Handbook of Clinical
Biochemistry

Second Edition

Handbook of **Clinical Biochemistry**

Second Edition

R Swaminathan

St Thomas' Hospital, UK

NEW JERSEY · LONDON · SINGAPORE · BEIJING · SHANGHAI · HONG KONG · TAIPEI · CHENNAI

Published by

World Scientific Publishing Co. Pte. Ltd.

5 Toh Tuck Link, Singapore 596224

USA office: 27 Warren Street, Suite 401-402, Hackensack, NJ 07601

UK office: 57 Shelton Street, Covent Garden, London WC2H 9HE

British Library Cataloguing-in-Publication Data
A catalogue record for this book is available from the British Library.

Cover photo courtesy of Smith, G.D., Pangborn, W.A. & Blessing, R.H., The Structure of T6 Human Insulin at 1.0 Å Resolution. Acta Cryst., D59, 474–482 (2003).

HANDBOOK OF CLINICAL BIOCHEMISTRY

ISBN-13 978-981-283-737-0
ISBN-10 981-283-737-X

Typeset by Stallion Press
Email: enquiries@stallionpress.com

Printed in Singapore by Mainland Press Pte Ltd.

Acknowledgement

My sincere gratitude to all who commented and suggested changes. This include Miss A. Sankaralngam, Dr. Rashim Slota, and many students who did a special study module (also called student selected component) with me on the subject of chemical pathology (clinical biochemistry). Special thanks to Ms Julie Clayton who helped with the manuscript.

Ms. Jihan Abdat of World Scientific Publishing helped in many ways to get this book published. My sincere thanks to her. Finally, this book would not have been possible without the help, encouragement and support from my family (Kamala, Suresh, Abirami and Ramesh).

Any errors and inaccuracies in this book are no one's fault but mine.

Preface

In writing this revised edition, the overall aim of the book remains the same as the first edition. The text in this second edition has been thoroughly revised and where necessary, new material has been added.

One of the new features of this edition is the addition of summary/key points at the end of each chapter to facilitate revision. Self-assessment questions including traditional multiple choice questions and extended matching questions are introduced at the end to help with revision and examination preparation.

At the end of each chapter some key references for further reading are given to help readers.

Contents

chapter 1

Interpretation of Biochemical Tests

Introduction

Biochemical tests are now an important part of investigation and management of patients. Understanding the factors which influence laboratory results and how laboratory results could be used for the diagnosis and treatment, is important for the rational and effective use of laboratory tests. As with other investigations, biochemical results should be taken in the context of patients' clinical features (signs and symptoms) and other relevant findings.

Biochemical tests are performed for four main reasons — diagnosis, management, prognosis and screening. When used appropriately, biochemical tests can contribute substantially to the overall care of the patient. However, when used inappropriately, it can lead to unnecessary further investigations, pain and suffering to the patient and increased costs to the health service.

Diagnosis

Clinical diagnosis is usually based on history, physical examination and results of investigations. History and examination are the important elements in arriving at a diagnosis and studies show that up to 80% of cases can be diagnosed from history and clinical findings alone. Biochemical tests very often help to confirm the diagnosis or identify a metabolic syndrome. Seldom are they diagnostic except in a few instances, such as inherited disorders of metabolism.

1

Management

Biochemical tests are most often used in the management of patients. Approximately 60–70% of all biochemical tests are used for monitoring treatment or to follow the progress of the disease. Serial measurements are valuable in management, e.g. in patients with diabetic ketoacidosis, frequent measurements of blood glucose help to assess the response to insulin and to adjust dosage; in hypothyroid patients on thyroxine replacement therapy, regular measurements of thyroid function tests guide the adequacy of thyroxine replacement.

Biochemical tests are also useful in assessing the severity of the disease. The degree of abnormality in biochemical tests is (usually but not always) related to the severity of the disease. For example, in renal failure, the greater the plasma urea and creatinine, the more severe the reduction in renal function.

Prognosis

Biochemical tests, either individually or in combination, can give an indication of the prognosis, e.g. in patients with malignant tumours, serial measurements of tumour markers are of value in assessing the response to treatment and the possibility of recurrence.

Screening

When tests are done to detect the presence of a disease before clinical features are evident, it is described as screening. Screening may be applied to a population (population screening), to a selected sub-group of a population (selective screening), individuals (individual screening) or it could be opportunistic.

Population screening

These tests are done in an apparently healthy population to identify those who may have subclinical disease or those who are at risk of developing a disease. Population screening programmes should satisfy the following criteria:

1. The disease should have a significant effect on the quality of life or life expectancy, e.g. screening for Gilbert's syndrome, an inherited disorder of bilirubin metabolism, is of no value as it has no long-term effect on the health of the patient.
2. The screening test can detect the disease before irreversible damage has occurred.
3. Effective treatment is available and is acceptable for asymptomatic patients.
4. The screening test should be effective (specific and sensitive) and acceptable to the population to be screened.
5. The prevalence of the disease and the benefits of treatment should justify the cost of screening.
6. The population at risk can be defined.

Unequivocal benefits of screening have only been established for a few conditions. These include screening for phenylketonuria and hypothyroidism in the newborn and cervical screening for the detection of cervical carcinoma.

Selective screening

Screening can also be applied to a subgroup of the population known to be at risk of developing that disease, e.g. family members of a patient with hypercholesterolaemia or premature coronary heart disease could be screened for high cholesterol.

Individual screening

Here an individual is screened for a particular disease or diseases based on the individual's history. An example is antenatal screening of a foetus for inherited disease when a previous child of the parents has been found to have that disease or when there is a strong family history of that disease.

Opportunistic screening

Opportunistic screening is when a patient is screened for certain diseases when he presents to the doctor with an unrelated condition, e.g. detection of hypertension.

Other Uses of Laboratory Tests

Biochemical profiling

With the availability of multichannel analysers, it is now possible to analyse a small blood sample for a large number of biochemical tests. When a group of tests is applied to otherwise healthy individuals or to all admitted to hospitals, it is termed biochemical 'profiling'. In general, this type of approach has caused more harm than good as the efficiency of detection of a disease is low. Unnecessary investigations may follow when non-specific abnormalities in test results are found. It has been argued that admission profiling can detect potentially treatable diseases at an early stage. Diseases that can be detected by screening hospital patients include hyperparathyrodism, hypothyroidism, diabetes mellitus, renal disease and liver disease (alcoholism). However, the value of such admission screening is yet to be established as this approach can detect diseases, which may not manifest in the lifetime of the patient.

Baseline

Biochemical tests are often also used as a baseline before starting treatment to detect any harmful effects of treatment or to monitor the treatment.

Collection of Specimens

Biochemical investigations are done in body fluids, most often in plasma or serum. It is essential that blood samples are collected appropriately to prevent artefacts. In patients receiving intravenous therapy, blood should be collected from a different site to avoid contamination of the blood sample with the infusion fluid. Tourniquet is often used to obstruct venous blood flow in order to make the venepuncture easy. However, if the tourniquet is applied too long, the increased pressure will cause transfer of water and small molecular weight constituents into the interstitial compartment. This will often result in an increase in the concentration of large molecular weight substances such as proteins and in protein-bound substances such as calcium in the serum sample.

While transferring the blood from the syringe into the bottle, care should be taken to avoid haemolysis.

Once the blood is taken, it should be put into appropriate bottles containing the appropriate preservatives. Most investigations are now done in serum. However, there are some investigations for which plasma is required, e.g. measurement of fibrinogen requires plasma. Inappropriate use of anticoagulants has often led to spurious results, e.g. taking blood sample for electrolytes into an EDTA tube will cause very high potassium and low calcium concentrations. Blood for the measurement of glucose concentration should be taken into a fluoride tube; otherwise the blood glucose will artificially decrease due to continued glycolysis by red cells. Once the blood is collected, it should be transported to the laboratory within a specified time to prevent artificial results, e.g. if the blood sample is left at room temperature for several hours, it will lead to high potassium concentration. Storing samples below room temperature has similar effects.

Some investigations are done in urine — a random or a 24-hour urine collection. Appropriate preservatives should be used to avoid artefactual results, e.g. urine for calcium should be collected in a container with acid preservative to prevent precipitation of calcium phosphate. When 24-hour samples are required, patients should be given appropriate instructions on how to collect the urine samples. In investigations done on 24-hour samples, the most important source of error is incomplete urine collection. Measurement of urine creatinine concentration is sometimes used to check whether urine collection is complete.

Identification of Patient Specimens

It is essential that the sample is collected from the correct patient. Many errors have occurred due to improper identification of the patient. Once collected, the sample should be correctly labelled and accompanied by a properly completed request form.

Interpretation of Laboratory Results

In interpreting laboratory results, one of two questions is usually asked:

(i) Is the result normal or abnormal?
(ii) Has the result changed significantly from a previous result?

In answering the first question, the result is compared to a range (reference range).

Reference Ranges

Reference ranges can be either population-based or risk-based.

Population-based reference ranges

To determine the population-based reference range, blood samples are taken from a defined population — usually healthy individuals, but it can be from any defined population. If the analyte concerned is known to be affected by sex and age, the population should be divided according to these two factors. Once the blood is analysed, the results are examined to see whether it follows a Gaussian (symmetrical) distribution. If the result fails to follow a Gaussian distribution, trans-formations such as a log transformation can be done. The reference range is calculated as mean ± 2 standard deviation (SD), but often the range is taken as the value that represents 95% of the population (2.5–97.5%) when the values are ranked. This type of reference range excludes 5% of the population who are apparently healthy. It is also possible that some subjects with an undiagnosed 'disease' may be included in the population used to establish the reference range (for example, undiagnosed diabetes mellitus).

The probability of finding abnormal results in a healthy population increases if multiple tests are done at the same time, especially if the tests are not dependent on each other. If 20 independent tests are done, the probability of finding at least one result outside the reference value is 64%.

Risk-based reference range

Reference range is sometimes based on disease risk. The reference values for cholesterol are based on the risk of developing coronary heart disease. Epidemiological studies have shown that a cholesterol value of 4.0 mmol/L or lower carries a low risk of coronary heart disease.

It is important to remember that results within the reference range do not exclude disease and results outside the reference range do not always indicate the presence of a pathological disease. However, the more abnormal the result, the greater the chance that there is a disease process. The diagnosis is seldom based solely on biochemical results. Test results should be taken in conjunction with clinical findings and usually there is no absolute demarcation or cut-off values between disease and normal.

Detection of a Significant Change

The second question asked about a test result is whether there is a significant change from a previous result. In deciding whether a significant change has occurred, several factors need to be taken into account. These include analytical variation, biological variation related to the time of sampling and the procedure used. Ideally biological variation should be minimised by taking the sample under identical conditions of time, posture, etc. Other preanalytical factors should be minimised by using exactly the same techniques. When these factors are minimised, the variation between two results depends on the imprecision of the assay. If the difference between two results is equal to or more than 2.8 times the standard deviation (SD)

Table 1.1   Calculation of total variation

Serum free thyroxine concentration measured in a subject 2 months apart were 12 and 18 pmol/L. Is this difference significant?

The analytical variation for free thyroxine measurement is 1 pmol/L and the biological variation 1.2 pmol/L.

$$\text{Total variation} = \sqrt{(SD_A^2 + SD_I^2)} = \sqrt{(1^2 + 1.2^2)},$$
$$= \sqrt{(1 + 1.44)} = \sqrt{2.44} = 1.6.$$

For a difference between two values to be significant at 5% level, the two results should be greater than 2.8 × the total variation.

In this case, 2.8 × 1.6 = 4.48.

The difference observed is 6 pmol/L and therefore this change is clinically significant.

---

of the method, the difference is significant at 5% confidence limit. For example, if the analytical SD for serum sodium is 2 mmol/L, a change less than 5.6 mmol/L is within the limit of the analytical variation.

In order to decide whether the change is clinically significant, biological variation should be taken into account. Total variation (biological and analytical) is calculated from the analytical and biological variations using the formula:

$$SD_{Total}^2 = SD_A^2 + SD_I^2$$

where SD_{Total} is the total variation, SD_A is the SD of analytical variation and SD_I is the SD of biological variation (Table 1.1).

Factors Affecting Test Results

Test results can be affected by preanalytical, analytical and postanalytical factors. Preanalytical factors may be biological factors or factors related to the collection of specimen. The latter has already been discussed.

Biological Factors (Table 1.2)

Age

Many biochemical variables vary with age, and an appropriate age-related reference range should be used to interpret results of these tests. Examples of tests which vary with age include alkaline phosphatase, phosphate and gonadotrophins.

Sex

Tests such as sex hormones, serum creatinine, urate and GGT show differences between sexes.

Body composition

Body fat and lean body mass can influence some results. Creatinine and creatine kinase, which are derived from muscle, are said to be related to muscle mass. Triglycerides tend to be higher in obese individuals.

Table 1.2   Biological factors affecting biochemical results

Factors	Biochemical tests
Age	Alkaline phosphatase, uric acid, creatinine
Sex	Gonadotrophins, gonadal steroids, creatinine
Body composition	Creatinine, creatine kinase, triglycerides
Race/Ethnicity	Creatine kinase, prostate specific antigen
Time	Cortisol
Posture	Protein, renin, aldosterone
Stress	Prolactin, cortisol
Food intake	Glucose, triglycerides
Alcohol	Triglycerides
Exercise	Creatine kinase
Drugs — *in vivo* effects	Phenytoin — gamma-glutamyl transferase
	Thiazides — potassium
	Oestrogens — sex hormone-binding globulin

Ethnicity/race

The range found in healthy individuals from different ethnic groups vary for some analytes. Prostate specific antigen (PSA) is higher in African Americans and lower in Japanese compared to Caucasians. Serum creatine kinase (CK) values in African Americans are higher than in Caucasians.

Time of day

Cortisol, osteocalcin, parathyroid hormone, etc. show a circadian rhythm. Some analytes show seasonal changes, e.g. plasma 25-hydroxycholecalciferol.

Stress

Stress causes the release of cortisol, ACTH, prolactin, growth hormone, catecholamines, etc. Thus, it is very important to avoid stress when taking samples for these measurements.

Posture

Posture increases aldosterone and renin activity. Plasma proteins and protein-bound compounds tend to be higher on attaining a standing posture. This is due to the movement of fluid from the vascular compartment to the interstitial compartment.

Food intake

Glucose, triglycerides and insulin are examples of substances affected by food.

Drugs

Drugs can influence results by either interfering with the analysis or by physiological mechanisms. For example, in patients taking phenytoin, serum gamma-glutamyl transpeptidase (GGT) is higher due to enzyme induction.

Exercise

Exercise and trauma release CK and myoglobin.

Intrinsic biological variation

Although many biochemical variables are tightly controlled, there is variation within an individual. This individual variation is small for some analytes and large for others. For example, serum iron concentration fluctuates rapidly within the same individual whereas serum sodium, creatinine and calcium concentrations show less variation (Figure 1.1). For analytes which have a low intraindividual variation, the serum concentration may change within the reference range and

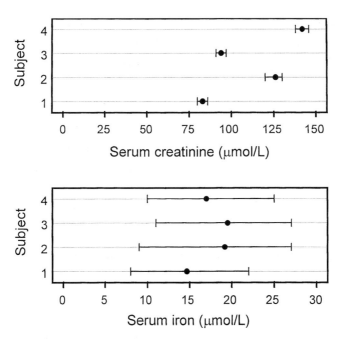

Figure 1.1   Within individual variation for (a) serum creatinine and (b) serum iron in four individuals. The circle represents the mean for the individuals and the lines indicate the range found in that individual (adapted from Fraser and Stevenson, 1998 with permission).

become abnormal for that individual. For analytes with low intraindividual variation, population reference ranges are less useful.

Analytical Factors

All analytical methods have errors — inaccuracy and imprecision (Figure 1.2). Inaccuracy refers to how close the result is to the true value. Imprecision refers to the reproducibility of the result and is usually expressed as coefficient of variation. To estimate imprecision, a sample is analysed several times and the mean and standard deviation (SD) are calculated. Coefficient of variation (CV) is derived using the formula:

$$CV = \frac{SD}{Mean} \times 100.$$

The Diagnostic Value of an Investigation

The diagnostic value of a test is described in terms of sensitivity, specificity and predictive value. The sensitivity of a test is the frequency with which a positive result is found in patients known to have the disease i.e. true positive (TP) rate. The specificity of a test is a measure of the frequency of negative results in patients (or persons) known to be free of the disease, i.e. true negative (TN) rate. A sensitivity of 90% implies that 90% of patients with the disease will have a positive result and 10% of people with disease will not show a positive result; a false negative (FN) result. A specificity of 95% means 95% of people without the disease will show a true negative result and 5% of the

Precise but inaccurate   Imprecise and inaccurate   Precise and accurate

Figure 1.2   Diagram illustrating accuracy and imprecision.

population without the disease will have a positive result, a false positive (FP) result. An ideal test should be 100% sensitive and 100% specific. However, in practice, no such test exists. Specificity and sensitivity can be calculated from the following formulae:

$$\text{Specificity} = \frac{\text{true negative}}{\text{false positive} + \text{true negative}} \times 100.$$

$$\text{Sensitivity} = \frac{\text{true positive}}{\text{true positive} + \text{false negative}} \times 100.$$

The predictive value of a positive test is a measure of the likelihood of having the disease under consideration when the test is positive. The predictive value of a negative result is the likelihood of not having the disease when the test result is negative. In determining the specificity, sensitivity and predictive values, it is important to be able to assign subjects to the right categories, i.e. patients should be allocated to a particular disease based on independent diagnosis. For some diseases, histological confirmation may be the only way of confirming the diagnosis. Calculation of predictive values is illustrated in Table 1.3.

In the preceding discussion, it has been assumed that the prevalence of the disease is 50%. When establishing the diagnostic value of a new test, it is not unusual to apply the test to selected groups of equal sizes — a group with the disease and a control group — a prevalence of 50%. In clinical practice however, the test will be applied to a larger number of people where the prevalence will be lower. The example given in Table 1.4 illustrates the effect of prevalence on the predictive value of a test, which has a sensitivity of 99% and specificity of 99%. When the prevalence is 50%, the predictive value of a positive result is 99%. However, when the prevalence is 1%, the predictive value falls to 50%. The predictive value of a test falls with decreasing prevalence. One way of improving the efficacy of a test would be to use the tests more selectively, i.e. by applying the test only on sound clinical grounds, thus increasing the prevalence of the disease.

Table 1.3   Calculation showing predictive value of a test

$$\text{Sensitivity} = \frac{TP}{TP+FN} \times 100.$$

$$\text{Specificity} = \frac{TN}{TN+FP} \times 100.$$

$$\text{Predictive value of a positive result} = \frac{TP}{TP+FP} \times 100.$$

$$\text{Predictive value of a negative result} = \frac{TN}{TN+FN} \times 100.$$

$$\text{Efficiency} = \frac{TP+TN}{(TP+FP+TN+FN)} \times 100.$$

Example

	Positive	Negative	Total
Disease	(TP) 90	(FN) 10	100
Health	(FP) 5	(TN) 95	100
	95	105	200

*TP = True positive; TN = True negative;
FD = False positive; FN = False = negative.

$$\text{Predictive value of a positive result} = \frac{90}{(90+5)} \times 100 = 94.7\%.$$

$$\text{Predictive value of a negative result} = \frac{95 \times 100}{(95+10)} = 90.5\% = \frac{95}{(95+10)} \times 100 = 90.5\%.$$

$$\text{Efficiency} = \frac{(90+95)}{200} \times 100 = 92.5\%.$$

Cut-off Value

In the discussion thus far, the test result was designated as positive or negative. As biochemical results are usually quantitative, the cut-off value, the concentration at which the test is considered as positive, can be varied.

Selection of the cut-off value will depend on the purpose of the test. In circumstances where the consequences of not diagnosing the diseases is great, it is important to select a cut-off value so as not to

Table 1.4 Effect of prevalence on the predictive value

Sensitivity of the test = 99%
Specificity of the test = 99%

Population of 1000 with a prevalence of 50%

	+ve results	−ve test	Total
Disease	4950	50	5000
Without the disease	50	4950	5000
Total	5000	5000	10,000

$$\text{Predictive value of a +ve result} = \frac{4950}{4950+50} \times 100 = 99\%$$

Prevalence of 1% — Population 10 000 with a prevalence of 1%

	+ve results	−ve test	Total
Disease	99	1	100
Without the disease	99	9801	9900
Total	198	9802	10,000

$$\text{Predictive value of a +ve result} = \frac{99}{99+99} \times 100 = 50\%$$

miss any individual with the disease, i.e. to keep false negatives as low as possible. Hence, a lower cut-off value to ensure a sensitivity of 100%, should be chosen. An example of this situation is in the screening for phenylketonuria (PKU), when missing the diagnosis of PKU has grave consequences. This approach however, will reduce specificity and will give a larger number of false positive results. In circumstances where it is important not to cause unnecessary anxiety and investigations, a high specificity is required and a high cut-off value should be selected.

To compare the performance of two tests, receiver operating characteristic (ROC) curves can be used. The sensitivity and specificity at

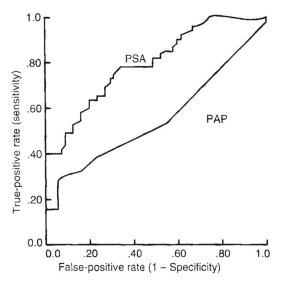

Figure 1.3   Receiver operating characteristic (ROC) curve for prostate-specific anti-gen (PSA) and prostatic acid phosphatase (PAP) in the diagnosis of prostatic carcinoma. The area under the curve for PSA is greater than PAP showing the PSA is a better test.

different cut-off values for each test are calculated and plotted as shown in Figure 1.3. The area under the curve is a measure of the per-formance — the greater the area under the curve, the more specific and sensitive the test.

Likelihood Ratios

Likelihood ratios are an alternative way of summarising the useful-ness of a diagnostic test. The ratio tells us how many more (or less) times patients with the disease are likely to have that particular result than patients without the disease. Likelihood ratios can be calculated for a positive (LR+) or a negative (LR−) results by the following equations:

The likelihood for a positive result is calculated as

$$LR+ = \frac{\text{sensitivity}}{1 - \text{specificity}}.$$

The likelihood for a negative result is calculated as

$$LR-=\frac{1-\text{sensitivity}}{\text{specificity}}.$$

The higher the likelihood ratio of a positive result, the greater the chances of finding the disease. Likelihood ratios above 10 and below 0.1 are considered strong evidence to rule in or rule out diagnoses respectively.

General Principles

Steady State vs. Transient State

In order to understand the pathophysiology of diseases, it is important to explain the difference between steady state and transient state. This principle can be illustrated by the following case example:

In a patient with chronic renal failure, the following blood results are found:

Sodium (mmol/L)	138	135–145
Potassium (mmol/L)	5	3.5–5.0
Bicarbonate (mmol/L)	18	23–32
Chloride (mmol/L)	102	90–108
Urea (mmol/L	35	3.5–7.2
Creatinine (μmol/L)	565	40–60

If students are asked about the urinary excretion of creatinine in this patient, most students will say that the excretion of creatinine will be low due to the low glomerular filtration rate. Similarly, if students were asked about the excretion of carbon dioxide in a patient with chronic obstructive airways disease, most would answer that it would be decreased and the arterial partial pressure of carbon dioxide is increased. However, in both these examples, it is likely the excretion will not be low as these patients are in a steady state. This can be explained using the diagram below (Figure 1.4).

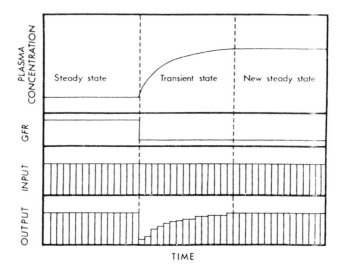

Figure 1.4   Theoretical changes in plasma concentration, GFR, production rate and urinary excretion of creatinine before and after a reduction in GFR. (Reproduced with permission.)

When the patient is healthy, the amount of creatinine excreted is equal to the amount produced (which depends on the person's muscle mass). As creatinine is excreted by filtration without reabsorption or significant secretion, the amount excreted is equal to the amount filtered (Equation 1).

$$\text{Amount produced} = \text{amount excreted} = \text{amount filtered,}$$
$$= \text{GFR} \times \text{plasma creatinine.} \qquad (1)$$

If the GFR falls, the amount filtered will fall resulting in less excretion. This will increase the plasma concentration. As the plasma concentration increases, the filtered amount will increase thus excretion will increase. Eventually, the amount excreted will be equal to the amount produced, and the plasma concentration will be steady at a higher value. In this new steady state, input (production of creatinine in this case) will be equal to the output (urinary excretion).

In the steady state if there is a transient increase in input (e.g. sudden increase in potassium intake), the plasma concentration and the

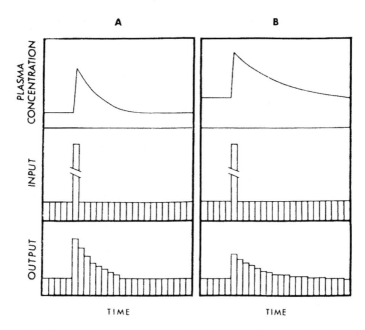

Figure 1.5   Changes in plasma concentration, input and output with time of potassium following a transient increase in input, (a) in the normal steady state and (b) in an abnormal steady state; chronic renal failure with reduced glomerular filtration rate.

excretion will increase until the load is excreted and the plasma concentration will come back to the previous level (Figure 1.5a). If the steady state is an abnormal one (like the patient with chronic renal failure described above), the increase in plasma concentration will be greater and it will take a longer time before the steady state is re-established (Figure 1.5b).

The following general principles will be of help in the understanding of changes in plasma concentration:

1. A change in plasma concentration of a substance is the result of either a change in input rate (production or intake) or due to a change in the output rate (metabolism or excretion).
2. When the steady state is disturbed, plasma concentration and output will change until input and output are equal — a new steady state.

3. In the steady state, irrespective of weather, this is a normal or abnormal state, input and output are equal.
4. A transient increase in input when the steady state is abnormal will be handled at a different rate than when the steady state is normal.

Concentration vs. Amount

Most clinical biochemistry investigations are done in plasma or serum and what is measured is the concentration. A change in concentration of an analyte can be due to a change in the amount of substance in plasma or due to a change in the volume of fluid. It is a common mistake to assume that when the plasma sodium concentration is low, the amount of sodium in the body is low. However, very often a low plasma sodium concentration is due to an increase in water content.

Concentration vs. Activity

In many circumstances, we measure the total concentration of a substance in serum or plasma. However, the substance may be partly bound to proteins, and very often, it is the unbound or free analyte which is the physiologically important fraction. For example, calcium in serum is bound to proteins (mainly albumin) and the physiologically active form is ionised calcium. The total concentration however, may change due to changes in the binding protein concentration, e.g. in pregnancy, the concentration of binding protein for thyroxine increases due to increased synthesis and this will cause the total thyroxine concentration to be high with normal free thyroxine concentration. Therefore, in interpreting the total concentration of an analyte, it is important to bear in mind that the total concentration may change without a change in the active fraction.

Urinary Excretion

When the excretion or concentration of a substance is measured in the urine, it is important to remember that there is no reference range. This is particularly true for urine electrolytes, urea and osmolality. The concentration of electrolytes in serum is highly regulated such that, if there is a change in serum concentration, homeostatic mechanism(s) will return the serum concentration back to the original value. Urinary excretion, on the other hand, is a method of regulating plasma concentration. Therefore, one cannot determine a reference value for it. In the steady state, urinary excretion is a reflection of input. In the case of electrolytes, it is a reflection of intake. For example, a urinary excretion of 100 mmol/d of sodium implies that the person is taking 100 mmol per day of sodium . On the other hand, if we know that the person is taking 50 mmol of sodium per day, a urinary excretion of 100 mmol/d of sodium tell us that he is losing sodium through his kidneys. Interpretation of urinary values should be done in relation to the input (intake) and/or in relation to the clinical state. For example, if a person is volume depleted (extracellular volume is low), the body's response is to increase sodium reabsorption and the urinary exertion should be very low (usually < 20 mmol/L or d). Any value higher than this implies that the person is losing sodium through his kidneys.

Further Reading

1. Fraser CA, Stevenson HP. Production and use of data on biological variation in laboratory medicine. *CPD Bulletin in Clin Biochem* 1998; 1: 14–17.
2. Fraser CA. *Biological Variation: From Principles to Practice* 2001. AACC Press.
3. Payne RB, Morgan DB. Sodium, water and acid–base balance: Teaching transient and steady states. *Med Educ* 1977; 11:133–135.

Summary/Key Points

1. Biochemical tests are useful in the diagnosis, monitoring, screening and prognosis of disease. Most biochemical tests are done for monitoring treatment or to detect complications of treatment. Biochemical screening of healthy subjects are of little value except in a few well-defined situations.

2. A test result can vary due to biological and analytical variation. Analytical variation or imprecision is assessed by the standard deviation (or coefficient of variation).
3. When interpreting results, they are usually compared to a reference range, which encompasses 95% of values in a healthy population. For analytes such as serum cholesterol and serum 25-hydroxyvitamin D concentration, this approach is not valid and risk-based value is used to compare results.
4. A test result within the reference range does not necessarily imply that there is no disease and a test result outside the reference range does not indicate disease.
5. Tests results can vary due to many physiological factors such as age, gender, ethnicity, time of day, body size, etc.
6. Even when these physiological factors are controlled, there is intrinsic variation within an individual. This intraindividual variation is large for some analytes (e.g. iron, cortisol) and low for others (e.g. sodium, calcium and thyroxine).
7. For an analyte with low within person biological variation, a change within the reference range is potentially clinically significant.
8. The diagnostic value of a test is described by sensitivity (percentage of positive results in a group with the disease) and specificity (percentage of negative results in a group without the disease). The predictive value of a test is influenced by the prevalence of the disease in the population. For a disease with low prevalence, even a test, which is highly specific and sensitive, will give a large number of false positives.
9. The value of two tests can be compared using the receiver operating characteristic curves (ROC) in which specificity is plotted against sensitivity.

chapter 2

Disorders of Fluid Balance

Introduction

Distribution and Composition of Body Fluids

Water accounts for approximately 60% of body weight in young men and 55% in young females. In newborn infants, the total body water (TBW) may be as high as 75% of body weight and it progressively reduces to 60% of body weight by 10 years of life. TBW, as a percentage of body weight, decreases steadily with age and in obesity. In obese subjects, it can be as low as 45%. Expressed as a percentage of lean body mass, total body water accounts for 75% of lean body mass in adults.

Water is distributed between intracellular (ICF) and extracellular (ECF) compartments separated by the cell membrane. The extracellular compartment is further divided into plasma water, interstitial water (outside the circulation) and a small amount of transcellular water which is water in specialised spaces such as cerebrospinal fluid space, pleural space, synovial space, intra-ocular spaces and the lumen of the gut (Table 2.1).

Composition of Extracellular Fluid

The main extracellular cation is sodium (Table 2.2). Concentration of ions in plasma is usually expressed as mmol/L of plasma. However, these ions are distributed in plasma water which forms only 93% of plasma. Hence, the concentration of ions in plasma water is higher (sodium concentration in the plasma water is 153 mmol/L). If the amount of solids (proteins and lipids) is grossly elevated, as in hyperlipidaemia or hyperproteinaemia, the concentration of ions expressed per litre of plasma may

Table 2.1   Distribution of body water in a 70-kg adult man

	Volume (L)	% of body weight
Total Body Water	42	60
• Intracellular Water	28	40
• Extracellular Water	14	20
— Plasma Water	3	5
— Interstitial Water	10	14
— Transcellular Water	1	1

Table 2.2   Composition of intracellular and extracellular compartments (mmol/L)

	Intracellular fluid	Extracellular fluid Plasma/Serum	Interstitial
Cations			
• Sodium	10	140*	145
• Potassium	160	4	4
• Calcium (ionised)	1	1.2	1.2
• Magnesium (ionised)	13	1.0	1.0
Anions			
• Chloride	3	102	117
• Bicarbonate	10	27	27
• Phosphate & others	106	1.0	1.0
• Protein	65	16	0
Total	368	292	302

*Concentration in plasma water is 153 mmol/L.

be artificially low (i.e. pseudohyponatraemia). The concentration of sodium in plasma and interstitial fluids are very similar — the small difference is accounted for by the Gibbs-Donnan equilibrium due to the presence of a higher protein concentration in plasma.

The total volume of transcellular fluid at any one time is small — about one litre. However in the gastrointestinal tract, a much larger volume (5–8 litres) is secreted and reabsorbed each day (Table 2.3). In pathological states, e.g. diarrhoea, fistula and vomiting, the loss of these fluids can produce serious fluid and electrolyte abnormalities.

Table 2.3 Composition of transcellular fluids (mmol/L)

	Saliva	Gastric juice	Bile fluid	Pancreatic fluid	Ileal fluid	Colonic fluid	Sweat
N^+	20–80	20–100	150	120	140	140	65
K^+	10–20	5–10	5–10	5–10	5	5	8
Cl^+	20–40	120–160	40–80	10–60	105	85	39
HCO_3^-	20–60	0	20–40	80–120	40	60	16

Composition of Intracellular Fluid

The main intracellular cations are potassium and magnesium while the main intracellular anions are phosphate and proteins (Table 2.2). The sodium–potassium pump maintains low intracellular sodium and high potassium concentrations. The higher total ion concentration in ICF compared to ECF (368 vs. 300 mmol) is due to the Gibbs-Donnan effect of high concentration of non-diffusible anions — proteins and organic phosphate.

Distribution of Water between ICF and ECF

Osmotic forces chiefly determine the movement of water across the cell membrane, which are freely permeable to water. Cell volume, which is essential for normal cell functions, is thus regulated by the regulation of plasma osmolality.

Osmolality

Osmolality is the number of osmoles per kg of water and osmolarity is number of osmoles per litre of solution. In clinical practice, the difference between osmolality and osmolarity is negligible.

The number of particles dissolved in body fluids determines its osmolality, which is held within narrow limits. In the ECF, sodium and its anions are the major contributors to osmolality. Plasma osmolarity can be calculated from plasma electrolytes, glucose and urea:

$$\text{Calculated plasma osmolarity} = 2[Na^+ + K^+] + glucose + urea.$$
$$\text{(All in mmol/L)}$$

In most circumstances, the calculated osmolarity and measured osmolality differ very little. However, in the presence of unmeasured osmotically active molecules (e.g. alcohols, mannitol, etc.) or in the presence of excessive lipids or proteins, which may cause pseudohyponatraemia, calculated values will be lower than measured osmolality. This difference is sometimes referred to as an osmolar gap.

Although sodium, glucose and urea contribute to the osmolality of ECF, not all of these molecules are 'physiologically' active. For example, urea is diffusible across the cell membrane and therefore an increase in osmolality produced by high concentration of urea does not cause movement of water. Effective osmolality or tonicity of ECF is mainly due to sodium and it anions.

If water is added to the ECF, osmolality of ECF will decrease and water will move into cells, resulting in an increase in the volume of ICF and ECF until the osmolalities are equal (Figure 2.1a). On the other hand, if isotonic saline is added to ECF, osmolality of the ECF will not change and the ECF volume will increase without changing ICF volume (Figure 2.1b). Loss of water will cause a decrease in ICF and ECF, while loss of isotonic saline will cause a decrease in ECF only. If sodium (without water) is added to ECF, osmolality of the ECF will increase and this will result in water moving out of cells, causing an increase in ECF and a decrease in ICF volume. These experiments illustrate that the plasma sodium concentration does not reflect ECF volume changes. Plasma sodium concentration is a reflection of the water content. On the other hand, the amount of sodium determines the ECF volume.

Regulation of Total Body Water

Total body water is regulated by the intake and excretion of water, which are in turn determined by osmolality. Typical values for water intake and output in a normal adult man are illustrated in Table 2.4.

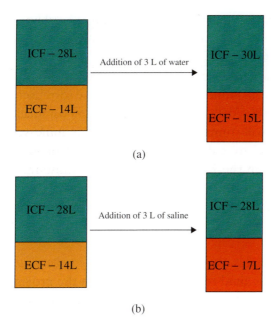

Figure 2.1   (a) Addition of 3 L of water. (b) Addition of 3 L of isotonic saline.

Table 2.4   Typical water balance in a normal adult man

	Intake (ml/d)		Output (ml/d)
Water in diet	850	Urine	1500
Ingestion of water	1400	Skin	500
Water of oxidation	400	Respiratory tract	400
		Faeces	200
Total	2600		2600

Water loss through the skin and lungs (insensible loss) is determined by thermoregulation. Evaporation of water accounts for approximately a quarter of the heat lost from the body. Of the 1500 mls of urine, a small proportion (500 ml) is obligatory loss, which is required to excrete the osmolar load. This obligatory amount of urine volume increases with the demand for solute excretion (e.g. during intake of a high protein diet). Thus approximately 1.5 litres of water is obligatory loss. As 500 ml of water is produced as a result of metabolism, about 1 litre of water intake is obligatory.

Water intake and output are controlled by osmoreceptors in the hypothalamus via influences on thirst and the secretion of antidiuretic hormone (ADH/AVP). An increase in plasma osmolality (sodium) of as little as 1% will stimulate thirst and increase ADH secretion. Other stimuli to thirst are a decrease in ECF volume and angiotensin II. A small decrease in the ECF volume does not cause thirst or increased ADH secretion, whereas a decrease greater than 10% to 15% in blood volume or ECF volume will. Angiotensin II is also an important stimulus for thirst and it acts directly on the brain, causing an increase in water intake.

ADH regulates water reabsorption at the collecting tubules of the kidney. Isotonic fluid reabsorption occurs in the proximal segment of renal tubules. This is followed by the production of a hypotonic fluid as a result of the countercurrent mechanism. In the collecting tubules, in the presence of ADH, water will move out due to the high osmolality of the medulla. By this mechanism, osmolality of the urine can be altered from 50 to 1400 mosmol/kg. The ability of the kidney to alter the osmolality of the urine depends on several factors and these are listed in Table 2.5. Osmotic diuresis, e.g. due to the presence of glucose in urine, prevents the development of medullary hyperosmolality thus, urine cannot be concentrated or diluted (Figure 2.2). When the osmolar load is high, the ability to dilute as well concentrate the urine is limited and the urine osmolality is fixed close to that of plasma (Figure 2.2).

Table 2.5 Factors essential for the production of hypertonic urine

Adequate fluid reaching the loop of Henle
Countercurrent mechanism
Presence of ADH
Responsiveness of collecting tubular cells to ADH
Medullary hyperosmolality

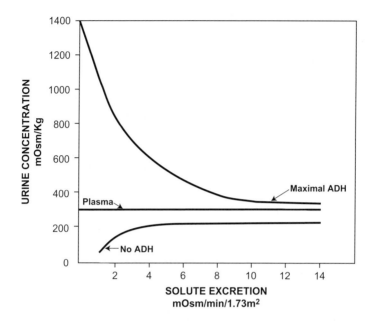

Figure 2.2 Effect of osmotic diuresis on urine concentrating and diluting abilities.

Table 2.6 Factors influencing the secretion of antidiuretic hormone

Osmolality
ECF volume
Nausea
Pregnancy
Hypoglycaemia
Drugs — alcohol, nicotine, morphine

Secretion of ADH is also stimulated by non-osmotic factors such as volume depletion and stress (e.g. surgery or trauma) (Table 2.6). One of the important non-osmotic stimuli for ADH secretion is ECF volume depletion. Figure 2.3 shows the ADH response to volume depletion and changes in ECF osmolality. A small change in ECF volume does not cause an increase in the secretion of ADH but a large decrease in ECF volume (greater than 15%) will

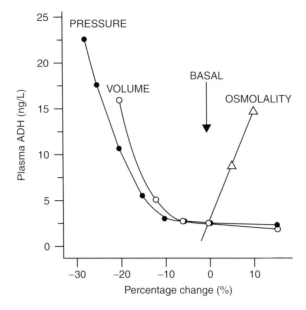

Figure 2.3   Effect of changes in ECF volume, blood pressure and osmolality on the secretion of ADH.

result in a dramatic increase in ADH secretion. Hypotension has a similar effect to that of hypovolaemia (Figure 2.3). Surgical stress is a common cause of an increase in ADH secretion in hospitalised patients. Nausea, but not vomiting, is a powerful stimulus for ADH excretion.

Regulation of ECF Volume

ECF volume is determined by the amount of body sodium, which in turn is regulated primarily through renal sodium excretion. A normal 70-kg adult man has approximately 4000 mmol of sodium, most of which is in the ECF. About 30% of the body's sodium is found in the bone, but only 60% of this is exchangeable. Sodium intake, which is determined by habit, varies widely (50–300 mmol/day) and sodium balance is maintained by renal excretion with small amounts excreted in the stools and sweat. In a healthy person, large amounts of sodium are filtered daily (25,200 mmol — approximately six times the total body sodium) and only 100–150 mmol is excreted: a reabsorption rate of 99.4%. The renal reabsorption of sodium is finely regulated to maintain sodium balance and consequently to maintain ECF volume. As the maintenance of ECF volume and hence blood volume is crucial, there are several sensitive and powerful homeostatic mechanisms involved in regulating renal sodium excretion.

Changes in the ECF volume are sensed by volume receptors in the carotid sinus, aortic arch and afferent glomerular arterioles as changes in the effective circulating volume (ECV). ECV refers to the part of ECF in the vascular space that effectively perfuses the tissues. ECV is not a measurable entity and under normal circumstances, ECF varies directly with ECV volume. However, in disease states, e.g. heart failure, ECV may be reduced with increased ECF volume. Stimulation of these volume receptors will cause the stimulation of several systems, including the sympathetic system and the renin–angiotensin system. Stimulation of the sympathetic system will cause haemodynamic changes in order to maintain the ECV and indirectly increase sodium reabsorption. Other factors influencing sodium reabsorption are given in Table 2.7.

Aldosterone, a steroid hormone secreted by the zona glomerulosa of the adrenal cortex is controlled by angiotensin II. Renin, a proteolytic enzyme secreted by the macula densa cells in the kidney acts on circulating angiotensinogin to convert it to angiotensin I, which in turn, is converted to angiotensin II by angiotensin-converting enzyme. Renin secretion is increased by sodium deficiency and

Table 2.7   Factors influencing sodium reabsorption

Hormonal factors
- Aldosterone
- Atrial natriuretic peptide (ANP)
- Sodium transport inhibitor (ouabain-like factor)
- Dopamine
- Kinins
- Prostaglandins

Haemodynamic factors
- GFR
- Peritubular capillary haemodynamics

volume depletion. Hyperkalaemia can also stimulate the secretion of aldosterone. Aldosterone acts by increasing the reabsorption of sodium in exchange for potassium and/or hydrogen ion secretion. Angiotensin II, in addition, has vasoactive actions, thereby minimising the effects of a decrease in plasma volume.

A variety of natriuretic factors can increase sodium excretion. Atrial natriuretic peptide (ANP), secreted by the myocytes is one of the well-characterised natriuretic factors. ANP increases sodium excretion by inhibiting sodium reabsorption in the inner medullary collecting tubules via a cyclic GMP-mediated mechanism. ANP, by a direct effect on vascular smooth muscles, has a hypotensive effect. In addition to ANP, two other natriuretic peptides have been described: brain natriuretic peptide (BNP) and C-type natriuretic peptide (CNP). BNP, first described in the brain, is synthesised and secreted from the ventricles of the heart. The biological effects of BNP are similar to that of ANP. CNP appears to be a neuropeptide rather than a cardiac peptide. In addition to these natriuretic peptides, many other natriuretic factors have been described; one of these is a sodium–potassium–ATPase inhibitor.

As the mechanisms responsible for ECF volume regulation are many and powerful, it is impossible to get clinical sodium depletion by decreased intake. Sodium depletion almost always follows abnormal losses. As the amount of body sodium determines the ECF volume, sodium depletion manifests as volume depletion without necessarily changing plasma sodium concentration.

Distribution of Fluid between Plasma and Interstitial Fluid

The movement of fluid between plasma and interstitial fluid (IF) is important as it supplies nutrients and remove waste products from the cells. This movement of fluid is controlled primarily by capillary hydrostatic pressure and plasma oncotic pressure, as capillaries are permeable to water and solutes (Figure 2.4).

At the arteriolar end of the capillary, fluid passes out into the IF as the hydrostatic pressure is higher than the oncotic pressure. At the venous end, plasma oncotic pressure is greater than the hydrostatic

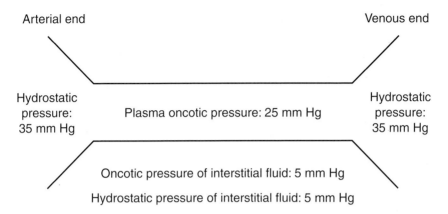

Figure 2.4   Haemodynamic factors controlling fluid movement across capillary wall.

pressure and fluid is reabsorbed. Some fluid is returned by lymphatics. This movement of fluid in and out of IF is rapid — about 75% of plasma is exchanged every minute.

Although the fluid moving through the capillary wall is relatively protein-free, albumin does slowly pass into the interstitial space and return to the circulation by the lymphatics. Permeability of capillaries varies in different organs, with the hepatic sinusoids being more permeable. It is important to note that although albumin concentration in IF is low, about 50% of total body albumin is extravascular since the volume of IF is 3–4 times greater than the plasma volume.

A small increase in hydrostatic pressure or decrease in oncotic pressure will be expected to cause accumulation of IF. However, this does not happen until there are gross changes due to the following safety factors:

1. Increased lymphatic flow.
2. When fluid enters the interstitial compartment, it will decrease the oncotic pressure of the IF. This will minimise the gradient for further fluid entry.
3. Increase in IF volume will raise interstitial hydrostatic pressure and oppose the entry of fluid.

Disorders of Water Metabolism

Any excess or deficit of water in the body is shared between ICF and ECF and can be detected by changes in plasma sodium concentration. For example, if 6 litres of water is lost, 2 litres will be lost from ECF and the plasma sodium concentration will increase by 20 mmol/L to 160 mmol/L. Similarly a gain of 6 litres of water will result in a decrease in the plasma sodium concentration by 20 mmol/L. Symptoms of water depletion and water excess are caused by changes in the hydration of the cells.

Water Depletion

Causes of water depletion are given in Table 2.8.

A decreased intake of water, especially in elderly subjects or unconscious patients, is not an uncommon finding in hospital practice. When there is a lack of ADH (neurogenic diabetes insipidus) or resistance to the action of ADH (nephrogenic diabetes insipidus), water can be lost in the urine causing water depletion. In osmotic diuresis, there is a greater loss of water than sodium leading to water depletion. Sweat is

Table 2.8   Causes of water depletion

Decreased water intake
- Elderly
- Very young
- Unconscious subjects

Increased water loss
- Renal loss
  — Central (neurogenic) diabetes insipidus
  — Nephrogenic diabetes insipidus
    • Congenital
    • Acquired — hypokalaemia, hypercalcaemia
  — Osmotic diuresis — glucose
- Insensible loss
  — Increased sweating
  — Hyperventilation

a hypotonic fluid, thus if there is excessive sweating, water depletion may result. If the humidity of the external environment is low, water can be lost without visible sweating. It is important to note that in conditions where there is increased loss of water, water depletion will not develop unless there is a failure to increase water intake.

Symptoms of water depletion are non-specific such as confusion, thirst and dry mouth. As the loss of water is shared between ICF and ECF, the decrease in ECF volume is relatively small and signs of volume depletion are minimal in water depletion. If the loss of water is non-renal, there will be increased reabsorption of water by the renal tubules and a small amount of concentrated urine will be produced; plasma sodium concentration will be high. As water depletion develops slowly, brain cells adapt to the high ECF osmolality by producing osmotically active organic compounds (called osmolytes) in order to prevent cerebral dehydration.

Water depletion is treated by the administration of water orally or by 5% glucose intravenously. It is important to reduce the deficit slowly if the water depletion has occurred over a period of time, in order to prevent cerebral oedema. Furthermore, when replacing the lost water, it is important to give additional amounts of water, sufficient to compensate for obligatory loss. The amount of water deficit can be calculated from the increase in plasma sodium concentration.

Water Excess

The normal system can cope with a water intake of up to 20 L/day without significant gain of total body water. Therefore, water excess is usually observed with impaired excretion. The excess water accumulates both in the ICF and ECF. Swelling of the brain cells causes symptoms such as confusion, behavioural disturbances, headache, convulsions and coma. The most important laboratory finding is a decrease in plasma sodium concentration with normal or low plasma urea concentration. Causes of water excess are given in Table 2.9.

During bladder irrigation, a decrease in plasma sodium concentration is common due to the absorption of irrigation fluid. Occasionally,

Table 2.9   Causes of water excess

Increased intake
- Psychogenic polydipsia — very rare
- Bladder irrigation

Decreased excretion
- Renal failure
- ADH secretion
    - Stress — post-operative states (common)
    - Pulmonary disease — pneumonia, tuberculosis, etc.
    - Neurological disease — infections, vascular accidents
    - Ectopic secretion of ADH — oat cell carcinoma
    - Drugs — cyclophosphamide, etc.
- Potentiation of ADH effects
    - Carbamazepine, chlorpropamide
- Exogenous ADH/analogues
    - Oxytocin

Others
- Cortisol deficiency

---

symptomatic hyponatraemia develops. The most common cause of water excess in hospital practice is the secretion of ADH due to stress or trauma. When these patients are given hypotonic fluids, i.e. 5% dextrose or a dextrose saline, hyponatraemia will develop.

Syndrome of inappropriate ADH secretion (SIADH) is a frequent finding in hospital populations. This syndrome is described under hyponatraemia. Cortisol deficiency due to adrenal failure or anterior pituitary failure can lead to water retention (see under hyponatraemia). Note that hyponatraemia of Addison's disease is due to a combination of water excess, as a result of cortisol deficiency, and sodium depletion as a consequence of aldosterone deficiency.

The treatment of water excess is described under hyponatraemia.

Disorders of ECF Volume

Changes in volume, i.e. ECF volume, are mainly due to changes in the amount of sodium in the body. As sodium is normally lost or gained with water, changes in body content of sodium results in an increase or decrease in extracellular fluid.

Volume Deficit

Extracellular fluid volume depletion is one of the commonest fluid disorders (Table 2.10). A decrease in the ECF volume will stimulate the sympathetic system, causing tachycardia and vasoconstriction. If the loss of fluid is gradual, then postural hypotension will be the only manifestation. The renal response to volume depletion is to increase the reabsorption of sodium and water. As ECF volume depletion is a strong stimulus for ADH secretion, water reabsorption will be increased and a small amount of concentrated urine with very little sodium will be produced. As the urea in the tubular fluid is reabsorbed passively, a decrease in urine flow rate would cause increased reabsorption of urea and plasma urea concentration will increase. High plasma urea and a high urea/creatinine ratio is a useful method of detecting volume depletion. In severe volume depletion, GFR may decrease and "acute tubular necrosis" may result. The clinical features of volume depletion include weakness, apathy, postural dizziness, syncope, thirst and muscle cramps. The patient will have reduced skin turgor, reduced

Table 2.10   Causes of volume depletion

Gastrointestinal losses
- Gastric — vomiting, nasogastric suction
- Intestinal — fistula, tube drainage, diarrhoea

Renal losses
- Diuretics
- Mineralocorticoid deficiency
- Osmotic diuresis

Sequestration
- Ileus
- Peritonitis

Others
- Loss via skin — e.g. burns

ocular pressure, hypotension (may be only postural) and oliguria. Significant laboratory findings are an increase in plasma urea concentration and an increase in plasma urea/creatinine ratio. If the sodium loss is via a non-renal route, urine sodium concentration will be very low, typically less than 20 mmol/L and the urine osmolality will be high. Plasma sodium concentration in patients with volume depletion may be normal, low or high depending on the composition of the fluid lost and on the composition of fluids administered. When the fluid loss is isotonic, the concentration of plasma sodium will remain within normal limits. But it may decrease if hypotonic fluid is administered (either orally or intravenously), as hypovolaemia is a powerful stimulus for the secretion of ADH. If the fluid lost is hypotonic, the plasma sodium concentration may be high.

Treatment of volume depletion is by infusion of isotonic (0.9%) saline.

Volume Excess

Oedema

Oedema is defined as a palpable swelling produced by expansion of the IF volume. It is associated with many clinical conditions such as congestive heart failure, nephrotic syndrome and cirrhosis of the liver. There are two important steps involved in the formation of oedema: (1) alteration in capillary haemodynamics, and (2) renal retention of sodium and water. Initially, as a result of altered capillary haemodynamics, fluid moves out of the vascular space into the interstitial compartment. This results in a reduction of plasma volume and tissue perfusion, which triggers homeostatic mechanisms to increase the renal sodium reabsorption in an attempt to return the plasma volume to normal. However, because of the alteration in capillary haemodynamics, most of the fluid will enter the interstitial compartment, eventually resulting in oedema. The net effect is an expansion of the total ECF volume with plasma volume maintained at near normal levels. It is important to recognise that in this case, the increased sodium and water retention is an appropriate response to the decrease in tissue perfusion produced by altered capillary haemodynamics. Thus rapid removal of fluid, for example by diuretic therapy, may cause diminished tissue perfusion.

Mechanism of oedema

1. Increased capillary hydrostatic pressure

Although capillary hydrostatic pressure at the arteriolar end is important in determining the movement of fluid out of the capillaries, it is not sensitive to changes in arterial pressure due to autoregulation of the precapillary sphincter. In contrast, the capillary pressure at the venous end is not well regulated and changes in venous pressure can lead to accumulation of interstitial fluid. Venous pressure can be increased either by an increase in the volume of blood in the venous system or when there is venous obstruction. The former is seen in heart failure and the latter is seen in deep vein thrombosis.

2. Decreased plasma oncotic pressure

A decrease in plasma albumin concentration, as a result of loss of albumin in the urine or due to reduced synthesis as in cirrhosis, will cause decreased reabsorption of interstitial fluid and hence oedema.

3. Increased capillary permeability

Capillaries are normally impermeable to large molecules such as proteins and albumin. If the capillary permeability is increased, albumin moves out into the interstitial compartment and more fluid is retained in the interstitial compartment, as a result of increased oncotic pressure. In certain forms of angioneurotic oedema (e.g. after replacement with interleukin II) and in conditions associated with adult respiratory distress syndrome, capillary permeability is increased.

4. Lymphatic obstruction

As some of the fluid accumulated in the interstitial compartment is normally removed by the lymphatics system, obstruction to the lymphatic system will also result in oedema.

5. *Renal sodium retention*

When capillary haemodynamics are altered, increased renal reabsorption of sodium and water is the appropriate response to the decrease in ECV. However, primary renal retention of sodium may also cause oedema, as seen in chronic renal failure and acute glomerular nephritis. The increased reabsorption of sodium is probably a consequence of the decreased filtration rate.

In some situations, multiple factors are involved in the accumulation of fluid. In hepatic diseases, for example, hypoalbuminaemia caused by decreased albumin synthesis, splanchnic pooling and increases in sinusoidal pressure all contribute to the accumulation of fluid.

Common causes of oedema are congestive heart failure, nephrotic syndrome and cirrhosis of the liver.

Pulmonary oedema

In diseases of the left ventricle, increased left ventricular end-diastolic pressure and left atrial pressure are transmitted to the pulmonary capillaries, and fluid accumulates in the pulmonary compartment. Hypoalbuminaemia does not lead to oedema in the lungs because alveolar capillaries are more permeable to proteins and the difference in oncotic pressure between the capillaries and interstitial fluid is small.

Pleural effusion and ascites

In healthy people, the amount of fluid in the pleural and peritoneal spaces at any one time is small (about 10–50 ml), but there is a continuous turnover. In the pleural cavity, 5–10 litres of fluid are formed and reabsorbed daily. Accumulation of fluids in these spaces depends on similar haemodynamic factors as those outlined above.

Fluid accumulating in these spaces is sometimes classified as exudates or transudates. This classification is based primarily on protein concentration: Exudate has a protein concentration greater than 30 g/L and transudate has a protein concentration of less than

30 g/L. Serum/fluid lactate dehydrogenase ratio will also aid in this classification. A transudate is believed to occur when there is excess IF formation. This results from a decrease in oncotic pressure or a reduction in the reabsorption of IF (caused by elevated hydrostatic pressure). Examples of transudates are pleural effusion, ascites and pericardial effusion of nephrotic syndrome, and ascites of cirrhosis. An exudate results from increased capillary permeability (e.g. in infection of the pleura or peritoneal cavity), or because of obstruction to the lymphatic flow.

Hyponatraemia

Hyponatraemia is a common electrolyte disorder — 15–20% of hospital patients have serum sodium < 135 mmol/L and 3–5% patients have serum sodium < 130 mmol/L. Severe hyponatraemia is associated with increased mortality. When the serum sodium is 105 mmol/L or lower, the mortality rate is > 50%. However, this increased mortality may be due to the severity of the underlying cause of hyponatraemia rather than due to the hyponatraemia itself. Hyponatraemia is more common in the elderly due to the increased incidence of comorbidity and the use of drugs which impair water excretion.

Causes

Causes of hyponatraemia are listed in Table 2.11. Patients with pseudohyponatraemia have low sodium concentration in the plasma, as measured by indirect ion selective electrode (ISE), because of the high protein or lipid content of plasma. As the sodium concentration in plasma water is normal, measured serum osmolality will be normal.

Hyponatraemia due to a shift of water from the ICF is seen when the osmolarity of the ECF is high due to osmotically active substances such as glucose or mannitol in the ECF.

When hyponatraemia is accompanied by hypoosmolality, it is usual to classify hyponatraemia according to the ECF volume status — hypo-, normo- or hypervolaemia.

Table 2.11   Causes of hyponatraemia

Increased osmoles in ECF
  • Mannitol or glucose
Pseudohyponatraemia
  • Increased protein or lipids in plasma
Hyponatraemia with reduced ECF volume
  • Volume depletion (as in Table 2.10)
Hyponatraemia with 'normal' ECF volume
  • Water excess (as in Table 2.9)
Hyponatraemia with increased ECF volume
  • Congestive heat failure
  • Nephrotic syndrome
  • Cirrhosis

Hypovolaemic hyponatraemia

Hypovolaemic hyponatraemia is when there is a loss of sodium and water from the body, causing ECF volume depletion, which stimulates the secretion of ADH. If the patient is given water or hypotonic fluids, hyponatraemia will result. All causes of sodium and water loss from the body (e.g. gastrointestinal or renal loss) can cause such hyponatraemia. One of the most common causes of hyponatraemia is diuretic-induced hyponatraemia. Thiazides diuretics are more likely to cause hyponatraemia than loop diuretics. Loop diuretics inhibit sodium chloride reabsorption in the thick ascending limb of the loop of Henle. The reabsorption of sodium chloride in this segment is important for the generation of hyperosmolality of the medullary interstitum which is essential for increased water reabsorption. As loop diuretics interfere with this process, the ability to reabsorb water is reduced despite high ADH levels due to volume depletion. Thiazides, on the other hand, act on the early part of the distal tubule and do not interfere with the generation of medullary hyperosmolality. Therefore, water reabsorption is not impaired.

Cerebral salt wasting: Lesions of the CNS, such as subarachnoid haemorrhage and meningitis, have been associated with loss of sodium in the urine and is sometimes described as salt wasting syndrome.

However, many authorities question the existence of such a condition. It is believed that the loss of sodium in such situations is part of the SIADH syndrome.

Salt wasting nephropathy is seen occasionally in many renal diseases such as interstitial nephropathy, medullary cystic disease, polycystic disease of the kidney and toxic nephropathy due to cisplatin or chronic analgesic abuse. In this condition, there is an inappropriate loss of sodium in the urine leading to hypovolaemic hyponatraemia.

Adrenal failure is a rare cause of hypovolaemic hyponatraemia. Due to the absence of aldosterone, there is increased sodium loss accompanied by retention of potassium and hydrogen ions. Hyponatraemia is associated with hyperkalaemia and metabolic acidosis.

Hypervolaemic hyponatraemia

In oedematous states such as congestive heart failure, nephrotic syndrome and cirrhosis, hyponatraemia is observed, especially when the underlying condition is severe. Although the total ECF volume is increased in these situations, the effective circulating volume is decreased, leading to increased ADH secretion and water retention. Increased proximal tubular reabsorption of sodium leads to the decreased delivery of fluids to the distal segments of the nephron, further reducing the diluting ability of the kidney, thus contributing to the hyponatraemia. Diuretic therapy further aggravates this.

Euvolaemic hyponatraemia

Hyponatraemia due to increased water with normal total body sodium is described as euvolaemic hyponatraemia. This can arise either as a result of increased water intake or a reduced capacity to excrete water or a combination of the two. Increased water intake leading to hyponatraemia is rare as the kidney has the capacity to excrete more than 20 litres of free water per day. Psychogenic polydipsia is a rare disorder leading to hyponatraemia.

One of the common causes of euvolaemic hyponatraemia in hospital patients is the inappropriate infusion of hypotonic fluids after surgery. ADH is secreted in response to the pain, anxiety and surgical stress. When hypotonic fluid is given (4% dextrose saline or 5% dextrose), hyponatraemia results. In patients undergoing transurethral prostatectomy, acute hyponatraemia may develop due to the absorption of irrigation fluid (1.5% glycine) from the prostatic venous bed. When the glycine is metabolised, free water is generated. Factors contributing to the hyponatraemia include the duration of the operation, the hydrostatic pressure of the irrigation fluid and blood loss. With the introduction of newer techniques for prostatectomy, this is rarely seen now.

Syndrome of inappropriate ADH secretion (SIADH) is another common cause of euvolaemic hyponatraemia or dilutional hyponatraemia. In this syndrome, ADH is secreted in spite of low plasma sodium (low plasma osmolality), hence the term. There are a large number of causes of SIADH (Table 2.12). With the increased ADH, water is retained leading to the expansion of ECF and ICF. The former causes the release of factors leading to natriuresis, which is a feature of SIADH. The criteria for the diagnosis of SIADH are given in Table 2.13.

Patients with HIV may develop hyponatraemia due to a combination of factors: increased ADH secretion due to chest infection, CNS infection or malignancy; volume depletion due to diarrhoea, adrenal insufficiency due to infection with *Cytomegalovirus* or *Mycobacterium avium-intracellulare*.

Hyponatraemia due to water excess is seen in marathon runners and other endurance athletes. Hyponatraemia is due to a combination of excess water intake (intake greater than losses due to sweating), decreased excretion (as a result of reduced renal blood flow and stress-induced ADH release) and loss of sodium due to sweating. Non-steroidal anti-inflammatory drugs are thought to increase the risk of hyponatraemia in athletes due to the inhibition of prostaglandins. In the absence of prostaglandins, sodium reabsorption in the thick ascending loop of Henle is increased, thereby enhancing the action of ADH causing water retention.

Table 2.12   Causes of syndrome of inappropriate ADH secretion (SIADH)

Increased/inappropriate secretion from hypothalamus

- Pulmonary diseases
  - Pneumonia
  - Tuberculosis
  - Acute respiratory failure
  - Mechanical ventilation

- Neuropsychiatric diseases
  - Infections — meningitis, encephalitis
  - Head injury
  - Vascular — subarachnoid haemorrhage
  - Tumours

- Drugs
  - Cyclophosphamide
  - Carbamazepine
  - Ecstasy (MDMA)
  - Antidepressants

- Others
  - Post-operative
  - Exercise-induced hyponatraemia
  - Glucocorticoid deficiency
  - Hypothyroidism

Ectopic secretion of ADH

- Carcinoma of lungs — oat cell tumour
- Others — thymus

Potentiation of the effect of ADH

- Chlorpropamide
- Carbamazepine

Administration of ADH or ADH-like compounds

- Oxytocin
- Desmopressin

Table 2.13   Criteria for the diagnosis of SIADH

- Hyponatraemia with hypoosmolality
- Absence of ECF volume depletion
- Absence of ECF excess — oedema
- Urine osmolarity inappropriate for the plasma osmolarity (urine not maximally dilute)
- Normal renal, adrenal and thyroid status
- Urine sodium excretion >20 mmol/L (continued natriuresis)

Case 2.1

A 65-year-old man was admitted with a history of cough, increasing disorientation and confusion. He had been a heavy smoker for 50 years. Admission blood results were:

Serum		Reference Range
Sodium (mmol/L)	116	135–145
Potassium (mmol/L)	4.7	3.5–5.0
Bicarbonate (mmol/L)	25	23–32
Chloride (mmol/L)	87	90–108
Urea (mmol/L	3	3.5–7.2
Creatinine (μmol/L)	90	60–112
Glucose (mmol/L)	5	
Osmolality (mOsm/kg)	233	285–295
Urine		
Osmolality (mOsm/kg)	698	
Sodium (mmol/L)	43	

Hyponatraemia in this patient is due to too much water in his body (dilutional hyponatraemia). It is unlikely to be due to loss of sodium as he had no features of volume depletion and his serum urea was normal. With a history of chronic smoking most likely diagnosis is bronchogenic carcinoma causing SIADH.

Hyponatraemia, often symptomatic, is seen in young adults taking ecstasy (methylenedioxy methamphetamine or MDMA). A combination of excessive water intake and reduced excretion due to inappropriate ADH secretion is the cause of hyponatraemia.

Desmopressin, an ADH analogue, which is used in the treatment of diabetes insipidus, von Willibrand disease and enuresis, can cause hyponatraemia. Oxytocin, which has some ADH-like activity when given in large amounts with hypotonic fluids, has been reported to cause hyponatraemia.

Hyponatraemia is also a recognised complication of antidepressants: selective serotonin reuptake inhibitors (SSRI), tricyclics antidepressants and monoamine oxidise inhibitors. Incidence of hyponatraemia with SSRIs is greater than with other classes of antidepressants. The mechanism of hyponatraemia is thought to be due inappropriate ADH secretion. Risk factors for the development of hyponatraemia include old age, female gender, use of diuretics and low body weight.

Glucocorticoid insufficiency (ACTH deficiency or secondary adrenal failure) causes hyponatraemia. Glucocorticoids are essential for the excretion of water load. In the absence of glucocorticoids, there is non-osmotic stimulation of ADH secretion, the mechanism of which is unclear. A direct action of glucocorticoids on the renal tubules in excreting water load has also been suggested to contribute to hyponatraemia.

Case 2.2

A 50-year-old lady complained of tiredness and weakness. Her family noticed that her skin had become darker although she had never been abroad. Investigations by her GP showed the following results:

Serum		Reference Range
Sodium (mmol/L)	124	135–145
Potassium (mmol/L)	6	3.5–5.0
Bicarbonate (mmol/L)	17	23–32
Chloride (mmol/L)	96	90–108
Urea (mmol/L	18	3.5–7.2
Creatinine (µmol/L)	100	60–112
Glucose (mmol/L)	3.2	

> *History of tiredness, dark skin and hyponatraemia with hyperkalaemia is suggestive of Addison's disease. She underwent a short Synacthen test which confirmed the diagnosis. She was treated with hydrocortisone and fludrocortisone and her tiredness and weakness improved rapidly.*

Hyponatraemia that is observed in sick patients is sometimes described as sick cell syndrome. However, many have questioned the existence of such a syndrome. The postulated mechanism is leakage of non-diffusible solutes into the ECF with movement of sodium into cells.

Hypothyroidism is often described to cause hyponatraemia. The exact mechanism is unclear but is thought to be due to increased ADH secretion as a result of reduced cardiac output.

Hyponatraemia can develop in subjects who have a very low intake of osmolar load to excrete. This may be seen in people consuming large volumes of beer (beer potomaina) or those who eat a very low protein diet. A normal adult eating an average diet produces an osmolar load between 600 to 900 mosmol/day. If the minimum urine osmolality that can be achieved (maximum diluting capacity) is 50 mosmol/kg, then the water intake has to exceed 12 litres before water retention occurs. If the osmolar load is low (100 mosmol/day), as in beer drinkers, water retention and hyponatraemia develops if the intake of water exceeds 2 litres per day. Contributing factors to beer potomaina include low electrolyte content of beer and the consumption of a low protein diet.

Table 2.14 shows a list of drugs causing hyponatraemia and their possible mechanisms.

Clinical Features of Hyponatraemia

The symptoms and signs related to hyponatraemia depend on the severity and rapidity of development of hyponatraemia. Features include nausea, vomiting, lethargy, confusion, headache, loss of consciousness and fits. These symptoms are thought to be due to the swelling of brain cells caused by the influx of water. Symptoms are mild or absent if the hyponatraemia develops slowly, even if it is severe

Table 2.14　Drugs causing hyponatraemia with possible mechanisms

---

Drugs affecting sodium and water balance
- Diuretics

Drugs causing increased ADH secretion

- Antidepressants — selective serotonin reuptake inhibitors (SSRI), tricylics and monoamine oxidase (MAO) inhibitors
- Antipsychotic drugs — phenothiazine
- Antiepileptic drugs — carbamazepine, sodium valporate
- Anticancer drugs — vincristine, cisplatin, cyclophosphamide
- Others — ecstasy (MDMA)

Drugs potentiating the action of ADH

- Antiepileptics — carbamazepine, lamotrigine
- Antidiabetics — chlorpropamide
- Anticancer drugs
- Non-steroidal anti-inflammatory drugs (NSAIDs)

Drugs mimicking ADH

- Desmopressin
- Oxytocin

---

due to neuronal adaptation. During slowly developing hyponatraemia, neuronal cells reduce their osmolality by excreting osmoles (called osmolytes), thereby preventing water movement into the cells. These osmolytes have been identified as myo-inositol, scyllo-inositol, N-acetyl aspartate and choline.

Investigation of Hyponatraemia

When investigating the cause of hyponatraemia, which is not clinically apparent, measurement of plasma osmolality is often helpful to identify pseudohyponatraemia and hyponatraemia due to osmotically active substance (Figure 2.5). If the osmolarity is normal, it is indicative of pseudohyponatraemia and if it is high, it is due to the presence of

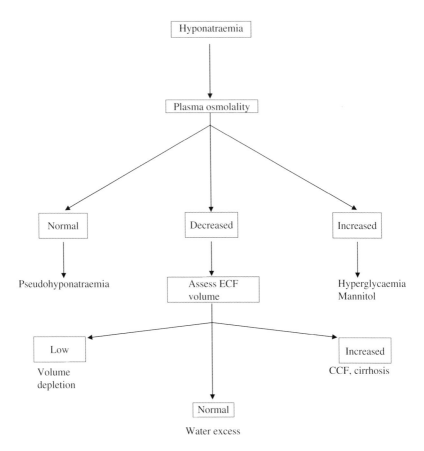

Figure 2.5   A scheme for investigation of hyponatraemia.

osmotically active substances. When hyponatraemia is accompanied by hypoosmolality, the volume status of the patient should be assessed. Measurement of urine osmolarity and urine sodium may be helpful. Urine that is not maximally dilute (inappropriate osmolarity), in the presence of euvolaemic hyponatraemia, is indicative of SIADH. A urine sodium of 20 mmol/L or greater is also seen in SIADH. Urine sodium < 20 mmol/L in the presence of hypovolaemic hyponatraemia suggests non-renal loss of sodium. Urine sodium >20 mmol/L, in the presence of hypovolaemic hyponatraemia, suggests renal loss of sodium. In normovolaemic hyponatraemia, serum urea is normal or low due to dilution and reduced reabsorption. Serum urate is also often low in these patients.

Other tests, which may be indicated, are thyroid function tests and short ACTH stimulation (Synacthen) tests to exclude thyroid and adrenal failure respectively.

The two main categories of hyponatraemia, water excess and volume depletion, can be distinguished by examining the volume status, plasma urea and creatinine concentrations (Table 2.15).

Management of Hyponatraemia

The underlying disorder causing hyponatraemia should be treated. In hypo- and hypervolaemic hyponatraemia, treatment of the underlying disorder is usually sufficient to correct the hyponatraemia. In euvolaemic hyponatraemia, treatment of the low sodium may be required if it is accompanied by clinical features of hyponatraemia. In severe acute symptomatic hyponatraemia, correction of serum sodium by infusion of hypertonic saline may be required. However, serum sodium should not be corrected rapidly as there is a risk of development of central pontine myelinolysis. It is recommended that the maximum increase in plasma sodium should not be more than 12 mmol/L/24-hr or to a value of

Table 2.15 Differentiation between hyponatraemia of volume depletion and hyponatraemia of water excess (dilutional)

	Volume depletion	Dilutional hyponatraemia
Recent weight loss	+	−
Thirst	+	
Orthostatic hypotension & tachycardia	+	−
Decreased CVP	+	−
Poor skin turgor & dry mucous membranes	+	−
Haemo concentration	+	−
Plasma urea	High	Normal
Plasma urea/creatinine	High	Normal

125 mmol/L. Diuretics such as furosemide may be required to remove the large amounts of sodium given.

In mild cases or in chronic hyponatraemia, no specific treatment may be necessary. Restriction of water intake to 500–800 mL/day will result in slow rise in plasma sodium concentration. Drugs, which oppose the action of ADH, may be used to reduce the water content of the body. Until recently, demeclocycline (a tetracycline) and lithium, both of which block the action of ADH, have been used. More recently, AVP receptor antagonists, conivaptan, lixivaptan, and tolvaptan are available. Of these, conivaptan blocks both V1a and V2 receptors whereas the other two antagonists only block V2 receptors.

Hypernatraemia

Hypernatraemia is not uncommon in hospital populations, occurring in up to 2% of patients. Many of these cases develop hypernatraemia after admission to the hospital and it has been shown that up to 60% of hypernatraemia is hospital acquired.

Hypernatraemia is most commonly caused by a lack of water and rarely due to excess sodium. Inadequate water intake in hospital patients, especially the elderly, is a common cause of hospital-acquired hypernatraemia. This is exacerbated if there is increased losses of water, e.g. fever or hyperventilation. Loss of hypotonic (sodium concentration less than that of plasma) fluid can also cause hypernatraemia as in osmotic diuresis (e.g. hyperosmolar non-ketotic diabetic coma, HONK, see Chapter 10). Hypernatraemia can also result from the inability to concentrate urine because of ADH deficiency or reduced responsiveness to ADH. These patients usually present with polydipsia and polyuria.

A very rare cause of hypernatraemia is excess sodium/salt administration orally or intravenously, without adequate amounts of water. In adults, it is difficult to cause hypernatraemia with oral sodium administration unless they have a faulty thirst mechanism. Parenteral sodium-containing drugs, when given in large quantities, can rapidly cause hypernatraemia. In children, oral salt administration can cause hypernatraemia due to the inability to access water and the limited capacity to excrete sodium load.

The cause of hypernatraemia is most often clinically obvious and very little further investigation is required. In the rare case of hyper-natraemia due to excess sodium intake, measurement of urine sodium may be helpful. If the hypernatraemia is due to water deficit, the urine sodium is likely to be low due to reduced ECF volume and activation of renin–angiotensin–aldosterone system. In salt poisoning, urine sodium excretion is high.

Management of Hypernatraemia

In addition to correcting the underlying cause, water deficit should be corrected with oral or intravenous fluids (5% dextrose). Hyper-natraemia due to water deficit especially if chronic, should be corrected slowly to prevent cerebral oedema. In chronic hypernatraemia, brain cells adapt by generating osmolytes (as described above) to increase cellular osmolarity to prevent neuronal cell dehydration. If the hyper-natraemia is corrected rapidly, water will move rapidly into brain cells causing cerebral oedema. Slow correction will allow neuronal cells to adapt and reduce intracellular osmolarity. Serum sodium should not be corrected by more than 12 mmol/L/24-hr.

Case 2.3

A 50-year-old man was found to have fracture of the base of skull following a road traffic accident. Two days after the accident, he was pass-ing large volumes of urine and his blood results were:

Serum		Reference Range
Sodium (mmol/L)	165	135–145
Potassium (mmol/L)	3.5	3.5–5.0
Bicarbonate (mmol/L)	23	230–32
Chloride (mmol/L)	126	90–108
Urea (mmol/L	13	3.5–7.2
Creatinine (µmol/L)	140	60–112
Glucose (mmol/L)	4.5	

> *Hypernatraemia in this patient is due to loss of water from the body. Large volumes of urine suggest an inability to concentrate the urine. With a history of skull fracture, the most likely diagnosis is inadequate ADH secretion due to damage to the pituitary following the trauma. He was given intravenous dextrose as well as DDAVP, an analogue of ADH, and his urine volume and plasma sodium decreased.*

Polyuria

Polyuria defined as a urine output of 2.5 litres or greater in 24 hours is caused by many conditions and these are listed in Table 2.16. Excess fluid administration and diuretics are not discussed further as they are obvious.

When the osmolar load is high, the ability to concentrate urine is diminished (Figure 2.2), causing polyuria. This is due to interference in the counter current mechanism, leading to reduced medullary osmolarity and as a result, inability to concentrate. Diabetes mellitus is an example of osmotic diuresis caused by high glucose concentration. In chronic renal failure, each remaining nephron receives a high osmolar load causing polyuria; however, the polyuria is not severe as the number of nephrons are reduced. Very high protein intake can cause an osmotic diuresis due to the increased urea load. Diabetes insipidus is a condition due to deficiency of ADH (cranial or neurogenic diabetes insipidus) or failure of the renal tubules to respond to ADH (nephrogenic diabetes insipidus), causing the passage of large amounts of dilute urine (polyuria). Cranial diabetes insipidus can be due to trauma, infection, infiltration, tumours or occasionally idiopathic and very rarely due to a genetic defect.

Acquired nephrogenic diabetes insipidus, due to hypercalcaemia and hypokalaemia is a relatively common cause. The polyuria however is not severe, urine volume rarely exceeding 4 litres per day.

Investigation of Polyuria

Initial investigation of a patient presenting with polyuria should include simple blood tests such as glucose, potassium, calcium

Table 2.16   Causes of polyuria

Osmotic diuresis
- Diabetes mellitus
- Chronic renal failure
- High intake of protein or solute

Inability to concentrate urine
- Lack of ADH — neurogenic diabetes insipidus
  — Hereditary
  — Acquired
  - Trauma
  - Tumours
  - Surgery
  - Granulomatous disease — sarcoidosis
  - Meningitis and encephalitis
  - Idiopathic/autoimmune
  - Genetic
- Reduced responsiveness to ADH — nephrogenic diabetes insipidus
  — Hereditary
  — Acquired
  - Hypokalaemia
  - Hypercalcaemia
  - Drug induced — lithium
  - Renal disease
  - Post-obstructive uropathy

Increased intake
- Psychogenic polydipsia

and creatinine. These will exclude diabetes mellitus, hypokalaemia, hypercalcaemia and chronic renal failure. Further investigations include plasma sodium and urine osmolality (either a random or early morning urine). Diabetes insipidus is excluded if the urine osmolality is > 750 mOsmol/kg. In diabetes insipidus (cranial as well as nephrogenic), urine osmolality will be low, and plasma sodium and osmolality are high or near the upper end of the reference range. In psychogenic polydipsia, urine osmolality is low, together with low or normal plasma sodium and osmolality. A water deprivation test may be required if these tests are equivocal. In this test, the patient is deprived of water for 8 hours under supervision, urine osmolality is measured every 2 hours, and the patient is weighed regularly. The test is stopped if the patient loses more than 3% of body weight. At the end of 8 hours, an injection of 20-μg desmopressin (DDAVP), an analogue of ADH, is given and urine osmolality measured hourly for a further 2 hours. During this part of the test, the patient is allowed to drink water. If urine osmolality during the first 8 hours reaches 750 mOsmol/kg or higher, diabetes insipidus is excluded. If urine osmolality increases only after desmopressin, then cranial diabetes insipidus is diagnosed. Failure to achieve this urine osmolality even after desmopressin is indicative of nephrogenic diabetes insipidus.

Water deprivation test may give equivocal results because the patient may be drinking surreptitiously or adding water to urine. In patients who had been drinking excessive amounts of water prior to the test and in partial diabetes insipidus, the results may be equivocal as well. In patients who have been drinking excessively, urine osmolarity may not rise because the plasma osmolarity has not gone above the osmotic threshold for ADH release and/or due to washout of the renal medulla, resulting in inability of the renal tubules to reabsorb water.

If the results are equivocal, some centres recommend hypertonic saline infusion test. In this test, plasma ADH response to an intravenous infusion of 5% saline is measured. In patients with nephrogenic diabetes insipidus or in psychogenic polydipsia, the response will be normal, but those with cranial diabetes insipidus will have an abnormal response.

Further Reading

1.   Hannon MJ, Thompson CJ. The syndrome of inappropriate antidiuretic hormone: Prevalence, causes and consequences. *Eur J Endocrinol* 2010; 162 Suppl 1:S5–12.
2.   Liamis G, Milionis H, Elisaf M. A review of drug-induced hyponatremia. *Am J Kidney Dis* 2008; 52:144–153.
3.   Rose BD, Post TW. *Clinical Physiology of Acid–base and Electrolyte Disorders.* 2001 McGraw-Hill: New York, 5th Edition.

4. Sherlock M, Thompson CJ. The syndrome of inappropriate antidiuretic hormone: Current and future management options. *Eur J Endocrinol* 2010; 162 Suppl 1:S13–8.

5. Zenenberg RD, Carluccio AL, Merlin MA. Hyponatremia: Evaluation and management. *Hosp Pract (Minneap)* 2010; 38:89–96.

Summary/Key Points

1. Sodium is the main extracellular cation. Amount of sodium in the body determines the ECF volume.

2. Sodium concentration in the ECF is determined by the amount of water in the body.

3. Sodium balance is maintained by the regulation of renal excretion which in turn, is regulated by renin–angiotensin–aldosterone system and other factors such as atrial natriuretic peptide.

4. Thirst and ADH maintain water balance. ADH secretion is increased by reduced ECF volume and a decrease in BP in addition to a rise in osmolality.

5. Disorders of sodium and water are common. Loss of sodium results in ECF volume depletion and gain of sodium results in oedema. Loss or gain of water is shared by ICF and ECF and is easily detected by changes in serum sodium concentration.

6. Hyponatraemia is a common electrolyte problem and is usually divided according to the ECF volume status: hypovolaemic, euvolaemic and hypovolaemic hyponatraemia. Common causes are diuretic therapy, inappropriate fluid therapy and syndrome of inappropriate ADH secretion (SIADH).

7. SIADH is a diagnosis of exclusion and the urine osmolality is inappropriately high (not maximally dilute).

8. Treatment of hyponatraemia should be slow to avoid central pontine myelnosis.

9. Hypernatraemia is usually due to inadequate body water.

10. Polyuria can be caused by excess osmolar load (diabetes mellitus) or inability to concentrate urine (diabetes insipidus).

chapter 3

Potassium

Introduction

Potassium is the major intracellular cation and the concentration in cells is about 100–150 mmol/L of cell water depending on the cell type. Total body potassium is dependent on muscle mass and in an average 70-kg man, there is approximately 3000 mmol of potassium. Most (95–98%) of this is in the cells. The amount of potassium in the extracellular fluid (60 mmol) is small and is similar to the amount of potassium taken daily in the diet. The ratio of intracellular to extracellular potassium is a critical determinant of the cell membrane polarization which influences nerve conduction and muscle contraction. Any deviation in plasma potassium from normal (3.6–5.0 mmol/L), either low or high, can cause significant clinical effects.

The plasma potassium concentration depends on intake, excretion and the distribution between ECF and ICF. Dietary potassium intake varies between 50 and 150 mmol per day. Plasma potassium is regulated by homeostatic mechanisms controlling internal as well as external balance. The kidney largely controls external balance, which determines total body potassium, with a small (5–15 mmol) but sometimes important contribution by the gut. Internal balance determines the ratio of intracellular to extracellular potassium.

Internal Balance

Several factors modulate the distribution of potassium between the ICF and ECF (Table 3.1). The high intracellular potassium

Table 3.1   Factors modifying the distribution of potassium between intracellular and extracellular fluids

---

Sodium–potassium pump
Acid–base status
- H^+ concentration
- Bicarbonate concentration

Insulin
Catecholamines
Aldosterone

---

concentration (150 mmol/L) relative to the extracellular concentration (5 mmol/L) is primarily regulated by the sodium–potassium–ATPase pump, which is present in all cell membranes. Three sodium ions are moved out of the cell and two potassium ions are transported into the cell by this pump for each molecule of ATP hydrolysed. Stimulation of the pump will cause hypokalaemia and inhibition (e.g. digoxin toxicity) will cause hyperkalaemia.

The concentration of hydrogen and bicarbonate ions separately influence the internal balance of potassium. An increase in H^+ ions or a decrease in bicarbonate will increase the extracellular potassium concentration due to the shift of potassium from the cells to the ECF in exchange for H^+ ions. However, hyperkalaemia seen with organic acids, where the anions are permeable across the cell membrane is less than after mineral acids, because the organic anions follow the H^+ ions into the cell, thereby maintaining electroneutrality.

Insulin stimulates the sodium–potassium pump and promotes the uptake of potassium ions into liver and muscle cells, thereby lowering plasma potassium levels. This effect is independent of the effect of insulin on glucose uptake and is important in regulating plasma potassium concentration. An increase in plasma potassium is also known to be a stimulus for insulin release.

Catecholamines are known to cause hypokalaemia by the redistribution of potassium into the intracellular compartment. This effect is thought to be mediated by the activation of the sodium–potassium pump via stimulation of β-adrenergic receptors. Hypokalaemia found frequently in acutely ill patients might be explained by this effect.

β-Adrenergic agonists increase potassium uptake while β_2 blockers and α-adrenergic agonists decrease potassium uptake by cells causing hypo and hyperkalaemia respectively.

External Balance

Potassium rich foods are fruits, meat, vegetables (spinach, tomatoes, etc.) and dried fruits. Potassium is almost completely absorbed in the gastrointestinal tract. External potassium balance, which determines total body potassium is regulated mainly by the kidney, as excretion by other routes, such as the skin and gastrointestinal tract, are relatively small, usually 5–15 mmol/day. The contribution of the gastrointestinal tract in maintaining potassium balance is increased in renal failure where potassium excretion by the colon is increased. Potassium is freely filtered at the glomerulus and almost all of it is completely reabsorbed before the distal convoluted tubule. Urinary excretion of potassium depends on potassium secretion in the distal tubule. A number of factors influence potassium secretion and therefore, potassium excretion (Table 3.2). A chronic high potassium intake increases potassium secretion. The exact mechanism by which this is brought about is unclear. Plasma potassium influences the secretion of potassium by the distal renal tubules and thereby maintains plasma potassium concentration. An acute increase in sodium delivery to the distal tubule will increase sodium reabsorption, which will enhance urinary potassium excretion by increasing the electrical gradient. Chronic increases in sodium delivery, such as during chronic high salt intake has no effect because of the influence of aldosterone secretion.

Table 3.2   Factors influencing potassium secretion in the distal tubule

- Potassium intake
- Plasma potassium
- Sodium delivery to the distal tubule
- Aldosterone
- Hydrogen ion concentration
- Flow rate of tubular fluid
- Presence of nonabsorbable anions in the tubular fluid, e.g. carbenicillin

Aldosterone has a major role in potassium homeostasis. Aldosterone acts at the principal cells of the distal tubule and collecting ducts and increases the reabsorption of sodium and secretion of potassium and hydrogen ions. High potassium levels directly stimulates aldosterone secretion. Aldosterone also increases hydrogen ions secretion by the intercalated cells of the collecting ducts.

Alkalosis increases intracellular potassium, thus enhancing distal tubular potassium loss. The reverse is true in acidosis.

Increased flow of fluid maintains a favourable concentration gradient, thereby enhancing potassium secretion.

Presence of nonabsorbable anions such as bicarbonate or carbenicillin in the distal tubular fluid increases potassium secretion by increasing the transepithelial potential difference.

The kidney adapts to changes in potassium intake. However, when the potassium intake is very low or absent, there is an obligatory loss of 10–15 mmol/day. This is in contrast to sodium whose excretion closely matches the intake. Therefore, potassium depletion can arise if potassium intake is low.

Assessment of Potassium Status

Plasma potassium concentration may not necessarily reflect total body potassium status as it may be affected by redistribution between the ICF and ECF. Both red and white cells may be used to measure intracellular potassium concentration to assess the potassium status, however these methods are cumbersome. Total body potassium can be assessed by counting the radioactivity from the naturally occurring isotope (^{40}K) using a whole body counter. This measurement should be related to lean body mass as the amount of potassium in an individual depends on the size of the individual. However, these methods are only useful as research tools.

Of these methods, plasma potassium is the only readily available method. The relationship between plasma potassium and total body potassium is curvilinear so that there is relatively little potassium deficit until the plasma potassium falls below 3 mmol/L. For measurements below 3 mmol/L, a drop of 1 mmol/L indicates a deficit of about 200–300 mmol.

Urinary potassium

Measurement of urine potassium levels may be useful in the differential diagnosis of potassium disorders. The amount of potassium excreted in 24 hours reflects dietary intake in healthy individuals. In the presence of hypokalaemia, a urine potassium excretion of 20 mmol/day or greater suggests renal potassium loss as the kidneys can reduce the excretion of potassium to 10–15 mmol/day. A potassium excretion of less than 20 mmol/day suggests non-renal causes of hypokalaemia. As 24-hour urine is cumbersome and difficult to collect accurately, a spot urine potassium is sometimes used. A value greater than 20 mmol/L is suggestive of renal loss of potassium. However, the concentration of potassium in spot urine may be influenced by the urine flow rate. If the urine is very dilute, the urine potassium concentration will be lower than 20 mmol/L even though the 24-hour excretion is >20 mmol/day (e.g. if the urine volume is 5 litres, a urine potassium of 10 mmol/L will be misleading). To avoid this problem, the transtubular potassium gradient (TTKG) is sometimes calculated by simultaneously measuring potassium and osmolality in urine and plasma.

TTKG = (urine potassium × serum osmolality)/(plasma potassium × urine osmolality)

TTKG gives an estimate of the serum to tubular fluid potassium ratio at the cortical collecting tubules. A value of 3 or less for TTKG suggests renal conservation of potassium and a value of greater than 7 suggests renal loss.

Hypokalaemia (Table 3.3)

Hypokalaemia is common in hospital patients. The frequency of hypokalaemia varies depending on the population studied. In hospitalized patients, the incidence has been reported to vary from 8% to 22%. In patients on diuretics (excluding potassium-sparing diuretics), up to 50% may have low serum potassium. This incidence is higher in people of African origin and in women. In patients with

Table 3.3   Causes of Hypokalaemia

Reduced intake

Transcellular shift
- Insulin
- Refeeding syndrome
- Metabolic alkalosis
- Catecholamines and other sympathomimetics
- Rapid proliferation of cells — treatment of pernicious anaemia with vitamin B_{12}
- Periodic paralysis

Increased losses
- Renal losses
  - Diuretics — a common cause due to increased delivery of sodium to distal tubules
  - Osmotic diuresis — glycosuria
  - Metabolic alkalosis (e.g. vomiting)
  - Increased mineralocorticoid activity
  - Primary and secondary hyperaldosteronism
  - Cushing's syndrome
  - Liquorice ingestion
  - Bartter's Syndrome/Gitelman's Syndrome/Liddle Syndrome
  - Renal tubular acidosis
  - Magnesium depletion
- Gastrointestinal losses
  - Diarrhoea
  - Vomiting/Nasogastric suction
  - Laxative abuse
  - Ureterosigmoid anastamosis
  - Villus adenoma
  - Zollinger-Ellison syndrome
  - VIPoma

eating disorders and in AIDS patients, the incidence is relatively high (up to 20%). The risk of hypokalaemia is also higher in patients after bariatric surgery. Severe hypokalaemia (plasma potassium <2.4 mmol/L) is seen in approximately 1% of acute admissions. Hypokalaemia can occur with or without a deficit of total body potassium.

Causes

Hypokalaemia can be caused by decreased intake, shift of potassium into cells or abnormal losses (gastrointestinal tract or kidney). Occasionally pseudohypokalaemia is seen when the measured plasma potassium is low due to *in vitro* uptake into cells when there are large number of cells, e.g. in acute myeloid leukaemia.

If potassium intake is very low, potassium depletion and hypokalaemia will develop as there is an obligatory loss of about 10–15 mmol/day. This type of hypokalaemia is occasionally seen in:

(a) elderly subjects who consume a diet poor in potassium,
(b) starving patients who can develop secondary hyperaldosteronism and urinary potassium loss due to sodium depletion, and
(c) in patients on intravenous fluids with inadequate potassium supplements.

Transcellular shift

Hypokalaemia seen in hospital patients is explained by increased uptake of potassium into cells by stress-induced catecholamine release. β-Adrenergic agonists such as salbutamol, fentolol, and terbutaline can cause hypokalaemia. Rare causes of transcellular shift include glue sniffing (toluene) and ingestion of barium salts. Severe hypokalaemia may occur due to rapid uptake of potassium together with phosphate and magnesium during refeeding syndrome which occurs when malnourished patients are treated with entral or parenteral feeding.

During treatment of megaloblastic anaemia, hypokalaemia develops due to the rapid uptake of potassium by the newly formed red cells.

Familial hypokalaemic periodic paralysis is a rare inherited disorder due to a defect in the potassium channel and is characterised by episodes of muscle weakness and hypokalaemia. Thyrotoxic periodic paralysis is seen in oriental males, who present with hypokalaemia and paralysis, which usually occurs after a heavy carbohydrate meal and/or exercise. Paralysis and hypokalaemia manifest only when the patients are thyrotoxic.

Gastrointestinal losses

Causes of potassium depletion due to gastrointestinal loss include diarrhoea, vomiting, nasogastric suction and loss via fistula. Diarrhoeal fluid is rich in potassium and severe diarrhoea is accompanied by loss of bicarbonate. These patients tend to have metabolic acidosis (hyperchloraemic) and hypokalaemia. Contributing factors to the hypokalaemia in vomiting or nasogastric suction are loss of potassium, metabolic alkalosis, secondary hyperaldosteronism, and lack of intake. Metabolic alkalosis increases potassium uptake into cells and causes potassium loss in the urine. Loss of sodium and water leads to secondary hyperaldosteronism, which in turn, leads to further loss of potassium in the urine. Diversion of the urinary tract to the colon, as in ureterosigmoid anastomosis, leads to potassium loss as a result of the colonic mucosa reabsorbing sodium and chloride in exchange for potassium and bicarbonate. Rare causes of gastrointestinal loss of potassium include villous adenoma of the rectum which causes loss of large volumes of mucus with potassium, sodium and chloride, pancreatic tumours secreting vasoactive intestinal peptide (VIPoma) and gastrin-secreting tumours (Zollinger-Ellison syndrome).

Renal losses

Diuretics are one of the most common causes of hypokalaemia. The incidence of hypokalaemia in patients taking diuretics (non-potassium–sparing diuretics) varies from 20–50% (see below). Osmotic diuresis is another cause of potassium loss. Metabolic alkalosis can lead to renal potassium loss due to increased secretion of potassium in the distal tubules in exchange for sodium reabsorption.

Aldosterone excess (primary or secondary) causes increased loss of potassium and hypokalaemia. Hypokalaemia is also seen in subjects taking large amounts of liquorice or patients treated with carbenoxolone.

Active ingredients in these products such as, glycyrrhizic acid, glycyrrhetinic acid and their metabolites inhibit the enzyme, 11β-hydroxysteroid dehydrogenase type 2 that converts cortisol to cortisone, allowing cortisol to bind to the renal mineralocorticoid receptor, causing hypokalaemia.

In Cushing's syndrome, the concentration of cortisol exceeds the capacity of 11β-hydroxysteroid dehydrogenase type 2. Therefore, cortisol binds to mineralocorticoid receptors causing a loss of potassium. Hypokalaemia is more likely to be seen in Cushing's syndrome due to ectopic ACTH syndrome as the level of cortisol is usually high. In magnesium depletion, hypokalaemia is often seen due to the underlying cause, e.g. gastrointestinal loss causing loss of magnesium and potassium. In addition, hypomagnesaemia causes loss of potassium in urine by a mechanism not fully understood.

Case 3.1

A 62-year-old lady was referred by her GP to the hospital because of muscle weakness and hypertension. Her serum potassium was found to be 2.8 mmol/L and a 24-hour urine potassium was 60 mmol/d.

A urine potassium excretion of more than 20 mmol/day, in the presence of hypokalaemia, suggests renal loss of potassium. In a hypertensive patient, diuretic treatment should first be excluded. If this is excluded, excess mineralocorticoid is likely. Plasma renin and aldosterone were measured. This showed low renin and high aldosterone, confirming a diagnosis of primary hyperaldosteronism.

Bartter's syndrome and Gitelman's syndrome are two inherited disorders causing hypokalaemia. Bartter's syndrome, characterised by hypokalaemic metabolic alkalosis with high renin and aldosterone, is due to mutation in the genes coding for the sodium–potassium–chloride cotransporter (NKCC2), apical potassium channel (ROMK1) or the basal chloride channel (CLC-K) in the thick ascending limb. In Gitelman's syndrome, in addition to hypokalaemic alkalosis, there is hypocalciuria and hypomagnesaemia. The defect in this syndrome has been localised to thiazide-sensitive sodium–chloride cotransporter (NCCT) in the distal tubule. Patients with Bartter's Syndrome may remain asymptomatic until adult life. Gitelman's syndrome can present in childhood or adult life, with symptoms such as muscle weakness, tetany and fits.

Liddle's syndrome is a rare inherited disease characterised by hypertension, hypokalaemia and suppressed renin activity and aldosterone concentration. This is due to a mutation in the amiloride-sensitive epithelial sodium channel leading to increased sodium reabsorption and potassium secretion.

In type I (distal) renal tubular acidosis, the impaired sodium–hydrogen exchange in the distal tubule leads to increased sodium–potassium exchange. In type II (proximal) renal tubular acidosis, increased bicarbonate delivery to the distal tubule causes increased potassium secretion.

Other rare causes of potassium wasting due to increased mineralocorticoid activity include rare forms of congenital adrenal hyperplasia (11β-hydroxylase deficiency and 17α-hydroxylase deficiency), causing increased production of deoxycorticosterone (DOC).

Case 3.2

A 21-year-old female with an eating disorder was admitted to hospital with generalised weakness. Initial investigation showed:

Serum		Reference Range
Sodium (mmol/L)	135	135–145
Potassium (mmol/L)	1.8	3.5–5.0
Bicarbonate (mmol/L)	40	23–32
Chloride (mmol/L)	2.2	3.5–7.2
Urea (mmol/L)	83	90–108
Creatinine (µmol/L)	65	60–112

The cause of severe hypokalaemia in this patient is likely to be multifactorial. Reduced intake, increased loss due to abuse of diuretics and purgatives and bulimia are all contributing factors. Repeated vomiting causes metabolic alkalosis, which in turn can reduce plasma potassium and the accompanying secondary hyperaldosteronism due to volume depletion, further aggravates the hypokalaemia.

Drugs (Table 3.4)

Drugs are major causes of hypokalaemia with diuretics being the commonest among them. Both loop diuretics (furosemide and ethacrynic acid) and thiazides, which act on the distal tubules, cause hypokalaemia due to increased delivery of sodium to the collecting ducts where the sodium is reabsorbed in exchange for potassium. Loss of sodium and water from the body stimulates the renin–angiotensin–aldosterone system causing secondary hyperaldosteronism. Increased flow rate in the tubules further enhance potassium secretion due to the increase in potassium gradient. Carbonic anhydratase inhibitors, which cause loss of bicarbonate, increase potassium loss as a result of the bicarbonate reaching the distal tubule and increasing the electronegativity of the lumen facilitating the secretion of potassium.

Amphotercin an anti-fungal drug and cisplatin cause loss of potassium in the urine and is usually accompanied by hypomagnesaemia. Presence of carbenicillin in the distal tubular fluid increases potassium secretion by increasing the transepithelial potential difference.

Table 3.4   Drugs causing hypokalemia

Diuretics
  • Loop diuretics — Furosemide, ethacrynic acid
  • Thiazides — Hydrochlorthiazide
  • Carbonic anhydrase inhibitors — Acetazolamide

Amphotericin B
Pencillins
  • Carbenicillin

Cisplatin
Beta agonists
  • Theophylline

Aminoglysoides

Effects of Hypokalaemia and Potassium Depletion (Table 3.5)

It is unusual to see symptoms or signs until the serum potassium falls below 3.00 mmol/L. Potassium depletion affects several tissues including the cardiac muscle, skeletal muscle, smooth muscle and the kidney. Hypokalaemia causes hyperpolarisation of membrane potential resulting in increased myocardial excitability, which can precipitate cardiac arrhythmias ranging from atrial tachycardia, ventricular tachycardia or fibrillation. Development of these features depends on the degree of hypokalaemia as well as the rate of change in serum potassium. ECG changes seen in hypokalaemia are depressed ST segment, flattened T waves and prominent or biphasic U waves (Figure 3.1). Toxicity of digoxin and similar cardiac glycosides is also enhanced.

Patients with hypokalaemia have muscle weakness and in severe cases, paralysis with absent reflexes can be seen. An increase in plasma CK is relatively common. Occasionally, acute lysis of muscle cells (rhabdomyolysis) develops. The smooth muscle of the gut is affected causing reduced motility or paralytic ileus. The effect of hypokalaemia on the smooth muscle of blood vessels can lead to hypotension.

Table 3.5   Consequences of hypokalaemia

Cardiac arrhythmias
- ECG changes — flattening of T waves and depression of ST segments
  - Development of U waves
  - Tall P waves and prolonged PR interval
  - Widening QRS complex

Neuromuscular effects
- Skeletal muscle — weakness to paralysis, rhabdomyolysis in severe cases
- Smooth muscle — paralytic ileus — urinary retention

Renal effects
- Metabolic alkalosis — increased H^+ secretion.
- Polyuria — due to decreased responsiveness to ADH

ECG Pattern of Hypokalemia

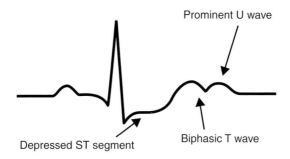

Figure 3.1 Electrocardiographic changes in of hypokalaemia.

In the kidney, hypokalaemia and potassium depletion cause a defect in urine concentrating ability due to resistance to the action of ADH, acquired nephrogenic diabetes insipidus, resulting in polyuria, nocturia and polydipsia. This is due to a decrease in the expression of aquaporin-2 water channels in the collecting ducts. Hypokalaemia also leads to sodium retention, decreased reabsorption of chloride and increased reabsorption of bicarbonate leading to metabolic alkalosis. The latter is due to increased secretion of hydrogen ions instead of potassium ions during sodium reabsorption in the distal tubules.

Investigation of Hypokalaemia

Detection of potassium depletion depends on history, examination and measurement of plasma potassium. The assessment of the size of potassium deficit is difficult; the only available guide is plasma potassium measurement.

The cause of hypokalaemia, in many cases, is obvious from the history and clinical signs. A detailed drug history should be taken to exclude the use of diuretics, purgatives, etc. and to exclude causes of

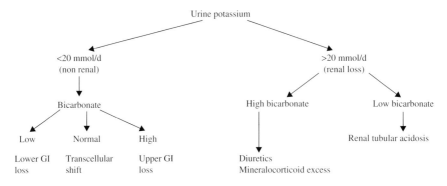

Figure 3.2   Investigation of hypokalaemia.

transcellular shift. When the cause is not apparent from the clinical picture, urinary potassium measurement is helpful. Urine potassium excretions of greater than 20 mmol/d or a transtubular potassium gradient (TTKG) of more than 7 in the presence of hypokalaemia is suggestive of renal losses of potassium. Measurement of plasma bicarbonate will be of help at this stage (Figure 3.2). A low urine potassium and low serum bicarbonate is suggestive of lower gastrointestinal loss such as diarrhoea. A high serum bicarbonate and low urine potassium is indicative of upper gastrointestinal loss such as vomiting. When the urine potassium and plasma bicarbonate levels are high, causes such as diuretic therapy and/or mineralocorticoid excess should be thought of and further investigations, such as measurements of renin and aldosterone may be necessary. A low serum bicarbonate with low serum potassium and high urine potassium suggest renal tubular acidosis. Urine chloride measurement is occasionally helpful to differentiate surreptitious vomiting (chloride < 20 mmol/L) from rare diseases such as Bartter's syndrome (chloride > 20 mmol/L).

Management

All patients likely to lose potassium should be given potassium supplements. To replace potassium deficit, potassium salts can be given orally in the form of slow release potassium preparations such as Slow K (8 mmol per tablet) or Sando K (12 mmol of potassium as chloride or bicarbonate per tablet) or intravenously, the latter only in cases of severe hypokalaemia. Great care must be exercised in giving potassium by intravenous infusion, especially when renal function is inadequate. A general guide is not to infuse potassium at a concentration exceeding 40 mmol/L, at a rate exceeding 20 mmol per hour or more than 200 mmol per day. Plasma potassium concentration and ECG should be monitored closely.

Hyperkalaemia

Hyperkalaemia is less common than hypokalaemia occurring in 1% to 10% of hospital patients and severe hyperkalaemia is a medical emergency. As the amount of potassium in the extracellular fluid is small, hyperkalaemia represents a very small increase in extracellular potassium.

Causes (Table 3.6)

As the intracellular potassium concentration is around 150 mmol/L, lysis of cells or leakage from cells during or after venepuncture can lead to artefactual hyperkalaemia. Leakage of potassium may occur during venepuncture, when there is prolonged venous occlusion or haemolysis. In patients with very high white cell or platelet count, pseudohyperkalaemia is seen due to lysis of white cells and platelets during the clotting process. This can be excluded if the blood sample is taken into a heparin tube and plasma separated immediately. When blood samples are left unseparated for more than 4–6 hours or kept at low temperatures, potassium will leak out of red cells due to reduced activity of the sodium–potassium pump (as a result of depletion of glucose or low temperature). Plasma potassium concentration may be high when blood samples are taken from the same arm as an intravenous infusion containing potassium or if the sample is contaminated with potassium EDTA.

Table 3.6   Causes of hyperkalaemia

Pseudohyperkalaemia
- Efflux from cells during venepuncture (e.g. prolonged application of tourniquet)
- Stored blood (either too long or inappropriately at low temperature)
- Thrombocytosis or leucocytosis
- Haemolysis (during venepuncture — use of small needles)
- Wrong anticoagulant (potassium EDTA)

Reduced renal exertion
- Renal failure
- Potassium-sparing diuretics
- Mineralocorticoid deficiency (Addison's disease)
- ACE inhibitors

Transcellular shift
- Insulin deficiency
- Acidosis
- Rapid cell breakdown — trauma/tumour lysis syndrome
- Beta blockers
- Digoxin overdose

Increased intake — especially with poor renal function
- Transfusion of stored blood

Case 3.3

A blood sample was taken by a GP from a patient on diuretics and sent to the laboratory the following morning. Serum potassium was reported as 6.8 mmol/L. There was no visible haemolysis and her renal function (urea and creatinine) were within the reference range. A repeat sample taken in A&E that day was 3.5 mmol/L.

The first sample was left at room temperature for nearly 10 hours and this causes potassium to diffuse out of the cells. Potassium is maintained within the cells by the sodium–potassium pump which uses glucose for energy. When the sample is left for prolonged periods, the glucose in the sample is utilised and the pump slows down, allowing potassium to leak out of cells into the plasma. Keeping the sample at 4°C will have the same effect because low temperature decreases the activity of the sodium–potassium pump.

A shift of potassium from cells can cause hyperkalaemia, especially if there is increased tissue breakdown as in rhabdomyolysis or tumour lysis syndrome. Metabolic acidosis especially due to mineral acids leads to a shift of potassium out of cells in exchange for hydrogen ions. In diabetic patients, hyperkalaemia is seen due to lack of insulin and due to hypertonicity induced by hyperglycaemia. Hypertonicity leads to the release of potassium from cells due to a shift of water out of cells, causing an increase in intracellular potassium concentration, which increases the concentration gradient across the cell membrane, thereby increasing potassium movement. It is also thought that during the movement of water, potassium moves out due to 'solvent drag'. β-Adrenergic blockers and digoxin toxicity cause hyperkalaemia due to reduced entry of potassium into cells. A rare cause of transcellular shift is hyperkalaemic periodic paralysis, which is an autosomal dominant disorder due to activating mutation of the skeletal muscle voltage-gated sodium channel.

True hyperkalaemia is usually due to a combination of increased load and diminished renal function. Acute renal failure especially in hypercatabolic states where increased potassium is released from tissues can lead to life-threatening hyperkalaemia. The metabolic acidosis of renal failure further aggravates the hyperkalaemia. Increased intake of potassium, either orally or intravenously, is unlikely to cause hyperkalaemia if kidney function is normal. However, if the infusion rate is high, or the renal function is poor as for example in old people or saline-depleted patients, then hyperkalaemia can occur.

Potassium-sparing diuretics such as spironolactone, triamterene and amiloride cause reduced potassium secretion by the renal tubules and can cause hyperkalaemia if there is increased potassium intake and/or reduced renal function.

Case 3.4

A 26-year-old man was admitted to hospital after a road traffic accident and he underwent surgery for open reduction of his fractured femur and repair of damaged liver. The day after the operation his urine output had decreased and his plasma results were:

Serum		Reference Range
Potassium (mmol/L)	6.0	3.5–5.0
Urea (mmol/L)	22.5	3.5–7.2
Creatinine (μmol/L)	365	60–112

This patient had developed acute oliguric acute renal failure due to hypovolaemia (blood loss). Hyperkalaemia in this patient is due to reduced renal function and increased release of potassium from cells damaged by trauma.

In mineralocorticoid deficiency (Addison's disease, hyporeninemic hypoaldosteronism, or congenital adrenal hyperplasia), decreased sodium–potassium exchange in the distal tubules results in hyperkalaemia. Heparin, which inhibits aldosterone synthesis and ACE inhibitors, may cause hyperkalaemia especially if there is increased input of potassium. Non-steroidal anti-inflammatory drugs inhibit intrarenal prostaglandins, which are important in the intrarenal production of renin may cause hyperkalaemia. Table 3.7 summarises the list of drugs which can cause hyperkalaemia.

Case 3.5

A 65-year-old man with type 2 diabetes was treated with amiloride for hypertension. His GP added ramipril as the blood pressure remained high. When his blood was taken a week after starting ramipril, it showed a serum potassium of 5.9 mmol/L. His renal function (serum creatinine) remained within the reference range.

Ramipril is an ACE inhibitor and the rise in potassium of this patient is due to the reduced aldosterone secretion.

Case 3.6

A 16-year-old type 1 diabetic patient was admitted as an emergency with a diagnosis of diabetic keto acidosis. On admission, his serum potassium was 6.0 mmol/L and the glucose was 35 mmol/L. He was treated with insulin and intravenous saline infusion. A repeat sample, taken 60 minutes after starting the treatment, showed a serum potassium of 3.1 mmol/L.

Before treatment, his potassium was high because of insulin deficiency, hypertonicity and metabolic acidosis, although his total body potassium was low. With insulin treatment, potassium moves into cells rapidly, causing hypokalaemia.

Table 3.7   Drugs causing hyperkalaemia

Redistribution
- Beta blockers
- Digoxin
- Suxamethonium

Reduction in excretion due to decreased aldosterone
- Heparin
- ACE inhibitors
- Angiotensin II receptor blockers
- Non-steroidal anti-inflammatory drugs (NSAIDs)
- Anti-fungal — ketaconazole
- Ciclosporin
- Tacrolimus

Blockers of aldosterone action
- Spironolactone
- Eplerenone

Inhibition of sodium channel
- Amiloride, triamterene (potassium-sparing diuretics)
- Trimethoprim
- Pentamidine

Effects of Hyperkalaemia

Clinical effects of hyperkalaemia depend on the severity of hyperkalaemia and the rapidity of onset. Hyperkalaemia affects the heart, nerve and muscle. Most patients are asymptomatic until potassium rises above 6 or 6.5 mmol/L. Hyperkalaemia can cause muscle weakness

ECG Pattern of Hyperkalemia

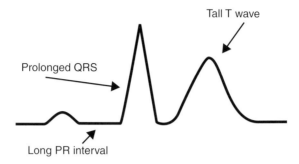

Figure 3.3   Electrocardiographic changes in hyperkalaemia.

with loss of tendon reflexes, mental confusion, paresthesia, fatigue, numbness, nausea and palpitation. Typical ECG changes are tall tented T waves, widening of the QRS complex and diminution or absent P waves (Figure 3.3). Severe cases can cause ventricular arrhythmias that can be fatal.

Investigation of Hyperkalaemia

Artefactual increase in plasma potassium concentration should first be excluded, if necessary, by analysis of a second sample. Careful history should be taken to exclude drug-induced hyperkalaemia. Measurements of plasma sodium, bicarbonate, urea and creatinine will exclude common causes such as renal failure. Further investigation with measurements of renin and aldosterone may be required if the sodium is low and potassium is high to exclude Addison's disease. A TTKG value less than 7 is suggestive of mineralocorticoid deficiency.

Management

Severe hyperkalaemia (serum potassium of 6.0 mmol/L with ECG changes or serum potassium of 6.5 or greater, irrespective of ECG changes) is a medical emergency and should be treated immediately. Infusion of 10 mL of 10% calcium gluconate given over 3–5 minutes will antagonise the cardiotoxic effects of hyperkalaemia and stabilise the myocardium. Hyperkalaemia can be corrected by shifting potassium into cells by infusion of 50 mL of 50% intravenous glucose and 10 units of insulin given over 5 minutes. Alternatively, 50–100 mL of 1.26% sodium bicarbonate or 10–20 mg salbutamol via a nebulizer may be given to reduce serum potassium. Dialysis may be required if the plasma potassium continues to rise even after these measures, if the load of potassium to be excreted is high or if the renal function is poor.

In mild to moderate hyperkalaemia, potassium intake should be reduced and drugs likely to cause hyperkalaemia should be withdrawn. A loop diuretic to increase potassium excretion may be adequate if renal function is not poor. Cation exchange resins like sodium polystyrene sulphonate (Kayexalate) are also helpful in reducing plasma potassium.

Further Reading

1. Davison AM, Cameron TS, Grunfield J-P, Kerr DNS, Ritz E, Winearls GG. *Oxford Textbook of Clinical Nephrology* 1998. Oxford: Oxford University Press.
2. Halperin ML, Kamel KS. Electrolyte quintet: Potassium. *Lancet* 1998; 352:135–140.
3. Nejirenda M, Tang T, Padfield PL, Seek JR. Hyperkalaemia. *Br Med J* 2009; 339:1019–1024.
4. Rose BD, Post TW. *Clinical Physiology of Acid–base and Electrolyte Disorders* 2001. McGraw-Hill: New York, 5th Edition.
5. Smellie WSA. Spurious hyperkalaemia. *Br Med J* 2007; 334:693–695.

Summary/Key Points

1. Potassium is mainly intracellular. Plasma potassium is not a good guide to total body potassium.
2. Acute changes in plasma potassium can occur due to movement of potassium into and out of cells.

3. Hypokalaemia is a common electrolyte disorder and it can be caused by transcellular shift or loss of potassium via gastrointestinal tract or kidneys. Diuretic therapy is a common cause.

4. Urine potassium excretion >20 mmol/day in the presence of hypokalaemia is suggestive of renal loss of potassium.

5. Plasma potassium may be spuriously high due to artefacts such as lysis of cells or leak of potassium after venepuncture.

6. Hyperkalaemia can be caused by transcellular shift or reduced excretion and/or increased intake. Renal failure is a common cause.

7. Severe hyperkalaemia is a medical emergency. Effects of hyperkalaemia can be reduced by infusion of calcium gluconate. Infusion of glucose and insulin will help to move potassium into cells. If these methods fail, or the rise of plasma potassium is rapid, renal dialysis may be required to correct the hyperkalaemia.

chapter 4

Disorders of Acid–Base Balance

Introduction

Concentration of H$^+$ ions (pH) in the extracellular fluid in health ranges from 35–45 nmol/L (pH 7.45–7.35) and in the intracellular fluid it is slightly higher. H$^+$ ion concentration in body fluids is maintained within narrow limits because of its effects on the activity of enzymes, and on function of organs e.g. at extremes of H$^+$ ion concentration, blood vessels do not respond to catecholamines, resulting in vasodilatation and hypotension. The range of pH over which cardiac function, metabolic activity and central nervous system function can be maintained is narrow. The widest range of pH compatible with life is from 6.80 to 7.80 (H$^+$ ion concentration: 16–160 nmol/L).

Acid–Base Homeostasis

Under physiological circumstances, large amounts of acids (compared to normal H$^+$ ion concentration) are produced daily. H$^+$ ions are produced during metabolic processes such as ATP hydrolysis, respiratory chain reactions and the production of nucleotides. Acid produced in the body can be conveniently divided into volatile and non-volatile acids. During oxidative metabolism, about 15,000–20,000 mmol of carbon dioxide is produced per day. As the carbon dioxide is excreted by the lungs, it is considered a volatile acid. The process of complete combustion of carbon involves the intermediate generation and metabolism of several thousand millimoles of relatively strong organic acids such as lactic acid, tricarboxylic acids, and ketoacids. The type of acid produced depends on the type of fuel metabolised. These organic acids do not accumulate in body fluids and their concentration normally remains in the millimolar range as these organic acids are almost fully metabolised under normal circumstances. Thus, there is no net acid accumulation. However, if there is a mismatch between production and metabolism, organic acids will accumulate causing metabolic acidosis, e.g. during severe strenuous exercise, lactic acid will accumulate causing acidosis. In addition, gut bacteria produce about 300 mmol of organic acids each day such as acetic, propionic and butyric acids. These acids are formed from anaerobic metabolism of neutral foodstuff remaining in the intestinal lumen. They are completely absorbed and join the larger pool of metabolic organic acids and are metabolised. Thus, these acids also do not contribute to net acid production in the body.

Metabolism of some components such as proteins, nucleic acid and small fractions of some carbohydrates and lipids generate specific organic acids that cannot undergo full metabolism to carbon dioxide. These organic acids include uric acid, oxalic acid, glucuronic acid, hippuric acid, etc. In addition, certain inorganic acids such as sulphuric acid and phosphoric acid are also produced from metabolism of sulphur-containing amino acids and organic phosphorous-containing compounds. This process generates about 40–80 mmol of acid each day. These acids, called non-volatile acids, contribute to the net acid load of the body. Acid production of the body can therefore be summarised as 15,000–20,000 mmol of volatile acid (carbon dioxide) and 40–80 mmol of non-volatile acids (sulphuric, phosphoric acids, etc). In spite of the large amount of acid production, pH is maintained within narrow limits by several mechanisms. Strong buffers in the extracellular and intracellular fluids buffer the hydrogen ions. Important buffer systems in the ECF are carbonic acid/bicarbonate, proteins and phosphate buffer systems. In the ICF, haemoglobin, proteins and phosphates are important buffers. Of the extracellular buffers, the most important buffer system is the bicarbonate/carbonic acid system. The relationship between pH, bicarbonate and carbonic acid concentrations in the ECF is given by the following equation (Henderson–Hasselbalch equation):

$$pH = pK + \frac{\log[HCO_3]}{\log[H_2CO_3]}$$

where pK is the log of the dissociation constant of carbonic acid, $[HCO_3^-]$ is the plasma bicarbonate concentration and $[H_2CO_3]$ is the plasma carbonic acid concentration. The carbonic acid concentration is calculated as $\alpha P_a CO_2$ where α is the solubility constant of carbon dioxide, and $P_a CO_2$, is the partial pressure of CO_2 in kPa. At normal $P_a CO_2$ of 5.3 kPa and bicarbonate concentration of 24 mmol/L, the ratio of $[H_2CO_3]$ to $[HCO_3^-]$ is 1/20. The importance of this buffer system is that the lungs regulate carbonic acid while the kidney, as discussed below, regulates the bicarbonate concentration.

Intracellular proteins are important cellular buffers because of their high concentration. Phosphate is an important buffer in the urine and ICF but has a small part to play in the ECF. Haemoglobin plays an important role in the transport of carbon dioxide and the maintenance of H^+ ion concentration.

Transport of Carbon Dioxide and the Role of Haemoglobin (Figure 4.1)

Carbon dioxide produced during metabolism in cells diffuses out of the cells into the ECF, thereby increasing the $P_a CO_2$. Some carbon dioxide combines with water to form carbonic acid. Most of the carbon dioxide, however, diffuses into the red cells down a concentration gradient. Within the red cells, carbon dioxide combines with water to form carbonic acid. This is facilitated by the enzyme carbonic anhydrase (carbonate dehydratase). Carbonic acid, in turn, dissociates to H^+ ions and bicarbonate. The H^+ ion released from the above reaction is buffered by deoxyhaemoglobin, which has a greater buffering capacity than oxyhaemoglobin. The bicarbonate diffuses

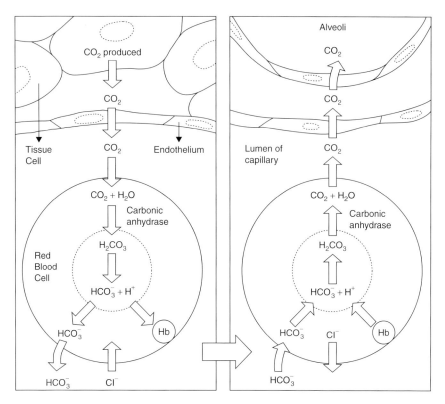

Figure 4.1　Transport of carbon dioxide and oxygen between lungs and tissue.

out into the plasma and chloride ion moves into the cells to maintain electrical neutrality. This is termed the 'chloride shift'. The net effect is that carbon dioxide is converted to bicarbonate and is carried in the plasma. A small amount of carbon dioxide is carried as carbonic acid and as carbamino compounds, which is produced as a result of carbon dioxide combining directly with nitrogen groups in the protein. In the lungs, the process is reversed. Carbon dioxide is produced from the bicarbonate by reversal of the reaction catalysed by carbonic anhydrase and it diffuses into the alveoli because of low partial pressure in the alveoli. At the same time, haemoglobin gets oxygenated.

Under normal circumstances, the rate of cellular carbon dioxide production equals the rate of carbon dioxide excretion by the lungs. The rate of carbon dioxide excretion is governed primarily by the rate of minute ventilation. When there is hyperventilation or hypoventilation, there is a transient loss or gain of carbon dioxide. However, in the steady state, the rate of carbon dioxide removal by the lung equals the rate of carbon dioxide production irrespective of ventilation.

One of the factors regulating the rate of ventilation is a change in CSF or arterial hydrogen ion concentration. Chemoreceptors in the medullary centres or in the carotid body are activated by changes in hydrogen ion concentration in the CSF or arterial blood respectively. The increase in hydrogen ion concentration causes stimulation of the chemoreceptors resulting in an increase in ventilation.

Role of Kidney in the Maintenance of Acid–Based Balance

Kidneys are responsible for the excretion of non-volatile acids produced during metabolism. In addition, they play an another important role in reabsorbing the filtered bicarbonate.

Bicarbonate reabsorption (Figure 4.2)

Large quantities of bicarbonate, approximately 4300 mmol or about four times the total body buffering capacity, are filtered daily. In order to maintain H^+ ion balance, this filtered bicarbonate has to be reabsorbed.

About 90% of the filtered bicarbonate is reabsorbed in the proximal tubule and the rest in the intercalated cells of the distal tubules and collecting ducts. The luminal surface of the proximal tubular cells is impermeable to bicarbonate ion, thus reabsorption of bicarbonate is achieved indirectly. Within the tubular cells, carbon dioxide combines with water to form carbonic acid and then to H^+ ion and bicarbonate, facilitated by the enzyme carbonic anhydrase (Figure 4.2). The hydrogen ion formed diffuses from the tubular cell

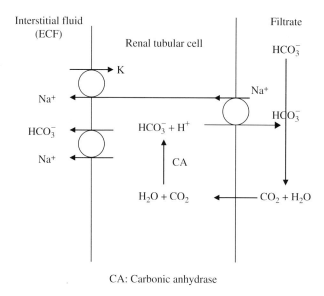

CA: Carbonic anhydrase

Figure 4.2   Reabsorption of bicarbonate in the renal tubule.

into the lumen via a Na^+/H^+ antiporter (major part) and via an H^+–ATPase. In the lumen, the secreted H^+ combines with the filtered bicarbonate to form carbonic acid and then to carbon dioxide and water. This process is facilitated by carbonic anhydrase, which is present in large amounts in the tubular brush border. The secretion of hydrogen ion across the luminal membrane is in exchange for sodium ions. The carbon dioxide produced within the lumen enters the tubular cell and contributes to the formation of bicarbonate within the cell. The bicarbonate formed within the cell is transported via Na^+/HCO_3^- symporter into the circulation. This symporter transports three bicarbonate ions for every sodium ion. The net effect is the return of one mole of bicarbonate to the circulation for every mole of H^+ secreted. At normal bicarbonate concentration, all the filtered bicarbonate is reabsorbed. Note that at this stage there is no net secretion of H^+ in the proximal tubules. The importance of the luminal carbonic anhydrase in bicarbonate reabsorption is illustrated by the use of carbonic anhydrase inhibitors such as acetozolamide, which is filtered by the kidney but does not diffuse into the tubular cells and thus acts on the luminal side. Acetazolemide administration results in the excretion of large amounts of bicarbonate in the urine.

As the H^+ ion secretion in the proximal tubule is associated with reabsorption of sodium, the rate of reabsorption of bicarbonate in the proximal tubule is influenced by factors that regulate reabsorption of sodium in the proximal tubule. ECF volume expansion reduces bicarbonate reabsorption while contraction increases bicarbonate reabsorption. Alkalosis produced by ECF volume depletion is sometimes referred to as 'contraction alkalosis'. Other factors that affect bicarbonate reabsorption include luminal bicarbonate concentration, luminal flow rate, arterial pCO_2 and angiotensin II.

Net hydrogen ion excretion (Figure 4.3)

During the reabsorption of filtered bicarbonate, there is no net excretion of H^+ ions, which occurs in the distal nephron segments, where hydration of carbon dioxide to carbonic acid and then to hydrogen ions

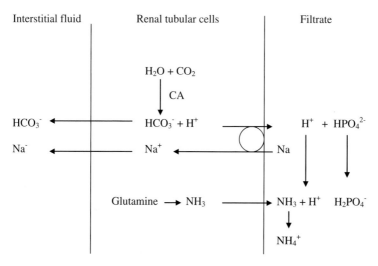

Figure 4.3   Hydrogen ion excretion by renal tubules.

and bicarbonate occurs as in the proximal tubule. The hydrogen ions produced are secreted into the lumen by a H^+–ATPase pump. The bicarbonate ions produced enters the circulation via Cl^-/HCO_3^-– exchanger. The H^+–ATPase pump is stimulated by aldosterone. Hydrogen ions secreted into the lumen are buffered by phosphate or ammonia. The amount of hydrogen ion buffered by phosphate is about 30 mmol/day or approximately 30–40% of the total hydrogen ion excretion and the rest as ammonium ion. Phosphate which is filtered as the divalent ion (HPO_4^{2-}) combines with H^+ ion to form $H_2PO_4^-$.

The renal handling of ammonia is complex. Ammonia is produced in the proximal tubular cells from glutamine which is pro- duced in the liver. Glutamine is converted to α-ketoglutarate and NH_4^+, which is secreted into the lumen via Na^+/H^+ exchanger (NH_4^+ is substituted for H^+). The α-ketoglutarate is metabolised and in this process, two bicarbonate ions are produced. Ammonium ion produced in the proximal tubule is reabsorbed in the thick ascending limb by the $Na^+/K^+/Cl^-$ carrier. The reabsorbed NH_4^+ dissociates to NH_3 and H^+ because of the less acidic environment of the tubular cell. The luminal membrane of the thick ascending limb is imperme- able to NH_3 and the NH_3 diffuses into the medullary interstitum. The NH_3 then re-enters the proximal tubule where it is converted to NH_4^+ and recycled (Figure 4.4). Some of the NH_3 enters the collect- ing ducts where it becomes NH_4^+ by taking up H^+ ions secreted by the collecting duct. The collecting duct is not permeable to NH_4^+ hence, this is described as 'ammonia trapping'. NH_4^+ formed is then exerted in the urine. Approximately 40–50 mmol of hydrogen ions, i.e. 60–70% of total H^+ is excreted as ammonium ions. The amount of hydrogen ion excreted as phosphate is measured as titratable acid- ity, which is the amount of alkali necessary to raise the pH of the urine to 7.4. The amount of titratable acidity depends on the excretion of phosphate, which in turn is dependent on dietary intake of phosphate and PTH. The excretion of phosphate is not regulated in response to the needs to maintain acid–base regulation. Acid–base regulation is primarily achieved via NH_4^+ production which responds to changes in acid–base status. In response to acidosis, ammonia production can increase up to ten times the normal.

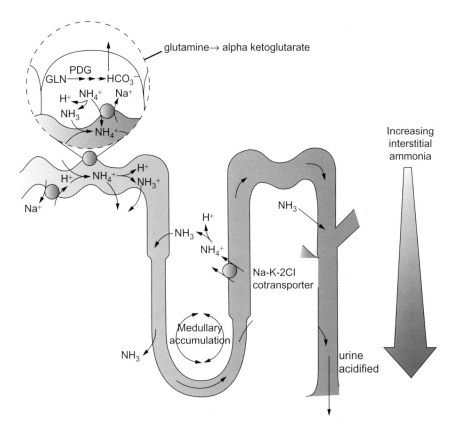

Figure 4.4 Ammonia recycling in the renal tubules.

1. *Factors regulating hydrogen ion secretion (Table 4.1)*

Distal tubular sodium reabsorption is linked to secretion of H^+ ions or potassium ions. It is estimated that approximately half the sodium reabsorption is linked to excretion of H^+ ions and the other half to secretion of potassium ions. Thus, factors which affect distal tubular sodium reabsorption will influence H^+ secretion. Aldosterone, which increases sodium reabsorption, will increase the excretion of H^+ ions causing alkalosis. In potassium depletion, more H^+ ion is secreted in exchange for sodium reabsorption. The presence of non-reabsorbable anions such as sulphate or carbenicillin will facilitate the secretion of H^+ ions because of the increased

Table 4.1   Factors influencing hydrogen ion secretion

Aldosterone and cortisol
Hypokalaemia
Extracellular fluid H^+ concentration
Non-absorbable anions in the filtrate
Chloride deficiency
ECF volume
Arterial pCO_2

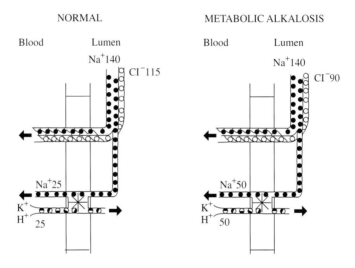

Figure 4.5   The importance of chloride in hydrogen ion secretion.

electronegativity within the lumen. ECF volume depletion increases sodium reabsorption which also increases bicarbonate reabsorption. Arterial pCO_2 causes an increase in H^+ ion secretion and hence, an increase in plasma bicarbonate.

2. *Importance of chloride in hydrogen ion secretion (Figure 4.5)*

Sodium reabsorption in proximal tubule is accompanied by chloride or bicarbonate. If there is chloride deficiency, there will be less reabsorption of sodium in the proximal tubule and more of it will reach the distal tubule, where it will be reabsorbed in exchange for H^+ ions. This will result in metabolic alkalosis.

Disorders of Acid–Base Balance

Acid–base disorders are classified into respiratory and metabolic. If the primary abnormality or change is in pCO_2, it is called 'respiratory'. If the primary change is in the bicarbonate concentration, it is termed 'metabolic'.

Compensation

When there is an acid–base disturbance, there is an attempt at compensation to reduce the magnitude to this change in pH. For example, if an acidosis develops due to an accumulation of lactic or keto acids (metabolic acidosis), there is respiratory compensation, a compensatory respiratory alkalosis. If there is accumulation of CO_2 (respiratory acidosis), this is followed by an increase in bicarbonate generation by the kidneys, a compensatory metabolic alkalosis.

The importance of the buffers and respiratory compensation is shown by the classical experiment performed by Pitts. In this experiment, acid was infused into two identical groups of dogs. In one group, respiration was controlled by a ventilator and the other group was allowed to breathe normally. The same amount of acid was added to a volume of water equal to the volume of total body water of the dogs. The pH of the water decreased to less than 1.0. In the ventilated group, the pH dropped to 6.7 whereas in the group which breathed on their own, the pH was 7.25. This experiment illustrates the major role of buffering (50% of which is intracellular) and the relatively small but important role of hyperventilation.

Summary of response to a change in hydrogen ion concentration

When there is a disturbance in H^+ ion concentration, several mechanisms try to minimise the change in ECF H^+ ion concentration.

1. Distribution and ECF buffering: H^+ ion load is distributed and buffered by the ECF buffers.
2. Intracellular buffering: During acidosis, H^+ ions move into the cells where it is buffered. Movement of H^+ ion into the cells is accompanied by a compensatory movement of potassium out of the cells to maintain electrical neutrality. Hyperkalaemia is, therefore, not an uncommon finding in metabolic acidosis. Extra-cellular buffering takes place immediately whereas intracellular buffering takes several hours to be completed (Figure 4.6).
3. Compensation: Metabolic acidosis or alkalosis is compensated by respiratory mechanisms and the kidney compensates respiratory acid–base disturbances. Respiratory compensation happens very quickly whereas renal compensation takes several days (Figure 4.6). An increase in H^+ ion concentration as a result of a metabolic problem stimulates the respiratory centre either via central or peripheral receptors, causing an increase in ventilation. This lowers P_aCO_2 and minimises the change in H^+ ion concentration. When acidosis is due to a respiratory problem, the increase in

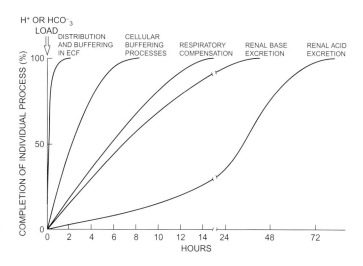

Figure 4.6   Time course of response to changes in hydrogen concentration. Time required to reach the maximum response is shown.

P_aCO_2 causes an increase in the renal tubular secretion of H^+ ions. This generates more bicarbonate in the ECF.

4. The final mechanism dealing with a change in H^+ ion concentration is by the kidneys. Figure 4.6 shows the relative time scale for these processes to reach maximum effectiveness.

Metabolic Acidosis

Metabolic or non-respiratory acidosis can arise due to one of three mechanisms: increased acid in the body (either due to increased production or ingestion), decreased H^+ ion excretion or loss of bicarbonate. Increases in H^+ ion concentration is also minimised by respiratory compensation, which happens fairly rapidly.

Causes of metabolic acidosis (Table 4.2)

1. *Increased production of acid*

Common causes of increased production of acid are diabetic ketoacidosis and lactic acidosis. In diabetic ketoacidosis, due to a lack of insulin, fatty acids are metabolised to ketone bodies and the accumulation of ketone bodies leads to acidosis. Lactic acidosis is commonly caused by hypoxia such as, after cardiac arrest, severe haemorrhage, shock, hypotension and congestive cardiac failure. Lactic acidosis without hypoxia occurs in liver disease, alcoholic intoxication, inherited metabolic diseases such as glucose-6-phosphatase deficiency, and with the use of some drugs like metformin.

Ingestion of salicylate in toxic doses can cause acid–base disturbance. Salicylate uncouples oxidative phosphorylation and inhibits various enzymes leading to increased production of organic acids. Salicylate itself will stimulate the respiratory centre to cause respiratory alkalosis. In adults with salicylate intoxication, the picture is either a respiratory alkalosis or a mixture of respiratory alkalosis and metabolic acidosis. In children and the elderly, metabolic acidosis is more predominant. In any age group, severe salicylate intoxication will result in a predominant metabolic acidosis.

Table 4.2　Causes of metabolic acidosis

Increased acid load
- Increased endogenous production
  — Diabetic ketoacidosis
  — Lactic acidosis
- Ingestion
  — Salicylate
  — Methanol
  — Ethylene glycol

Decreased excretion
- Renal failure
- Distal renal tubular acidosis (Type I)

Loss of bicarbonate
- Loss via gastrointestinal tract
  — Fistula — biliary, pancreatic or intestinal,
  — Severe diarrhoea
  — Ureterosigmoidostomy
  — Cholestyramine
- Renal loss
  — Proximal renal tubular acidosis — (Type II)

Case 4.1

A 60-year-old man was admitted to hospital following the onset of severe abdominal pain radiating to the back. He was shocked; his abdomen was rigid; his femoral pulses were impalpable. Ultrasound examination showed a greatly enlarged abdominal aorta.

		Reference Values
pH	7.05	7.35–7.45
pCO_2 (kPa)	3.5	5.0–6.5
$[HCO_3]$ (mmol/L)	7	22–28

This is severe acidosis. pCO_2 is reduced, indicating that this must be metabolic (non-respiratory) in origin. The low bicarbonate is characteristic of non-respiratory acidosis (bicarbonate is consumed in buffering hydrogen

ions). The low pCO$_2$ reflects respiratory compensation. This man had a rup-
tured abdominal aortic aneurysm, causing hypotension, shock and restricting
tissue oxygen supply. There is a switch from aerobic to anaerobic metabolism,
producing a lactic acidosis. The pCO$_2$ in this case is not as low as one would
expect for this pH, probably because the pain and rigidity of the abdomen may
have restricted respiratory movements.

Ingestion of methanol leads to metabolic acidosis due to the formation of formic acid from methanol by the action of the enzyme
alcohol dehydrogenase. Ethylene glycol, which is the main component of antifreeze, is metabolised by the same enzyme to oxalic acid,
resulting in severe metabolic acidosis.

2. *Decreased excretion of hydrogen ions*

Acidosis is common in both acute and chronic renal failure. In
chronic renal failure, metabolic acidosis develops when net H$^+$ ion
excretion rate falls below the production rate. More specific disturbance of H$^+$ ion secretion is seen in Type I renal tubular
acidosis (Type I RTA), where there is impairment in H$^+$ ion secretion by the distal tubules.

3. *Loss of bicarbonate*

Loss of intestinal fluid, especially pancreatic and small intestinal
secretions rich in bicarbonate, can lead to metabolic acidosis.
Such loss can occur in severe diarrhoeal states such as in cholera,
ileal fistula, ureterosigmoidostomy and ileal conduits. In these
two latter situation the chloride reaching the colonic/ileal
mucosa is reabsorbed in exchange for bicarbonate resulting in a
net loss of bicarbonate.

Failure to reabsorb all the filtered bicarbonate at the proximal
tubule leads to a loss of bicarbonate and consequently, proximal
renal tubular acidosis is seen (Type II RTA). This can be due to a

widespread tubular dysfunction (the Fanconi syndrome) or due to an isolated defect (see Chapter 5). In Type II RTA, the distal acidification mechanism is intact. When patients with Type II RTA are in a steady state, the urine pH can be acidic with very little urine bicarbonate. This is because the defect in bicarbonate reabsorption is partial and all the filtered bicarbonate can be reabsorbed in the steady state (Figure 4.7). When serum bicarbonate concentration is

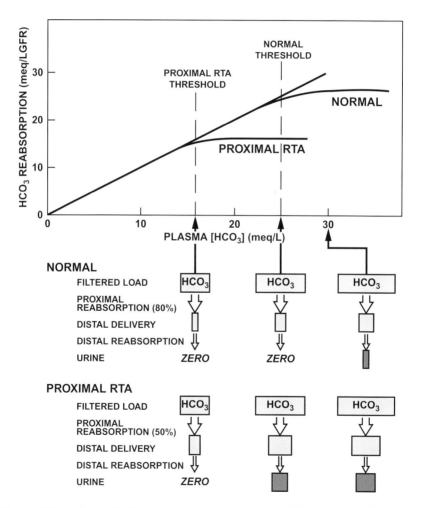

Figure 4.7   Relationship between plasma bicarbonate and bicarbonate reabsorption in normal subjects and in patients with proximal RTA.

raised by either oral or intravenous administration, bicarbonate will appear in the urine before the serum bicarbonate concentration reaches normal values. Bicarbonate can also be lost when carbonic anhydrase inhibitors are administered.

Features and effects of metabolic acidosis (Table 4.3)

In addition to the features of the disease causing metabolic acidosis, features common to all metabolic acidotic states are listed in Table 4.3. Hyperkalaemia is usually present in metabolic acidosis, except in renal tubular acidosis where hypokalaemia is more likely to occur. Acidosis can lead to lethargy and in severe acidosis, coma may result. In the

Table 4.3   Features and effects of metabolic acidosis

Metabolic
- Hyperkalaemia
- Increased protein breakdown
- Reduction in ATP synthesis
- Insulin resistance

Cardiovascular
- Cardiac arrhythmias
- Decreased cardiac contractility
- Vasodilatation
- Hypotension
- Increased pulmonary vascular resistance

Central nervous system
- Lethargy
- Confusion
- Coma

Renal and bone
- Osteopenia — in chronic metabolic acidosis
- Hypercalciuria — nephrocalcinosis
                — renal stones

Retardation of growth
Respiratory
- Hyperventilation
- Dyspnoea

cardiovascular system, cardiac arrhythmias, decreased cardiac contractility and reduced response to catecholamines, resulting in vasodilatation and hypotension are seen. In chronic metabolic acidosis, osteopenia develops due to mobilisation of bone calcium as a result of bone buffering. Acidosis can reduce renal tubular calcium reabsorption leading to hypercalciuria, nephrocalcinosis and renal stones. In children, chronic metabolic acidosis can lead to retardation of growth.

Investigation of the cause of metabolic acidosis

The cause of metabolic acidosis is usually obvious from history and examination. Calculation of anion gap may be of some help. In plasma, the total anions must equal the total cations. However, the commonly measured cations sodium and potassium exceed the commonly measured anions chloride and bicarbonate. The difference between these ions is called 'anion gap' and is due to proteins, mainly albumin, and other unmeasured anions such as lactate, ketones, etc. Table 4.4 shows the causes of metabolic acidosis divided according to whether the anion gap is normal or high. The metabolic acidosis with normal anion gap is also called 'hyperchloraemic metabolic acidosis' as the chloride concentration is increased in these situations.

Table 4.4 Causes of metabolic acidosis classified according to anion gap

High anion gap acidosis
- Lactic acidosis
- Ketoacidosis
- Renal failure
- Ingestion
  — Salicylate
  — Methanol
  — Ethylene glycol

Normal anion gap acidosis
- Gastrointestinal loss of bicarbonate
- Renal loss
  — Proximal renal tubular acidosis
  — Distal renal tubular acidosis

Another useful investigation is osmolar gap, which is the difference between the measured osmolality and calculated osmolarity. When the osmolar gap is high (measured osmolality exceeds the calculated osmolarity), presence of unmeasured osmoles such as methanol, ethylene glycol or alcohol is likely.

Management of metabolic acidosis

The cause of the metabolic acidosis must first be treated whenever possible. Administration of bicarbonate to correct metabolic acidosis is not usually recommended unless the metabolic acidosis is severe. If the acidosis is severe (pH < 7.10), or when the cause of the metabolic acidosis cannot be corrected (such as renal failure or renal tubular acidosis), bicarbonate is administered. It is recommended that the bicarbonate be given in small amounts so as not to bring the pH back to normal quickly. The usual recommendation is not to raise the pH above 7.20. If large amounts of bicarbonate are given, it can cause adverse effects such as hypokalaemia, hypocalcaemia, paradoxical increases in ventilation, and the shifting of the oxygen dissociation curve to the left, which impairs oxygen delivery to the tissues.

Metabolic Alkalosis

Metabolic (non-respiratory) alkalosis results from loss of H^+ ions or from administration of alkali. The causes of the metabolic alkalosis are outlined in Table 4.5. In metabolic alkalosis, H^+ ions move from the ICF into the ECF and potassium ions move into the cells leading to hypokalaemia. Compensatory hypoventilation will result in an increase in P_aCO_2. The high serum bicarbonate concentration in metabolic alkalosis is maintained by the accompanying chloride deficiency, potassium depletion or volume depletion. Vomiting due to pyloric stenosis or gastric aspiration can lead to metabolic alkalosis. Unlike vomiting from other causes where acid and bicarbonate are lost, in pyloric stenosis, only acid is lost. The resulting chloride deficiency maintains the metabolic alkalosis by the mechanism outlined in Figure 4.5. ECF volume contraction due to the loss of fluid causes

Table 4.5   Causes of metabolic alkalosis

Loss of hydrogen ions
- Gastrointestinal
  — Gastric aspiration
  — Vomiting due to pyloric stenosis
- Renal losses
  — Mineralocorticoid excess
  — Cushing's syndrome
  — Drugs with mineralocorticoid-like activity, e.g. carbenoxolone
  — Diuretics
  — Potassium depletion
  — Chloride depletion

Excessive alkali administration
- Milk alkali syndrome

secondary hyperaldosteronism, which causes loss of H^+. Potassium depletion causes increased H^+ ion secretion in the distal tubules, in exchange for sodium leading to metabolic alkalosis.

Ingestion of large amounts of alkali as in milk alkali syndrome or administration of large amounts of bicarbonate can cause metabolic alkalosis.

Case 4.2

A 50-year-old man was referred to the casualty department by his GP because of a history of repeated vomiting over the past month. He vomited undigested food after almost every meal, lost weight and was dehydrated.

Serum		Reference Range
pH	7.55	7.35–7.45
pCO2 (kPa)	7.2	5.0–6.5
Bicarbonate (mmol/L)	45	22–28
Sodium (mmol/L)	146	135–145
Potassium (mmol/L)	2.8	3.5–5.0
Urea (mmol/L)	34.2	2.5–7.5
Creatinine (μmol/L)	150	60–120

> *He is alkalotic and the pCO$_2$ is raised which indicates that he has a metabolic (non-respiratory) alkalosis partially compensated by retention of carbon dioxide. The alkalosis is due to loss of gastric acid. Note that respiratory compensation in non-respiratory alkalosis is often poor, since carbon dioxide is a respiratory stimulant.*

Effects of metabolic alkalosis (Table 4.6)

Metabolic alkalosis, when acute, can cause a reduction in ionised calcium due to increased protein binding of calcium. In chronic states, increased mobilisation of calcium from the bone mediated by PTH, will restore the ionised calcium concentration. Hypokalaemia is usually seen because of increased cellular uptake of potassium. Other effects are listed in Table 4.6.

Table 4.6   Features and effects of metabolic alkalosis

Metabolic
- Hypokalaemia
- Decreased plasma ionised calcium concentration
- Hypomagnesaemia
- Hypophosphataemia
- Stimulation of anaerobic glycolysis

Cardiovascular
- Vasoconstriction
- Reduction in coronary blood flow
- Predisposition to ventricular and supraventricular arrhythmias

Respiratory
- Hypoventilation — hypoxia and hypercapnia

Central nervous system
- Reduction in cerebral blow flow
- Lethargy
- Delirium
- Stupor
- Fits

Management of metabolic alkalosis

Appropriate treatment consists of infusion of isotonic sodium chloride solution and potassium replacement to correct the accompanying hypokalaemia and volume contraction. The primary cause should always be treated first. Very rarely it may be necessary to correct the metabolic alkalosis itself, especially when it is very severe and accompanied by significant hypoventilation. In these circumstances, administration of dilute hydrochloric acid or other acidifying salts such as lysine or arginine hydrochloride may be required.

Respiratory Alkalosis

The causes of respiratory alkalosis are given in Table 4.7. Anxiety-induced hyperventilation is the most common cause of acute respiratory alkalosis. In chronic respiratory alkalosis, there is renal compensation resulting in a decrease in bicarbonate concentration. In acute hyperventilation, the patient might have lightheadedness, parasthesiae, numbness and tingling of the extremities. Tetany may occur in severe cases, due to a reduction in ionised calcium concentration as a result of increased protein binding of calcium.

Table 4.7   Causes of respiratory alkalosis

Hypoxia
- High altitude
- Severe anaemia
- Pulmonary disease — e.g. pneumonia, intestinal fibrosis, emboli

Respiratory stimulation
- Drugs — e.g. salicylate
- Cerebral causes — e.g. trauma, infections
- Hepatic failure
- Gram-negative septicaemia
- Voluntary/hysterical hyperventilation

Mechanical ventilation

Pulmonary disease

Case 4.3

As part of a class experiment in physiology, a medical student volunteered to have a sample of arterial blood taken. During the time the demonstrator was explaining the procedure to the class, the student became very nervous and experienced a tingling sensation in her fingers and toes.

Serum		Reference Range
pH	7.52	7.35–7.45
pCO_2 (kPa)	3.5	5.0–6.5
[HCO_3] (mmol/L)	22	22–28

She is alkalotic, with a low pCO_2 indicating acute respiratory alkalosis. Anxiety-induced hyperventilation is the likely cause. The tingling reflects hyperaesthesiae, a consequence of a reduction in ionised calcium concentration.

The treatment of acute respiratory alkalosis involves correction of the underlying cause. In severe anxiety provoked hyperventilation syndrome, breathing into a paper bag generally terminates an acute attack.

Respiratory Acidosis

Respiratory acidosis is caused by impaired CO_2 excretion. When chronic, more bicarbonate is generated by the renal tubules as compensation. Causes of respiratory acidosis are listed in Table 4.8. Acute respiratory acidosis is seen when there is a sudden reduction of ventilation (e.g. depression of respiratory centre by narcotic overdose), airway obstruction or cardiac arrest. Chronic respiratory acidosis occurs in patients with chronic obstructive airways disease.

Clinical manifestations of respiratory acidosis depend on the severity of the disorder and the rate at which carbon dioxide retention occurs. Acute increase in P_aCO_2 results in somnolence, confusion and ultimately carbon dioxide narcosis. As carbon dioxide is a cerebral vasodilator, the blood vessels in the optic fundi are often dilated, engorged, torturous and frank papilloedema may occur.

Table 4.8   Causes of respiratory acidosis

Depression of respiratory centre
- Drugs — sedatives, anaesthetics, opiates
- Cerebral trauma, tumours
- Extreme obesity

Airway obstruction
- Aspiration of foreign bodies
- Laryngospasm
- Obstructive sleep apnoea

Neuromuscular disease
- Poliomyelitis, motor neurone disease.
- Myasthenia gravis

Pulmonary disease
- Pulmonary fibrosis
- Pneumonia (severe)
- Respiratory distress syndrome
- Severe asthma
- Pneumothorax
- Adult respiratory distress syndrome (ARDS)

Case 4.4

A young man sustained a chest injury in a road traffic accident, which resulted in a large flail segment of his thoracic cage, which made it difficult for him to breathe.

Serum		Reference Range
pH	7.24	7.35–7.45
pCO_2 (kPa)	8.0	5.0–6.5
$[HCO_3]$ (mmol/L)	25	22–28

He is acidotic, and the elevated pCO_2 indicates that this is respiratory in origin. The cause is the impaired respiratory gas exchange as a result of the flail segment of his chest. In this short time, no renal compensation would be expected.

Case 4.5

An elderly man was admitted to hospital with an exacerbation of his long-standing chronic obstructive pulmonary disease, itself a consequence of his cigarette smoking habit.

Serum		Reference Range
pH	7.30	7.35–7.45
pCO$_2$ (kPa)	9.5	5.0–6.5
[HCO$_3$] (mmol/L)	35	22–28

He has respiratory acidosis, but the pH is not as low as one would have been predicted from the pCO$_2$. The high bicarbonate gives a clue to the cause. In long-standing carbon dioxide retention, increased renal hydrogen ion excretion compensates for the acidosis, resulting in increased renal bicarbonate generation. The fact that he is now acidotic probably reflects a further increase in his pCO$_2$ as a result of an acute respiratory infection, in addition to his longstanding problem.

Management

In acute respiratory acidosis, treatment of the underlying cause should be the prime objective. Hypoxia is the immediate danger thus, oxygen therapy is required. In chronic respiratory acidosis, however, hypoxia is an important stimulus for the respiratory drive and as such, rapid correction of hypoxia may be dangerous. Oxygen therapy in these patients should be started with caution as the correction of hypoxia may reduce the ventilatory drive, leading to severe carbon dioxide narcosis. Arterial P$_a$CO$_2$ should be regularly monitored. Mechanical ventilation may be required in extreme cases.

Assessment of Acid–Base Status

Measurements made in the laboratory to assess acid–base status are pH and pCO$_2$, which are usually measured in arterial blood using pH and pCO$_2$ electrodes. Bicarbonate can either be calculated from the pH and pCO$_2$ based on the Henderson–Hasselbalch equation or measured directly. Measured bicarbonate includes not only bicarbonate ions but also carbon dioxide dissolved in the plasma as well as carbamino compounds. Thus, the correct terminology for measured bicarbonate is total CO$_2$ but for clinical purposes, total CO$_2$ and bicarbonate can be used synonymously as the difference between the two is 1–2 mmol/L. Some instruments for blood gas analysis generate other derived parameters such as base excess and standard bicarbonate. These are no longer used in clinical practice and are not discussed any further.

Recognition of Acid–Base Disturbance

Acid–base disturbances can be easily classified by examining pH and pCO_2 and/or bicarbonate. Table 4.9 below shows the changes in pH, pCO_2 and bicarbonate in acid–base disorders. Occasionally, when there is a mixed disturbance, i.e. a combination of respiratory and metabolic disturbance, a metabolic map may be helpful (Figure 4.8).

Table 4.9   Recognition of acid–base disturbance

	H^+	pCO_2	HCO_3^-
Metabolic acidosis	↑	↓*	↓
Metabolic alkalosis	↓	↑*	↑
Respiratory acidosis	↑	↑	↑*
Respiratory alkalosis	↓	↓	↓*

*Denotes compensatory response.

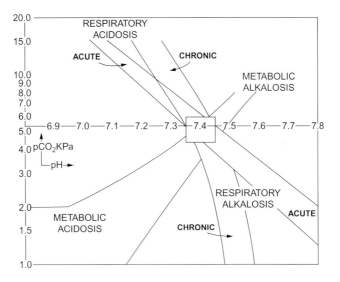

Figure 4.8   Metabolic map showing the relationship between pH and pCO_2.

This figure shows the range of values for pCO_2 and pH in simple acid–base disturbances. To identify an acid–base disorder, pH and pCO_2 are plotted on the graph and if these fall within the band shown, the disturbance is a simple one. If they fall between the bands, there is a mixed disturbance.

Oxygen Transport and its Disorders

The transfer of oxygen from atmospheric air to the mitochondria involves several processes: alveolar ventilation, diffusion of oxygen into the blood, transport of oxygen in the blood and oxygen delivery and uptake by tissues.

The partial pressure of oxygen in arterial blood, P_aO_2, depends on partial pressure of oxygen in alveoli, P_AO_2, which is determined by oxygen content of inspired air. P_AO_2 can be calculated by the equation:

$$P_AO_2 = P_IO_2 - kP_aCO_2,$$

where P_IO_2 is partial pressure of oxygen in inspired air, P_aCO_2 is partial pressure of CO_2 in arterial blood (hence alveolar, as CO_2 rapidly diffuses across alveolar membrane) and k is the reciprocal of respiratory quotient.

P_aO_2 is less than P_AO_2 by about 1 kPa due to some blood entering the arterial system without going through ventilated areas of the lungs and the fact that oxygen diffuses more slowly. The difference between P_AO_2 and P_aO_2 increases with age and in many pulmonary diseases. Factors causing this increase in pulmonary diseases include impairment of diffusion, ventilation–perfusion mismatch, and shunting of blood, blood reaching the arterial system without passing through ventilated areas of the lung.

Oxygen entering the blood is carried mainly by haemoglobin with small amounts as dissolved oxygen. The relationship between the saturation of haemoglobin and P_aO_2 is sigmoid (Figure 4.9), such that haemoglobin is 90% saturated at a P_aO_2 of 8 kPa and 97% saturated

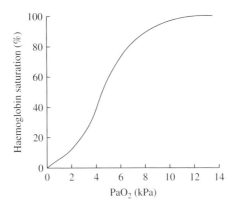

Figure 4.9   Oxygen dissociation curve of haemoglobin.

at normal P_aO_2 of 13.3 kPa. Increasing the P_aO_2 any further will not increase the saturation of haemoglobin and therefore the amount of oxygen carried by the blood. Factors which affect the oxygen dissociation curve include $[H^+]$, P_aCO_2 and red cell intermediate 2,3-diphosphoglycerate (2,3-DPG). An increase in 2,3-DPG or $[H^+]$ will shift the curve to the right and this will facilitate the release of oxygen to tissues, i.e. at a given P_aO_2, haemoglobin is less saturated and therefore more oxygen is released. A shifting of the curve to the left has the opposite effect.

The delivery of oxygen to tissues depends not only on the P_aO_2 (oxygen saturation of haemoglobin) but also, the amount of haemoglobin, cardiac output and blood flow to the tissues.

Hypoxaemia, reduced P_aO_2, can arise in many conditions (Table 4.10). In ventilation–perfusion imbalance, some ventilated areas of lungs are not perfused and some perfused areas are not ventilated. This results in a reduction in P_aO_2 but not always an increase in P_aCO_2. Carbon dioxide is rapidly diffusible and hyperventilation can remove the carbon dioxide. However, hyperventilation cannot correct the low P_aO_2 as the haemoglobin in the perfused and ventilated alveoli is already fully saturated and cannot carry more oxygen. If the V/Q imbalance is moderate, P_aCO_2 will be normal or low and

Table 4.10   Causes of hypoxaemia

Low inspired oxygen
- High altitude

Hypoventilation
- Depression of respiratory centre

Impairment in diffusion
- Pulmonary fibrosis

Shunting of blood
- Extrapulmonary shunts — atrial or ventricular septal defect
- Intrapulmonary shunt — arteriovenous fistula

Ventilation–Perfusion imbalance
- Chronic obstructive airways disease
- Pulmonary embolism

P_aO_2 will be low. In more advanced diseases, hyperventilation cannot compensate and P_aCO_2 will also rise. In such patients, hypoxia become the dominant respiratory driver.

Further Reading

1.   Adrogue HJ, Madias NE. Management of life-threatening acid base disorders. *New Engl J Med* 1998; 338:26–34, 107–111.
2.   Judge, BS. Metabolic acidosis: Differentiating the causes in the poisoned patient. *Med Clin North Am* 2005; 89:1107–1124.
3.   Rose BD. *Clinical Physiology of Acid–base and Electrolyte Disorders* 2001. McGraw-Hill: New York, 5th Edition.

Summary/Key Points

1.   Hydrogen ion concentration (pH) of the body fluids is maintained within very narrow limits, in spite of the production of large amounts (15–20 mol) of volatile acids (carbon dioxide) and non-volatile acids (80 mmol) (produced from the metabolism of proteins).
2.   The hydrogen ions produced are buffered by extracellular and intracellular buffers before the acids are excreted by the lungs

(volatile acids) and the kidneys (non-volatile acids). An important extracellular buffer is the bicarbonate buffer system. The components of this buffer system, namely carbonic acid (carbon dioxide) and bicarbonate are regulated by the lungs and kidney respectively. Extracellular pH is directly proportional to bicarbonate concentration and inversely proportional to arterial partial pressure of carbon dioxide (P_aCO_2).

3. Kidneys play an important role in maintaining acid–base balance. Large amounts of bicarbonate are filtered at the glomerulus and most of this is reabsorbed in the proximal tubules.

4. Net hydrogen ion secretion takes place in the distal tubules and these hydrogen ions are excreted as titratable acid (buffered by phosphate) or as ammonium ions.

5. Titratable acid excretion depends on the excretion of phosphate which is not regulated by the acid–base status. Increased hydrogen ion excretion in response to an acid load is via increased ammonium ion excretion.

6. Factors influencing the secretion of hydrogen ions in the distal nephron include extracellular pH, aldosterone, potassium status, and ECF volume.

7. Acid–base disturbance of one system (respiratory or metabolic) is compensated by the other, e.g. when there is metabolic acidosis, there is respiratory compensation and vice versa.

8. Metabolic acidosis can be caused by increased acid load (diabetic keto acidosis, lactic acidosis; or ingestion of acid producing substances such as methanol), decreased excretion of hydrogen ions (renal failure or distal renal tubular acidosis) or due to loss of bicarbonate via the gut (diarrhoea) or via the kidneys (proximal renal tubular acidosis).

9. Metabolic alkalosis is often caused by loss of gastric acid and the loss of chloride maintains metabolic alkalosis.

chapter 5

The Kidneys

Introduction

Kidneys are the major organs involved in maintaining the composition of the internal environment. In addition, kidneys have
excretory, metabolic and endocrine functions (Table 5.1).

Each kidney has approximately one million nephrons and each
nephron contains a renal corpuscle and a tubule which is divided into
proximal convoluted tubule (PCT), loop of Henle, distal convoluted
tubule (DCT) and collecting duct. The renal corpuscle consists of a
glomerulus, a tuft of capillaries that is surrounded by the Bowman's
capsule which is the blind end of the tubule. The tuft of capillaries in
the glomerulus is supplied by the afferent arteriole and drained by the
efferent arteriole. Blood is filtered at the glomerulus and the composition of the filtrate is modified as it passes along the various segments of
the tubule before it becomes urine. The kidneys receive approximately
20–25% (about 1.2 L/min) of the cardiac output. From this, about
120 mL/min (180 L/day) of filtrate is formed at the glomerulus. The
composition of the ultra filtrate formed at the glomerulus is similar to
that of plasma except for large molecular weight substances like proteins
and cellular components. These are retained by the glomerular filtration
barrier, which consists of the capillary endothelial cells, between which
there are numerous pores (fenestrations), the basement membrane and
the epithelial cells of the nephron called podocytes. Filtration of proteins depends on the size and the electrical charge. Proteins with a
molecular weight of 68,000 or greater and molecules with negative
charge are not readily filtered. Small molecular weight proteins such as

Table 5.1   Functions of the kidney

Excretory
  • Urea, creatinine and urate
Regulatory
  • Maintenance of ECF volume and composition
Metabolic
  • Gluconeogenesis
Endocrine
  • Renin production
  • Erythropoietin production
  • Synthesis of 1,25-dihydroxycholecalciferol
  • Catabolism of hormones — insulin, parathyroid hormone

β_2 microglobulin (molecular weight 11,000) are freely filtered and these are almost completely reabsorbed and catabolised by the proximal tubular cells.

Production of the ultrafiltrate, which is a passive process, depends on two opposing forces. The pressure difference between the afferent and efferent arterioles forcing fluid out. The oncotic pressure difference between plasma and ultra filtrate and the hydrostatic pressure within the lumen of the nephron oppose this effect. The net filtration pressure is about 17 mmHg. As the glomerular filtrate passes down the tubules water, electrolytes and other constituents are reabsorbed, some substances are secreted leaving a urine volume of about 1–2 L/day. Summary of how substances are handled in the nephron is given in Table 5.2. In the proximal tubule, most of the glucose, amino acids, potassium and bicarbonate and 65–70% of the sodium accompanied by water and chloride are reabsorbed. The fluid entering the next segment, the loop of Henle, is isotonic. This segment is important for the generation of the counter current mechanism. The descending limb of the loop of Henle is relatively impermeable to solutes but permeable to water. Water moves out and the fluid becomes hypertonic. The thin section of the ascending limb of loop of Henle is relatively impermeable to water but not to solutes — sodium and chloride. As sodium and chloride move out, the tubular fluid becomes hypotonic. The

Table 5.2   Renal handling of substances

Substance	Amount filtered per day	Amount excreted per day	Percentage reabosrbed
Water	180 L	1.5 L	99.2
Sodium	24,000 mmol	100 mmol	99.6
Chloride	200,000 mmol	100 mmol	99.5
Bicarbonate	5000 mmol	<2 mmol	>99.9
Glucose	900 mmol	0	100
Albumin	1.3 g	15 mg	98.8
Potassium	900 mmol	50 mmol	94.5

thick ascending loop of Henle is also impermeable to water but not to solutes and this makes the fluid even more hypotonic. The net result of the counter current multiplication mechanism is the production of a hypotonic fluid leaving the loop of Henle and a hypertonic medullary interstitium (Figure 5.1). In the distal tubule, sodium reabsorption continues, which is mediated by aldosterone in exchange for H^+ ions or potassium ions. Water reabsorption in the collecting duct depends on the hydration state of the individual. In the presence of antidiuretic hormone (ADH), the cells of the collecting ducts become permeable to water and water moves out of the collecting duct into the interstitial fluid, down the osmotic gradient produced by the counter current mechanism. The highest urine osmolarity achievable by this process is about 1200–1400 mOsmol/kg.

Assessment of Renal Function

The excretory and regulatory functions of the kidney depend on three processes, namely filtration, reabsorption and secretion. When testing the functions of the kidney, one or more of these processes can be assessed. Filtration can be assessed by measuring the glomerular filtration rate (GFR). In addition, the permeability of the glomerular membrane can also be assessed. The most common renal function test in clinical practice is the assessment of GFR.

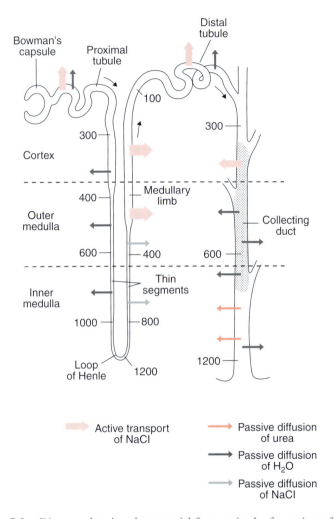

Figure 5.1   Diagram showing the essential features in the formation of urine.

Glomerular filtration rate is measured to detect mild renal impairment, to calculate the dosage of drugs which are eliminated by the kidney, to monitor changes in renal function and to assess the severity of renal impairment.

GFR can be assessed either by measuring a serum constituent which is excreted by filtration (e.g. urea and creatinine) or by a clearance technique.

Measurement of GFR

GFR can be determined by measuring the clearance of an endogenous or exogenous substance, which is freely filtered at the glomerulus and not reabsorbed, secreted or metabolised by the tubules. Classically, inulin has been used. An alternative is ^{51}Cr-labelled ethylenediaminetetra acetic acid (EDTA). The clearance of these can be calculated from the following equation:

$$\text{Clearance} = \frac{UV}{P} \text{ mls/min}$$

where U is the urinary concentration of the substance, V is the urine flow rate in mls/min and P is plasma concentration of the substance. Measurement of inulin or ^{51}Cr-EDTA clearance requires a constant infusion therefore, it is not used in routine practice. Instead, plasma clearance is used. In this, a bolus intravenous injection of a marker is given and blood samples are collected at timed intervals. The concentration of the marker is measured in blood and the rate of disappearance of the marker is calculated. The GFR is worked out from the amount of marker injected and the disappearance rate. The markers used in practice include ^{51}Cr-EDTA, ^{125}I-iothalamate, inulin or ^{99m}Tc-labelled diethylene triamine pentaacetic acid (DTPA).

Clearance of creatinine, which is an endogenous product, is an alternative. To measure creatinine clearance, urine is collected over a period of time, usually 24 hours, and a blood sample is taken during this period. Creatinine is measured in plasma and urine and the clearance is calculated from the above formula. However, as there is some secretion of creatinine in the tubules, creatinine clearance overestimates the true GFR. This discrepancy between creatinine clearance and true GFR increases as renal function deteriorates.

One of the major errors involved in measuring creatinine clearance is the difficulty in collecting 24-hour urine accurately, even in highly motivated subjects. Supervised collection of urine over a shorter period may be an alternative, but as the time period gets shorter, the errors in measurement also increases. As creatinine from food can be absorbed, it is best that the subject be on a meat-free diet when collecting urine samples for creatinine clearance. Because of these potential errors, creatinine clearance is seldom used in clinical practice.

Serum markers of GFR

Serum markers of GFR include urea, creatinine, and cystatin C. An ideal marker of GFR is one that is produced endogenously not protein bound, freely, filtered and is neither reabsorbed nor secreted.

1. *Urea*

    Urea, a metabolic end product of protein metabolism is produced in the liver from deamination of amino acids. Urea is not protein bound, is freely filtered by the glomerulus and is excreted in the urine. However, urea is passively reabsorbed and the rate of reabsorption depends on urine flow rate; the lower the urine flow rate, the higher the reabsorption. Serum urea is not a good index of GFR because production rate and reabsorption, in addition to GFR, affects its concentration (Table 5.3). Increased production due to high protein diet, bleeding into the gastrointestinal tract and hypercatabolic states will cause increase in serum urea concentration without a change in GFR. In volume depletion, the rate of urine flow decreases due to compensatory mechanisms, urea reabsorption increases and serum urea rises.

2. *Creatinine*

    Creatine is produced in the liver, kidneys and pancreas from arginine and glycine. It is transported to other organs such as the muscle and brain, where it is converted to the high-energy compound, creatine phosphate. In muscle, creatine phosphate non-enzymatically breaks down to form creatinine at a constant rate of 1–2% per day. The total amount of creatinine produced per day is dependent on the amount of creatine phosphate, which in turn is determined by the muscle mass. Once released from the muscle, creatinine enters the circulation. Creatinine in plasma is not protein bound, is freely

Table 5.3   Factors affecting serum urea concentration

Production
- Protein intake
- Hypercatabolic states — trauma, burns, surgery
- Gastrointestinal bleeding
- Liver disease

Increased reabsorption
- Volume depletion
- Congestive heart failure

Glomerular filtration rate
- Renal disease

filtered at the glomerulus and is excreted without reabsorption. There is some secretion of creatinine by the renal tubules in normal subjects. Serum creatinine concentration is commonly used in clinical practice to assess GFR. However, it is not a sensitive marker, as serum creatinine does not rise above the reference range until the GFR falls well below 50% of normal (Figure 5.2). Furthermore, muscle mass, diet and exercise can also affect serum creatinine. Serum concentration of creatinine is higher in men than in women and this is attributed to high muscle mass in men. However, the contribution of muscle mass to variation in serum creatinine is relatively small under normal circumstances. A diet rich in meat can increase serum creatinine transiently by up to 30% due to the absorption of creatinine from the diet. Blood samples for serum creatinine should preferably be taken in the fasting state. The reference range of serum creatinine also varies with age and sex. These disadvantages limit the use of creatinine. However, serum creatinine within an individual shows very little variation (low intraindividual variation) (Figure 1.1). Therefore, a change in serum creatinine in an individual is a sensitive index of a change in GFR in that individual and serial measurements are useful in following the progress of renal function. The use of serum creatinine can be summarised as follows: Serum concentration of creatinine within the reference range does not indicate normal GFR, serum creatinine concentration outside the reference range is indicative of reduced GFR, and serum creatinine in an individual is a useful marker to follow changes in GFR.

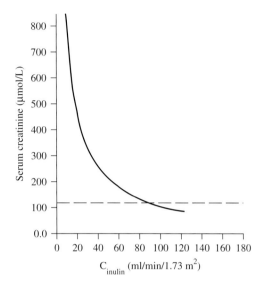

Figure 5.2   Relationship between creatinine and GFR.

3. *Cystatin C*

Cystatin C is a small peptide (molecular weight 13,000), produced by all nucleated cells and is a cysteine protease inhibitor. It is freely filtered by the glomerulus completely reabsorbed, and catabolised in the proximal tubular cells. As it is cleared by glomerular filtration, its serum concentration is a reflection of GFR. Serum concentrations of cystatin C is less affected by diet, age, gender, race and muscle mass. Serum cystatin C is, therefore, advocated as a better marker of GFR. However, this assay is expensive compared to serum creatinine. Recent studies show that serum cystatin C can be affected by factors other than GFR and these include thyroid status, glucocorticoid therapy, malignancy and HIV.

Estimated GFR

GFR can be estimated from serum creatinine using a formula. The Cockcroft and Gault formula is one of the earliest equations to be used to estimate creatinine clearance.

Creatinine clearance (mL/min) = F [(140-age) × weight (kg)]/serum creatinine (μmol/L)

where F is a constant. For males it is 1.224 for females it is 1.074.

This formula is used commonly in oncology to calculate the dosage of cytotoxic drugs. However, this formula is less reliable in patients who are obese or in whom there is oedema or ascites and this has not been validated for different ethnic groups.

More recently, a formula developed by the Modification of Diet in Renal Disease (MDRD) study group is widely used. Serum creatinine (S cr), age, race and gender are included in the four variable formula.

$$eGFR = 175 \times [Scr/88.4]^{-1.154} \times (age)^{-0.203}$$

where eGFR is estimated GFR in mL/min/1.73m^2, S cr is serum creatinine in μmol/L and age in years. For females, the result is multiplied by 0.742 and for African Americans, the result is multiplied by 1.212.

Validation studies have shown that this formula is fairly reliable when the GFR is <60 mL/min. This formula, however, has not been validated for races other than Caucasians and African Americans. This formula is unreliable in several clinical situations which are listed in Table 5.4.

The Chronic Kidney Disease Epidemiology Collaboration (CKD-EPI) formula was recently published as an improvement over the MDRD formula especially when GFR is >60 mL/min.

In children, the Schwartz formula is used, where in addition to serum creatinine, height and a constant (k) is applied depending on the age of the child:

Estimated GFR (mL/min/1.73m^2) = k × height (cm)/Serum creatinine (μmol/L)

Assessment of Glomerular Permeability

Normal glomerulus is impermeable to large molecules. Presence of large molecular weight substances such as proteins in the urine

Table 5.4   Situations where MDRD formula is not applicable

Rapidly changing GFR
Pregnancy
Patients with muscle wasting disorders
Paraplegia
Spina bifida
Assessment of living donors
To calculate dosage of drugs excreted by the kidney (e.g. cancer chemotherapeutic agents)
When GFR is > 60 mL/min/1.73m^2

suggests impairment of the glomerular permeability. Presence of red cells would indicate severe glomerular damage. However, blood can appear in the urine from anywhere in the urinary tract and the presence of red cell casts, which are red cells embedded in a proteineous matrix in the urinary sediment, is highly suggestive of increased glomerulus permeability. The presence of protein in the urine is also suggestive of glomerular damage but there may be other causes of proteinuria (see later). Comparing the clearance of two proteins with differing molecular weights such as IgG (molecular weight 150,000) and transferrin (molecular weight 68,000) is useful in assessing the degree of permeability.

Tests of Renal Tubular Function

Tests of tubular function are required less often in clinical practice. In proximal tubular dysfunction, glucose, amino acids and bicarbonate may appear in the urine. However, amino acids can appear in the urine in other conditions, such as in inborn errors of amino acid metabolism and glycosuria may occur in diabetes mellitus.

As proximal tubules also reabsorb and catabolise small molecular weight proteins such as β_2 microglobulin, increased excretion of such proteins is also an indication of proximal tubular dysfunction.

Distal tubular function can be assessed by the ability to secrete H$^+$ ions and to concentrate the urine.

Urinary concentrating ability

This is classically assessed using the water deprivation test. This is described in Chapter 2.

Tests of urinary acidification

Assessment of urinary acidification can be done by the ammonium chloride loading test. The basis of this test is to examine the effect of a mild degree of metabolic acidosis induced by administration of ammonium chloride on urinary pH. Administration of ammonium chloride is not necessary if the patient is already acidotic. Ammonium chloride (0.1 g/kg) is administered orally and the urine pH is measured hourly for 6–8 hours. A normal response in indicated by a urine pH of 5.5 or less. Failure to reach this pH indicates a defect in acidification capacity.

Renal Disorders

Acute Renal Failure/Acute Kidney Injury

Acute kidney injury (AKI) is a rapidly progressive loss of renal function which is frequently reversible and is usually associated with oliguria or anuria. Oliguria is defined as urine output of < 400 mL/day. An international body has recently established a uniform definition of AKI. The criteria for the diagnosis of AKI is an abrupt (within 48 hours) reduction in kidney function defined as an absolute increase in serum creatinine $\geq 26\ \mu mol/L$, or a $\geq 50\%$ increase in serum creatinine or a reduction in urine output (oliguria of < 0.5 mL/kg/hr) for more than six hours. However, urine volume may be normal or even increased in a substantial percentage of patients with AKI. AKI is a fairly common condition in hospitals, especially in the elderly and in acutely ill patients. Five percent to 20% of critically ill patients may develop AKI and in severe sepsis, the incidence may be as high as 50%. Severe AKI is associated with significant mortality of up to 50% in the elderly. Causes of AKI are usually divided into prerenal, renal and postrenal causes (Table 5.5). Prerenal AKI is caused by a decrease in renal blood

Table 5.5   Causes of acute renal failure

Prerenal
- Absolute decrease in effective blood volume
  — Haemorrhage
  — GI loss of fluids
  — Burns
- Relative decrease in effective blood volume
  — Sepsis
  — Anaphylactic shock
  — Congestive heart failure
  — Myocardial infarction

Renal causes
- Nephrotoxins
  — Aminoglycosides
  — Cephalosporins
  — Toxins
  — Radiocontrast media
  — Pigments — myoglobin (rhabdomyolysis)
- Intrinsic renal disease
  — Following prerenal failure leading to acute tubular necrosis
  — Glomerulonephritis
  — Acute interstitial nephritis
- Vascular
  — Vasculitis
  — Malignant hypertension
- Metabolic
  — Hypercalcaemia
  — Immunoglobin light chain

Postrenal causes
- Obstruction
  — Bladder neck obstruction
  — Prostatic hypertrophy
  — Fibrosis
  — Ureteric stones
  — Bence-Jones proteinuria

flow caused by circulatory insufficiency. Causes of prerenal AKI include haemorrhage, diarrhoea, vomiting, and burns. Hypotension, induced by congestive heart failure may also cause AKI. Reduction in effective circulatory volume usually leads to reduction in urine flow rate, thereby causing an increased reabsorption of urea causing serum urea to increase. This state is sometimes referred to as 'prerenal uraemia'. With further reduction in circulatory volume, the GFR decreases and the plasma creatinine will increase. At this stage, the increase in urea is greater than the increase in creatinine, i.e. urea/creatinine ratio will be high. Further reduction in circulatory volume will lead to damages to the tubular cells, which is described as acute tubular necrosis (ATN). Prior to this stage, if circulatory volume is restored, renal function rapidly returns to normal. Once ATN has developed, renal function will not return to normal until several weeks or months later. ATN is a form of AKI where there is damage to the renal tubule cells. Hypoperfusion and nephrotoxins (myoglobin, antibiotics, etc.) are the main causes of ATN. Presence of brown casts of epithelial cells in the urine is indicative of ATN. Renal causes of AKI are many and include various forms of glomerulonephritis, nephrotoxins and renal vascular disease such as renal artery occlusion due to embolism or thrombosis. Postrenal AKI is due to urinary obstruction, from renal stones, fibrosis, or obstruction of the bladder neck by prostatic hypertrophy.

Pathophysiology of acute renal failure

The pathophysiology of ATN due to hypoperfusion is not fully understood. It is not clear why renal hypoperfusion does not return to normal when the circulation is restored. Factors thought to be involved in the pathogenesis of ATN and their interactions are outlined in Figure 5.3.

Features of acute kidney injury

Acute renal failure is usually described in three phases — oliguric phase, diuretic phase and recovery phase. During the oliguric phase, urine output falls below 400 mL/day but up to 40% of cases may have no oliguria. The duration of oliguria varies considerably. During this phase the GFR is low and urinary excretion of nitrogenous waste products, electrolytes and hydrogen ions are impaired. Characteristically, there is metabolic acidosis, increased plasma urea and creatinine, hyperkalaemia and hyperphosphataemia. The rate of

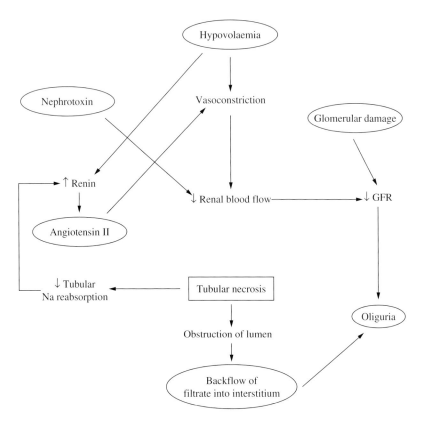

Figure 5.3   Pathogenesis of acute renal failure.

rise in plasma potassium and urea depends on the metabolic state of the individual patient. In hypercatabolic patients, the increase will be rapid. Hyponatraemia is a common finding during this phase due to the inability of the kidneys to excrete water. Hyperkalaemia is due to a combination of metabolic acidosis and the inability of the kidney to excrete potassium ions released from tissues. Hyperphosphataemia is caused by decreased excretion of phosphate released from cells. Hyperphosphataemia will lead to a decrease in plasma calcium concentration by a direct effect on the solubility of calcium phosphate as well as through inhibition of the production of 1,25-dihydroxycholecalciferol (calcitriol).

Case 5.1

A 65-year-old man was admitted with severe acute abdominal pain. He underwent an emergency laparotomy where he was found to have a thrombosis of the superior mesentric artery. He had a resection of most of his small bowel and part of his large bowel. On the second day after his operation, he was found to be hypotensive and his urine output was 250 mL. Results of biochemical investigations were as follows:

Serum		Reference range
Sodium (mmol/L)	130	135–145
Potassium (mmol/L)	4.9	3.5–5.0
Urea (mmol/L)	27.5	2.5–7.2
Creatinine (μmol/L)	212	66–112
Urine		
Sodium (mmol/L)	10	
Urea (mmol/L)	655	

This patient had, by definition, AKI and the likely cause in his case is pre-renal uraemia due to ECF volume depletion. His urine sodium excretion was < 20 mmol/L and the urine was concentrated more than twenty times (urine to plasma urea ratio). The increase in urea was more than that of creatinine due to increased reabsorption of urea as well as increased catabolism. He was given intravenous saline and blood. After some time, his urine output increased and his serum urea and creatinine returned to normal.

The start of the diuretic phase usually indicates recovery and during this period, the urine volume may increase to several litres a day. The diuretic phase is characterised by an increase in GFR without an improvement in tubular function thus, electrolytes and water can be lost.

In the recovery phase, tubular function returns to normal and tubular cells regenerate. The diuresis decreases and various abnormalities in biochemistry resolve.

Case 5.2

A 25-year-old man was rescued from an underground cave where he was trapped in a narrow gap for nearly 24 hours. On admission, he was dehydrated and had severe pain in his thighs. He was hypotensive and investigations showed the following results:

Serum		Reference range
Sodium (mmol/L)	132	135–145
Potassium (mmol/L)	6.3	3.5–5.0
Urea (mmol/L)	43.5	2.5–7.2
Creatinine (μmol/L)	375	66–112
Bicarbonate (mmol/L)	15	23–32
Urine		
Sodium (mmol/L	50	
Urea (mmol/L)	212	
Osmolality (mmol/kg)	345	

This man has developed AKI as a result of crush injury to his leg muscles, due to the fact that he was trapped in the cave. His urine sodium results were high and the urine was not concentrated (urine:serum urea ratio < 5). These are all features of acute tubular necrosis. His serum potassium was high and bicarbonate concentration was low, indicating metabolic acidosis.

Diagnosis of acute kidney injury

After a careful history and clinical examination, an analysis of urine and serum biochemistry will be helpful in the diagnosis of AKI and in the detection of aetiology. Biochemical investigations are especially useful in differentiating oliguria due to volume depletion from ATN (Table 5.6). In volume depletion states, serum urea is disproportionately higher than serum creatinine; the urine will be concentrated (urine osmolarity > 500 mOsmol/kg) and there will be very little sodium in the urine (fractional excretion of sodium is typically < 1%) (Table 5.6). However, it must be stressed that none of these tests can invariably differentiate between the two conditions. Furthermore, frequent use of diuretics in these patients makes the urine test unreliable or difficult to interpret.

Table 5.6   Biochemical tests to distinguish oliguria due to volume depletion from acute tubular necrosis

	Volume depletion	Acute tubular necrosis
Urine sodium concentration	< 20 mmol/L	> 40 mmol/L
*Fractional excretion of sodium (%)	< 1	> 2
Urine:plasma urea ratio	> 8	< 3
Urine:plasma osmolality ratio	> 1.5	< 1.1
Urine osmolality (mOsmol/kg)	> 500	< 350
Urine:serum creatinine ratio	> 40	< 20
Urine microscopy	Hyaline casts	Abnormal
Fractional excretion of urea	< 35	> 35
Fractional excretion of uric acid	< 7	> 15
Low molecular weight proteins	Low	High
Brush border enzymes	Normal	High

*Fractional excretion of sodium (%) = $\frac{\text{urine sodium}}{\text{plasma sodium}} \times \frac{\text{plasma creatinine}}{\text{urine creatinine}} \times 100$.

Recently, low molecular weight proteins, brush border enzymes or cytokines have been investigated as possible biomarkers of AKI. Kidney injury molecule-1 (KIM-1), neutrophil gelatinase-associated lipocalin (NAGL), N-acetylglucosaminidase (NAG), alanine amino peptidase, interleukin-18 and fatty acid-binding protein are some of these.

Management of AKI

The best form of management is prevention by proper fluid balance, avoidance of nephrotoxins agents, etc. Principles of management of AKI are (i) correction or treatment of the cause, (ii) support the kidney until its recovery (iii) prevention of treatment of complications. Correction of any fluid loss, correction of any postrenal factors, optimising cardiac output and reduction or elimination of nephrotoxic drugs are important. It is common practice to use furosemide, dopamine and mannitol in the hope of delaying or preventing AKI. The available evidence shows that these agents offer no significant benefit and may be even harmful. Volume depletion from any cause should be treated immediately since longstanding renal hypoperfusion will lead to AKI.

Treatment of hypocalcaemia and hyperuricaemia should be undertaken as soon as possible.

In established AKI, measures should be taken to minimise the adverse effects of electrolyte and acid–base balance. Plasma potassium should be monitored carefully and if hyperkalaemia develops, it should be treated. Sodium and water intake should be carefully controlled to prevent overload. Adequate amounts of carbohydrates should be given to minimise endogenous protein breakdown, which can be substantial in patients with sepsis, shock or rhabdomyolysis. Increased protein breakdown may increase the rate of rise of potassium, hydrogen ions and phosphate. Measures should be taken to prevent and control infection. In many patients, however, these conservative measures may not be adequate and haemodialysis, haemofiltration or peritoneal dialysis may be required (renal replacement treatment). Indications for such treatment include rapidly rising potassium and urea concentration, severe acidosis, fluid overload, complications such as pericarditis, encephalopathy or gastrointestinal bleeding, or a general deterioration of the patient. During the diuretic phase, adequate amounts of fluid and electrolytes should be given to compensate for the losses. Plasma sodium, potassium, creatinine, bicarbonate, calcium, phosphate and magnesium should be monitored regularly. In addition to these general measures, specific treatment such as immunosuppressive drugs, treatment of hypertension, etc. may be required in certain diseases.

In spite of intensive treatment, mortality due to renal failure still remains high especially in the elderly, critically ill patients and patients with multiple organ failure. Hence, this reflects the importance of prevention of AKI. It is estimated that 5% of patients with AKI (this may be as high as 16% in the elderly) develop chronic renal failure.

Chronic Kidney Disease/Chronic Renal Failure

Chronic kidney disease (CKD) is characterised by a progressive and generally irreversible decline in GFR. The prevalence of CKD has been estimated to be around 10% in the general population and majority of them are undetected. Hence, the recommendation that whenever serum creatinine is measured, eGFR should be reported in order to detect

Table 5.7　Causes of chronic renal failure

Common
- Glomerulonephritis
- Hypertension
- Diabetes mellitus
- Infections of the kidney
- Urinary tract disease
- Congenital abnormalities — polycystic disease of kidney

Uncommon
- Medullary cystic disease
- Cystinosis
- Oxalosis
- Systemic vasculitis
- Systemic lupus erythematosus
- Myeloma
- Haemolytic uraemic syndrome
- Nephrocalcinosis
- Gouty nephropathy

Table 5.8　Stages of chronic kidney disease

Stage	GFR (mL/min/1.73m^2)	
1	>90	Abnormalities in urine, structural abnormalities, genetic trait to kidney disease
2	60–89	As for stage 1
3A	45–59	Moderate
3B	30–44	
4	15–29	Severe, prepare for renal replacement therapy
5	<15	End stage renal failure, on renal replacement therapy

these patients early. CKD is four times more common in African Americans compared to Caucasians and the incidence increases with age. It is caused by a large number of diseases (Table 5.7) and some of the common causes of chronic renal failure are glomerulonephritis, diabetic nephropathy and hypertension. CKD is classified into five stages (Table 5.8). The rate of progression of chronic renal failure varies enormously from weeks to years. Although these diseases have different aetiology and different histological features, they all produce a syndrome which has common clinical features. In most patients with chronic renal failure, compensatory and adaptive mechanisms maintain homeostasis until the GFR falls below 15 mL/min. When GFR falls below this level, some form of renal replacement therapy in the form of dialysis or renal trans-plantation is required.

Pathophysiology

The ability to maintain near normal health is explained by the intact nephron hypothesis which states that in CKD, a proportion of nephrons are destroyed but the remaining nephrons are fully functional.

1. *Sodium and water*

   The kidney has a remarkable capacity to maintain total body sodium content within normal limits until the very end stages of CKD. In healthy subjects, sodium balance is maintained by regulation of urinary sodium excretion. In CKD, to maintain sodium balance, the remaining nephrons must excrete a higher proportion of the filtered sodium compared to a normal subject. The mechanism by which this is achieved is not clear. It has been suggested that a natriuretic substance may be involved. In healthy subjects, sodium excretion can vary over a wide range depending on the dietary intake. However, in CKD, this ability is restricted, and if sodium intake is reduced or increased there may be loss or retention of sodium. Rarely, some patient's exhibit inappropriate natriuresis and this is sometimes referred to as 'salt-losing nephropathy'. However, most patients retain sodium when the sodium intake is increased, leading to volume expansion.

   Patients with chronic renal failure have polyuria. This is due to the osmotic diuretic effect of urea. With a reduced number of nephrons, each nephron has to handle more osmoles, causing an osmotic diuresis. However, the polyuria in chronic renal failure is not severe and the urine volume seldom exceeds 4 L/day. As the renal failure progresses, the ability to vary the osmolarity of the urine progressively decreases and urine osmolality may be fixed. Therefore, patients with CKD are liable to become water-depleted or water-intoxicated very easily. The inability to concentrate the urine is particularly noticeable during the night and CKD patients often complain of nocturia.

2. *Potassium*

Potassium balance is maintained until the end-stage by the ability of remaining nephrons to secrete more potassium. The exact mechanism by which this is brought about is not clear and until the very late stage, patients can respond to changes in potassium intake. In the late stages, hyperkalaemia develops and this can be precipitated by factors, which increase potassium input (from the cells due to trauma, surgery or infection or dietary intake) or drugs that interfere with potassium excretion such as ACE inhibitors or β-adrenergic blockers.

Case 5.3

A 64-year-old lady was seen at the pre-admission unit for assessment prior to an ear operation, stapedectomy. Pre-operative examination showed her blood pressure was 150/90 and she had proteins in her urine. She was otherwise healthy and her blood results were as follows:

Serum		Reference Range
Sodium (mmol/L)	138	135–145
Potassium (mmol/L)	4.7	3.5–5.0
Urea (mmol/L)	43.5	2.5–7.2
Creatinine (μmol/L)	375	66–112
Bicarbonate (mmol/L)	22	23–32
Calcium (mmol/L)	2.15	2.25–2.65
Phosphate (mmol/L)	1.5	0.8–1.3

This lady has CKD. She was relatively symptom-free as the remaining nephrons were able to maintain homeostasis. Her serum urea and creatinine indicate that her GFR is low and the eGFR puts her in CKD category 3a.

3. *Acid–base balance*

Initially, acid base balance is maintained by increased secretion of H^+ ions and ammonia by each remaining nephron. However, this capacity is limited and metabolic acidosis eventually develops and plasma bicarbonate concentration decreases. The rate of progression of metabolic acidosis (i.e. the rate of decrease in bicarbonate concentration) is relatively slow compared to the rate of deterioration in GFR due to the buffering action of bone. This buffering action increases bone reabsorption and may contribute to renal osteodystrophy (see later). Even in end-stage renal disease, the plasma bicarbonate concentration seldom falls below 12–15 mmol/L. If a patient with chronic renal failure is seen with a bicarbonate concentration below this value, a search should be made for causes of superimposed metabolic acidosis. This steady state metabolic acidosis with low bicarbonate is well tolerated by most patients with CKD and probably reflects the slow development of the acidosis.

4. *Calcium and phosphate*

Patients with CKD are hypocalcaemic and hyperphosphataemic. The pathogenesis of this is illustrated in the Figure 5.4. As serum phosphate is mainly determined by renal excretion, hyperphosphataemia develops in CKD. This leads to reduce synthesis of 1,25-dihydroxycholecalciferol (calcitriol). Reduction in renal mass also contributes to reduced synthesis of calcitriol. Decreased synthesis of calcitriol leads to decreased absorption of calcium from the gut leading to hypocalcaemia. Hyperphosphataemia causes soft tissue calcium deposition when the calcium/phosphate product exceeds the solubility product further aggravating the hypocalcaemia. Hypocalcaemia, in turn, leads to increased parathyroid hormone secretion (secondary hyperparathyroidism), leading to increased bone resorption. Hyperphosphataemia can directly stimulate the secretion of PTH, contributing to secondary hyperparathyroidism. Decreased calcitriol gives rise to a picture of osteomalacia. Metabolic acidosis of CKD (described earlier) causes demineralisation of bone due to the buffering action of bone. A combination of these effects on the bone leads to the complex picture of renal osteodystrophy.

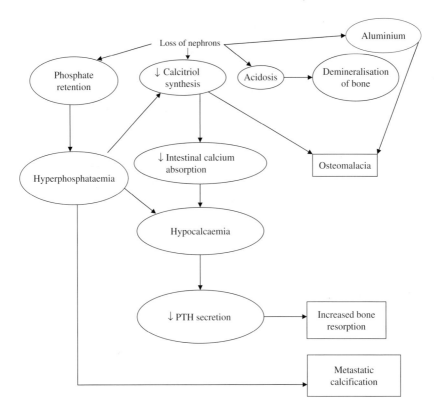

Figure 5.4 Pathogenesis of hypocalcaemia and renal osteodystrophy in chronic renal failure.

Aluminium, which is excreted by the normal kidney, is retained in CKD. Aluminium comes from oral phosphate binders used to prevent hyperphosphataemia or from aluminium salt used as water softeners. Retention of aluminium is an important factor in the pathogenesis of osteomalacia in CKD.

5. *Other abnormalities*

Patients with CKD tend to have a slight elevation of serum magnesium concentration, especially when the GFR is below 20% of normal, due to decreased excretion. Hypermagnesaemia itself does not cause symptoms. Patients with chronic renal failure are more prone to get cardiovascular complications and this may be partly due to changes in lipid metabolism; increased triglyceride concentration and reduced HDL cholesterol. Chronic renal failure is also associated with decreased testosterone and oestrogen synthesis and may lead to decreased libido and impaired fertility in men and women. Mild glucose intolerance is not uncommon in patients due to insulin resistance. However, in severe CKD, hypoglycaemia will develop as the kidney is an important contributor to gluconeogenesis.

Hyperuricaemia is a consistent finding in CKD when the GFR is below 20% of normal levels. Anaemia is one of the most consistent manifestations of uraemia. The haematocrit in end-stage CKD is in the region of 20–25%. Anaemia is most often normocytic-normochromic. The major cause of anaemia is deficiency of erythropoietin due to decreased renal mass. In addition, uraemic 'toxins' may depress bone marrow and shorten the lifespan of red blood cells. Patients with CKD also have a haemorrhagic tendency related to abnormalities in platelets, which are often reduced in number due to increased peripheral destruction. Platelets also show functional defects such as decreased adhesiveness and aggregation.

Clinical features of CKD (Table 5.9)

Clinical features of CKD are summarised in Table 5.9. This collection of features is termed 'uraemic syndrome' and it is due to a combination of the effects of retained 'toxins' on organs and the failure of homeostatic and endocrine functions of the kidneys. Substances which accumulate in CKD and which are thought to be toxic, include purine metabolites, amines, indoles, phenols, myo-inositol, polyol, and middle molecules (molecular weight between 500 and 5000).

Management of patients with CKD

In all patients, attempts should be made to identify the cause of renal failure and treat it in order to prevent or delay the progression of CKD. In all patients, factors which aggravate the development of renal failure

Table 5.9   Clinical features of chronic renal failure

Nervous system
- Lethargy
- Sleep disturbance
- Peripheral neuropathy
- Stupor, fits

Cardiovascular system
- Hypertension
- Ventricular hypertrophy
- Calcification of valves
- Cardiac failure
- Pericarditis
- Pericardial effusion

Gastrointestinal system
- Anorexia
- Hiccup
- Nausea and vomiting
- Gastrointestinal bleeding

Musculoskeletal system
- Growth failure (children)
- Weakness — myopathy
- Bone pain — osteodystrophy
- Carpal tunnel syndrome

Haemopoitic system
- Anaemia
- Bleeding tendency

Genitourinary system
- Nocturia, polyuria
- Impotence, reduced libido
- Infertility

Skin
- Pruritis
- Pallor/Pigmentation
- Purpura
- Bruising

such as hypertension, should be sought out and treated immediately. However, most patients will progress to end-stage renal failure, requiring dialysis or transplantation. Until that stage, conservative measures can reduce symptoms. Sodium and water balance should be carefully maintained and in those patients who have a tendency to retain salt, reduced salt intake and/or diuretics may be necessary. Oral administration of bicarbonate may be required to treat the acidosis. Hyperkalaemia is usually not a problem until the late-stage and it can be controlled if present, by oral ion exchange resin. Hyperphosphataemia, which can lead to osteodystrophy, can be controlled by oral phosphate binding agents. However, aluminium salts should be used carefully as this might aggravate osteodystrophy. Hypocalcaemia and the consequent secondary hyperparathyroidism should be controlled by the administration of 1α-hydroxyvitamin D or calcitriol. Calcimimetics such as cinacalcet which acts on the calcium-sensing receptor is also used to treat secondary hyperparathyroidism. It is usually recommended that patients with CKD restrict protein intake to reduce the formation of nitrogenous waste products. It has also been suggested that a reduction in protein intake will reduce the rate of progression of renal failure as a high protein intake increases GFR, causing subsequent glomerular damage. However, trials on the role of protein intake on the rate of progression of renal failure have not been conclusive. The diet should not be too low in protein as the patient will then go into a negative nitrogen balance and this will lead to endogenous protein breakdown. Treatment with HMG-CoA reductase inhibitors and aspirin should be considered to reduce the risk of cardiovascular disease. Treatment with recombinant erythropoietin is given to correct anaemia. However, there is some evidence that raising the haemoglobin above 120 g/L may be harmful. In patients with proteinuria and CKD with or without diabetes, ACE inhibitors and angiotensin II receptor blockers are recommended to slow the progression of CKD.

Patients with CKD should be monitored regularly. Regular monitoring of electrolytes, bicarbonate, urea, creatinine, calcium, phosphate, magnesium, haemoglobin as well as parathyroid hormone is required in all patients, including those on dialysis.

When the conservative management outlined above is no longer adequate to maintain homeostasis, renal replacement therapy is required. Ideal renal replacement therapy is renal transplantation. Until a kidney is available or if there are contraindications for surgery, dialysis, haemo or peritoneal, is necessary to maintain the internal environment. Dialysis involves diffusion of solutes across a semiper-meable membrane. In the case of haemodialysis, a synthetic membrane is used in the dialysis machine and this separates blood from the dialy-sis fluid in the extracorporeal circulation. In peritoneal dialysis, the peritoneal membrane acts as the semipermeable membrane. The ade-quacy of dialysis is usually assessed by calculating the function KtV where K is clearance of urea, t is time of dialysis and V is the volume of distribution of urea, which is essentially the patient's total body water. Although there are some disadvantages to this function, most renal units use this as it correlates with outcome. Patients on dialysis should have regular monitoring as described above and in addition, aluminium levels should be checked.

Renal transplantation from a cadaver or a living person is required in patients with CKD. Unless the donor kidney is from an identical twin, lifelong immunosuppressive therapy is required after transplan-tation. Regular monitoring of renal function (serum creatinine) and blood levels of immunosuppressive agents like cyclosporine and tacrolimus is required in these patients.

Proteinuria and Nephrotic Syndrome

The glomerular membrane consists of the capillary endothelium, which is fenestrated, a basement membrane and the podocyte. This membrane carries a negative charge. Due to the size of the pores, some proteins are filtered at the glomerulus. The degree of filtration of a protein depends on the size and the charge. The normal kidney filters about 7–10 g of protein a day due to the fact that the perme-ability of the glomerular capillaries is about 50 times that of the capillaries in the skeletal muscle. The extent and nature of protein fil-tered depends on its molecular weight and charge. Neutral (uncharged) molecules with an effective molecular diameter of < 4 nm are freely filtered and the filtration of neutral molecules > 8 nm diameter approaches zero. Between these values, filtration is inversely proportional to diameter. However, the endothelium and basement membrane proteins are negatively charged therefore, nega-tively charged molecules are relatively impermeable. Studies with

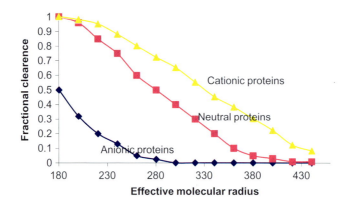

Figure 5.5   Fractional clearance of neutral, anionic and cationic proteins plotted as function of molecular size.

anionically charged and cationically charged dextrans show that filtration of anionic molecules of 4-nm diameter is less than half that of neutral molecules of the same size and filtration of cationic molecules is slightly greater (Figure 5.5). As albumin with a diameter of 7.2 nm is negatively charged, its filtration is less than expected from its molecular size. Filtered protein is reabsorbed in the proximal tubules by endocytosis and catabolised within the tubular cells, leaving < 150 mg/24-hr to be excreted in urine. Less than half of this is from plasma and the rest from the tubular cells. The amount of albumin in normal urine is < 30 mg/day. Most abundant protein in the urine is Tamm-Horsfall protein, which is a glycoprotein secreted by the loop of Henle and distal convoluted tubule.

Protein excretion greater than 150 mg/24-hr is seen after strenuous exercise, in a benign condition called 'orthostatic proteinuria', in fever and in other inflammatory conditions such as burns, sepsis, trauma and surgery. The mechanism of increased protein excretion in fever and other inflammatory conditions is probably an increase in vascular permeability induced by cytokines. The mechanism of proteinuria in strenuous exercise is both glomerular and tubular. During severe exercise, blood is diverted to the skeletal muscle from the kidney, thereby reducing the reabsorption of proteins normally filtered and in addition, proteins may be released from the tubular cells.

Orthostatic proteinuria is a benign condition, which is seen in approximately 5% of young adults. In this condition, proteinuria develops only when the subjects are upright. When the subject assumes an erect posture, the increase in hydrostatic pressure in the renal veins due to pressure of the liver on the inferior vena cava causes increased protein excretion. This condition has no long-term clinical consequences.

Mechanism of Proteinuria

Table 5.10 shows the mechanism and causes of proteinuria.

Table 5.10 Causes of proteinuria

Overflow proteinuria
- Normal proteins
  — Haemoglobin
  — Myoglobin
- Abnormal proteins
  — Bence-Jones proteins

Glomerular proteinuria
- Altered renal haemodynamics
  — Exercise
  — Orthostatic proteinuria
- Increased glomerular permeability
  — Fever and other acute inflammatory conditions such as trauma, sepsis
  — Glomerulonephritis

Tubular proteinuria
- Tubular damage
  — Heavy metal poisoning
  — Drug toxicity
- Interstitial disease
  — Interstitial nephritis
  — Pyelonephritis

Secretory proteinuria

Overflow proteinuria

When the concentration of proteins small enough to pass through the normal barrier is increased in plasma, the amount of proteins reaching the tubular cells will exceed the tubular reabsorptive capacity, leading to proteinuria. An important pathological cause is the presence of immunoglobulin light chains (Bence-Jones protein) in multiple myeloma.

Glomerular proteinuria

Glomerular proteinuria is caused by the increased filtration of proteins as a result of increased permeability. Important pathological causes include primary renal disorders such as glomerular nephritis and a number of systemic conditions. Glomerular proteinuria is divided into selective and non-selective. In non-selective proteinuria, large molecular proteins are filtered whereas in selective proteinuria, only low molecular weight proteins are excreted.

Tubular proteinuria

Proteins, which are normally filtered because of their small molecular weight, are reabsorbed by the tubules. Tubular proteinuria develops when there is impaired reabsorption of these proteins or if resorption is saturated. β_2 microgloblin is an example of tubular proteins. Tubular proteinuria is seen in conditions such as heavy metal poisoning.

Secretory proteinuria

Secretory proteinuria is when the proteins are produced from the cells lining the renal tract. An example of such a protein is the Tamm-Horsfall protein.

Detection of Proteinuria

Urine protein is usually detected at the bedside using a dipstick. The principle of this method is the pH effect of proteins. The dipstick incorporates a pH indicator dye. Proteins present in the urine buffer the H^+ ions, leading to a change in pH. The detection of proteinuria by dipstick depends on the type of protein present — albumin is detected better than globulin. These dipsticks usually reliably detect albumin when the concentration exceeds 150 or 200 mg/L. As the concentration of protein in the urine depends on urine flow rate, early morning urine is preferable. The dipstick method will give false positive result if the urine is alkaline, if the urine is contaminated by antiseptics, or if the patient had been given X-ray contrast media. In the laboratory, urine protein is determined by a dye binding method. As these dyes bind differentially to different proteins, it has been recommended that that urine total protein measurement should be replaced by albumin measurement. It has been the practice to measure protein excretion in 24-hour urine collection. As 24-hour urine collection is cumbersome and often inaccurate, it is now recommended to use a spot urine and measure protein:creatinine ratio (PCR). A PCR value of < 15 mg/mmol is normal. Recently it has been recommended that PCR should be replaced with albumin:creatinine ratio (ACR).

Albuminuria (Microalbuminuria)

Albumin can be detected in normal urine by sensitive immunological methods and in healthy subjects, albumin excretion is < 30 mg/24-hr. Increased excretion of albumin detectable by such methods is called 'microalbuminuria'. Increased albumin excretion is of prognostic significance in diabetes (see Chapter 10) and has been shown to be an independent risk factor for cardiovascular disease in non-diabetic patients. It is conveniently measured in a spot urine and expressed in relation to creatinine excretion. An albumin:creatinine ratio (ACR) value of more than 3.5 mg/mmol in females and 2.5 in males is abnormal.

Investigation of Patients with Proteinuria

A flow chart describing the investigation of proteinuria is given in Figure 5.6. If protein is detected by a dipstick method, proteinuria

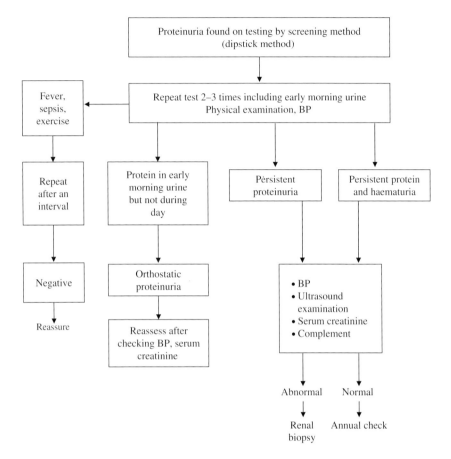

Figure 5.6   Investigation of a patient found to have proteinuria.

should be confirmed by the measurement of total protein or albumin excretion by quantitative laboratory methods. After excluding causes such as fever, orthostatic proteinuria or exercise-induced proteinuria, simple renal function tests (serum urea and creatinine) followed by electrophoresis of urine should be done. Electrophoresis of the urine sample will detect Bence-Jones proteins as well as the type of protein excreted. Further investigations include radiological or ultrasound examination of the kidney.

If the protein:creatinine ratio (PCR) is persistent and greater than 30 mg/mmol, further investigations including blood glucose, renal function, serum lipids, anti-nuclear factor, complement C3 and C4 are necessary. If these are not diagnostic, renal biopsy may be required. If the PCR is between 15 and 30 mg/mmol, consider other types of proteinuria such as tubular proteinuria.

Nephrotic Syndrome

Nephrotic syndrome is characterised by heavy proteinuria (> 3 g/day or > 300 mg/mmol creatinine), hypoalbuminaemia and oedema. Typically, in patients with nephrotic syndrome, proteinuria exceeds 3.5 g/day, usually 5 g/day. Although the proteinuria is compensated by increased protein synthesis, this is insufficient to prevent the development of hypoalbuminaemia. The decrease in albumin concentration may lead to a reduction in plasma oncotic pressure and subsequently to oedema as described in Chapter 2. Reduction in plasma oncotic pressure is also thought to stimulate hepatic lipoprotein synthesis, and hyperlipidaemia is a typical finding in patients with nephrotic syndrome. A large number of conditions can manifest as nephrotic syndrome (Table 5.11) and the causes of nephrotic syndrome can be classified according to clinical presentations or in histological terms. Clinically, they can be divided into primary (idiopathic) or secondary. Nephrotic syndrome is said to be primary when it occurs in association with one of the primary glomerular diseases whose cause is unknown. The syndrome is said to be secondary when

Table 5.11   Causes of nephrotic syndrome

Primary — renal disease
- Minimal change glomerulonephritis
- Focal segmental glomerulonephritis
- Membranous nephropathy
- Proliferative glomerulonephritis

Secondary — in systemic conditions
- Diabetic nephropathy
- Amyloidosis
- Systemic lupus erythematosus
- Henoch-Schönlein purpura

it occurs in association with conditions such as diabetes mellitus, systemic lupus erythematosus or by exposure to toxins or heavy metals. In morphological terms, nephrotic syndrome can be classified according to the underlying glomerular alteration. One of the primary nephrotic syndromes of importance is the minimal change disease or lipoid nephrosis, where minimal or no changes are seen on light microscopy. Under electron microscopy, fusion of epithelial cell foot processes are seen. This condition is common in children and is responsive to steroid treatment. In minimal change disease, there is no antecedent illness and the exact cause is unknown. The proteinuria is highly selective. The diagnosis of the cause of nephrotic syndrome usually relies on renal biopsy and immunocytochemistry. Protein selectivity index can be used to identify this condition. However, the sensitivity of this method is inadequate.

Case 5.4

A 7-year-old-boy was taken by his parents to the GP as they noticed his shoes was getting tight and he had visible swelling of his legs and face which was especially noticeable in the morning. His GP examined his urine and found that there was protein in his urine. The GP referred the patient to the hospital where further investigations were done and the results were as follows:

Serum		Reference Range
Sodium (mmol/L)	134	135–145
Potassium (mmol/L)	4	3.5–5.0
Urea (mmol/L)	34.2	2.5–7.2
Creatinine (μmol/L)	52	66–112
Bicarbonate (mmol/L)	28	23–32
Calcium (mmol/L)	1.75	2.25–2.65
Albumin (g/L)	16	36–48
PCR (mg/mmol)	320	

This boy had all the features of nephrotic syndrome; proteinuria, oedema and low serum albumin. His serum was reported by the laboratory to be lipaemic

> *and further analysis showed his cholesterol and triglycerides were high. He underwent a renal biopsy which showed he had minimal change disease. He was treated with steroids and made a good recovery.*

The oedema in nephrotic syndrome is classically attributed to hypoalbuminaemia. However, recent studies failed to find a correlation between serum albumin concentration and the degree of oedema, leading to the suggestion that other mechanisms must be involved in its pathogenesis.

Other abnormalities seen in nephrotic syndrome include increased α_2 globulin in electrophoresis, low circulating levels of transport proteins such as thyroxine-binding globulin (TBG), increased cholesterol, LDL and VLDL. Total calcium concentration is decreased due to low albumin. Complications of nephrotic syndrome include malnutrition, increased susceptibility to infection due to loss of immunoglobulins, accelerated atherosclerosis and disorders of coagulation.

Management of nephrotic syndrome includes treating the cause whenever it can be identified and general measures to counteract the effects of protein loss. These include a high protein diet, low sodium intake and the use of diuretics to reduce oedema. However, a rapid reduction in extracellular fluid volume by diuretics may lead to hypovolaemia and deterioration in renal function.

Renal Tubular Disorders

Renal tubular disorders may be an isolated defect or may affect many aspects of tubular function and they can be congenital or acquired.

Isolated Defects of Tubular Function

Glycosuria

Glucose is normally completely reabsorbed from the glomerular filtrate. Glycosuria may occur when plasma glucose rises (diabetes mellitus) or when the renal threshold for reabsorption is reduced. The latter may occur as part of a general tabular disorder or as an isolated defect due to a mutation in the sodium–glucose cotransporter.

Aminoaciduria

Amino acids are normally completely reabsorbed by specific transporters — both mono-specific and group-specific transporters. Aminoaciduria can result if the plasma amino acid concentration increases and the reabsorptive capacity is exceeded; overflow aminoaciduria. Examples include various inherited disorders of amino acid metabolism (homocysteinuria, phenylketonuria). Aminoaciduria may also be part of a general tubular disorder or a specific defect in amino acid transport. Cytinuria, Hartnup disease and histidinuria are examples of specific defects.

Cystinuria is due to an inherited defect in the neutral and basic amino acid transporter. The incidence of cystinuria has been reported to vary between 1 in 100,000 to 1 in 2500 in different races. In addition to cystinuria, there is an increase in the excretion of ornithine, arginine and lysine with an associated defect in intestinal absorption of these amino acids. The main manifestation of the disease is renal stones. Diagnosis depends on the demonstration of increased excretion of cystine. Management involves high fluid intake and alkalinisation of urine. Treatment with penicillamine, which reacts with cystine to form a soluble product is useful.

Hartnup disease is due to a defect in the absorption of neutral amino acids. This is a rare autosomal recessive disorder with an incidence of 1 in 20,000. Its main features are caused by a deficiency of nicotinamide, which is partly derived from dietary tryptophan.

Familial X-linked hypophosphatemic rickets

The characteristic abnormality in this rare inherited disorder is hypophosphataemia and decreased renal tubular reabsorption of phosphate as shown by a low renal threshold for reabsorption of phosphate (T_m P/GFR) (see Chapter 8). This is due to a mutation in the PHEX gene which results in loss of function, leading to reduced breakdown of the phosphaturic factor FGF-23. (See Chapter 8). As the name implies, these patients have rickets that does not respond to vitamin D. Treatment with oral phosphate and calcitriol is recommended.

Urine concentration defect

Defects in urinary concentration can be due to a tubular abnormality (nephrogenic diabetes insipidus) or due to the lack of ADH. This is discussed in Chapter 2.

Generalised Tubular Defect

Fanconi syndrome

This is a generalised disorder of tubular function as shown by glycosuria, aminoaciduria, phosphaturia and acidosis. It can occur as a primary disorder without a recognisable cause or secondary to other disorders — both inherited and acquired (Table 5.12). Cystinosis is the most common inherited disorder causing Fanconi syndrome. Cystinosis or cystine storage disease is a rare metabolic disease, inherited as an autosomal recessive trait and characterised by intracellular

Table 5.12   Causes of Fanconi syndrome

Primary
Secondary
- Inborn errors of metabolism
  - Cystinosis
  - Tyrosinaemia type I
  - Galactosaemia
  - Fructose intolerance
  - Wilson decease
  - Glycogen storage disease type I
- Acquired
  - Paraproteinaemia
  - Amyloidosis
  - Malignancy
- Toxins/Drugs
  - Heavy metals — mercury, cadmium, lead
  - Organic compounds — toluene
  - Drugs
    o Outdated tetracycline
    o Cisplatin

lysosomal accumulation of cystine in many organs, including kidneys. The defect is an abnormality in the transport of cystine out of lysosomes. Clinical features include failure to thrive, dehydration, rickets and renal failure. Measurement of leucocyte content of cystine is necessary for diagnosis. There is no specific treatment except renal transplantation.

Renal tubular acidosis (RTA)

These are a group of disorders in which there is normal anion gap (hyperchloraemic) metabolic acidosis. RTA is classified into type I (distal), type II (proximal) or type IV (secondary to hypoaldosteronism).

1. Distal renal tubular acidosis (Type I)

In type I RTA, there is a defect in the secretion of hydrogen ions by the intercalated cells of the cortical collecting duct. Causes of distal renal tubular acidosis include hereditary defects, autoimmune diseases such as Sjogren's syndrome and SLE, nephrocalcinosis (can be a cause as well as an effect), renal transplantation, sickle cell disease, toxins such as toluene, lithium carbonate, amphotericin C and ifosamide. These patients cannot lower their urine pH and usually have hypokalaemia. Hypokalaemia is due to the increased exchange of potassium for sodium in the distal tubules. Clinical features of distal renal tubular acidosis include normal anion gap metabolic acidosis, hypokalaemia, urinary stone due to alkaline urine, hypercalciuria and low urinary citrate (see below), nephrocalcinosis (deposition of calcium in the kidney) and bone demineralisation (can cause rickets in children). Features vary widely from no symptoms to failure to thrive, rickets and possibly renal failure.

Diagnosis is based on the inability to acidify the urine pH to 5.5 or lower in the presence of systemic acidosis (pg. 121). In the incomplete or partial variety, ammonium chloride loading test is useful. Treatment involves correction of any underlying disorder and treatment with alkali and potassium.

2. *Proximal renal tubular acidosis (type II)*

Type II RTA is due to a defect in the reabsorption of bicarbonate by the proximal tubules. This can be an isolated defect or part of a generalised proximal tubular disorder (Fanconi syndrome). These patients have normal distal tubular hydrogen ion secretion and the urine can be acidified to a pH value of less than 5.5. Proximal renal tubular acidosis can be a primary disorder of the tubule or secondary to other diseases which include familial disorders such as cystinosis, galactosaemia, glycogen storage disease type I, tyrosinemia, Wilson's disease or acquired diseases such as amyloidosis, multiple myeloma, or due to toxins (lead, highly active antiretroviral therapy (HART), ifosfamide). The acidosis in proximal renal tubular acidosis is usually less severe than in distal renal tubular acidosis. The increased sodium delivery to the distal tubules causes hyperaldosteronism which leads to hypokalaemia. Hypercalciuria is not a feature of proximal renal tubular acidosis and is not associated with renal stones. Diagnosis depends on the demonstration of bicarbonaturia when serum bicarbonate is raised to normal values (see Chapter 4). Treatment involves the correction of any under-lying disorder and administration of bicarbonate to correct the acidosis.

3. *Type IV renal tubular acidosis*

Type IV renal tubular acidosis is due to aldosterone deficiency or resistance to aldosterone, causing the failure of hydrogen and potassium ions secretion. Hyperkalaemia is a feature of this condition. The most common cause is hyporeninamic hypoaldosteronism (diabetic nephropathy, ACE inhibitors or AIDS). Treatment with fludrocortisone is indicated.

Renal Stones

Renal stone disease is a common condition. Prevalence varies from 2–5% in Asia, 8–15% in Europe and North America and up to 20%

in Saudi Arabia. Prevalence is affected by genetic, nutritional and environmental factors. Renal stones are four times more common in men than in women. They tend to recur and the recurrence rate is 75% during 20 years. In most industrialised countries, 80% of the stones are composed of calcium salts, usually calcium oxalate and less commonly calcium phosphate or a mixture of calcium oxalate and phosphate. The type of stones and their frequency are listed in Table 5.13. Struvite stones, which are due to infection, are becoming less common with the widespread use of antibiotic treatment. Uric acid stones account for about 5% of stones in most industrialised countries but may account for 30% in Mediterranean countries.

Pathogenesis

Normal urine is supersaturated with calcium oxalate and this will precipitate out of solution as crystals if the saturation increases further. Stone formation is normally inhibited in most subjects due to the presence of crystallisation inhibitors. If there is abnormal urinary composition due to metabolic or environmental causes, crystallisation tends to occur. Factors that may contribute to stone formation are listed in the Table 5.14.

Patients with calcium stones have hypercalciuria (40–60%), hyperuricosuria, hyperoxaluria or hypocituria. Pure calcium oxalate stones are usually associated with hyperoxaluria with or without hypercalciuria.

Table 5.13   Types of renal stones

Calcium oxalate, phosphate or both	60–80%
Struvite (magnesium, ammonium phosphate)	10–20%
Calcium phosphate	5%
Uric acid	5%
Cystine	<1%
Xanthine	very rare
2,8-dihydroxyadenine	very rare

Table 5.14 Factors contributing to stone formation

Hypercalciuria
- With normocalcaemia
  — Idiopathic hypercalciuria
  — High sodium intake
  — Renal tubular acidosis
- With hypercalcaemia
  — Primary hyperparathyroidism
  — Malignancy
  — Vitamin D intoxication

Hyperoxaluria
- Increased absorption or intake
  — Low calcium/high oxalate diet
  — Idiopathic hyperoxaluira
  — Small bowel disease
- Increased synthesis
  — Excess vitamin C intake
- Inherited disorders
  — Primary hyperoxaluria

Hyperuricosuria
- High purine diet
- Excessive production
  — Alcohol, obesity
  — Renal tubular dysfunction
  — Drugs

Low urine volume

Alkaline urine

Hypocitraturia
- Renal — renal tubular acidosis (type I)
- Small bowel disorders — malabsorption, bowel resection
- Metabolic acidosis
- Urinary infection

Deficiency of inhibitors

Hypercalciuria, defined as a calcium excretion of greater than 0.1 mmol/kg/24-hr, is a common abnormality and can be due to several mechanisms:

(a) Absorptive hypercalciuria as a result of increased calcium absorption in the gut. This is thought to be due to an inherited hyperresponsiveness to calcitriol. In some patients, non-calcitriol–mediated mechanisms may play a role.

(b) Renal hypercalciuria due to a primary defect in the renal tubular reabsorption of calcium.

(c) Resorptive hypercalciuria due to increased bone reabsorption such as primary hyperparathyroidism.

(d) Hypercalciuria associated with hypercalcaemia, e.g. primary hyperparathyroidism.

Hyperoxaluria is an important factor in the development of calcium oxalate stones. Hyperoxaluria increases calcium oxalate supersaturation and subsequent calcium oxalate stone formation. In normal subjects, 90% of dietary oxalate binds to calcium in the gut and is excreted in stools as calcium oxalate, leaving 10% to be absorbed. Hyperoxaluria can arise due to excessive intake or increased intestinal absorption. Foods rich in oxalate include spinach, rhubarb, chocolate, nuts, tea, wheat bran and strawberries. The amount of oxalate absorption is inversely related to calcium intake. This may explain the higher risk of stone formation in those on low calcium diets. Increased absorption of oxalate occurs due to ileal disease such as Crohn's disease and short bowel syndrome. In the latter, increased free fatty acids in the gut binds to calcium, thereby increasing the availability of oxalate for absorption. Primary oxaluria is a rare, inherited disorder of oxalate metabolism leading to high oxalate excretion.

The causes of increased uric acid excretion are discussed under purine metabolism (Chapter 27).

Hypocitraturia not only contributes to uric acid stone formation but also to the formation of calcium oxalate stones. Citrate in the urine forms soluble complexes with calcium, reducing calcium oxalate and calcium phosphate supersaturation and thereby inhibiting stone formation. Acidosis is the major cause of hypocitraturia. Acidosis leads to a decreased citrate synthesis and increased tubular reabsorption. Hypocitraturia is found with increased intake of diet rich in meat which causes increased acid production. In urinary tract infection, citrate excretion may be low due to destruction of citrate by bacteria in the infected urine.

Certain proteins produced in the kidney inhibit the growth of stones. These include Tamm-Horsfall protein, nephrocalcin, uropontin and inter-α-trypsin inhibitor. A genetic polymorphism leading to abnormal forms of these proteins may increase the risk of stone formation.

Investigation

The aim of the investigation is to identify abnormal risk factors for stone formation. After a careful history, an analysis of the stone, if passed, should be performed to identify the type of stone. After the first stone event, urine pH, serum calcium, phosphate, bicarbonate, creatinine, and uric acid should be measured in individuals. In recurrent stone formers, 24-hour urine should be collected for volume, pH, calcium, phosphate, sodium, uric acid, oxalate, citrate and creatinine. Serum PTH should be measured if the plasma calcium is high or at the upper end reference of range (Table 5.15).

Management

Prevention of subsequent stone formation depends on identifying and reducing any risk factors. Measures, which will decrease the frequency of stone formation, include an increase in fluid intake, avoidance of high oxalate diet, reduction in intake of proteins, high purine foods (to reduce the excretion of uric acid) and salt and avoidance of high doses of vitamin C. Specific forms of treatment may be required if an

Table 5.15 Investigation of urinary calculi

Stone analysis
Blood — Calcium, phosphate, bicarbonate, chloride, urate, creatinine, PTH*
Random urine — pH, amino acids (cystine), phosphate**, creatinine
　　　　　　 — Mid-stream sample for microscopy and culture
24-hour urine — Volume
　　　　　　 — Calcium, oxalate, creatinine, citrate**, magnesium**
　　　　　　 — Uric acid, creatinine

* Only indicated when serum calcium is high.
** Only in recurrent stone formers.

abnormality is found, e.g. allopurinol in uric acid stone formers. Thiazide diuretics may be helpful in reducing calcium excretion. Citrate or potassium bicarbonate may be required in stone formers with low urine citrate excretion.

Further Reading

1. Davison AM, Cameron TS, Grunfield J-P, Kerr DNS, Ritz E, Winearls GG. *Oxford Textbook of Clinical Nephrology* 1998. Oxford University press.
2. Hilton R. Acute renal failure. *Br Med J* 2006; 333:786–790.
3. Klahr S, Miller SB. Acute oliguria. *N Engl J Med* 1998; 338: 671–675.
4. Lamb E, Delaney M. *Kidney Disease and Laboratory Medicine* 2009. ACB Venture Publications: London.
5. Lameire P, VanBlesen W, Vanholder R. Acute renal failure. *Lancet* 2005; 365:417–430.
6. Orth SR, Ritz E. The nephrotic syndrome. *N Engl J Med* 1998; 338: 1202–1211.
7. Parmar MS, Kidney stones. *Br Med J* 2004; 328:1420–1424.

Summary/Key Points

1. Kidneys play a vital role in maintaining the volume and composition of body fluids. Additional functions are excretion of waste products, hormone production and metabolism.
2. Assessment of glomerular filtration rate gives an overall assessment of renal function. In clinical practice, when an accurate measurement of GFR is required, plasma clearance of an exogenous substance such as ^{51}Cr-EDTA is done. Creatinine clearance is not used very often in clinical practice because of difficulty in collecting urine accurately.
3. GFR can be assessed by measurement of serum creatinine. However, serum creatinine within the reference range does not imply normal renal function. Serum creatinine above the reference range indicates reduced GFR. Intraindividual variation of creatinine is small and changes even within the reference interval is indicative of changes in GFR.

4. GFR can be estimated from serum creatinine using the MDRD formula. Estimated GFR is not usually reliable when the GFR is > 60 mL/min/1.73 m^2 and when it is changing rapidly.

5. Acute kidney injury is common especially in hospital patients, elderly and those who are critically ill. Common causes of AKI are volume depletion and nephrotoxins. Examination of urine and blood may be helpful in differentiating prerenal AKI from intrinsic acute tubular necrosis. Mortality in AKI is high especially in the elderly.

6. Management of AKI depends on treating the cause, correcting any complications such as acidosis, hyperkalaemia, managing fluid balance and nutrition. Renal replacement therapy may be required in severe cases. Treatment with loop diuretics, dopamine or osmotic diuretics has very little benefit.

7. CKD is common in the population and is divided into five stages. Due to adaptation by the remaining nephrons, homeostasis is maintained until the GFR falls to < 15 mL/min/1.73 m^2.

8. Biochemical abnormalities in CKD include hypo- or hypernatremia, hyperkalaemia, high urea and creatinine, low bicarbonate, hypocalcaemia, hyperphosphataemia, secondary hyperparathyroidism and normocytic normochromic anaemia.

9. Conservative management of CKD include dietary manipulation, administration of erythropoietin, calcitriol and phosphate binders, controlling blood pressure and treatment of proteinuria with ACE inhibitors to reduce cardiovascular risk.

10. Renal replacement therapy, dialysis (haemo or peritoneal) or transplant is required eventually. Regular biochemical monitoring of CKD patients is required.

11. Small amounts of proteins are excreted daily in urine. Increased protein excretion indicate renal disease. It is now common practice to measure protein:creatinine ratio (PCR) to assess protein excretion. It has been recently recommended that this should be replaced by albumin:creatinine ratio (ACR).

12. Nephrotic syndrome defined as heavy proteinuria, hypoalbuminaemia and oedema can be caused by many diseases. Clinical

and biochemical effects of this syndrome are due to the loss of proteins in the urine.

13. Renal tubular disorders can be a general or an isolated defect. The net result is a decrease in tubular function leading to either excretion of substances like amino acids and glucose which are normally reabsorbed or retention of substances which are normally secreted, e.g. H^+ ions.

14. Renal calculi is a common condition and is caused by a super-saturation of urine. Factors important in the development of renal stones are hypercalciuria, hyperoxaluria, hypocituria, hyperuricosuria and low urine volume.

chapter 6

Calcium Metabolism

Introduction

Calcium, which is the most abundant mineral in the human body, has structural, neuromuscular, enzymatic and signalling functions (Table 6.1). Most of the calcium (99%) is present in bone (Table 6.2). The extracellular fluid content of calcium is relatively small (<0.2%) but it is important in regulating neuromuscular excitability and as a cofactor in the clotting cascade.

The intracellular calcium concentration is relatively low and the gradient between ECF and ICF free calcium is around 10,000:1. However, inside the cell, calcium is also present as complexes. Intracellular free calcium ion regulates the activity of various enzymes and is also an important second messenger (Table 6.1).

Plasma Calcium Concentration

Calcium in the circulation is present in three forms: ionised, protein-bound and complexed forms (Figure 6.1). Ionised calcium concentration is the physiologically active form and is tightly regulated. Calcium is mainly bound to albumin and to some extent to globulin. The amount of protein-bound calcium depends on protein concentration, protein type and hydrogen ion concentration. The binding of calcium to protein is inversely related to hydrogen ion concentration, binding increases in alkalosis and decreases in acidosis. Acutely, this will cause a change in ionised calcium concentration without affecting the total calcium concentration. In chronic states, however, ionised calcium

Table 6.1   Functions of calcium

Type of function	Function
Neuromuscular	Control of excitability
	Release of neurotransmitters
	Initiation of muscle contraction
Structural	Bone
	Teeth
Enzymic	Cofactor for coagulation factors
Signalling	Intracellular second messenger

Table 6.2   Distribution of calcium, phosphate and magnesium in the body (percentage)

	Calcium	Phosphate	Magnesium
Bone	99	85	55
Soft tissue	1	15	45
ECF	<0.2	<0.1	1
Total (mol)	25	19.4	1.0

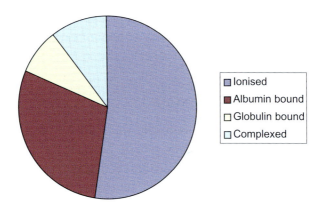

Figure 6.1   Fractions of plasma calcium.

concentration will return to normal as a result of homeostatic regulation and total calcium concentration will change accordingly. In acute respiratory alkalosis (hyperventilation) for example, plasma ionised calcium will decrease and may lead to tetany, whereas in chronic alkalosis, the ionised calcium will be normal but the total calcium will be high. In acidosis the opposite will be seen.

As albumin-bound calcium accounts for nearly 40% of the total calcium concentration changes in albumin concentration will lead to a change in the total calcium concentration, without producing any change in ionised calcium concentration. In order to interpret total plasma calcium concentration in patients with low albumin concentration, various formulae have been published to correct the total plasma calcium concentration to an albumin concentration, of 40 g/L. One such formula is:

Corrected calcium = measured total calcium + 0.02 (40 −measured albumin concentration).

The corrected calcium is only an approximation for ionised calcium. For an accurate assessment, ionised calcium should be determined directly by ion selective electrodes.

Calcium Balance

Normal calcium turnover in the body is illustrated in Figure 6.2. Some of the calcium in the bone (1–2%) is rapidly exchangeable with ECF and this exchangeable pool of calcium is important in plasma calcium homeostasis. In addition, there is a constant release of calcium from the bone and constant uptake of calcium by bone due to bone remodelling. In healthy young adults, bone formation and bone resorption are equal, resulting in no net change.

Calcium Absorption

The average calcium intake in the western diet is about 25 mmol/d (1000 mg/d). In many societies, however, calcium intake is much lower (for example in the Chinese diet, the average intake is about 10 mmol/d (400 mg/d)). The main sources of dietary calcium are dairy products and green vegetables. About 40% (10 mmol) of the dietary calcium is absorbed in the gastrointestinal tract. However, about 5 mmol of calcium is secreted into the gastrointestinal tract with the digestive juices. Thus, the net absorption of calcium is

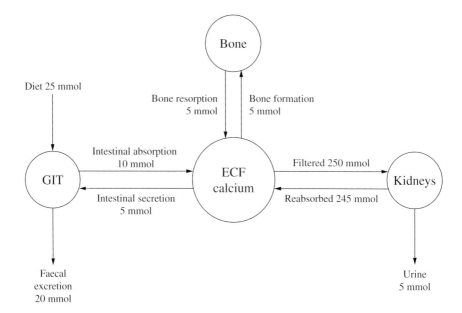

Figure 6.2   Calcium turnover in body.

approximately 5 mmol/d (20% of dietary intake). The capacity for calcium absorption is highest in the upper part of the intestine — duodenum and jejunum, and lowest in the ileum. However, majority of the calcium is absorbed in the lower small intestine as the lower part of the small intestine is longer and the food stays for longer periods of time in this segment. A small percentage of calcium is also absorbed in the colon. Calcium absorption is mediated by an active process at low dietary intakes of calcium and by a passive process at higher intake. The active component is influenced by the active metabolite of vitamin D; 1,25-dihydroxycholecalciferol (calcitriol). The passive component of absorption is directly proportional to the intraluminal calcium concentration. The intestinal absorption of calcium can be decreased by high alkalinity, by dietary phytates and oxalates which complexes with calcium within the lumen.

Efficiency of dietary calcium absorption is also related to dietary intake and body requirement. When the demand for calcium is high or when the dietary calcium intake is low, the efficiency (percentage of dietary calcium absorbed) increases.

Renal Handling of Calcium

The ionised and complex fractions of calcium are freely filtered at the glomerulus and 98% of the filtered calcium is reabsorbed. The kidney is an important regulator of plasma calcium concentration, as the turnover of calcium through the kidneys is nearly 10 times the calcium present in ECF and about 10–20 times the fluxes across the other organs. Majority of the filtered calcium (65%) is reabsorbed by an active process in the proximal tubules. This process is closely linked to sodium reabsorption and is not hormonally regulated. Changes in proximal sodium reabsorption, therefore, affect calcium reabsorption. Volume expansion and high salt intake reduce calcium reabsorption while hypovolaemia increases calcium reabsorption. Of the calcium reabsorbed in the distal segments (33% of filtered), only about 10% is subjected to hormonal regulation. Alkalosis and thyroid diseases increase while acidosis decreases calcium reabsorption.

Interpretation of urinary calcium excretion

Urine calcium excretion should be interpreted in relation to the serum calcium concentration. The relationship between calcium excretion and serum calcium is shown in Figure 6.3 and this was established in healthy subjects by infusing calcium. Calcium excretion is measured in the second void urine sample in the fasting state and expressed in relation to GFR (CaE).

$$CaE = [Urine\ Ca] \times [plasma\ creatinine]/Urine\ creatinine.$$

In the investigations of hypo- and hypercalcaemia, measurements of CaE is useful. This value should be plotted against the serum calcium. In patients with increased calcium reabsorption, this value will be to the right of the shaded area in the figure, showing that the excretion of calcium is less than expected for serum calcium. This is seen in hyperparathyroidism, humoral hypercalcaemia of malignancy and familial hypocalciuric hypercalcaemia. In patients with decreased calcium reabsorption, the value will be to the left of the shaded area and this is seen in hypoparathyroidism.

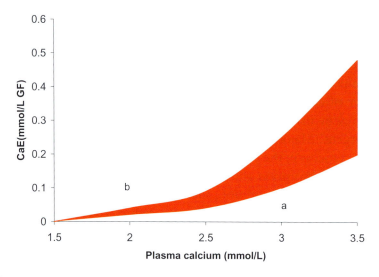

Figure 6.3   Relationship between plasma calcium concentration and calcium excretion (corrected for glomerular filtration). The shaded area represents the range found in healthy subjects. Symbol *a* represents a patient with a plasma calcium concentration of 3 mmol/L and CaE of 0.05 mmol/L GF is lower than expected (between 0.1 and 0.2 mmol/L GF). Symbol *b* represents a patient with hypocalcaemia (2 mmol/L) with a CaE of 0.1 mmol/L GF which is higher than expected.

Bone

In normal adults, the movement of calcium into and out of the bone is balanced. Fluxes of calcium across bone can be either due to bone remodelling (see Chapter 9) or due to an exchange between the ECF and the bone without any change in the bone matrix. In normal adults, about 5% of bone is remodelled per year. In contrast, 1–2% of total body calcium is exchanged between the ECF and the bone within a few days. This rapid exchange is important for regulation of plasma calcium concentration. This exchange process takes place across the osteocyte cell membrane, but the mechanism regulating this exchange is unclear. The rapid increase in plasma calcium concentration due to PTH is believed to be due to this exchange between the ECF and the bone.

Calcium Regulating Hormones

Extracellular fluid calcium concentration is tightly controlled by a complex homeostatic mechanism involving fluxes of calcium between ECF, kidney, bone and the gut. These fluxes are regulated by parathyroid hormone and 1,25-dihydroxycholecalciferol (calcitriol). The role of calcitonin in calcium homeostasis is small.

Parathyroid hormone

Parathyroid hormone (PTH) is an 84 amino acid peptide synthesised and secreted by the chief cells of the parathyroid gland. It is synthesised as a larger precursor of 115 amino acids, pre-pro-PTH which is converted to pro-PTH of 90 amino acids. A further loss of 6 amino acids leads to the formation of PTH (Figure 6.4). The secretion of PTH is controlled by ionised calcium concentration via a simple negative feedback loop. Serum PTH concentration decreases as the

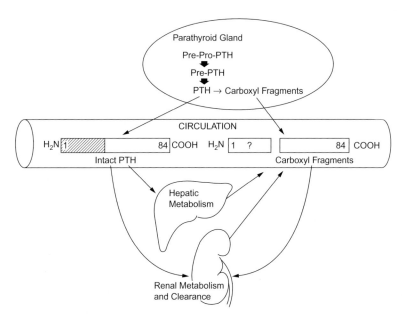

Figure 6.4 Synthesis and metabolism of PTH.

serum ionised calcium concentration increases and vice versa. Extracellular ionised calcium concentration regulates PTH over a narrow range (1.0–1.5 mmol/L) with little effect outside this range. Ionised calcium concentration regulates PTH secretion via the calcium sensing receptor (CaSR) located in the cell surface membrane. This receptor, a 1078-amino acid protein, is a member of the G-protein coupled receptor superfamily. It is unique in that activation of this receptor causes inhibition of PTH secretion. Mutation in this receptor can lead to disorders such as familial hypocalciuric hypercalcaemia and neonatal severe hyperparathyroidism. When the calcium concentration decreases, PTH secretion increases within minutes, followed by increased synthesis and eventually proliferation of the parathyroid chief cells.

In addition to calcium levels, PTH secretion is affected by calcitriol, magnesium, phosphate and possibly catecholamines. Calcitriol and acute hypermagnesaemia inhibit PTH secretion. In chronic low magnesium states, PTH secretion is inhibited (see Chapter 7). There is some evidence that phosphate has a direct stimulatory effect on PTH, a contributing factor to the secondary hyperparathyroidism of renal failure.

Major biological actions of PTH are
(i)   stimulation of osteoclast-mediated bone resorption,
(ii)   stimulation of renal calcium reabsorption,
(iii) inhibition of tubular phosphate reabsorption, and
(iv)   stimulation of renal production of calcitriol.

The amino terminal end of the PTH molecule binds to the PTH receptor to elicit these biological effects. The PTH receptor, like the calcium-sensing receptor, acts via a G-protein coupled mechanism.

Serum PTH shows diurnal variation, highest in the early morning and lowest at 9 a.m. Metabolism of PTH is complex and produces several fragments of varying biological and immunological reactivity (Figure 6.4). The intact peptide has a half-life of less than 4 minutes in the circulation and is then rapidly cleared by the kidney and liver. Hepatic Kupffer cells take up intact PTH and degrade it into very small peptides, as well as into discrete fragments. C-terminal fragments produced by such metabolism have a half-life of 2–3 hours and are cleared by glomerular filtration. Thus, in renal failure, the concentration of these fragments in the circulation will be high. Immunoassays for PTH in the past have produced variable results depending on the specificity of the antibody against fragments of PTH. Highly sensitive and specific immunometric assays are now widely available to measure the intact PTH molecule. However, recent studies show that even these assays measure some (up to 15%) non-intact PTH fragments.

Vitamin D

Vitamin D is a fat soluble vitamin and there are two physiologically important forms, vitamin D_2 (ergocalciferol) and vitamin D_3 (cholecalciferol). Fungus, plants and invertebrates produce vitamin D_2 and vertebrates produce vitamin D_3. Most vitamin D is produced in the skin, from 7-dehydrocholesterol by exposure to UV light (290–310 nm). Production of vitamin D occurs daily in the tropics and during spring and summer seasons in temperate climates. Major dietary sources of vitamin D are fish oils, fatty fish, egg yolk, liver and margarine that have added vitamin D. The amount of vitamin D produced in the skin is about 15% of the 7-dehydrocholesterol and this depends on the amount of exposure to sunlight, the surface area exposed and the colour of the skin. Production of vitamin D by Caucasians is greater than Asians, who in turn produce more vitamin D than people of African origin. Amount of vitamin D produced in the skin is regulated; prolonged exposure to UV light does not increase the production any further as inactive metabolites are produced. There is a large seasonal variation in the vitamin D status of individuals living in temperate climates. It has been estimated that majority of vitamin D is derived from sunlight.

- ## Metabolism of Vitamin D

  The metabolism and actions of D_2 and D_3 are similar. Vitamin D is transported in the serum, bound to vitamin D-binding protein (DBP),

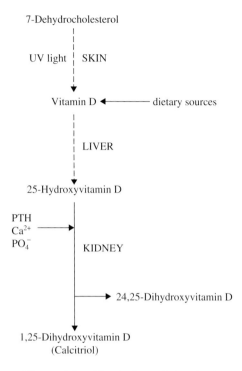

Figure 6.5   Metabolism of vitamin D.

to the liver where it is converted to the 25-hydroxyvitamin D (calcid-iol) by hydroxylation at the C25 position. 25-Hydroxyvitamin D is transported bound to vitamin D binding protein to the kidney where it is converted to either the active metabolite 1,25-dihydroxyvitamin D (calcitriol) by hydroxylation at the C1 position (Figure 6.5) or to 24,25-dihydroxyvitamin D which is an inactive metabolite. The enzyme responsible for the formation of calcitriol (1α-hydroxylase) is a complex cytochrome P450 mitochondrial enzyme system, located in the proximal nephron. This enzyme is under tight regulation primarily by PTH, ionised calcium, phosphate, hydrogen ion and calcitriol itself. Low calcium, low phosphate, low calcitriol and high PTH concentrations stimulate the production of calcitriol.

Calcitriol acts via vitamin D receptor to inhibit the hydroxy-lase enzyme activity. When the requirement for calcitriol is low,

25-hydroxyvitamin D is converted to 24,25-dihydroxyvitamin D by enzyme 24-hydroxylase. Until recently, this metabolite has been considered as an inactive metabolite. However, recent studies suggest that 24,25-dihydroxyvitamin D may play an important role in normal intramembranous bone formation.

Calcitriol is also produced outside the kidney by the placenta and activated macrophages in granulomatous diseases. Recent evidence shows that the enzyme 1α-hydroxylase is present in many tissues and that calcitriol produced by these tissues acts locally in a paracrine fashion. The production of calcitriol by these tissues is not regulated by PTH or calcium and is dependent on circulating calcidiol concentration. Calcitriol produced by these peripheral tissues does not enter the circulation. This locally produced calcitriol is now believed to have effects on tissues such as skeletal muscle, pancreatic β cells, immune cells and cell proliferation in general.

The half-life of calcitriol in the circulation is about 5 hours and it is either excreted in the urine or faeces as inactive metabolites.

- *Actions*

Calcitriol increases calcium and phosphate absorption from the gastrointestinal tract, increases bone resorption and enhances the effect of PTH in the renal tubule to increase renal tubular calcium reabsorption. Calcitriol binds to the cytosolic vitamin D receptor (VDR), a member of the steroid/thyroid/retinol superfamily. The VDR forms a heterodimer with retinoid X receptor (RXR). This binds to vitamin D responsive element (VDRE) in target genes, thereby influencing target gene expression. In the intestine, calcitriol stimulates the synthesis of calbindin, a calcium-binding protein, as well as other proteins such as calcium ATPase. Calcitriol stimulates the synthesis of osteocalcin by osteoblasts. It is a powerful agent stimulating the differentiation of committed osteoclast precursors, causing their maturation to form multinucleated cells that are capable of resorbing bone. Calcitriol is essential for mineralisation of bone. This is thought to be due to its effect on calcium and phosphate absorption from the gut, thereby providing adequate calcium and phosphate for mineralisation. Some suggest that there may be a direct effect of calcitriol on bone mineralisation. Calcitriol also increases reabsorption of calcium by the renal tubules via stimulation of calbindin in the distal tubules. Calcitriol has also been shown to stimulate the production of phosphate regulating factor, FGF-23 by osteocytes.

Calcitriol also has an effect on the muscle as shown by the fact that myopathy of vitamin D deficiency improves with vitamin D treatment. Calcitriol probably acts in the muscle by promoting calcium transfer across the sacroplasam. In addition to these effects, calcitriol affects the differentiation and proliferation of cells and has an immunomodulatory function.

It is now believed that vitamin D has important functions in many tissues and vitamin D deficiency may play a role in several diseases including diabetes mellitus, ischemic heart disease, hypertension and cancer.

- *Assessment of vitamin D status*

The major circulating form of vitamin D is calcidiol, which is bound mainly to DBP (85%) and to some extent to albumin (15%). DBP has high affinity for vitamin D metabolites and in healthy subjects only 0.03% of calcidiol and 0.4% of calcitriol are free.

Vitamin D status is assessed by measuring the concentration of calcidiol in serum. As yet, there is no international agreement as to what constitutes an adequate level of serum calcidiol. As serum calcidiol is not physiologically regulated a conventional reference range cannot be applied. Several studies have used the physiological effects of low serum calcidiol on PTH, bone markers and calcium absorption to determine the optimal concentration of calcidiol. This was defined as the concentration of calcidiol above which there were no changes in the parameters studied. A value greater than 75 nmol/L has been suggested as the optimal serum concentration for health, although some suggest a value as high as 125 nmol/L. A value less than 20 nmol/L is considered as severe deficiency, a value between 20 and 60 nmol/L as deficiency and a value between 60 and 75 nmol/L as insufficiency.

Calcitonin

Calcitonin, a 32 amino acid peptide, is synthesised and secreted by the parafollicular cells of the thyroid gland. Ionised calcium concentration is the most important regulator of calcitonin secretion; an increase in ionised calcium stimulates calcitonin secretion. Gastrointestinal hormones like gastrin are also potent secretagogues and these hormones facilitate and enhance the secretion of calcitonin in response to a given calcium concentration, accounting for the increased calcitonin secretion seen after meals. Calcitonin directly inhibits osteoclastic bone resorption and decreases plasma calcium concentration rapidly within minutes. However, the precise physiological role of calcitonin in calcium homeostasis is unclear. It has been suggested that calcitonin may be important in minimising the postprandial rise in calcium. The fact that calcitonin deficient patients (after thyroidectomy) and patients who have excess calcitonin production (medullary thyroid carcinoma) have no disturbance in calcium homeostasis suggests that calcitonin may not be an important regulator of calcium homeostasis in normal subjects. Calcitonin has a half-life of a few minutes and is metabolised predominantly by the kidney. Exogenous calcitonin is used in the treatment of Paget's disease and in severe hypercalcaemia.

Summary of Response to Hypocalcaemia or Hypercalcaemia

A fall in ionised calcium concentration is immediately sensed by the parathyroid glands, which respond by increasing PTH secretion. PTH will increase osteoclastic bone resorption, releasing calcium and phosphate from the bone into the ECF. PTH also increases renal calcium reabsorption, inhibits phosphate reabsorption and stimulates the synthesis of calcitriol which increases calcium and phosphate absorption from the gut. When these mechanisms are intact, the extracellular calcium concentration will return to normal.

A rise in ionised calcium concentration causes a decrease in PTH secretion. This will reduce renal tubular calcium reabsorption, osteoclastic bone resorption, and synthesis of calcitriol, which in turn reduces the absorption of calcium and phosphate from the gut.

Other Hormones

In addition to these hormones, a number of other hormones influence calcium metabolism (Table 6.3). Oestrogens are important regulators of bone turnover. In the absence of oestrogen, menopause, bone turnover and serum calcium concentration tend to increase (see Chapter 9). Testosterone in men has a similar effect to that of oestrogen. Glucocorticoids reduce intestinal calcium absorption, osteoblastic activity and increase bone resorption. Thyroid hormones stimulate bone resorption that can lead to hypercalcaemia.

Parathyroid Hormone Related Peptide (PTH-rP)

Parathyroid hormone related peptide (PTH-rP) is a 141 amino acid peptide produced by several cells. The first 13 amino acids of PTHrP are similar to that of PTH. It binds and activates the PTH receptor, mimicking the biological effects of PTH in the bones, kidney and gut. PTH-rP was first discovered in tumours derived from lung, breast and kidney. It is now known that PTH-rP is produced by nearly 50% of primary breast cancers and its production may be enhanced at the

Table 6.3 Hormones influencing calcium homeostasis (other than PTH, vitamin D and calcitonin)

Oestrogens
Testosterone
Glucocorticoids
Growth hormone
Insulin
Thyroid hormones
Prostaglandins
Parathyroid hormone-related peptide
(cytokines)

bone site by bone-derived factors such as transforming growth factor-β. PTH-rP has a pathophysiological role in humoral hypercalcaemia of malignancy as well as in hypercalcaemia due to bony secondaries. Immunohistochemical studies have shown the expression of PTH-rP in bone secondaries, suggesting its involvement in osteolytic bone lesions. It has now been shown that PTH-rP has a much wider physiological role via several receptors. It is produced by healthy skin cells, amniotic cells, and by lactating breast tissue. It is present in large amounts in breast milk. It may be involved in the transport of calcium from the mother to the foetus, across the placenta against a concentration gradient. PTH-rP has been shown to be important for the proliferation, differentiation and death of many cell types. PTH-rP is involved in cartilage differentiation and its absence leads to abnormalities in growth plate. It is important for the development of the mammary gland and teeth, as well as for the functioning of smooth muscles. PTH-rP and its receptors are widely expressed in the central nervous system and skin.

Disorders in Calcium Metabolism

Hypercalcaemia

Hypercalcaemia is a common finding in hospital patients, especially in ill patients (up to 3% of hospital patients may have hypercalcaemia). The prevalence of hypercalcaemia in healthy ambulatory populations varies from 1–3%.

Clinical features of hypercalcaemia

Hypercalcaemia itself can cause several clinical features (Table 6.4). Its effects on the kidney can lead to polyuria, volume depletion and renal failure. The extent to which hypercalcaemia produces clinical features depends on the plasma calcium concentration and the rate of rise. Symptoms are infrequent when the plasma calcium is less than 3.0 mmol/L. Severe hypercalcaemia is associated with neurological and renal effects. Hypercalcaemia may also potentiate digoxin toxicity.

Table 6.4   Clinical features of hypercalcaemia

	Features
Neuropsychiatric	General weakness, tiredness, lassitude
	Impaired concentration, drowsiness, personality changes, coma
	Muscle weakness
Gastrointestinal	Anorexia, nausea, vomiting, constipation, abdominal pain
	Pancreatitis
Renal	Polyuria, polydipsia, volume depletion, renal failure
	Renal calculi, nephrocalcinosis
Cardiovascular	Cardiac arrhythmias
	Hypertension
	Short QT interval
	ECG changes
Others	Corneal and vascular calcification
	Weight loss

Causes of hypercalcaemia

The causes of hypercalcaemia are listed in Table 6.5. The most common causes of hypercalcaemia, primary hyperparathyroidism and malignant disease account for more than 90% of cases of hypercalcaemia in adults.

1. *Primary hyperparathyroidism*

    The prevalence of primary hyperparathyroidism is 1 in 1000. The incidence increases with age and it has a female preponderance — female: male ratio 3:1. Most cases are seen in patients over the age of 50. In majority of patients, a benign solitary adenoma is responsible. In 3% of patients, multiple adenomas may be present, while in a small proportion, nodular hyperplasia of all glands is seen and rarely, carcinoma of the parathyroid gland is found. Multiple adenomas may be associated with one of the multiple endocrine neoplasia syndromes (MEN). Two distinct patterns of MEN are recognised — MEN-1 in which parathyroid adenomas is associated with tumours of the pituitary gland and pancreas, while MEN-2 consists of parathyroid adenomas,

Table 6.5 Causes of hypercalcaemia

	Cause
Common	Primary hyperparathyroidism
	Malignancy — without osteolytic lesion
	• Humoral hypercalcaemia of malignancy (HMM)
	Malignancy — with osteolytic lesion
	• Multiple myeloma
	• Secondaries from breast, prostate lung and skin
	• T cell leukaemia — lymphoma
Less common	Granulomatous diseases
	• Sarcoidosis
	• Tuberculosis
	Vitamin D intoxication
	Milk alkali syndrome
	Familial hypocalciuric hypercalcaemia
	Tertiary hyperparathyroidism
	Thiazide diuretics
	Thyrotoxicosis
	Drugs
	• Increased intestinal absorption of calcium
	— Milk alkali syndrome
	— Vitamin D and its metabolites
	• Increased mobilisation from bone
	— Vitamin A and its derivatives (etretinate, acitretin, isotretinoin)
	— Vitamin D and its metabolites
	• Decreased renal excretion
	— Thiazide diuretics
	— Lithium
	— Calcium channel blockers

phaeochromocytomas and medullary carcinoma of the thyroid. More than 50% of patients with primary hyperparathyroidism are asymptomatic and hypercalcaemia is found on routine screening. Others may present with renal calculi or bone pain. Primary hyperparathyroidism accounts for 5% of first-time stone formers and 15% of recurrent stone formers. Renal stones contain calcium oxalate and calcium phosphate and are caused by hypercalciuria. Some patients with primary hyperparathyroidism may present with features of bone disease such as muscle pain, bone pain and cystic lesions of bone.

Hypophosphataemia is a feature in primary hyperparathyroidism due to the phosphaturic effect of PTH in the kidney. However, this may be masked by renal impairment. Renal phosphate clearance or renal threshold for reabsorption of phosphate (TmP/GFR) is occasionally useful in the differential diagnosis. However, with reliable assays for PTH, the diagnosis of primary hyperparathyroidism is now fairly straightforward. Typically, serum PTH concentration is high or near the upper end of the reference range in the presence of hypercalcaemia, inappropriately elevated PTH.

- Management
  If the patient is symptomatic or if the plasma calcium concentration is more than 3 mmol/L, surgical removal of the adenoma is recommended. In most patients, parathyroid adenoma can be located at surgery. However, imaging techniques such as high-resolution ultrasound or ^{99}Tc sestamibi scan may help to localise the tumour. Occasionally, the tumour may be localised by measuring PTH in blood samples obtained by selective venous catheterisation. Intraoperative selective venous catheterisation and localisation of tumour is a useful technique when re-exploring the neck after a previously unsuccessful parathyroidectomy.

  There is controversy as to whether patients without symptoms or with mild hypercalcaemia should undergo surgery. These patients should have regular monitoring of their plasma calcium concentration, renal function and bone mineral density. Increase in calcium concentration, decrease in renal function or the development of osteoporosis indicate the need for surgery. Criteria for parathyroidectomy in these patients are listed in Table 6.6.

Case 6.1

A 65-year-old consulted her GP feeling unwell and blood tests showed that her serum calcium level was high. She was referred to the endocrine clinic and investigations there revealed the following results:

Serum		Reference Range
Calcium (mmol/L)	2.85	2.25–2.55
Phosphate (mmol/L)	0.7	0.8–1.3
Albumin (g/L)	35	36–47
Alkaline phosphatase (IU/L)	100	40–125
PTH (ng/L)	75	10.0–65.0
Correctedcalcium (mmol/L)	2.95	2.25–2.55
Urea (mmol/L)	6	3.5–7.2
Creatinine (μmol/L)	100	60–112

This lady's corrected calcium is high with a low phosphate. These results indicate high PTH activity which was confirmed by the elevated PTH concentration. A diagnosis of primary hyperparathyroidism was made and a urine calcium excretion showed she was excreting 12 mmol/d of calcium. This prompted the endocrinologist to refer the patient for surgery and she underwent a successful parathyroidectomy and her serum calcium returned to normal postoperatively.

Table 6.6 Criteria for parathyroidectomy

1. Serum calcium concentration 0.25 mmol/L above the upper limit of reference range.
2. Urine calcium excretion greater than 10 mmol/day.
3. Reduction in creatinine clearance by more than 30% compared with age-matched controls.
4. Reduction in bone mineral density of the femoral neck, lumbar spine, or distal radius by more than 2.5 standard deviations below peak bone mass (T score lower than −2.5).
5. Subjects who are younger than 50 years.
6. Patients for whom medical surveillance is not desirable or possible.
7. Presence of any complications (e.g. nephrolithiasis, overt bone disease).
8. An episode of hypercalcaemic crisis.
9. Presence of symptoms such as depression, bone pain or gastric symptoms.

2. *Malignancy-related hypercalcaemia*

Hypercalcaemia may be found in up to 30% of patients with malignancy. Multiple myeloma and secondary deposits in bone from primary tumours such as lung, breast, kidney and prostate are commonly associated with hypercalcaemia. Hypercalcaemia is also seen in malignant tumour without secondary deposits in bone, humoral hypercalcaemia of malignancy (HHM). The high total plasma calcium concentration in multiple myeloma is due to an increased binding of calcium by globulins and an increased mobilisation of calcium from bone by osteolytic lesions mediated by osteoclast-activating cytokines, produced by the tumour cells. The mechanism of hypercalcaemia due to bony secondary deposits is deemed to be caused by the local secretion of cytokines, such as prostaglandins as well as PTHrP. In HHM, the tumour produces PTHrP, which enters the circulation and binds to PTH receptors causing hypercalcaemia. HHM is typically seen in squamous cell carcinoma of the bronchus.

Case 6.2

A 78-year-old man was brought to the emergency department with a short history of severe pain in his back and polyuria. He had a radical prostatectomy 2 years prior to this admission for prostate cancer. Admission blood tests showed the following results:

Serum		Reference Range
Calcium (mmol/L)	2.25	2.25–2.55
Phosphate (mmol/L)	1.3	0.8–1.3
Albumin (g/L)	32	36–47
Alkaline phosphatase (IU/L)	175	40–125
PTH (ng/L)	<6	10.0–65.0
Correctedcalcium (mmol/L)	3.4	2.25–2.55
Urea (mmol/L)	12	3.5–7.2
Creatinine (μmol/L)	135	60–112

This man has severe hypercalcaemia with a suppressed PTH which suggests a non-PTH related cause. Malignancy is the most common cause. With a previous history of prostate cancer, hypercalcaemia due to secondary deposits in the bone is the most likely diagnosis. Radiological examination of the spine showed lytic lesions and his serum prostate specific antigen was elevated. He was rehydrated and was given chemotherapy, but he succumbed to the disease within a week of admission.

3. *Other causes of hypercalcaemia*

- Granulomatous disease

  Hypercalcaemia is seen in granulomatous diseases such as tuberculosis and sarcoidosis. It is caused by the production of calcitriol by activated macrophages. The production of calcitriol is dependent on circulating calcidiol, therefore hypercalcaemia under these conditions is seasonal. Hypercalcaemia is often accompanied by hypercalciuria due to increased intestinal absorption of calcium.

- Excess of vitamin A or D

  Excessive intake of vitamin D or its metabolites can cause hypercalcaemia. Vitamin A in toxic doses increases bone resorption, leading to hypercalcaemia. Vitamin A analogues, such as etretinate, acitretin and isotretinoin, which are used in dermatological practice and in the treatment of some malignancies, can cause hypercalcaemia.

- Milk alkali syndrome

  Milk alkali syndrome is an uncommon cause of hypercalcaemia. It is caused by the ingestion of large quantities of milk and antacid for the control of dyspeptic symptoms. More recently, a number of such cases have been reported in the elderly who have been encouraged to drink milk and take calcium supplements. Ingestion of large amounts of antacids causes increased loss of bicarbonate in the urine and this may lead to hypovolaemia which increases calcium reabsorption in the renal tubules.

- Familial hypocalciuric hypercalcaemia

  Familial hypocalciuric hypercalcaemia (FHH), also called familial benign hypercalcaemia, is an autosomal dominant condition characterised by mild hypercalcaemia and relative hypocalciuria (i.e. calcium excretion that is low in relation to the hypercalcaemia). FHH is due to a mutation in the calcium-sensing receptor. The condition is benign and does not require treatment. These patients are asymptomatic, thus it is important to recognise them in order to avoid unnecessary surgery.

  In the parathyroid glands, due to a mutation in the CaSR, the feedback inhibition of PTH secretion by calcium concentration is set at a higher point, necessitating higher serum calcium levels to inhibit PTH. This results in hypercalcaemia with inappropriately normal or mildly elevated PTH concentration. In the kidneys, mutation in the CaSR prevents feedback inhibition of calcium reabsorption in response to hypercalcaemia. Since calcium continues to be reabsorbed despite hypercalcaemia, the calcium concentration in the urine is low relative to the serum calcium concentration (relative hypocalciuria).

  Diagnosis of FHH depends on the demonstration of low urinary calcium excretion relative to glomerular filtration (CaE). In FHH, this is less than 0.015 mmol/L GF while in primary hyperparathyroidism, it is higher.

  Individuals who are homozygous for this mutation develop severe neonatal hyperparathyroidism, which is often fatal.

- Tertiary hyperparathyroidism

  In patients with chronic hypocalcaemia, such as those with malabsorption-induced vitamin D deficiency or renal failure, there is hyperplasia of the parathyroid glands. This increase in PTH concentration is an appropriate response to hypocalcaemia. However, in a small minority of patients, hypercalcaemia develops due to the parathyroid gland becoming autonomous, presumably as a result of prolonged hypocalcaemic stimulus. This is termed tertiary hyperparathyroidism. Hypercalcaemia may manifest after treatment of the primary condition.

- Other causes

  Hypercalcaemia may also be seen in patients with hyperthyroidism but usually it is mild and of no clinical significance. Thiazide diuretics, which reduce renal calcium excretion, can cause mild hypercalcaemia.

  Treatment with lithium can cause mild hypercalcaemia and this is thought to be caused by the insensitivity of the CaSR to calcium due to lithium, thereby shifting the set point for PTH secretion by calcium.

Investigation of hypercalcaemic patients (Figure 6.6)

If the calcium concentration corrected for albumin concentration is high, a repeat measurement of calcium concentration in the fasting state without venous stasis may be advisable. At the same time, measurement of serum phosphate, renal function tests and alkaline phosphatase activity will be helpful.

Measurement of intact PTH by a reliable assay is the most important investigation. In hypercalcaemia of malignancy, PTH concentration will be low or undetectable whereas in patients with primary hyperparathyroidism, it will be high or near the upper end of the reference range. Further radiological investigations may help to determine the cause of malignancy. If there is suspicion of myeloma, serum and urine electrophoresis should by done. Urine calcium excretion may be required to exclude FHH. In cases where diagnosis is not apparent with these investigations, measurements of PTHrP or serum calcitriol may be required.

Management of hypercalcaemia

In all cases, the cause of hypercalcaemia should be identified and treated appropriately. In mild hypercalcaemia this may be sufficient. Urgent treatment is required in severe hypercalcaemia to prevent complications. Volume depletion should be corrected and furosemide may be given to promote diuresis, which will increase calcium excretion. In those with increased bone resorption, inhibitors of bone resorption such as calcitonin, bisphosphonates and mithramycin may

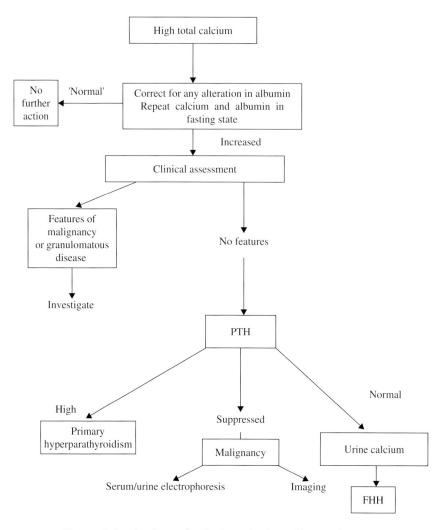

Figure 6.6   A scheme for the investigation of hypercalcaemia.

be used. Mithramycin can cause many side effects and is only used when hypercalcaemia is severe. Bisphosphonates, which are analogues of pyrophosphate, are the next step in the management of hypercalcaemia. These agents are powerful inhibitors of bone resorption. Infusion of pamidronate or zolendronic acid is commonly used in this situation. Calcitonin, which inhibits bone resorption, is another alter-native, but has only a modest and transient effect and therefore is not used in severe cases. Glucocorticoids are effective in hypercalcaemia associated with haematological malignancy, granulomatous disease and in vitamin D toxicity. Dialysis may be required in persistent life-threatening hypercalcaemia.

Phosphate infusion will lower plasma calcium concentration, but this is rarely used now because of the risk of metastatic calcification.

Hypocalcaemia

Hypocalcaemia is less common than hypercalcaemia and is relatively uncommon in hospital patients, except in patients with renal failure. The most common cause of apparent hypocalcaemia is a decrease in albumin concentration. This should be excluded by calculating 'corrected' calcium and/or ionised calcium measurements.

Clinical features

Hypocalcaemia is associated with increased neuromuscular excitabil-ity, such as tetany, paraesthesia and muscle cramps. Increased neuromuscular excitability can be demonstrated by Trousseau's sign (inflating the blood pressure cuff above systolic blood pressure causes carpal pedal spasm) or Chvostek's sign (tapping the facial nerve about 2 cm in front of the ear will cause twitching of facial muscles). Prolonged hypocalcaemia is associated with cataracts, mental retarda-tion, psychosis and increased intracranial pressure, which may lead to papilloedema (Table 6.7).

Causes of hypocalcaemia

Causes of hypocalcaemia are listed in Table 6.8. Artifactual hypocal-caemia due to taking blood into containers with EDTA or other chelating agents should first be eliminated. Calcium concentration should be corrected for low plasma protein concentration.

Table 6.7   Clinical features of hypocalcaemia

	Features
Acute	Chvostek's sign
	Trousseau's sign
	Tetany
	Numbness and paraesthesia
	Behavioural disturbance and stupor
	Muscle cramps and spasms
	Convulsions
	Prolonged QT interval
Chronic	Cataracts
	Basal ganglia calcification

Table 6.8   Causes of hypocalcaemia

Artefactual	• Blood taken into EDTA tube
	• Hypoalbuminaemia
Abnormalities in vitamin D	Vitamin D deficiency
	— Malabsorption
	— Lack of exposure to sunlight
	— Dietary
	• Defective vitamin D metabolism
	— Renal failure
	— Anticonvulsant therapy
	— Liver disease
	• Inherited disorders of vitamin D metabolism
	— Vitamin D-dependent rickets — Type I
	— Vitamin D-dependent rickets — Type II
Hypoparathyroidism	• Iatrogenic — surgical removal
	• Autoimmune
	• Congenital (DiGeorge syndrome)
Magnesium deficiency	
Acute pancreatitis	
Rhabdomyolysis	
Pseudohypoparathyroidism	

1. *Vitamin D deficiency*

Vitamin D deficiency can occur in patients with inadequate intake, malabsorption or those who do not go out into the sunlight such as the elderly or Asian women. Severe vitamin D deficiency leads to hypocalcaemia and secondary hyperparathyroidism. Reduced intestinal absorption of phosphate and phosphaturic effect of PTH leads to hypophosphataemia. Typically these patients have hypophosphataemia, hypocalcaemia and increased alkaline phosphatase activity (see Chapter 9). It is important to note that although vitamin D deficiency is relatively common, hypocalcaemia is only seen when the vitamin D deficiency is associated with low calcium intake.

2. *Abnormalities in vitamin D metabolism*

Renal failure is the commonest cause of impaired vitamin D metabolism. Factors contributing to hypocalcaemia in renal failure are discussed in Chapter 5. Vitamin D metabolism may be impaired in liver disease, causing failure of calcidiol formation. In these patients, other factors such as reduced absorption of vitamin D due to reduced bile salts may also contribute to low vitamin D status. Patients on chronic anticonvulsants therapy with phenytoin or phenobarbitone may develop vitamin D deficiency due to the induction of enzymes in the liver which convert vitamin D to inactive metabolites.

There are two forms of inherited disorders of vitamin D metabolism. In vitamin D-dependent rickets type I, there is a mutation causing deficiency of 1α-hydroxylase. In vitamin D-dependent rickets type II, there is a mutation causing an abnormal vitamin D receptor.

Case 6.3

A-25-year-old Asian lady saw her GP because of persistent aches in her limbs. On examination, her proximal muscles were weak. The GP also noticed that she had difficulty in getting up from her chair. Investigations showed the following results:

Serum		Reference Range
Calcium (mmol/L)	1.7	2.25–2.55
Phosphate (mmol/L)	0.5	0.8–1.3
Albumin (g/L)	35	36–47
Alkaline phosphatase (IU/L)	225	40–125
PTH (ng/L)	185	10.0–65.0
Correctedcalcium (mmol/L)	1.8	2.25–2.55
Urea (mmol/L)	3.8	3.5–7.2
Creatinine (μmol/L)	62	60–112

This lady had severe hypocalcaemia, accompanied by low phosphate, high ALP and high PTH. These features suggest vitamin D deficiency. In vitamin D deficiency, reduced absorption of calcium from the gut causes hypocalcaemia which stimulates PTH secretion. PTH in turn causes urinary loss of phosphate and hypophosphataemia. Her ALP is high due to the increased ostoblastic activity. Her serum calcidiol (25-hydroxyvitamin D) was < 10 nmol/L confirming the diagnosis of severe vitamin D deficiency. Her difficulty in getting up from the chair and her muscle weakness are also due to vitamin D deficiency.

3. *Hypoparathyroidism*

   Hypoparathyroidism is most commonly due to surgical removal of the parathyroid glands during thyroidectomy or other neck surgery and less commonly due to autoimmune disease. In infants, congenital absence of the parathyroid gland may be present as an isolated defect or in association with absence of the thymus, facial abnormalities and impaired cellular immune response (DiGeorge syndrome). In hypoparathyroidism, there is hyperphosphataemia, hypocalcaemia and undetectable PTH concentration.

4. *Pseudohypoparathyrodism*

   Pseudohypoparathyrodism is a rare disorder due to inherited abnormalities causing resistance to PTH. These patients have hypocalcaemia with high PTH. PTH binds to its receptor and activates cAMP via guanine nucleotide regulatory protein (GS protein). In type I pseudohypoparathyroidism, there is an abnormality in the α subunit of the GS protein; these patients do not show an increase in cAMP response to exogenous PTH. There are two subtypes, 1A and 1B. Type 1A has morphological features; short stature, mental retardation, short 3rd and 4th metacarpal. These features are absent in type 1B. In type II pseudohypoparathyroidism, cAMP response is normal but there is no effect on calcium or phosphate, suggesting that the defect is distal to the production of cAMP.

5. *Magnesium deficiency*

Hypocalcaemia is a feature of severe magnesium deficiency (see Chapter 7).

6. *Other causes*

Rhabdomyolysis or tumour lysis syndrome causes acute hypocalcaemia due to the release of intracellular phosphate which binds to calcium. Hypocalcaemia is also not uncommon in critically ill patients. Factors contributing to this include hypomagnesaemia, acute renal failure, blood transfusion (due to large amounts of citrate chelating calcium), vitamin D deficiency and hypoparathyroidism. Hypocalcaemia in pancreatitis is attributed to chelation of calcium by free fatty acids released by the action of lipase, forming calcium soaps in the peritoneum.

Many patients who have surgery of the parathyroid gland develop temporary hypocalcaemia. This may be due to atrophy of the remaining parathyroid tissues due to chronic hypercalcaemia or due to hungry bone syndrome where there is a rapid uptake of calcium by the bone after removal of the parathyroids.

Investigations

After excluding artefactual hypocalcaemia, renal function tests should be done (Figure 6.7). Hypocalcaemia accompanied by

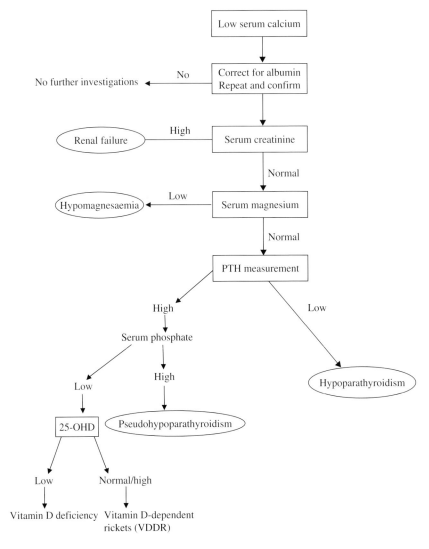

Figure 6.7   A scheme for the investigation of hypocalcaemia.

hypophosphataemia is typical of vitamin D deficiency whereas in renal failure and hypoparathyroidism, hyperphosphataemia is observed. Measurement of PTH levels may be required; PTH is high in vitamin D deficiency and low in hypoparathyroidism.

Plasma magnesium concentration is necessary to exclude magnesium deficiency.

Management

Symptomatic hypocalcaemia should be treated with intravenous calcium, 10 to 20 mL of 10% calcium gluconate infused slowly over 10 minutes. This may be repeated if necessary. The underlying cause should be investigated and treated. Patients with hypoparathyroidism presenting with hypocalcaemia are treated with calcitriol or alpha calcidiol. These patients should be monitored with serum and urine calcium as they are at risk of renal stones due to increased calcium excretion.

Further Reading

1. Bushinki DA, Mark RD. Calcium. *Lancet* 1998; 352:306–11.
2. Cooper MS, Gittoes NJJ. Diagnosis and management of hypocalcaemia. *Br Med J* 2008; 336:1298–1302.
3. Copper MS, Gittoes NJL. Hypocalcaemia. *Br Med J* 2008; 336: 1298–1302.
4. Hollis BW. Assessment and interpretation of circulating 25-hydroxy-vitamin D and 1,25-dihydroxyvitamin D in the clinical environment. *Endocrinol Metab Clin North Am* 2002; 39:271–286.
5. Mendy GR, Guise TA. Hormonal control of calcium homeostasis. *Clin Chem* 1999; 45:1347–1352.
6. Uttiger RD. Treatment of primary hyperparathyroidism. *N Engl J Med* 1999; 341:1301–1302.

Summary/Key Points

1. Calcium has many important functions and these include structural, neuromuscular, enzymatic and signalling functions.
2. Calcium in circulation is present in 3 forms: ionised (about 50%), protein-bound (mainly to albumin) (about 40%) and complexed (<10%). Of these, the ionised calcium is the

physiologically important fraction. Changes in concentration of proteins and hydrogen ions can affect protein binding of calcium. Interpretation of total calcium should take into account the concentration of serum proteins (albumin mainly) and the acid–base status.

3. Calcium homeostasis is maintained by parathyroid hormone (PTH) and vitamin D. PTH produced by the parathyroid glands is regulated by serum ionised calcium via calcium sensing receptor (CaSR). PTH acts on the renal tubules to increase the reabsorption of calcium, reduce the reabsorption of phosphate, increase bone resorption and stimulate the formation of calcitriol, which increases intestinal calcium absorption.

4. Vitamin D is mainly produced in the skin from 7-dehydrocholesterol by the action of UV light. Vitamin D is hydroxylated first in the liver and then in the kidneys to the active metabolite calcitriol, which acts on the intestine to increase calcium absorption.

5. Parathyroid hormone related peptide (PTH-rP) is a peptide which has structural similarities to PTH and can act at the same sites as PTH and cause hypercalcaemia. PTHrP is now known to be produced by a large number of tissues and is important for many functions including the differentiation, proliferation and death of many cell types.

6. Common causes of hypercalcaemia are primary hyperparathyroidism and malignancy.

7. Primary hyperparathyroidism is often asymptomatic and is discovered on routine screening. Features of primary hyperparathyroidism include renal stones, bone pain, polyuria, thirst and neuropsychiatric manifestations.

8. Hypercalcaemia in malignancy may be due to secondary deposits in bone or due to secretion of PTHrP (humoral hypercalcaemia of malignancy).

9. Familial hypocalciuric hypercalcaemia is a rare genetic disorder due to a mutation in CaSR. These patients have mild hypercalcaemia associated with PTH in the upper end of reference range ('inappropriately high'). It can be confused with

primary hyperparathyroidism and the demonstration of inappropriately low urine calcium excretion is helpful in diagnosing this disorder.

10. In the investigation of hypercalcaemia, measurement of PTH concentration is a vital investigation. If the PTH is low (suppressed) then malignancy including multiple myeloma is likely and investigations should be done to exclude these.

11. Hypocalcaemia in hospital population is often due to low albumin. If this is excluded, the common causes are renal failure (reduced calcitriol production), hypoparathyroidism or vitamin D deficiency.

chapter 7

Magnesium Metabolism

Introduction

Magnesium is the fourth most abundant cation in the body and within the cell it is the second most abundant cation, after potassium. Magnesium plays an essential physiological role in many cellular functions (Table 7.1). It is a cofactor for over 300 enzymes.

The normal human adult contains approximately 1000 mmol of magnesium (22–26 g) and the distribution within the body is given in Table 7.2. Most of the body's magnesium is found in the bone (53%) and skeletal muscles (27%). Intracellular magnesium concentration is approximately 20 mmol/L compared to a serum concentration of magnesium of about 0.9 mmol/L (range 0.70–1.10). Approximately 20% of total serum magnesium is protein-bound (60–70% to albumin and the rest to globulin). Most of the non-protein–bound (ultrafiltrable) fraction is ionised (65% of the total). The rest, about 15% of the total is complexed with anions such as phosphate and citrate (Figure 7.1). Intracellular free ionised magnesium constitutes only 0.5–5% of the total cellular magnesium; the remaining fraction is bound to anionic compounds or is sequestrated within the mitochondria and endoplasmic reticulum.

Magnesium Balance

The recommended daily allowance (RDA) for magnesium is 4.5 mg/kg/day for adults and it is higher in pregnancy, lactation and

Table 7.1 Physiological functions of magnesium

Enzyme function
- Enzyme substrate (ATPmg, GTPmg)
  — Kinases B, e.g. hexokinase
  — ATPases or GTPases, e.g. sodium–potassium–ATPase
  — Cyclases, e.g. adenylate cyclase
- Direct enzyme activation, e.g. phosphofructokinase, creatine kinase
Membrane function: cell adhesion
Calcium antagonist
- Muscle contraction/relaxation
- Neurotransmitter release
- Action potential conduction in nodal tissue
Structural function: protein, polyribosomes, etc.

Table 7.2 Distribution of magnesium in the adult human

Tissue	Weight (kg wet wt)	Concentration (mmol/kg wet wt)	Content (mmol)	% of total body magnesium
Serum	3.0	0.85	2.6	0.3
Red blood cells	2.0	2.5	5.0	0.5
Soft tissue	22.7	8.5	193	19.3
Muscle	30.0	9.0	270	27.0
Bone	12.3	43.2	530	52.9
Total	70.0		1000.6	100

following debilitating illnesses. Foods rich in magnesium are cereal grain, nuts, legume, chocolates and green vegetables that are rich in magnesium containing chlorophyll. Foods poor in magnesium are dairy products and beverages. Drinking water especially 'hard water' that contains up to 30 mg/L of magnesium is also an important source. The refining or processing of food and cooking, especially boiling will result in loss of magnesium.

The exact physiological mechanism regulating magnesium balance and plasma magnesium concentration is not fully understood. Figure 7.2 shows the metabolism of magnesium in a normal adult.

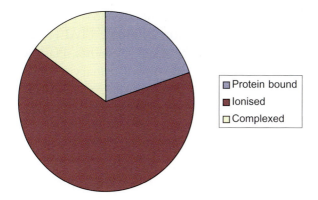

Figure 7.1   Distribution of serum magnesium.

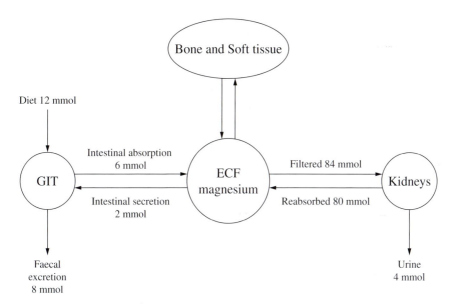

Figure 7.2   Magnesium metabolism.

In normal individuals, 30% to 50% of dietary magnesium is absorbed and absorption takes place throughout the intestine. Kidneys play a major role in the regulation of the magnesium homeostasis. Of the filtered magnesium, about 25% to 30% is reabsorbed in the proximal tubular segments, approximately 60% to 65% is reabsorbed in the thick ascending limb of the loop of Henle (TALH) while the rest (about 5%) is reabsorbed in the distal segments. Plasma magnesium concentration is a major determinant of urinary excretion. Changes in magnesium intake result in appropriate changes in magnesium excretion. How dietary magnesium influences excretion is not fully understood. Intestinal absorption of magnesium also shows adaptation to changes in dietary intake. Unlike calcium, there are no hormones regulating magnesium homeostasis. PTH has an effect on magnesium reabsorption and PTH secretion is influenced by serum magnesium. An acute decrease in magnesium concentration causes an increase in PTH whereas chronic low magnesium levels causes a reduced secretion. PTH can increase renal magnesium reabsorption but its importance in overall magnesium balance is little.

Assessment of Magnesium Status

Methods available to assess magnesium status are listed in Table 7.3. At present, there is no simple, rapid and accurate laboratory method to assess magnesium status and the most common method is by the measurement of serum or plasma magnesium concentration.

Magnesium concentration is preferably measured in serum rather than plasma as the anticoagulants may be contaminated with magnesium or could affect the assay. Haemolysis will increase serum magnesium as magnesium concentration in the cells is approximately three times greater than that of serum. Measurement of serum magnesium can also be affected by hyperbilirubinemia, lipaemia and high phosphate concentrations. Variation in serum magnesium concentration between individuals (interindividual) is between 5.9% and 7.5% while within individuals (intraindividual) it is between 3.4% and 4.7%. Total serum magnesium concentration is not the best method to evaluate the magnesium status, as changes in serum protein concentrations may affect serum magnesium. The measurement of ionised magnesium may be more meaningful.

Table 7.3   Tests used in assessing magnesium status

Serum magnesium concentration
- Total magnesium
- Ultrafiltrable magnesium
- Ionised magnesium

Intracellular magnesium content
- Red cells
- Mononuclear blood cells
- Skeletal muscle

Physiological test
- Metabolic balance studies
- 24-hour urinary excretion of magnesium
- Magnesium loading test

Intracellular free magnesium ion concentration
- Fluorescent dye
- Nuclear magnetic resonance spectroscopy

Others
- Magnesium balance
- Isotope studies
- Hair or tooth magnesium
- Functional assays

The measurement of magnesium concentration in red cells, white cells, platelets or muscle are not used in routine practice. Furthermore, the value of these tests in assessing the total body magnesium status has not been established. The 24-hour urine magnesium excretion reflects intestinal absorption. In the presence of hypomagnesaemia, a magnesium excretion higher than 1 mmol/day is suggestive of renal magnesium wasting.

Magnesium tolerance test appears to be an accurate means of assessing magnesium status. The percentage of magnesium retained after parental administration of magnesium (0.1 mmol of magnesium/kg of body weight in 50 mL of 5% dextrose over 4 hours) is measured and a high retention indicates deficiency. This test may be of limited value in patients with poor renal function or those in whom there is increased magnesium loss through the kidneys.

No single method is satisfactory to assess magnesium status. The simplest and readily available method is total serum magnesium concentration and the most useful test is the magnesium tolerance test.

Magnesium Deficiency and Hypomagnesaemia

The terms hypomagnesaemia and magnesium deficiency are usually used interchangeably, although total body magnesium depletion can be present with normal serum magnesium concentration and there can be significant hypomagnesaemia without total body deficit. The incidence of hypomagnesaemia in hospital patients is 6–11%. However, it is more common in critically ill patients (20–65%), alcoholics (30–80%) and patients with other electrolyte abnormalities.

Causes

Causes of hypomagnesaemia are listed in Table 7.4.

Redistribution

Hypomagnesaemia may develop in the postoperative period after total parathyroidectomy due to the entry of magnesium into cells as part of the 'hungry bone syndrome'. When starving subjects are fed (refeeding syndrome), hypomagnesaemia may be seen, in addition to hypokalaemia and hypophosphataemia, due to the uptake of magnesium into the cells.

Gastrointestinal causes

Reduced intake does not cause hypomagnesaemia but it is a contributing factor in other causes, e.g. alcohol-induced hypomagnesaemia. Loss of magnesium in chronic diarrhoea and malabsorption syndromes is not uncommon. Proton pump inhibitors such as omeprazole have been reported to cause severe hypomagnesaemia in a small percentage of patients, probably due to gastrointestinal loss of magnesium.

Table 7.4   Causes of hypomagnesaemia

Redistribution of magnesium
- Refeeding and insulin therapy
- Hungry bone syndrome

Gastrointestinal causes
- Reduced intake
- Reduced absorption: Malabsorption syndrome, chronic diarrhoea, and intestinal resection
- Proton pump inhibitors

Renal loss
- Reduced sodium reabsorption: Saline infusion, Diuretics
- Drugs
  — Diuretics
  — Cytotoxic drugs: e.g. Cisplatin
  — Antimicrobial agents: e.g. Aminoglycosides
  — Immunosuppressants: Cyclosporine, FK 506
  — β adrenergic agonists: Theophylline, Salbutamol, Rimiterol
  — Other drugs: e.g. Amphotericin B, Pentamidine, and Foscarnet
- Renal disease
  — Post-obstructive nephropathy, postrenal transplantation
  — Dialysis
- Inherited disorders: Gitelman's syndrome

Endocrine causes
- Hypercalcaemia, primary hyperparathyroidism, malignant hypercalcaemia
- Hyperthyroidism
- Hyperaldosteronism

Diabetes mellitus
Alcoholism

Renal loss

Proximal tubular magnesium reabsorption is reduced by chronic intravenous fluid therapy, particularly sodium-containing fluids, osmotic diuresis and loop diuretics. Cisplatin causes hypomagnesaemia in a large percentage of patients and the incidence increases with cumulative doses. Chronic hypomagnesaemia starts to develop

3 weeks after initiation of chemotherapy and persists usually for several months. Occasionally, hypomagnesaemia may persist for several years after completion of treatment. Carboplatin, an analogue of cisplatin, causes hypomagnesaemia in only 10% of patients. Hypomagnesaemia due to renal magnesium loss is seen with high doses of aminoglycosides. Hypomagnesaemia of cyclosporine therapy is usually mild, but occasionally, severe magnesium deficiency is seen.

Pentamidine may cause severe symptomatic hypomagnesaemia due to renal magnesium wasting and up to 70% of patients treated with foscarnet may develop hypomagnesaemia. Diabetes mellitus causes magnesium deficiency in 25–39% of patients. It is caused by increased excretion brought about by osmotic diuresis, but there may be a specific tubular defect as well. During treatment of diabetic ketoacidosis, hypomagnesaemia may be seen in up to 55% of such patients. Hypomagnesaemia, is seen in 30% of acute and chronic alcoholics, and this is due to multiple factors: poor nutritional status, vomiting and diarrhoea, malabsorption due to chronic pancreatitis or liver disease, acute alcoholic ketoacidosis, hyperaldosteronism secondary to liver disease and renal tubular dysfunction. Hypomagnesaemia is also seen in Gitelman's syndrome, an inherited disorder of sodium reabsorption (see Chapter 3).

Clinical Manifestation of Hypomagnesaemia and Magnesium Deficiency (Table 7.5)

Magnesium and potassium are closely related; hypokalaemia is a frequent finding in patients with hypomagnesaemia. This form of hypokalaemia cannot be corrected until the magnesium deficiency is corrected. The exact mechanism underlying this interrelationship is not clear. The hypokalaemia of magnesium deficiency contributes to the cardiac manifestation of hypomagnesaemia, but may delay the onset of tetany. Hypocalcaemia is a common manifestation in moderate to severe hypomagnesaemia, in up to one third of patients. Hypocalcaemia of magnesium deficiency cannot be corrected until magnesium deficiency is treated. Factors contributing to hypocalcaemia of magnesium deficiency are a decrease in PTH secretion,

Table 7.5 Clinical features of hypomagnesaemia

Electrolyte disturbance
- Hypokalaemia
- Hypocalcaemia

Neuromuscular and central nervous system
- Carpopedal spasm, convulsions
- Muscle cramps, muscle weakness, fasciculations, tremors
- Vertigo, nystagmus
- Depression, psychosis

Cardiovascular
- Atrial tachycardias, fibrillation
- Supraventricular tachycardia
- Ventricular arrhythmias
- Increased digoxin sensitivity

Complications of magnesium deficiency
- Atherosclerotic vascular disease
- Hypertension
- Myocardial infarction
- Osteoporosis

resistance to the action of PTH, a decrease in serum calcitriol and resistance to calcitriol.

The earliest manifestations of symptomatic magnesium deficiency are usually neuromuscular and neuropsychiatric manifestations. The most common clinical manifestation is hyperexcitablity manifested as positive Chvostek and Trousseau signs, tremors, fasciculations and tetany. Mechanisms contributing to these include a decreasing threshold of axon stimulation, increased nerve conduction velocity, increased neurotransmitter release, increased release of calcium from the sarcoplasmic reticulum in muscle and reduced re-uptake of calcium. The net effect is a muscle that is more readily contractible to a given stimulus and is less able to recover from the contraction, i.e. prone to tetany. The mechanism of effect of magnesium deficiency on the central nervous system is even more complicated and less well understood.

Case 7.1

A 67-year-old man was admitted to hospital with grand mal seizures. On admission his results were:

Potassium	2.5 mmol/L (3.5–5.0)
Calcium	1.3 mmol/L (2.25–2.55)
Magnesium	0.3 mmol/L (0.8–1.1)

He was on omeprazole, in addition to cizapal and hydrochlorothiazide. He was treated with intravenous potassium, calcium and magnesium and his electrolytes disturbances normalised. His magnesium excretion was low and his magnesium clearance related to creatinine clearance was < 2% showing that non-renal loss was the cause of hypomagnesaemia. This case illustrates hypomagnesaemia, causing hypocalcaemia and hypokalaemia due to omeprazole.

Cardiovascular manifestations of magnesium deficiency include effects on electrical activity, myocardial contractility and potentiation of digoxin toxicity. Electrocardigraphic changes include slight prolongation of conduction and depression of ST segment but these are nonspecific. Magnesium depletion increases the susceptibility to arrhythmogenic effects of drugs such as isoproterenol and cardiac glycosides. The effects of magnesium deficiency on the heart are further complicated by intracellular potassium depletion and hypokalaemia. Hypomagnesaemia and magnesium depletion may contribute to digoxin toxicity even when the concentration of digoxin is in the therapeutic range.

Epidemiological studies show an inverse relationship between magnesium intake and blood pressure and incidence of cardiovascular disease. Magnesium deficiency contributes to the progression of atherosclerosis by increased peroxidation of lipoproteins, increased platelet aggregation and hypertension.

Magnesium deficiency may be a risk factor for osteoporosis. Magnesium content of trabecular bone and magnesium intake is lower in osteoporotic subjects. Magnesium tolerance studies show increased retention of magnesium in osteoporotics. The mechanism whereby reduced magnesium intake exacerbates osteoporosis is unclear but is probably multifactorial.

Treatment

Symptomatic magnesium deficiency should be corrected promptly. In patients with ventricular tachycardia or convulsions, 12 mmol of magnesium as magnesium sulphate should be given over 2–3 hours, followed by another 12 mmol over the next 24 hours and if necessary, another 24 mmol may be administered over the next 24 hours.

Mild asymptomatic hypomagnesaemia can be treated by a diet rich in magnesium and/or oral magnesium supplementation as gluconate. An initial dose of 12 mmol per day, increasing to 48 mmol in divided doses (three or four times a day) is recommended to avoid diarrhoea. Administration of potassium and calcium together with magnesium may be necessary since associated loss of these cations is common in severe magnesium deficiency. Assessment of renal function before replacement therapy and monitoring of serum concentrations of magnesium, potassium and other major cations during therapy is recommended.

Hypermagnesaemia

In hospital populations, hypermagnesaemia is seen less frequently than hypomagnesaemia as the kidneys can readily excrete excess magnesium.

Causes

Causes of hypermagnesaemia are listed in Table 7.6. A common cause of hypermagnesaemia is excessive administration of magnesium salts or magnesium-containing drugs especially in patients with reduced renal function. Hypermagnesaemia due to redistribution from cells has been described in acute acidosis, e.g. in acidosis after massive theophylline overdose. Excessive usage of magnesium-containing medications such as laxatives, antacids and rectal enemas may lead to hypermagnesaemia. Hypermagnesaemia is not uncommon with the use of magnesium-containing cathartics especially during treatment of drug overdose. In patients with bowel disorders, the risk of hypermagnesaemia is higher. Urethral

Table 7.6   Causes of hypermagnesaemia

Redistribution
  • Acute acidosis

Excessive intake
  • Oral: Antacids, cathartics, swallowing salt water
  • Rectal: Purgatives
  • Parenteral
  • Urethral irrigation

Renal failure
  • Acute and chronic renal failure

irrigation with hemiacidrin has been reported to cause symptomatic hypermagnesaemia in patients with or without renal failure. Severe hypermagnesaemia follows the swallowing of salt water. Hypermagnesaemia is common in renal failure due to the administration of magnesium-containing medication.

Effects of Hypermagnesaemia

Signs and symptoms of hypomagnesaemia are not usually apparent until the serum concentration is in excess of 2 mmol/L (Table 7.7). Neuromuscular symptoms are the most common presentation of magnesium intoxication. These develop as a result of blockage of neuromuscular transmission and depression of the conduction system. Clinically, one of the earliest effects of magnesium intoxication is the disappearance of deep tendon reflexes. Somnolence may be observed at concentrations of 2 mmol/L or above. Other manifestations include muscle weakness proceeding to flaccid paralysis of voluntary and/or respiratory muscles leading to depressed respiration at concentrations in excess of 5 mmol/L. The effects on the neuromuscular junctions are antagonised by calcium, therefore the effects of hypermagnesaemia are exaggerated in the presence of hypocalcaemia. Severe symptomatic hypotension is a feature of severe hypermagnesaemia. Magnesium is also cardiotoxic and electrocardiographic findings include prolonged PR interval, increased QRS duration and QT

Table 7.7   Signs and symptoms of hypermagnesaemia

Neuromuscular
- Confusion
- Lethargy
- Respiratory depression
- Absent tendon reflexes
- Paralytic ileus
- Bladder paralysis
- Muscle weakness/paralysis

Cardiovascular
- Hypotension
- Bradycardia
- Inhibition of AV and interventricular conduction
- Heart block
- Cardiac arrest

Others
- Nausea, vomiting

intervals. Mild bradycardia may be seen and occasionally complete heart block and cardiac arrest may occur at concentrations greater than 7 mmol/L.

Management of Hypermagnesaemia

The possibility of hypermagnesaemia should be anticipated in any patient receiving magnesium treatment, especially if the patient has reduced renal function; and serum magnesium concentration should be monitored daily.

In patients with symptomatic hypermagnesaemia, all magnesium therapy should be withdrawn and 1 g of intravenous calcium gluconate should be given followed by infusion of 100–150 mg of calcium over 5–10 minutes. This usually causes dramatic improvement in the patient's clinical condition. Administration of glucose and insulin may also help to promote magnesium entry into cells. In patients with renal failure, peritoneal or haemodialysis against low dialysis magnesium fluid may be required.

Further Reading

1. Musso CG. Magnesium metabolism in health and disease. *Int Urol Nephrol* 2009; 41:357–362.
2. Swaminathan R. Magnesium and its disorders. *Clin Biochem Rev* 2003; 34:47–66.
3. Topf JM, Murray PT. Hypomagnesemia and hypermagnesemia. *Rev Endocr Metab Disord* 2003; 4:195–206.

Summary/Key points

1. Magnesium is an essential element for many physiological functions.
2. Magnesium homeostasis is largely regulated by renal excretion; however, the exact mechanism is not understood.
3. Assessment of magnesium status is difficult and magnesium tolerance test may be helpful.
4. Hypomagnesaemia is a common finding in hospital population.
5. Main causes of hypomagnesaemia are redistribution, gastrointestinal loss and renal loss.
6. Hypomagnesaemia can lead to hypocalcaemia, hypokalaemia, cardiac arrhythmias and neuromuscular manifestations.
7. In patients with hypocalcaemia, serum magnesium should be measured.
8. Hypermagnesaemia is less common and mainly caused by reduced excretion.

chapter 8

Disorders of Phosphate Metabolism

Introduction

Phosphorus is an essential element and is a major component of all tissues as cellular, organelle and membrane constituents. It is important for many vital functions including cellular metabolism and the regulation of intracellular processes. In addition, phosphate is a major component of bone and is an important buffer. Phosphorus is present as organic compounds such as phospholipids and phosphoproteins and in the inorganic form as phosphate.

Phosphate Homeostasis (Figure 8.1)

Most (85%) of the phosphate in the body (approximately 20 mol) is in the skeleton; about 14% is intracellular and less than 1% is in the extracellular fluid (Table 6.2). The average daily intake of phosphate is about 45 mmol and it comes mainly from dairy products, meat and vegetables, which are rich in phosphate. Absorption of phosphate is high from meat or dairy products where it is in a soluble form whereas absorption is poor from vegetables where it is in an insoluble form.

Like calcium, most of the phosphate is absorbed in the ileum although the efficiency of absorption is highest in the duodenum and jejunum. Mechanisms of phosphate absorption involve a vitamin D-dependent active transport mainly in the duodenum and jejunum and a passive saturable diffusion mechanism present throughout the intestine; the latter accounts for bulk of the absorption. Fractional intestinal absorption increases with low phosphate intake, demonstrating

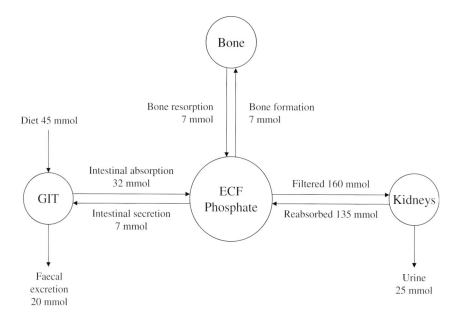

Figure 8.1   Phosphate homeostasis.

adaptation. Net absorption of phosphate is linearly related to dietary intake. Phosphate homeostasis is maintained mainly by the kidneys (Figure 8.1) and to a lesser extent by the intestine. Of the 160 mmol of phosphate filtered daily at the glomerulus, 80% is reabsorbed in the proximal tubule, about 10% in the distal parts of the nephron, leaving 10% to be excreted.

Phosphate reabsorption in the tubules is mediated by sodium–phosphate cotransporter, mainly Na-Pi-IIa. If plasma phosphate concentration is increased, urinary phosphate excretion will increase linearly above a threshold. This threshold of plasma phosphate, below which phosphate excretion is zero, is called the threshold phosphate concentration related to GFR (TmP/GFR) — the theoretical concentration of plasma phosphate, below which all phosphate is reabsorbed and above which phosphate is excreted in the urine. Determination of TmP/GFR is the best method to assess renal tubular reabsorption of phosphate. TmP/GFR can be easily determined from fasting urine and plasma concentrations of phosphate and

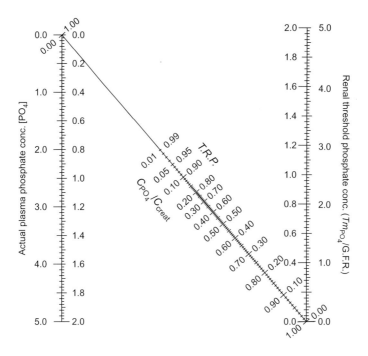

Figure 8.2 Nomogram to derive TmP/GFR.

Phosphate and creatinine concentrations in a random, fasting sample of urine and serum are measured and the ratio of clearance of phosphate to creatinine (TRP) is calculated from the following equation:

$$TRP = \frac{\text{urine phosphate} \times \text{serum creatinine}}{\text{serum phosphate} \times \text{urine creatinine}}$$

A straight line through the appropriate values of serum phosphate and TRP passes through the corresponding value of TmP/GFR.

creatinine using a nomogram (Figure 8.2). It is useful in the investigation of phosphate disorders.

Factors which influence phosphate reabsorption in the kidney include PTH, PTHrP, GH/IGF1 (Table 8.1). PTH and vitamin D are important regulators of phosphate reabsorption mediated by Na-Pi-IIa. High PTH inhibits while high calcitriol (1,25-vitamin D) increases

Table 8.1   Factors affecting renal tubular reabsorption

Parathyroid hormone
PTHrP
GH/IGF1
ECF volume
Acid–base status
Phosphatonins
Drugs

- Bisphosphonates
- Diuretics
- Cisplatin
- Acetazolamide
- Antiviral drugs (e.g. tenovir)
- Chinese herbal medicine (e.g. aristocholic acid)

Dietary phosphate
Vitamin D
Insulin
Epidermal growth factor
Thyroid
Calcitonin
Glucocorticoids
ANP

reabsorption. Recent studies have identified a group of compounds called phosphatonins which are inhibitors of phosphate reabsorption. These phosphatonins are thought to be responsible for the diet-induced changes in phosphate reabsorption. Phosphatonins include fibroblast growth factor-23 (FGF-23), matrix extracellular phospoglycoprotein (MEPE) and frizzled-related protein. FGF-23 also reduces the conversion of 25-hydroxyvitamin D to 1,25-vitamin D.

Plasma Phosphate

In healthy subjects, the total plasma phosphate concentration is about 3.9 mmol/L, and of this 0.8–1.3 mmol/L is in the inorganic form and the rest as phospholipids and other organic compounds. Most

(84%) of the plasma inorganic phosphate is free, approximately 10% is protein bound and about 6% is complexed with calcium or magnesium.

In healthy adults, there is a marked diurnal variation in plasma phosphate concentration, being lowest in the morning and highest during the night. Fasting abolishes this diurnal variation. The plasma phosphate concentration is high at birth and it remains high in children. During lactation, plasma phosphate is higher but is not altered in pregnancy. Postprandially, plasma phosphate concentrations tend to decrease, presumably because of insulin release. Severe exercise such as marathon running has varying effects on the plasma phosphate concentration, depending on the severity of exercise and carbohydrate ingestion. Serum concentration of phosphate is 0.06–0.10 mmol/L higher than that of plasma due to the release of intracellular phosphate from platelets and erythrocytes during clotting. This difference is especially marked in patients with thrombocytosis. It is recommended that samples for plasma phosphate estimation should be taken in the morning in the fasting state.

Hypophosphataemia

Mild hypophosphataemia (< 0.80 mmol/L) is a common finding among hospital patients and the prevalence varies from 14–39%. In some groups, such as patients in intensive care units and in patients with diabetic ketoacidosis (DKA), the incidence of hypophosphataemia is higher. Severe hypophosphataemia (< 0.30 mmol/L), however, is relatively uncommon and is seen in only 0.04–1.0% of hospital patients. Hypophosphataemia does not always indicate phosphate deficiency and phosphate deficiency may be present with normal or even increased plasma phosphate concentrations.

Hypophosphataemia may result from redistribution of phosphate into cells, reduced intestinal absorption or due to loss of phosphate. In most clinical situations, hypophosphataemia is often multifactorial.

Causes of Hypophosphataemia

Common causes of hypophosphataemia are listed in Table 8.2. In hospital populations, respiratory alkalosis and infusion of carbohydrates

Table 8.2   Causes of hypophosphataemia

Common
- Respiratory alkalosis
- Infusion of carbohydrates — glucose, fructose, parenteral nutrition
- Refeeding syndrome
- Alcoholism
- Antacids
- Diabetic ketoacidosis
- Postrenal transplantation
- Severe burns
- Hyperparathyroidism

Less common
- Neuroleptic malignant syndrome
- Hungry bone syndrome
- Heat stroke
- Oncogenic osteomalacia
- Chronic obstructive airways disease

Others
- Rapidly growing tumours
- Septicaemia
- Treatment of pernicious anaemia

(especially glucose) are the commonest causes. In respiratory alkalosis, there is an increased uptake of phosphate into the cells due to intracellular alkalosis stimulating phosphofructokinase activity, which leads to increased glycolysis and incorporation of phosphate into organic intermediates. Decreased renal tubular reabsorption is a contributing factor. Hyperventilation leading to respiratory alkalosis is the major cause of hypophosphataemia seen in severe liver disease, septicaemia, salicylate intoxication, head injury, heat stroke, acute gout, malignant neuroleptic syndrome and in mechanically ventilated patients.

Carbohydrate administration accounts for about 40% of hypophosphataemia cases. The degree of hypophosphataemia is related to the amount of carbohydrate infused. Even small amounts of intravenous glucose in the form of 4% or 5% dextrose can cause a significant decrease in plasma phosphate concentration. Infusion of carbohydrate leads to increased insulin concentration, which causes increased formation of intracellular phosphorylated glycolytic intermediates. This leads to an increased uptake of phosphate into the cells, mainly liver and skeletal muscle. A similar mechanism accounts for the severe life-threatening hypophosphataemia seen when severely malnourished subjects are refed rapidly with carbohydrates (refeeding syndrome) (see Chapter 17).

Septicaemic patients, patients with diabetic ketoacidosis during treatment and alcoholics are all at risk of hypophosphataemia. Drugs, which can cause hypophosphataemia, include aluminium containing antacids and other phosphate-binding agents. Patients in intensive care, including post-operative patients and those with severe burns, develop hypophosphataemia due to a combination of factors: respiratory alkalosis, carbohydrate infusion and drugs.

Rapid proliferation of cells, as may be seen with certain lymphomas and other haematological tumours can result in increased phosphate utilisation and hypophosphataemia. Renal loss of phosphate leading to hypophosphataemia is seen in primary hyperparathyroidism, Fanconi's syndrome, cystinosis, light chain nephropathy, oncogenic osteomalacia, X-linked hypophosphataemic rickets and paracetamol poisoning.

Consequences of Hypophosphataemia

Consequences of hypophosphataemia are listed in Table 8.3. It is unusual to see clinical manifestations in mild and moderate hypophosphataemia. In moderate hypophosphataemia of long duration, skeletal changes, i.e. rickets or osteomalacia is the only consistent abnormality. All other effects listed in Table 8.3 are seen when the hypophosphataemia is severe (phosphate concentration below 0.3 mmol/L) and long standing (several days). These consequences of hypophosphataemia are mainly due to impaired ATP production. This can result in abnormalities in cellular function such as reduced phagocytosis, haemolysis, and impaired muscle contraction, which may cause muscle weakness, cardiomyopathy and respiratory

Table 8.3   Consequences of hypophosphataemia

Haematological
- Red cells
  — ↓ 2,3-DPG and ↓ oxygen delivery
  — Haemolysis
- White cells
  — ↓ Phagocytosis
  — ↓ Chemotaxis
  — ↓ Bactericidal activity
- Platelets
  — ↓ Platelet number
  — ↓ Survival
  — ↓ Platelet ATP
  — ↑ Aggregation
  — ↓ Clot retraction

Muscular system
- Weakness/Myopathy — respiratory failure
- Rhabdomyolysis

Skeletal system
- Bone pain
- Osteomalacia or rickets
- Growth retardation

Nervous system
- Tremors, seizure, coma
- Parasthesiae

Cardiovascular system
- Cardiomyopathy

Respiratory system
- Respiratory insufficiency

insufficiency. Reduced erythrocyte 2,3-DPG may result in impaired delivery of tissues to oxygen.

Case 8.1

A 72-year-old woman was investigated for a 4-month history of dysphasia (difficulty in swallowing) and found to have a squamous cell carcinoma of the oesophagus. She underwent a pharyngolaryngeal oseophagectomy with neck dissection involving removal of her parathyroid glands. She was started on Frison feed postoperatively via a feeding jejunostomy. Postoperatively, she was given calcium and calcitriol to correct hypocalcaemia. Two days later, her serum phosphate was 0.17 mmol/L (0.8–1.3). Her urine phosphate excretion was negligible.

The cause of this patient's very low serum phosphate level was 'refeeding syndrome' that caused a rapid uptake of phosphate by cells.

Following addition of phosphate (48 mmol/day) to the feed, her serum phosphate increased to 0.70 mmol/L and then to 0.85 mmol/L in 2 days.

Investigation of Hypophosphatemia

Once the low phosphate is confirmed and if severe, further investigations may be required. A careful history and examination will identify many causes listed in Table 8.2. When the cause is not clear, investigations such as the assessment of renal phosphate threshold (TmP/GFR) will be useful. A low TmP/GFR in the presence of hypophosphatemia is an inappropriate response and a renal cause for the loss of phosphate should be sorted for.

Management of Hypophosphataemia

Withdrawal of precipitating causes (e.g. carbohydrate infusion) may be sufficient in mild to moderate hypophosphataemia. In severe hypophosphataemia, phosphate supplements in the form of sodium or potassium hydrogen phosphate can be given either orally (100 mmol/day in divided doses) or intravenously. Parenteral administration will be required in critically ill patients. Parenteral phosphate (0.10–0.20 mmol/kg body weight) is given until the plasma phosphate reaches 0.35 mmol/L or greater.

In patients at risk of developing hypophosphataemia, preventive measures should be taken, e.g. in patients given large amounts of carbohydrate, phosphate should be included in the infusion.

Oral phosphate administration may cause diarrhoea and intravenous phosphate can cause hyperphosphataemia, hypocalcaemia, hyperkalaemia and acidosis.

Hyperphosphataemia

In a general hospital population, the incidence of mild hyperphosphataemia (plasma phosphate > 1.6 mmol/L) is about 1.5%. The most common mechanism of hyperphosphataemia is reduced urinary excretion either due to a reduction in GFR or due to increased reabsorption. Increased intake and transcellular shifts of phosphate can also cause hyperphosphataemia especially when the GFR is compromised.

Causes (Table 8.4)

Pre-analytical and analytical artefacts are frequent causes of apparent 'hyperphosphataemia'. Haemolysis and prolonged storage of blood

Table 8.4   Causes of hyperphosphataemia

Artefact
- Haemolysed specimen
- Myeloma

Increased phosphate input
- Exogenous administration — oral, intravenous or rectal
- Release from cells
    - Tumour lysis syndrome
    - Rhabdomyolysis
    - Malignant hyperpyrexia
- Massive blood transfusion
- Haemolysis

Decreased excretion
- Renal failure — acute and chronic

Increased reabsorption
- Hypoparathyroidism
- Vitamin D toxicity
- Acromegaly
- Thyrotoxicosis

samples will cause spurious hyperphosphataemia due to the release of phosphate from red cells. High serum phosphate concentration noticed in some patients with multiple myeloma is due to analytical interference.

Renal failure accounts for more than 90% of cases of hyperphosphataemia in hospital patients. Plasma phosphate concentration in renal failure begins to rise when the GFR falls below 20–30% of normal. However, even in early renal failure, plasma phosphate concentration may rise during the postprandial period. A shift of phosphate from cells due to metabolic acidosis may also contribute to the hyperphosphataemia. Transcellular shift accounts for hyperphosphataemia of untreated diabetic ketoacidosis and lactic acidosis.

In tumour lysis syndrome and rhabdomyolysis, massive releases of intracellular phosphate can result in severe hyperphosphataemia. Excess phosphate administration, orally, intravenously or per rectum (as in enema) can cause of hyperphosphataemia.

Case 8.2

A 15-year-old boy was admitted for investigation of deteriorating health and was found to have acute lymphoblastic leukaemia. He was started on chemotherapy and 48 hours later, he developed oliguria. Investigation showed the following results:

Creatinine   600 μmol/L (60–120)
Phosphate   2.32 mmol/L
Calcium   1.20 mmol/L

This patient developed severe hyperphosphatemia due to tumour lysis syndrome. His hypocalcaemia is due to precipitation of calcium phosphate in soft tissues and he developed acute renal injury due to the deposition of calcium in the kidney.

Consequences of Hyperphosphataemia (Table 8.5)

Acute hyperphosphataemia can lead to hypocalcaemia, tetany and hypotension. Production of calcitriol is inversely related to serum

Table 8.5   Consequences of severe hyperphosphataemia

Hypocalcaemia
- Tetany
- Muscle cramps
- Paraesthesia
- Seizures
- Prolongation of the QT interval
- Behavioural changes

Tissue deposition of calcium phosphate salts

phosphate concentration and in chronic hyperphosphataemia, serum concentrations of calcitriol is low, leading to bone changes. The hypocalcaemic effect is due to the precipitation of calcium phosphate. In chronic hyperphosphatemia, the same mechanism leads to soft tissue calcification in the skin, myocardium, lungs, kidneys and muscle.

Management of Hyperphosphataemia

Acute effects of hyperphosphatemia are due to hypocalcaemia and management is aimed at the correction of this. In chronic hyperphosphataemia, serum phosphate should be reduced to prevent soft tissue calcification.

Serum phosphate should be controlled in patients with renal failure to prevent hypocalcaemia, and secondary hyperparathyroidism. This can be achieved by oral phosphate binders such as aluminium hydroxide, magnesium carbonate, calcium acetate and lanthanum carbonate.

In patients at risk of developing acute hyperphosphataemia such as those with tumour lysis syndrome, intravenous saline at the time of chemotherapy will reduce the risk.

Further Reading

1. Amanzedeh J, Reilly Jr. RF. Hypophospahatemia: An evidence based approach to clinical consequences and management. *Nat Clin Pract Nephrol* 2006; 2:136–148.

2. Brunelli SM, Goldfarb S. Hypophosphataemia: clinical consequences and management. *J Am Soc Nephrol* 2007; 18:1999–2003.
3. Crook M, Swaminathan R. Disorders of plasma phosphate and indications for its measurement. *Ann Clin Biochem* 1996; 33:376–396.
4. Peppers MP, Geheb M, Desai T. Hypophosphataemia and hyperphosphataemia. *Crit Care Clin* 1991; 7:201–214.

Summary/Key Points

1. Phosphate is important for many physiological functions.
2. Phosphate homeostasis is mainly regulated by renal reabsorption.
3. Factors regulating renal phosphate reabsorption include PTH, Vitamin D, PTHrP and phosphatonins.
4. Hypophosphataemia is common in hospital populations and is caused by a shift of phosphate into cells or due to increased loss.
5. Clinical effects are seen only when hypophosphataemia is severe and prolonged. The effects include muscle weakness, haemolysis, and reduced white cell and platelet function.
6. Severe and prolonged hypophosphatemia should be corrected by oral or IV phosphate.
7. The most common cause of hyperphosphatemia is reduced excretion.
8. The main effect of hyperphosphatemia is hypocalcaemia.
9. The management of high phosphate involves the correction of the underlying disorder, correction of hypocalcaemia and reduction of phosphate absorption by oral phosphate binding agents.

chapter 9

Metabolic Bone Disease

Bone Biology

Bone is a highly metabolically active, organised tissue consisting of a collagenous organic matrix, osteoid, and a mineral phase consisting of hydrated calcium salts of hydroxyapatite and amorphous calcium phosphate crystals. Bone has two main functions, structural (it forms a rigid skeleton) and homeostatic (bone has a central role in calcium and phosphate homeostasis). Bone can be classified into dense, compact bone and cancellous bone. Compact or cortical bone comprises the outer casing of long bones or vertebra and has a low surface area to volume ratio. In an adult, 80% of the skeleton is cortical bone and because of its rigidity and density, cortical bone gives the skeleton shape- and weight-bearing properties. Cancellous bone is located internal to the cortical bone and at the end of long bones. It is composed of trabeculae separated by marrow space. Trabecular bone is characterised by high surface:volume ratio which is important for calcium and phosphate metabolism. Cells in the bone include osteocytes, osteoblasts, which are bone-forming cells, and osteoclasts, which are bone-resorbing cells. Osteocytes has important function as mechanosensory cells. Osteocyte also produces fibroblast growth factor-23 (FGF-23), which is a factor regulating phosphate excretion. These cells occupy about 3–4% of the total volume of bone. In adults, bone collagen fibres are arranged in parallel or concentric sheets, embedded in a gelatinous glycosoaminoglycan ground substance. Nearly 90% of bone matrix is type 1 collagen that is synthesised by osteoblasts. Collagen has a trimeric helical structure formed from two $\alpha_1(1)$ and one α_2 (1) chains. Collagen is synthesised as a precursor, procollagen, which has a propeptide at the carboxyl (C) and amino (N) terminal ends. Once secreted, the propeptides are cleaved and collagen molecules are crosslinked. Lysine and hydroxylysine crosslinks develop between collagen molecules to stabilise the structure. As the collagen matures, the crosslinks are converted to deoxypyridinoline and pyridinoline.

Bone undergoes continuous remodelling throughout life. This remodelling is necessary to maintain the structural integrity of the skeleton and to serve its metabolic function. The bone remodelling cycle involves a series of highly regulated steps that depend on the interaction of osteoclasts and osteoblasts (Figure 9.1). The initial activation state involves the interaction of osteoclasts and the osteoblast precursor cells. This leads to the differentiation, migration and fusion of the large multinuclear osteoclasts. These cells attach to the mineralised bone surface and initiate bone resorption. This produces irregular scalloped cavities on the trabecular bone surface. Once the osteoclasts have completed the removal of bone, the reversal phase begins. In this phase, mononuclear cells, which may be of the macrophage lineage, are seen on the bone surface and these may release growth factors to initiate the formation of bone. During the final bone formation stage, the cavity created by resorption is completely filled in by successive layers of osteoblasts, which differentiate from the mesenchymal precursors, and deposit a mineralised matrix. It has been estimated that as much as 25% of the trabecular bone and 3% of cortical bone is resorbed and replaced each year. The sequence of resorption and formation has been referred to as a basic multicellular unit of bone turnover and the process of bone resorption followed by equal amount of bone formation has been termed 'coupling'. The process of coupling and remodelling involves communication between various cell types in bone. The process of remodelling starts with the migration of osteoclasts, resorption followed by apoptosis of osteoclasts and bone formation by osteoblasts. This complex process involves communication between osteoclasts and stromal cells (pre-osteoblasts) cells. This communication system has now been identified as the OPG/RANK/RANKL

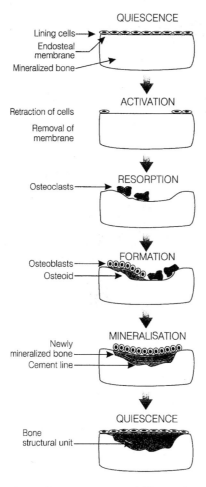

Figure 9.1   Bone remodelling cycle.

system (Figure 9.2). Receptor activator for nuclear factor κ B ligand (RANKL) is expressed on the surface of pre-osteoblasts/stromal cells and binds to RANK found on the surface of the osteoclastic precursor cells. This interaction is critical in the differentiation of osteoclasts. Osteoprotegerin (OPG) is a soluble decoy receptor for RANKL and blocks osteoclasts formation by inhibiting the binding of RANK to RNAKL (Figure 9.2). The effect of PTH, calcitriol, glucocorticoids

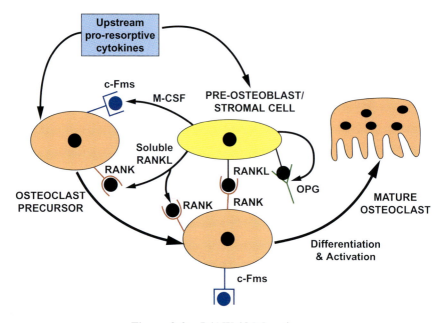

Figure 9.2   RANK/OPG pathway.

and oestrogens on bone is mediated via this pathway. Identification of the pathway has resulted in the development of novel therapeutic agents.

Bone remodelling is controlled by systemic hormones, as well as locally produced factors and by intracellular communication among the different bone cells (Table 9.1). Systemic factors include PTH and calcitriol. Calcitonin may play a minor role especially during skeletal development. Other systemic hormones, which may be important in regulating skeletal growth, are growth hormone via systemic and local IGF1 production, glucocorticoids that are necessary for bone cell differentiation, thyroid hormones and oestrogens. It has been suggested that androgens may also have a role in bone remodelling. Many cytokines and growth factors (Table 9.1) act locally as bone remodelling agents.

Abnormalities of bone remodelling can produce a variety of bone disorders and these abnormalities are summarised in the Table 9.2.

Table 9.1   Factors affecting bone remodelling

Systemic factors
- PTH
- 1,25(OH)$_2$ vitamin D
- Calcitonin
- Oestrogen
- Androgen
- GH/IGF1
- Thyroid hormone
- Glucocorticoids

Local factors
- Cytokines that cause bone loss — IL-1, IL-6
- Cytokines that prevent bone loss — IL-4
- Colony stimulating factors
- Prostaglandin, leukotrienes and nitric oxide
- Growth factors — IGF

Table 9.2   Abnormalities in bone remodelling in metabolic bone disease

	Bone resorption	Bone formation
Osteoporosis (type I)	↑↑	↑
Glucocorticoid-induced osteoporosis	↑	↓↓
Hyperparathyroidism	↑↑	↑↑
Hyperthyroidism	↑↑	↑↑
Paget's disease	↑↑	↑↑
Immobilization	↓	↓↓

Biochemical Markers of Bone Metabolism

Biochemical markers that reflect remodelling processes are: (1) enzymes or proteins that are secreted by cells involved in the remodelling process, (2) breakdown products generated during resorption of bone, and (3) by-products produced during synthesis of new bone. Table 9.3 shows the markers of bone formation and bone resorption.

Table 9.3   Markers of bone remodelling

Bone resorption markers
- Tartrate-resistant acid phosphatase
- Hydroxyproline
- Pyridinium crosslinks
  — pyridinoline
  — deoxypyridinoline
- Hydroxylysine
- Carboxyl terminal cross-linked telopeptides of collagen type I — CTx
- Amino terminal cross-linked telopeptides of collagen type — NTx

Bone formation markers
- Bone specific alkaline phosphatase
- Osteocalcin
- Carboxyl terminal propeptide of collagen type I (PICP)
- Amino terminal propeptide of collagen type I (PINP)

Markers of Bone Resorption

Acid phosphatase

Osteoclasts produce acid phosphatase, which is tartrate resistant and can be measured by a kinetic method or by immunoassays.

Urinary hydroxyproline and hydroxylysine

Urinary hydroxyproline, and hydroxylysine are breakdown products of collagen. Hydroxyproline is largely replaced by more sensitive and specific assays. These two compounds are produced *in situ* during collagen synthesis by hydroxylation of proline and lysine. Once collagen breaks down, hydroxyproline and hydroxylysine are not reutilised.

Pyridinium crosslinks

Posttranslational modification of lysine to hydroxylysine produces the non-reducible pyridinium crosslinks, pyridinoline and deoxypyridinoline

(DPD), that stabilises mature collagen. Both pyridinoline and DPD are released from bone during resorption. Of these, DPD is more specific for bone. Sixty percent of the crosslinks are released in protein or peptide bound form while the remainder is free. These crosslinks are not metabolised and are excreted in the urine. Total or free crosslinks can be measured by an HPLC method or immunoassay.

Crosslink telopeptides

In the process of bone resorption, amino and carboxyl terminal fragments of collagen are released with crosslinks attached. These fragments with attached crosslinks are called telopeptides — N telopeptide (NTx) and C telopeptide (CTx) — these are excreted in the urine. These can be measured in the urine or serum by immunoassays.

Other bone resorption markers are serum cathepsin K, RANK, RANL-L and OPG. The utility of these markers are still under evaluation.

Bone Formation Markers

Bone specific alkaline phosphatase

Alkaline phosphatase is present in osteoblasts and is released during bone formation. Alkaline phosphatase has several isoforms and the bone and liver isoforms are products of the same gene. The differences between these isoforms are in posttranslational modification; therefore, they are difficult to differentiate. Immunoassay methods are now available for bone alkaline phosphatase. Measurements of total alkaline phosphatase may reflect bone metabolism, but the changes will be masked by changes in other isoforms.

Osteocalcin

Osteocalcin, which is synthesised by osteoblasts, odontoblasts and some chondrocytes, is a good marker of bone formation. It is a small (49 amino acid) protein that is rich in glutamic acid. Osteocalcin binds to hydroxyapatite and gets deposited in bone matrix; the exact function of osteocalcin is not known. During the synthesis of bone matrix, some osteocalcin is released into the circulation. The half-life of osteocalcin in circulation is short and is cleared by the kidney. Serum osteocalcin can be measured by immunoassays. However, osteocalcin is labile in blood and fragments can be released during storage due to degradation by protelytic enzymes from red cells. Variability between assays due to the cross reaction of fragments can give rise to different results on the same sample. Antibodies that recognise both the intact molecule and the large N-terminal molecule appear to provide the best clinical information. During bone resorption, fragments of osteocalcin are released into the circulation and the concentration of these fragments reflect bone resorption.

Procollagen peptides

By-products of type I collagen synthesis are the amino and carboxyl terminal procollagen I extension peptides (PINP and PICP) which have molecular weights of 35 and 1000 kD, respectively. These peptides are cleared by liver and can be measured by immunoassays. Of these, PINP appears to be more useful in assessing bone formation.

Clinical Value of Bone Markers (Table 9.4)

Bone remodelling shows a diurnal rhythm, which is reflected in the biochemical markers of bone turnover, especially DPD and osteocalcin. The day-to-day variation of some of the markers can be as high as 20%. Furthermore, some of the markers, e.g. bone alkaline phosphatase, are only relatively specific for bone. Metabolism and

Table 9.4   Clinical usefulness of bone markers

Study the pathogenesis
Monitor treatment
Undertake short-term studies
Select method of treatment

clearance of the markers can also influence their concentration. Renal function can affect the concentration of osteocalcin, NTx, CTx and DPD, and liver clearance affects bone alkaline phosphatase.

Bone markers cannot be used to diagnose osteoporosis or to identify those with low bone mass. However, they may be useful in identifying those who are losing bone at a faster rate. Bone markers are useful in monitoring treatment and in helping to select the correct therapeutic agent. As changes in bone density caused by treatment take at least 12 months to be reliably measured by densitometry, changes in bone markers can be used to detect effectiveness of treatment within 2–3 months. The advantages of bone markers are that they are non-invasive, inexpensive and can be repeated.

Rickets and Osteomalacia

Osteomalacia is characterised by the impairment of mineralisation, leading to accumulation of unmineralised matrix or osteoid. When this occurs during childhood or infancy, while the bones are growing, it results in rickets with the widening of the epiphysis and impaired skeletal growth. Classical clinical features of osteomalacia include musculoskeletal pain, skeletal deformity, muscle weakness and hypocalcaemia. In osteomalacia, several distinctive radiological appearances such as the loser zone or pseudofracture, which consists of a large area of osteoid, are seen. In rickets, there is widening and irregularity of the metaphysis and skeletal deformity.

Causes of osteomalacia (rickets) and their mechanisms are listed in Table 9.5. The most common cause of osteomalacia and rickets is vitamin D deficiency. Vitamin D deficiency commonly occurs due to reduced cutaneous production of vitamin D due to the lack of exposure to sunlight. Elderly and Asian immigrants living in Europe are at risk of developing vitamin D deficiency. Vitamin D deficiency may also occur in malabsorption syndromes due to reduced absorptions of vitamin D, calcium and phosphate. Anticonvulsant treatment and hepatic disease may be associated with low plasma calcidiol and osteomalacia.

In vitamin D deficiency, serum calcium, phosphate and calcidiol are low, while serum PTH and ALP are high. Management of

Table 9.5   Causes of osteomalacia/rickets

Vitamin D deficiency
- Lack of dietary intake
- Lack of exposure to sunlight — elderly and Asian immigrants
- Malabsorption syndrome

Disorders of vitamin D metabolism
- Renal failure — reduced renal mass
  — hyperphosphataemia inhibits 1α-hydroxylase
- Anticonvulsant treatment — hepatic microsomal enzyme induction causing increased metabolism of vitamin D
- Hepatic disease
- Vitamin D-dependent rickets type I — lack of 1α-hydroxylase enzyme — an inherited disorder

Resistance to vitamin D
- Vitamin D-dependent rickets type II — a defect in the vitamin D receptor complex

Hypophosphataemia
- Severe phosphate depletion
- X-linked hypophosphataemia
- Oncogenic osteomalacia

Calcium deficiency

Drugs
- Bisphosphonates
- Aluminium
- Fluoride

osteomalacia and rickets involves adequate replacement of vitamin D, calcium and phosphate in the diet.

Osteoporosis

Osteoporosis is the most common metabolic bone disease and is characterised by low bone mass. Osteoporosis is defined as a disease characterised by low bone mass and microarchitectural deterioration of bone tissue leading to enhanced bone fragility and a consequent increase in fracture risk. The operational definition of osteoporosis is a bone mineral density that is 2.5 SD below the mean peak value in young adults. Osteoporosis can be divided into two major categories, primary and secondary. Primary osteoporosis can be further subdivided into type I and type II. Type I is seen in postmenopausal women while type II, sometimes called senile osteoporosis, is seen in both men and women. Secondary osteoporosis occurs when the decrease in bone density is due to other diseases (Table 9.6). One of the most important causes of secondary osteoporosis is glucocorticoid-induced osteoporosis.

Pathophysiology of Osteoporosis

Osteoporosis results when the bone reabsorption exceeds bone formation. An increase in bone turnover when uncoupled can lead to more rapid bone loss. The age-dependent bone loss (senile osteoporosis) is thought to result predominantly from impaired bone formation. The oestrogen deficiency related bone loss (type I

Table 9.6   Causes of secondary osteoporosis

Anorexia nervosa
Alcohol
Drug-induced osteoporosis
  • Glucocorticoids
  • Heparin
  • Warfarin
Hypogonadism
Exercise-induced amenorrhoea
Hyperprolactinaemia
Immobilisation
Coeliac disease
Crohn's disease
Malabsorption states
Myeloma
Organ transplantation
Hyperthyroidism and thyroxine replacement therapy
Rheumatoid arthritis

or postmenopausal) is associated with a higher rate of bone turnover, where the increase in bone resorption is relatively greater than the increase in bone formation. Trabecular bone is more sensitive to high turnover bone loss than cortical bone, while age-related loss affects both trabecular and cortical bone. Bone mineral density (BMD) at a given age is a result of the maximum BMD achieved during life and the rate of bone loss that occurs with ageing. BMD increases in childhood and early adult life, reaching the maximum during the second or early third decade in men and women; it then remains relatively stable until about the age of 40 to 45. Thereafter, there is an age-related decline in BMD that continues into extreme old age. In women, following menopause, there is a superimposed period of rapid bone loss predominantly due to oestrogen deficiency, lasting for 5–10 years (Figure 9.3). Up to 80% of the variability in peak bone mass between people is determined by genetic factors. Other factors which determine the peak bone mass and the rate of loss of bone during life are given in the Table 9.7.

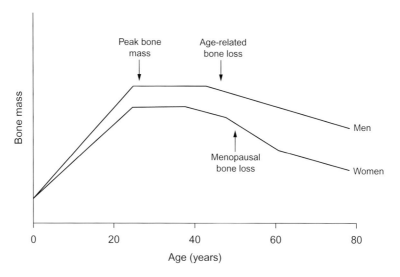

Figure 9.3   Changes in bone mineral density with age.

Table 9.7   Factors contributing to bone loss and peak bone mass

---

Genetic factors
Sex hormones
- Oestrogens
- Testosterone
- DHEA

Environmental and lifestyle factors
- Calcium intake
- Exercise
- Vitamin D
- Alcohol
- Sodium intake
- Protein intake
- Caffeine
- Cigarette smoking
- Drugs treatment
    — Thiazide diuretics
    — Glucocorticoids
    — Thyroid hormones
    — Anticonvulsants
    — Aluminium-containing antacids
- Medical conditions
    — Hyperthyroidism
    — Hyperparathyroidism
    — Cushing's syndrome
    — Diabetes
    — Rheumatoid arthritis

Epidemiology

Osteoporosis is an increasing health problem and leads to increased mortality and morbidity. The prevalence of osteoporosis increases rapidly with age and it is estimated that up to 50% of post-menopausal women will have low bone mass and 20–30% will have osteoporosis. Prevalence of osteoporosis ranges from < 15% in women aged 50–59 to 70% in women over 80 years of age. Osteoporosis is more common in Caucasian women compared to Asians or African Americans. One estimate is that over half of all women and up to one third of men will experience osteoporotic fractures during their lifetime. The problem is further compounded by the ageing population. Osteoporosis increases the risk of fractures, which is associated with a mortality of 20%. The economic cost of osteoporosis related fractures in the United States has been estimated to be over $6 billion.

Case 9.1

A 45-year-old lady was seen by a neurologist for her weak limbs. Investigation showed she had myasthenia gravis and was treated with prednisolone. Two years later, she presented to the metabolic bone clinic with back pain. Radiological examinations showed she had a crush fracture of her lower thoracic vertebrae. Biochemical investigations were all normal. Dual energy X-ray absorptiometry (DEXA) showed that her bone mineral density was 3.0 standard deviations below that of a young person (T score of –3.0), confirming the diagnosis of osteoporosis. Use of long-term steroids was the main contributing factor in this patient. She was treated with calcium, vitamin D and bisphosphonate to prevent further fractures.

Diagnosis

The diagnosis of osteoporosis depends on the measurement of bone mineral density (BMD) and the most reliable technique for measuring BMD is dual energy X-ray absorptiometry (DEXA). In osteoporosis, there are very few changes in the plasma concentration of calcium, phosphate or total alkaline phosphatase. The role of biochemical markers has been discussed earlier. As the economic consequence of osteoporosis is enormous, osteoporosis should be detected as early as possible. However, at the moment, there is no consensus as to the best way of screening the population.

Management

The management of osteoporosis consists of the removal of any associated risk factors, treatment with calcium and vitamin D supplements and one of the antiresorptive agents. In established osteoporosis, treatment with bisphosphonates will reduce the rate of bone loss. Bisphosphonates, which are stable analogues of pyrophosphate, are poorly absorbed and they bind to bone mineral, thereby reducing resorption. Calcitonin also has been used to reduce bone resorption. Other possible modes of treatment include treatment with raloxifene, which is a selective oestrogen receptor modulator. Strontium ranelate and teriparatide (recombinant human PTH 1-34) are new drugs which improve bone formation. Recently, monoclonal antibodies that block the action of RANKL (denosumab) have been successfully tried with good responses and is now available. Monitoring of treatment can be done either bio-chemically or by measurements of BMD at regular intervals.

Paget's Disease of Bone

Paget's disease of bone is a focal disorder of bone remodelling due to abnormally increased osteoclast-mediated bone resorption. It is rarely seen before the age of 35 and the prevalence increases with age, affecting 2–5% of the population above the age of 50. There is a wide geographical variation in the incidence of the disease. The disease is more common in Central and Western Europe particularly in the North of England, USA, Australia, New Zealand while it is rarely found in Scandinavia and Asia. The aetiology of Paget's disease is not known. Infection of genetically predisposed individuals with a paramyxovirus has been suggested as a possible causative agent. There is accelerated bone turnover, normal bone architecture is destroyed, and bone is laid down randomly, resulting in abnormal bone struc-ture. The affected bones change in shape, size and direction causing considerable morbidity, yet the majority of patients are asymptomatic. In Paget's disease, increased bone resorption is tightly linked to increased bone formation. This is reflected in the proportional increase in markers of bone formation and resorption. The most com-monly affected bones are the pelvis, spine, femur and the skull. As the majority of the patients are asymptomatic, the disease may be diagnosed incidentally during radiological examination or when finding unexplained increases in serum alkaline phosphatase activity. When symptoms are present, pain is the presenting complaint. Deformities affecting the weight-bearing bones, mainly the tibia and femur, may be present in 15% of patients at the time of diagnosis; the most frequent deformity being bowing of the legs. In addition, characteristic deformities and enlargement of the affected bones may be seen in the skull and jaw. Fracture as a presenting feature is seen in about 9% of patients and these occur more commonly in long bones. A rare complication of Paget's disease is malignant transformation and development of osteosarcoma. Neurological complications include irreversible hearing loss, spinal or brain stem compression and rarely hydrocephalus. When the disease is localised to the lower limbs, it can cause difficulty in walking. Paget's disease of the skull may give headaches.

Diagnosis depends on characteristic radiological appearance and increased biochemical markers of bone turnover. There is a proportional increase in indices of bone resorption and indices of bone formation. Despite these large changes in bone turnover, plasma concentrations of calcium and phosphate are normal. However, hypercalcaemia may develop in patients who are immobilised.

Treatment involves antiresorptive agents such as bisphosphonates and calcitonin. Less frequently used treatments include mithramycin and gallium nitrate. With adequate treatment, skeletal morbidity is reduced and complications can be prevented and this may lead to long-term clinical and biochemical remission.

Renal Osteodystrophy

Renal osteodystrophy is a bone disease complicating chronic kidney disease (CKD). As the bone disease is associated with systemic vascular disease, the term chronic kidney disease–mineral and bone disorder is preferred now. It is characterised by (a) abnormalities of calcium, phosphorus, parathyroid hormone, or vitamin D metabolism, (b) abnormalities in bone turnover, mineralisation, volume, linear growth, or strength (c) vascular or other soft-tissue calcification. Renal osteodystrophy can be classified on the basis of bone turnover (high, normal, or low), mineralisation (normal or abnormal) and bone volume (high, normal, or low). Abnormalities in bone range from osteomalacia, to osteitis fibrosa, and from adynamic bone (low bone turnover) to high bone turnover. The aetiology is complex. As discussed in Chapter 5, CKD is associated with hyper-phosphataemia, hypocalcaemia, secondary hyperparathyroidism and reduced production of calcitriol. Aluminium from dialysis fluid or from phosphate binders accumulate in the bone and inhibit bone mineralisation.

Treatment involves controlling secondary hyperparathyroidism by the administration of calcitriol, phosphate binders and the use of calcimimetic drugs such as cinacalcet. Parathyroidectomy may be required if the hyperparathyroidism is severe.

Further Reading

1. Khosla S. The OPG/RANKL/RANK system. *Endocrinology* 2001; 14:5050–5055.
2. Looker AC, Bauer DC, Chesnut CH *et al*. Clinical use of biochemical markers of bone remodeling: Current status and future directions. *Osteoporos Int* 2000; 11:467–480.
3. Raisz LG. Physiology and pathophysiology of bone remodelling. *Clin Chem* 1999; 45:1353–1358.
4. Vasikaran SD. Utility of biochemical markers of bone turnover and bone mineral density in management of osteoporosis. *Crit Rev Clin Lab Sci* 2008; 45:221–258.
5. Walths NB. Clinical utility of biochemical markers of bone remodelling. *Clin Chem* 1999; 45:1359–1368.

Summary/Key Points

1. Bone is an active tissue, which undergoes constant reabsorption and formation. This remodelling cycle is mediated by osteoblasts and osteoclasts. The RANK/RANKL/OPG pathway is involved in the regulation of this cycle.

2. Bone turnover can be assessed biochemically by bone formation markers (bone ALP, osteocalcin, and PINP) and bone resorption markers (tartrate-resistant acid phosphatase, pyridinium cross links, NTx and CTx). These markers, however, show marked diurnal variation and many are influenced by food intake.

3. Bone turnover markers are useful in following the treatment of metabolic bone diseases.

4. Osteomalacia (or rickets) is associated with defects in bone mineralisation and is commonly caused by low vitamin D status.

5. Osteoporosis is a major health problem worldwide. It is defined as bone mineral density, which is 2.5 standard deviations below the peak bone mass in a young person. The most common form of osteoporosis, primary osteoporosis, is due to age-related loss of bone together with accelerated bone loss during the early post-menopausal period.

6. Treatment with antiresorptive agent (bisphosphonates), strontium ranelate or teriparatide (PTH) will improve bone density and reduce the risk of fractures.

7. Paget's disease is a disorder of bone remodelling of unknown aetiology. Many patients are asymptomatic and discovered on routine radiology or blood tests (raised ALP).

8. Patients with Paget's disease may present with bone pain, deformity or fractures and are treated with bisphosphonates.

9. Renal osteodystrophy is multifactorial in origin. Abnormalities in PTH, calcitriol, phosphate and aluminium contribute to this disorder. Control of serum phosphate and the use of calcitriol and calcimimetics will help in the management of this disorder.

chapter 10

Carbohydrate Metabolism and Its Disorders

Introduction

Glucose and free fatty acids are important metabolic fuels to meet the energy requirements of the body. Under normal circumstances, some tissues such as the brain and red cells use only glucose. The body has a limited supply of carbohydrates, which is stored as glycogen in the liver (approximately 75 g) and in the muscles (approximately 400 g). Most of the energy in the body is stored as fat. An average 70-kg man has 15 kg of triglycerides stored in the adipose tissue. This is equivalent to 60–90 days of energy requirements. As the carbohydrate stores are limited, glucose has to be synthesised during fasting from other sources such as amino acids, glycerol and lactate. This takes place mainly in the liver and (to some extent) in the kidneys.

Normal Fuel Homeostasis

Postprandial (absorptive) State

Following a meal, carbohydrates in the food is digested to glucose and other monosaccharides. Some of the monosaccharides such as fructose are converted to glucose. Glucose and other monosaccharides are absorbed and reach the liver where it is either oxidised or converted to glycogen and fatty acids. Postprandially, glucose is the predominant fuel used by skeletal muscle. The glucose taken up by the muscle is also used for glycogen synthesis. Fat is absorbed as chylomicrons and they enter the circulation through the thoracic duct. The triglycerides in chylomicrons are hydrolysed by lipoprotein lipase, an enzyme present in the peripheral endothelium to free fatty acids and glycerol. The free fatty acids are taken up by the adipose tissue and are converted to triglycerides and stored. The fatty acids are also used as fuel in muscle. Protein in the diet is digested to amino acids and absorbed. Uptake of amino acids and protein synthesis in the liver, muscle and adipose tissue are stimulated by insulin.

Post-absorptive State

In the fasting state, the liver gradually becomes a glucose-producing organ from a glucose-consuming organ. There is a decrease in glycogen synthesis and an increase in glycogen breakdown, glycogenolysis. In the post-absorptive state, most of the glucose produced by liver is taken up by insulin-independent tissues such as brain and red cells, brain takes up 50% while the red blood cells take up 20%. Insulin-dependent tissues such as the muscle and adipose tissue account for only 20% of total glucose utilization. During the first 12 hours of starvation, up to 75% of glucose is derived from glycogen. As starvation progresses, more and more glucose is produced by gluconeogenesis. Glucogenic substrates include lactate, produced from glycolysis (Cori cycle), glycerol produced from the breakdown of triglycerides in adipose tissue and amino acids, mainly alanine, produced from muscle protein (glucose–alanine cycle). In the post-absorptive state, most of the body's energy requirement is met by fatty acids, which are released from triglycerides in the adipose tissue by hormone sensitive lipase, which is stimulated by glucagon. As starvation progresses, the free fatty acid concentration increases. Glycerol released from triglycerides is transported to the liver and converted to glucose.

Prolonged Starvation

As starvation progresses, large amounts of acetyl CoA are produced from β-oxidation of fatty acids, from which ketone bodies are produced. Serum ketones increase and they are oxidised in the muscles. Muscle releases amino acids, primarily alanine and glutamine, which are used for gluconeogenesis. As starvation progresses, the kidney contributes more to gluconeogenesis. Most of the energy requirement of the body during prolonged starvation is met by free fatty acids. As the concentration of the ketone bodies increases, the brain adapts to using them as fuel, thereby reducing the need for glucose and hence protein breakdown.

Glucose Homeostasis

Blood glucose concentration is maintained within fairly narrow limits throughout the day. Even in extreme starvation, it does not fall below 3 mmol/L and in the postprandial state, it seldom reaches more than 7.0 mmol/L. Plasma glucose concentration is a balance between intake and endogenous production (absorption from the gut, glycogenolysis and gluconeogenesis) on one hand and tissue utilisation (glycogen synthesis, tissue oxidation and lipogenesis) on the other. These processes are controlled by two sets of hormones: anabolic hormones, primarily insulin and insulin-like growth factor, and catabolic hormones or counterregulatory hormones such as glucagon, catecholamines, cortisol and growth hormone. Insulin, which is secreted from the pancreas in response to an increase in plasma glucose, promotes glucose uptake by tissues, thereby decreasing plasma glucose concentration. Counterregulatory hormones stimulate the release of glucose from glycogen and stimulate gluconeogenesis.

Insulin

Insulin is released by the β cells of islets of Langerhans of the pancreas. It is a 51 amino acid peptide of molecular weight 6000, contains two chains (A and B), joined by two disulphide bridges with a third disulphide bridge within the A chain (Figure 10.1). Insulin is synthesised as a preprohormone of 100 amino acids and is converted rapidly into proinsulin, an 86 amino acid peptide, which is stored in secretory granules of the β cells. In the secretory

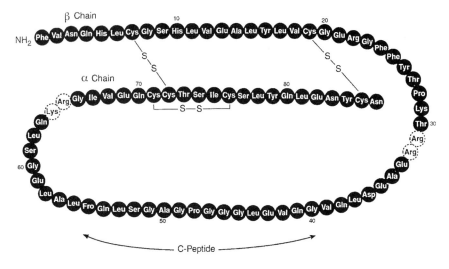

Figure 10.1   Structure of proinsulin.

granules, proinsulin is cleaved by two calcium regulated endopeptidases, prohormone convertases I and II (PCI and PCII) into insulin and C-peptide (Figure 10.1). In healthy individuals, insulin is secreted in a pulsatile fashion and the stimuli for secretion are glucose, amino acids, pancreatic and gastrointestinal hormones [glucagon-like polypeptide (GLP-1), gastric inhibitory peptide or glucose dependent insulinotrophic polypeptide (GIP)]. Some drugs such as sulphonylureas and β-adrenergic agonists will also stimulate insulin secretion. Insulin release is inhibited by hypoglycaemia, somatostatin and drugs such as diazoxide, phenytoin, phenothiazines, nicotinic acid, α-adrenergic agonists and β-adrenergic blockers.

The stimulation of insulin secretion by glucose is greater when it is given orally compared to an intravenous injection due to the secretion of incretins, gastrointestinal hormones particularly GLP-1 and GIP in response to a meal. GLP-1 and GIP are degraded by dipeptidyl dipeptidase-4 (DDP-4). The liver extracts and degrades approximately 50% of the insulin released into the circulation. The rest is filtered in the kidneys, reabsorbed and degraded in the proximal tubule. The half-life of insulin in the circulation is between 5 and 10 minutes.

Proinsulin, which is the major storage form of insulin, has relatively low biological activity. It is released into the circulation in small amounts and in the fasting state, proinsulin accounts for 10–15% of total insulin concentration. Hepatic clearance of proinsulin is lower than that of insulin and the half-life of proinsulin is 2–3 times longer than that of insulin.

When proinsulin is cleaved, C-peptide, which has no biological activity, is released together with insulin. C-peptide has a much longer half-life of about 35 minutes, therefore the fasting concentration of C-peptide is 5–10-folds higher than that of insulin. C-peptide is mainly removed by the kidneys, where it is filtered and degraded.

Mechanism of action of insulin

Insulin binds to specific receptors in the plasma membrane. The human insulin receptor is a heterotetramer comprising of two α- and two β-subunits. The α-subunit is located on the outer surface of the plasma membrane and contains the insulin binding site. The β-subunit extends intracellularly and contains an intrinsic tyrosine kinase. Binding of insulin to the α-subunit activates the tyrosine kinase that phosphorylates tyrosine residues and proteins, including autophosphorylation of the β-subunit, phosphorylation of intracellular proteins such as the insulin receptor substrate I and initiation of the phosphorylation cascade.

Glucose transport

Two families of proteins mediate the transport of glucose into cells. The sodium–glucose transporter present in the small intestine and the kidney promotes the uptake of glucose and galactose. The second family of transporters known as facilitative glucose transporters (GLUT) are located on the cell surface.

- GLUT 1 is widely expressed in all cells including foetal cells and provides many cells with the basal glucose requirement.
- GLUT 2 is present in hepatocytes, β cells of the pancreas, small intestine and basolateral membrane of the kidney. This is a low affinity high capacity transport system allowing non-rate limiting movement of glucose into and out of the cells.
- GLUT 3 is found mainly in the neurons, but can also be found in the placenta and testes, and provides a constant high level of glucose required by neurons.
- GLUT 4, present in muscle and adipose tissue, is responsible for insulin-stimulated glucose transport. When insulin levels are low, these transporters are localised in intracellular compartments and is inactive. Insulin stimulates the translocation of these to the plasma membrane, thereby promoting glucose uptake.
- GLUT 7 allows the diffusion of glucose out of the endoplasmic reticulum in the liver.

Actions of insulin

Insulin promotes an anabolic state, i.e. the storage of carbohydrates and lipids and protein synthesis. The main target organs for insulin are the liver, skeletal muscle and adipose tissue. These organs show differences in their responses to insulin. In the liver, insulin stimulates glycolysis and glycogen synthesis, suppresses lipolysis and promotes lipogenesis. In the peripheral tissues, insulin induces lipoprotein lipase, thus facilitating the uptake of fat into adipose tissue and muscle. It stimulates triglyceride synthesis from glycerol and fatty acids in the adipose tissue. In the muscle, insulin increases glucose transport, glucose metabolism and glycogen synthesis. It also increases the uptake of amino acids and stimulates protein synthesis.

Insulin-like Growth Factors

Insulin-like growth factors 1 and 2 (IGF1 and IGF2), are polypeptides, which are structurally related to insulin and have similar metabolic effects to that of insulin. IGF1 is produced predominantly in the liver and is growth hormone dependent. However, many cells produce IGF1 locally, which does not enter the circulation. IGF1 is a major regulator of cell growth and differentiation. The concentration of IGF1 in circulation is approximately 1000-fold higher than that of insulin. However, majority of IGF1 is bound to IGF binding proteins and < 10% of IGF1 is free. These binding proteins regulate the delivery of IGF1 to the target tissues. The action of IGF1 is mediated by specific IGFI receptors; IGF1 binds to insulin receptors as well but with less affinity. Certain neoplasms produce IGF2 which causes hypoglycaemia (see later). IGF2 functions in an autocrine fashion as a growth-promoting factor during gestation.

Glucagon

Glucagon is a 29 amino acid peptide secreted by the α cells of the pancreas. Glucagon acts mainly in the liver where it binds to specific receptors, stimulating cyclic AMP production and intracellular calcium. The main actions of glucagon are the stimulation of glucose production by increased glycogenolysis, gluconeogenesis and ketogenesis. To a much lesser extent, lipolysis is stimulated in the adipose tissue. Glucagon secretion is stimulated primarily by low plasma glucose concentration. Stress, exercise, amino acids and catecholamines may also increase while insulin inhibits glucagon release.

Other hormones, which affect glucose homeostasis, are adrenaline, cortisol, growth hormone and somatostatin. Cortisol stimulates gluconeogenesis and increases the breakdown of protein and fat. Growth hormone stimulates gluconeogenesis, lipolysis and antagonises the action of insulin. Adrenaline that is stimulated by hypoglycaemia and stress increases glycogenolysis.

Diabetes Mellitus

Diabetes mellitus is a heterogeneous group of disorders of carbohydrate metabolism characterised by hyperglycaemia due to defects in insulin secretion, insulin action or both. Patients with diabetes mellitus may develop acute complications such as ketoacidosis or hyperosmolar coma and chronic complications such as retinopathy, renal failure, neuropathy and atherosclerosis. Prevalence of diabetes is not fully known but it affects a large percentage of the population. The prevalence increases with age and is higher in certain populations. One estimate is that by the age of 65, 25% of the African American population of the United States and 17% of the Caucasians will have diabetes mellitus. The total economic cost of diabetes mellitus in the United States has been estimated to be over 92 billion dollars and it causes an estimated 150,000 deaths annually in the United States.

Classification of Diabetes (Table 10.1)

Two major types are recognised; type 1, insulin dependent and type 2, insulin independent.

Type 1 Diabetes Mellitus

In type 1 diabetes, there is deficiency of insulin because of the loss of pancreatic cells. These patients are dependent on insulin to sustain life and to prevent ketosis. Approximately 5–10% of the total diabetic

Table 10.1   Classification of diabetes mellitus

Type 1 diabetes — β cell destruction
Type 2 diabetes (aetiology unknown — varying degree of insulin
  resistance and insulin secretory deficit)
Other specific types
  • Genetic defects of β cell function — maturity onset diabetes
    of youth (MODY)
  • Genetic defects in insulin action
  • Diseases of exocrine pancreas
  • Endocrinopathies
  • Drug or chemical induced
  • Infections
Gestational diabetes mellitus

population belong to this category. Type 1 diabetes mellitus results from a cell-mediated autoimmune destruction of the pancreatic β cells. The aetiology of type 1 diabetes mellitus is not fully understood. Genetic as well as environmental factors play a role. Twin studies show that the concordant rate between twins is approximately 30%. The genetic susceptibility to type 1 diabetes mellitus is strongly associated with HLA-DQA, DRB and DQB. Other genetic factors such as the insulin gene and cytotoxic T-lymphocyte antigen may indicate the risk of diabetes. Several studies also support the view that environmental factors are important. Infections *in utero* may contribute to the initiation of β cell autoimmunity and later infections may accelerate this. It has been suggested that a combination of susceptibility genes and environmental factors initiate an autoimmune response to the insulin-producing cells. These patients are also more likely to develop other autoimmune diseases such as Addison's disease, Grave's disease, vitiligo and myasthenia gravis. The autoimmune reaction is reflected by the presence of antibodies against glutamic acid decarboxylase (GAD65), insulin and islet cell antigen 2 — a tyrosine phosphatase-like protein (IA-2 and IA-2β). The diagnostic sensitivity of these three antibodies varies from 40–80% with a specificity of 99%. The presence of these antibodies in first-degree relatives predicts the development of disease. Combined genetic and antibody testing may improve the prediction of development of type 1 diabetes mellitus in the general population.

In a small proportion of individuals, there is no evidence of autoimmunity and there is strong genetic tendency. This is sometimes termed idiopathic diabetes.

Type 2 Diabetes Mellitus

Type 2 diabetes is a heterogeneous disorder associated with insulin resistance together with either a relative or absolute insulin deficiency; it does not generally require insulin treatment for survival. It accounts for 90–95% of all diabetes. Most patients are obese. The development of type 2 diabetes requires both genetic as well as environmental factors. It has been suggested that type 2 diabetes mellitus is a consequence of a thrifty genotype, which evolved when food supplies were scarce and physical activity was high. In modern society, this thrifty genotype is a disadvantage leading to obesity, insulin resistance and type 2 diabetes. Another hypothesis is that insulin resistance in type 2 diabetes mellitus is a consequence of foetal malnutrition. The contribution of genetic factors is shown by the almost 100% concordance of type 2 in identical twins and the fact that type 2 diabetes mellitus is 10 times more likely to develop in an obese person with a diabetic parent than an equally obese person without a diabetic family history. However, the mode of inheritance is unknown and the genes responsible have not been identified. This form of diabetes can be present for several years before diagnosis and many of these undiagnosed subjects already have evidence of diabetic complications. Type 2 diabetes develops more frequently in women who have had gestational diabetes.

Prevalence of type 2 diabetes

Prevalence of type 2 diabetes varies widely from 50% or more in some populations, e.g. Pima Indians and Polynesians, to less than 3% in other populations, e.g. rural China. It is generally more prevalent in industrialised countries and in the emerging middle classes of developing countries. Recent estimates suggest that there will be 150 million diabetics worldwide and this figure is expected to double by 2025. Type 2 diabetes is also highly prevalent in South Asians, especially those who have moved to the West. It is predicted that type 2 diabetes will be a major public health problem especially in evolving industrialised societies. Type 2 diabetes is increasingly diagnosed in children and the trend parallels the increasing incidence of obesity in children.

Insulin resistance

In the majority of patients with type 2 diabetes, insulin resistance is present and precedes the development of diabetes. Insulin resistance is related to obesity, particularly to central or abdominal obesity. Genetic abnormalities may also contribute to this. It has been postulated that the direct release of free fatty acids and/or other products from visceral adipose tissue into the portal circulation and the liver may be an important contributor to the development of insulin resistance. Longitudinal studies have shown that insulin resistance leads to impaired glucose tolerance. Although the majority of obese individuals are insulin resistant, only a fraction progress to type 2 diabetes. This progression is related to the inability of the β cell to compensate for the insulin resistance with appropriate hyperinsulinaemia.

Metabolic syndrome

The metabolic syndrome is a cluster of factors: insulin resistance, hyperinsulinaemia, low HDL cholesterol and visceral obesity. Subjects with metabolic syndrome are more prone to get cardiovascular disease and type 2 diabetes. Prevalence of this syndrome is around 20% and it is rising with increasing incidences of obesity. This syndrome is also called syndrome X or Raven's syndrome. The definition of this syndrome is not fully agreed by international bodies. There are several versions and the one recommended by the International Diabetes Federation (IDF) includes central obesity and any two of the following: raised triglycerides, low HDL cholesterol, hypertension, and raised plasma glucose.

Maturity Onset Diabetes of Youth (MODY)

This subtype of diabetes is an inherited disorder, an autosomal dominant form of inheritance, and characterised by the onset of diabetes before the age of 25 and with normal insulin sensitivity. Abnormalities in six genes have been identified in these patients and the most common mutation is in the gene for hepatic transcription factor or hepatic nuclear factor (HNF-1α). This type of diabetes accounts for only a small proportion of diabetic subjects.

Other Forms of Diabetes

In a small group of subjects, diabetes mellitus may develop due to diseases of the pancreas (e.g. pancreatitis or malignancy), diseases of the endocrine glands (Cushing's syndrome, acromegaly, etc.), drugs or chemicals (glucocorticoids, α-interferon, etc).

Impaired Glucose Tolerance

Impaired glucose tolerance (IGT) is diagnosed on the basis of an oral glucose tolerance test when the values are intermediate between normal and diabetes. IGT is associated with increased risk of atherosclerosis and subjects with IGT progress to diabetes mellitus at the rate of 1–5% per year.

Diagnosis of Diabetes Mellitus

The diagnosis of diabetes depends on the demonstration of hyperglycaemia. The criteria for diagnoses of diabetes mellitus are listed in Table 10.2. In symptomatic patients, a random venous plasma glucose ≥ 11.1 mmol/L or fasting venous plasma glucose ≥ 7.0 mmol/L is adequate for the diagnosis. In patients without symptoms, diagnosis requires confirmation by at least one additional raised glucose on another day. If the glucose values (either random or fasting) are equivocal, an oral glucose tolerance test may be required. The

Table 10.2   Criteria for the diagnosis of diabetes mellitus

Symptoms of diabetes plus
- Random plasma glucose ≥ 11.1 mmol/L or
- Fasting plasma glucose ≥ 7.0 mmol/L or
- Plasma glucose ≥ 11.1 mmol/L at 2 hours after an OGTT

Impaired fasting glucose
- Fasting plasma glucose between 5.6 and 6.9 mmol/L

(a risk factor for future development of diabetes and cardiovascular disease)

Impaired glucose tolerance (both criteria must be met)
- Fasting plasma glucose < 7.0 mmol/L
- 2-hour post-OGTT plasma glucose concentration between 7.8 and 11.0 mmol/L

(Based on the latest WHO/ADA criteria)
(Blood glucose values are 10–15% lower than plasma glucose concentrations)

Table 10.3   Protocol for the performance of oral glucose tolerance test

Discontinue any medication that may affect glucose tolerance.
Normal diet containing at least 250 g of carbohydrate for 3 days prior to test.
Normal activity for 3 days prior to test.
Overnight fast (10–16 hours).
Basal blood sample.
Administer 75 g of glucose in 250 mL of water.
Remain seated throughout the test.
No smoking during the test.
Blood samples at 60 and 120 minutes after glucose load.

Table 10.4   Indications for OGTT

Diagnosis of gestational diabetes mellitus.
Equivocal fasting or random glucose concentration.
Unexplained nephropathy, neuropathy or retinopathy.
Epidemiological studies.
Diagnosis of acromegaly.

protocol/procedure for the glucose tolerance test is given in Table 10.3 and the indications are listed in Table 10.4. A large number of factors can affect the reproducibility of this test (Table 10.5), which has a coefficient of variation of 15–25%.

The American Diabetes Association (ADA) has now recommended that glycated haemoglobin may be used as a diagnostic test for diabetes mellitus as this is more convenient for the patient (no need to fast). However, there is no universal agreement on the use of glycated haemoglobin as a diagnostic test as it can be affected by factors other than glycaemia.

Gestational Diabetes Mellitus

During normal pregnancy, there is increased insulin resistance especially during the latter part of pregnancy. However, this is accompanied by increased insulin secretion and normal glucose concentration is maintained. Failure to increase insulin secretion leads to

Table 10.5   Factors that affect glucose tolerance test

Patient factors
- Duration of fast
- Previous carbohydrate intake
- Age
- Body weight
- Exercise
- Intercurrent illness
- Medication (e.g. thiazides, steroids, oral contraceptives)
- Posture
- Anxiety
- Smoking
- Time of day

Administration of glucose
- Form of glucose (anhydrous or monohydrate)
- Quantity of glucose
- Volume in which glucose is given
- Rate of ingestion

gestational diabetes mellitus (GDM), which is defined as carbohydrate intolerance of variable severity seen for the first time during pregnancy. Prevalence of this condition is not known but has been reported to vary from 1–20%. The risk factors for GDM include family history of diabetes in a first-degree relative, obesity, advanced maternal age, glycosuria and adverse outcomes such as stillbirth and macrosomia in a previous pregnancy (Table 10.6). GDM is usually asymptomatic and not life threatening to the mother but is associated with increased neonatal morbidity such as macrosomia, hypoglycaemia and maternal complications such as increased rate of caesarean section and pre-eclampsia. Maternal hyperglycaemia leads to increased secretion of insulin by the foetus, resulting in rapid stimulation of foetal growth and macrosomia. Early recognition of GDM and treatment will reduce the perinatal morbidity and mortality. GDM is also associated with increased risk of development of type 2 diabetes mellitus subsequently. The incidence is about 60% by 15 years after parturition. ADA recommends screening for GDM between 24 and 28 weeks of gestation in all pregnant women or at least in women who are at

Table 10.6   Risk factors for gestational diabetes mellitus

Family history of diabetes in a first-degree relative
Recurrent glycosuria (especially in the fasting state)
Previous macrosomic infant (> 4.5 kg)
Large for date babies in the current pregnancy
Polyhydramnios in the current pregnancy
Obesity
Previous still birth (with pancreatic β-cell hyperplasia)
Family origin with a high prevalence of diabetes e.g. South Asians,
    Black Caribbean, Middle Eastern

increased risk. Screening is usually performed with a 50 g glucose load without regard to time of day or time of last meal. Plasma glucose concentration at 1 hour of ≥7.8 mmol/L is indication for further diagnostic glucose tolerance test, using 75 g of glucose. A 2-hour value of ≥8.6 mmol/L is diagnostic of GDM. In the UK, it is recommended by National Institute of Clinical Excellence (NICE) that a 2-hour 75-g OGTT should be done at 24–48 weeks in those at high risk of developing it and the criteria for the diagnosis is a fasting glucose of ≥7.0 mmol/L or a 2-hour plasma glucose of ≥7.85 mmol/L. In women who had GDM in a previous pregnancy, OGTT is recommended at 16–18 weeks and if this is normal, a further OGTT at 28 weeks is recommended.

Management of diabetes mellitus

The aim of the treatment is to relieve symptoms and to achieve, as far as possible, normoglycaemia in order to prevent acute as well as long-term complications. In all diabetics, dietary control is important and in obese subjects, weight reduction is essential. Exercise, which increases insulin sensitivity, should be encouraged. Pharmacological treatment includes insulin treatment in type 1 diabetes and oral hypoglycaemic agents such as sulphonylureas, biguanides, glitazones and DDP-4 inhibitors (Sitaglaptin) and/or insulin in type 2 diabetics. Long-term studies have shown that adequate glycaemic control reduces the risk of long-term complications. However, intensive treatment may lead to periods of hypoglycaemia, which should be reduced or avoided. In addition to these measures, patients should be educated to look after themselves. Diabetic patients should be followed regularly and their glycaemic control monitored.

Monitoring of Treatment

The aim of treatment in diabetes is the achievement of near normal blood glucose values. Therefore, regular biochemical monitoring is required (Table 10.7).

Urine glucose

Excretion of glucose in the urine can be monitored using simple dipstick methods. However, glycosuria is an unreliable method to monitor treatment as the renal threshold for glucose varies between

Table 10.7 Biochemical tests for monitoring treatment in diabetes mellitus

Urine glucose
Urine ketones
Blood glucose
- Random
- Fasting
- Postprandial
- Home glucose monitoring

Glycosylated protein
- Haemoglobin — HbA_{1c}
- Albumin — fructosamine

Serum 1,5-anhydroglucitol
Serum lipids
- Total cholesterol
- HDL cholesterol
- Triglycerides

Urine protein
- Microalbuminuria

Renal function tests
- Serum creatinine

people. Furthermore, urine glucose measurement does not detect hypoglycaemia. Urine glucose, however, gives an estimate of glucose control over several hours as compared to the spot blood test.

Urine ketones

Detection of urine ketones is useful in monitoring unstable or labile type 1 diabetes. However, the method (dipstick) does not detect the quantitatively important ketone 3β-hydroxybutyrate and blood ketones may be increased 10–20-fold before the urine test becomes positive.

Blood glucose monitoring

Many patients monitor their blood glucose concentration at home using a glucose meter. Frequency of monitoring varies from once or twice a day to several times a day. Regular monitoring throughout the day including pre- and post-meal values are useful in adjusting the dosage of treatment especially in type 1 diabetics, in the pregnant, in GDM, in labile, brittle and in young diabetics. During the early stages of treatment, it also helps to educate patients.

Glycated haemoglobin

Blood glucose concentration changes very rapidly so taking measurements, even several times a day do not give a fully integrated value of glycaemic control. Measurements of glycated proteins can provide evidence of overall glycaemic control. Glycation is the non-enzymatic addition of sugar residues to amino groups of proteins. Haemoglobin in adults consists mainly of HbA_0 and a minor fraction of HbA_1 that can be further resolved into fractions HbA_{1a}, A_{1b} and A_{1c}. Condensation of glucose with the N-terminal valine residue of each β-chain of haemoglobin A gives rise to an unstable shift base. This may either dissociate or undergo Amadori rearrangement to form a stable ketomine, haemoglobin A_{1c}. Haemoglobin A_{1a} is haemoglobin A with fructose-1,6-diphosphate or glucose-6-phosphate adducts and haemoglobin A_{1b} is pyruvic acid linked to the amino terminal valine of the β-chain.

HbA_{1c} refers to a specific form of glycated haemoglobin, formed by the irreversible binding of glucose to the N-terminal valine residue of the haemoglobin β-chain. Total glycated haemoglobin refers to carbohydrate binding to all sites on the haemoglobin. The percentage of glycated haemoglobin formed is directly proportional to the glucose concentration and the time that the red cells have been exposed to glucose. Measurement of glycated haemoglobin, therefore, gives an integrated picture of the mean blood glucose concentration during the preceding 60 days. Haemoglobin A_{1c} results is abnormally low when the turnover of red cells is high, e.g. haemolytic anaemias, chronic sepsis, rheumatoid arthritis and chronic blood loss. Glycated haemoglobin values are free from day to day fluctuations in glucose and are unaffected by exercise or recent food ingestion. Haemoglobin variants can affect the value depending on the method used. Carbamylated haemoglobin, which is the binding of urea to haemoglobin is increased in renal failure patients and may cause falsely high A_{1c} value. Hypoglycaemia is not detected by this test.

Based on the long-term diabetes control and complication (DCCT) study, a multicentre randomised trial that compared intensive treatment with conventional treatment, it is now recommended that in type 1 diabetes patients, blood glucose should be controlled such that haemoglobin A_{1c} is around 7.2% (or 53 nmol/mol) or lower. Similar findings in type 2 diabetes has been demonstrated by the United Kingdom Prospective Diabetes Study (UKPDS).

Fructosamine

Fructosamine refers to glycated serum proteins and it reflects predominantly glycated albumin (80%). As the half-life of albumin is about 20 days, the concentration of fructosamine reflects glycaemic control over the preceding 2–3 weeks. Fructosamine assays can be automated, and are precise and cheaper than glycated haemoglobin assays. Fructosamine measurement may be useful in patients where short-term monitoring is required, e.g. in pregnant diabetics. It may also be useful in patients with haemoglobin variants where the life span of red cells may be reduced. Increased turnover of proteins as in liver disease and nephrotic syndrome may have an effect on fructosamine values.

Advanced glycation end products

Non-enzymatic attachment of glucose to molecules with long half-lives, such as collagen and DNA, produces a stable compound which undergoes a series of additional rearrangements, dehydration and fragmentation reactions, resulting in irreversible advanced glycation end products (AGE). The amount of these products does not return to normal when hyperglycaemia is corrected. They continue to increase throughout the life span of the protein. Hyperglycaemia accelerates the formation of AGE, and the measurement of AGE may have a role in providing a measure of diabetic control longer than that given by glycated haemoglobin. The value of AGE measurements in clinical practice remains to be established.

Urinary albumin excretion — Microalbuminuria

Increased albumin excretion is one of the earliest signs of diabetic renal disease. Albumin excretion in the range of 30–200 mg/24-hr is defined as microalbuminuria. This term refers to albumin excretion lower than that detected by conventional methods. The presence of increased albumin excretion signifies an increase in the transcapillary escape rate of albumin and is a marker of microvascular disease. Persistent microalbuminuria gives a 20-fold greater risk for the development of renal disease in patients with both type 1 and type 2 diabetes. Prospective studies have clearly demonstrated that increased albumin excretion precedes and is highly predictive of diabetic nephropathy, end-stage renal disease and proliferative retinopathy in type 1 diabetes. Furthermore, results of the DCCT trial have shown that intensive metabolic control can reduce the incidence of microalbuminuria. Microalbuminuria is also an independent predictor of cardiovascular mortality, both in diabetic and non-diabetic patients.

A 24-hour urine sample is the best way to demonstrate microalbuminuria. However, this is inconvenient and often incomplete. Expressing the results as albumin:creatinine ratio (ACR) in a sample collected over a shorter period (4 hours or overnight) is an alternative. However, albumin excretion shows diurnal variation — 30–50% lower at night and is affected by exercise, posture and diuresis. Sample for albumin excretion should be collected without prior exercise. It should not be collected during acute illness, urinary tract infection or after a fluid load. A convenient method to screen for microalbuminuria is to measure ACR in at least two separate random samples (preferably the first sample passed in the morning). ACR values of < 2.5 mg/mmol in males and < 3.5 mg/mmol in females are considered normal. Urine samples should be stored at 4°C or at –70°C and not at –20°. Albumin excretion can be measured by a sensitive immunoassay technique and several dipstick methods are also available to detect microalbuminuria.

1,5-anhydroglucitol

This pyranose form of glucose is of predominantly dietary origin. It is metabolically inactive, has a long half-life and serum concentrations are stable within a narrow range. It is excreted by the kidney where glucose competes with 1,5-anhydroglucitol (AG) for reabsorption. In poorly controlled diabetes mellitus, serum 1,5-AG is low. Serum 1,5-AG has been suggested as another marker of glycaemic control.

Complications of Diabetes

Acute complications

1. *Diabetic ketoacidosis*

    Diabetic ketoacidosis (DKA) is an important acute complication of type 1 diabetes but it can occur in type 2 diabetes during severe infection or other major illness. It may be the presenting feature of diabetes in some patients. It results from a combination of a lack of insulin, increased glucagon and other counterregulatory hormones, which leads to an inability to utilise glucose, increased glycogenolysis, lipolysis, ketogenesis and gluconeogenesis. Increased endogenous glucose production in the liver together with impaired glucose uptake leads to fasting hyperglycaemia. At the same time, lipolysis is unopposed due to a lack of insulin, leading to the production of excess acetyl CoA and increased ketogenesis (Figure 10.2). The increased production of hydroxybutyric acid and acetoacetate leads to a metabolic acidosis. Hyperglycaemia leads to an osmotic diuresis that causes volume depletion. Decreased ECF volume leads to decreased renal perfusion, prerenal uraemia and eventually oliguric renal failure. Hydrogen ions in the extracellular fluid is exchanged for potassium in the cells, leading to hyperkalaemia. However, the osmotic diuresis causes loss of potassium and there is a decrease in total body potassium. Increased ventilation caused by metabolic acidosis is described as Kussmaul breathing, a deep sighing hyperventilation.

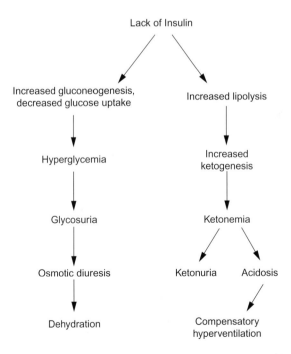

Figure 10.2   Pathogenesis of diabetic ketoacidosis.

Biochemical investigations in a patient suspected of diabetic ketoacidosis will show hyperglycaemia, high serum urea and creatinine, low pH, low pCO_2, low bicarbonate and urine ketones. Plasma sodium concentration is low or low normal due to a loss of sodium in the urine and due to the shift of water out of the ICF as a result of hyperglycaemia. The degree of ketonuria and ketonaemia may be underestimated as the reagent stick commonly used (Acetest) detects acetoacetate whereas the predominant ketone is β-hydroxybutyrate.

- Management

  DKA is a medical emergency and treatment should be prompt. Principles of treatment are replacement of insulin, fluid (isotonic saline), correction of acidosis if very severe and treatment of the precipitating cause. Administration of insulin will reverse the metabolic effects and reduce the production of ketones. Rate of insulin administration should be adjusted according to blood glucose concentration. Plasma potassium concentration should be monitored regularly as hypokalaemia may develop due to increased uptake by cells and potassium supplements may be required. If the acidosis is severe (pH < 7.0), sodium bicarbonate should be infused to raise the pH to 7.15 or 7.20. During treatment, if blood glucose decreases rapidly, hypernatraemia may develop due to a rapid shift of water from the ECF to ICF. Careful monitoring of the rate of decrease in blood glucose will avoid this complication. During treatment, blood glucose, electrolytes and occasionally blood gas should be monitored. Isotonic saline should be given to correct volume depletion and when the plasma glucose has fallen to 15 mmol/L, this may be changed to dextrose-saline.

Case 10.1

A 15-year-old girl was brought to casualty in a semi-comatose state. Her mother stated that her daughter had been feeling very tired over the previous 7–8 weeks and was also complaining of thirst and passing urine very often. Mother also noticed that her daughter had lost weight. On admission, she was drowsy with a pulse of 112/min and her blood pressure was 90/65 mmHg. Initial investigations showed:

Serum		Reference Range
Sodium (mmol/L)	137	135–145
Potassium (mmol/L)	5.5	3.5–5.0
Bicarbonate (mmol/L)	10	23–32
Chloride (mmol/L)	95	90–100
Urea (mmol/L)	15	3.5–7.2
Creatinine (μmol/L)	145	60–112
Glucose (mmol/L)	30	

These features suggest diabetic ketoacidosis. Her serum anion gap was high at 37 mmol/L ({sodium + potassium} — {chloride + bicarbonate}). Blood gas analysis confirmed that she had metabolic acidosis [pH 7.10 and a PaCO₂ of 3.0 kPa (4.6–6.0)]. She showed evidences of volume depletion (low blood pressure, raised pulse and high urea) due to the osmotic diuresis induced by high glucose. Although her serum potassium was high, she is likely to be potassium depleted. She was treated with isotonic saline, insulin and she made an uneventful recovery.

2. Hyperosmolar non-ketotic coma

Hyperosmolar non-ketotic coma (HONK) is seen in type 2 diabetics especially in the elderly. Severe hyperglycaemia with blood glucose concentration > 50 mmol/L develops, leading to osmotic diuresis, severe dehydration and high plasma osmolality. The exact pathogenesis of this condition is not clear. The absence of ketosis is thought to be due to sufficient insulin being present to prevent ketogenesis but not enough to prevent hyperglycaemia. Hypernatraemia is commonly seen due to a loss of water as result of osmotic dieresis. Severe hyperglycaemia causes movement of water out of the ICF, causing cellular dehydration.

- Management
Management consists of rehydration with isotonic saline and intravenous insulin treatment. Hypernatraemia may worsen as glucose

concentration decreases and careful monitoring of electrolytes and glucose is necessary. This condition has a mortality of nearly 50% and thromboembolic events are common. In some centres, heparin is given prophylacticaly. These patients usually do not require insulin once they recover from episode.

Case 10.2

A 65-year-old man of West Indian origin was brought to casualty department in a coma. His wife reported that he was on medication for hypertension and for the past 5 weeks, he was unusually thirsty and had consumed several litres of Coca Cola each day. The following results were obtained on admission:

Serum		Reference Range
Sodium (mmol/L)	159	135–145
Potassium (mmol/L)	4.9	3.5–5.0
Bicarbonate (mmol/L)	23	23–32
Urea (mmol/L)	25	3.5–7.2
Creatinine (μmol/L)	135	60–112
Glucose (mmol/L)	70	

This patient clearly has severe hyperglycaemia but he is not acidotic (bicarbonate is normal). His calculated osmolarity was 423 mmol/L (2{sodium + potassium} + urea + glucose) (2{159 + 4.9} + 23 + 70). The diagnosis of hyperosmolar non-ketotic coma was made and he was treated with isotonic saline, insulin and potassium. The severe hyperglycaemia had caused osmotic diuresis, leading to the loss of water, electrolytes and volume depletion. Serum sodium is high because of the increased loss of water relative to sodium by the osmotic diuresis. Loss of fluids has caused a prerenal acute renal failure.

---

3. *Hypoglycaemia*

Hypoglycaemia is a common acute complication in diabetes mellitus and is due to a high dose of insulin or sulphonylureas or missing a meal after a usual dose of these drugs. In any diabetic subject who presents with unusual features or coma, blood glucose should be checked. Some patients are unaware of hypoglycaemia due to the lowering of the hypoglycaemic threshold (see below).

4. *Lactic acidosis*

This is a complication seen in diabetes mellitus due to biguanide treatment — see acid–base balance in Chapter 4.

Long-term complications

Diabetes affects almost every organ in the body as a result of microvascular or macrovascular disease. Microvascular disease leads to retinopathy, nephropathy and neuropathy. The pathogenesis of macrovascular complications is similar to atherosclerotic vascular disease. Macrovascular disease leads to increased incidence of coronary heart disease, cerebrovascular disease, peripheral vascular disease and hypertension. Other complications include cataracts and skin thickening.

The exact mechanism of long-term complications of diabetes mellitus is unknown but is thought to be due to changes in the glycoprotein basement membrane, leading to increased permeability. Another contributing factor is the abnormalities in the polyol pathway (Figure 10.3). Glucose can be reduced to sorbitol by aldose reductase then to fructose by polyol dehydrogenase. The presence of hyperglycaemia causes an increase in the activity of the polyol pathway in insulin-independent tissues such as red blood cells, nerve and lens. The increased sorbitol causes an osmotic effect and this is thought to play a role in the development of diabetic cataract. In addition, the high concentration of sorbitol decreases cellular uptake of myoinositol, which in turn causes a decrease in the activity of plasma membrane sodium–potassium–ATPase. Inhibition of the sodium–potassium–ATPase together with hypoxia affects nerve function, contributing to the development of diabetic neuropathy. Drugs, which inhibit aldose reductase, improve the nerve function. Formation of AGE is

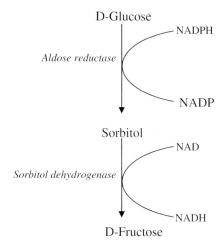

Figure 10.3   The polyol pathway.

another contributing factor in the development of long-term complications. Formation of AGE causes stiffening of the extracellular matrix, loss of elasticity and interferes with the function of endothelial cells, macrophages and smooth muscle of blood vessels. These in turn may contribute to the development of microangiopathy.

1. *Lipid abnormalities in diabetes mellitus*

Lipid abnormalities in type 2 diabetes are common; both qualitative and quantitative abnormalities are seen. These contribute to the development of atherosclerosis. Main quantitative abnormalities are raised triglycerides and low HDL cholesterol. Qualitative abnormalities include large VLDL particles, small dense LDL particles, increased triglyceride content of LDL and VLDL, glycation of apolipoproteins and increased oxidation of LDL. The exact pathogenesis of these lipids abnormalities is not fully understood.

In type 1 diabetes, there is increased triglycerides, VLDL and LDL with low HDL cholesterol especially if the glucose control is poor. Lack of insulin leads to increased free fatty acids as insulin inhibits hormone sensitive lipoprotein lipase. This increases triglyceride synthesis and VLDL. In severe insulin deficiency, there may be chylomicronaemia.

2. *Diabetic nephropathy*

Diabetic nephropathy (DN) is the leading cause of end-stage renal failure and contributes to premature death in diabetic subjects. Diabetic nephropathy develops in about 30% of type 1 diabetics and about 25% of type 2 diabetics. Approximately 3% of newly diagnosed type 2 diabetics have evidences of diabetic nephropathy. People of African origin and Pima Indians show greater incidence.

Diabetic nephropathy is one of the microvascular complications of diabetes and is characterised by a form of glomerulosclerosis. Diabetic nephropathy runs a predictable course. Early on, there is elevated GFR followed by increased albumin excretion (microalbuminuria) and subsequently, increased blood pressure. The next stage is the development of overt proteinuria, which progresses to renal failure. The time interval to reach these stages varies from 10–20 years from the onset of diabetes. Early on in this process, the glomerular basement membrane loses its anion charge (due to loss of heparin sulphate proteoglycan) and albumin excretion increases. GFR declines progressively once proteinuria is established and progress to end-stage renal failure. Poor glycaemic control, hypertension and hyperlipidaemia accelerate the progression of this condition. Adequate glycaemic control and aggressive management of hypertension at the microalbuminuria stage together with ACE inhibitors will delay or halt the progression of the disease. It is now recommended that all diabetic subjects have an annual ACR measurement.

- Pathogenesis of DN

One of the earliest changes in DN is the thickening of the basement membrane, accumulation of extracellular matrix and consequent expansion of mesangium. A possible pathogenetic mechanism is shown in Figure 10.4.

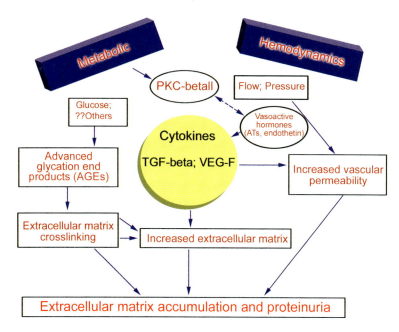

Figure 10.4   Pathogenesis of diabetic nephropathy.

Hypoglycaemia

Hypoglycaemia is a dangerous condition as the brain depends almost entirely on glucose for its energy. Hypoglycaemia is usually defined as fasting plasma glucose < 2.5 mmol/L. However, this definition is arbitrary as some patients do not develop symptoms at concentrations lower than this. Pathological hypoglycaemia is diagnosed when there are symptoms of hypoglycaemia accompanied by low plasma glucose concentration and when symptoms are relieved by glucose (Whipple's triad). Clinical features of hypoglycaemia are those produced by the effect on the central nervous system, neuroglycopenia, and those produced by the production   of counterregulatory hormones. Acute hypoglycaemia leads to increased catecholamines release, causing nervousness, weakness, headache, sweating, dizziness, tremor, tachycardia, palpitations, anxiety and hunger. Neuroglycopenic symptoms include visual symptoms, headache, blunted mental acuity, loss of

Table 10.8 Symptoms of hypoglycaemia.

Adrenergic
- Anxiety
- Restlessness
- Nausea
- Tremor
- Sweating
- Palpitations

Neuroglycopenia
- Behavioural changes
- Tiredness, confusion
- Blurred vision
- Amnesia
- Parasthesiae, weakness
- Personality changes
- Psychosis
- Seizures
- Coma

motor function, confusion, abnormal behaviour, fits and coma (see Table 10.8).

Symptoms of hypoglycaemia are non-specific; therefore, symptoms can only be confidently attributed to hypoglycaemia when the Whipple's triad is fulfilled. The concentration of glucose at which symptoms appear, the glycaemic threshold, varies within the same individual and between individuals. The glycaemic threshold can be varied by previous glucose concentrations; the threshold increases with previous hyperglycaemia and decreases with previous exposure to hypoglycaemic episodes. Hypoglycaemic unawareness (loss of warning) is due to the lowering of the glycaemic threshold.

Causes of Hypoglycaemia (Table 10.9)

Insulinoma

Insulinoma is a rare, benign neoplasm of the pancreatic β cells. They are usually solitary in 80% of cases. Multiple nodules may be

Table 10.9   Causes of hypoglycaemia

Insulinoma
Non-pancreatic neoplasms
Drugs
- Insulin
- Sulphonylureas

Alcohol
Liver disease
End stage renal failure
Endocrine diseases
- Hypopituitarism
- Addison's disease

Autoimmune insulin syndrome
Reactive hypoglycaemia
- Postgastrectomy
- Alcohol provoked

present in a small proportion (7–10%) often as part of the multiple endocrine neoplasia (MEN I) syndrome. Insulinoma can occur at any age although it is rare before the age of 10, and occurs equally in both men and women. Patients with insulinoma may present with behavioural changes or with the features of acute hypoglycaemia including faintness, ischaemic attacks or funny turns. Blood glucose should be measured in all patients who experience fits, transient ischaemic attacks or funny turns in order to exclude hypoglycaemia. The main cause of hypoglycaemia is the suppression of hepatic glucose output rather than increased peripheral glucose uptake by endogenous hyperinsulinaemia. Hypoglycaemia typically occurs in the fasting state. Plasma free fatty acids and β-hydroxybutyrate are normal or low despite the presence of hypoglycaemia due to persistent inhibition of lipolysis and ketogenesis by hyperinsulinaemia.

Diagnosis depends on demonstrating that the symptoms are due to hypoglycaemia, that symptoms can be provoked by fasting and relieved by glucose and that plasma insulin concentration is inappropriate. Measurable plasma insulin concentration and a C-peptide of

300 pmol/L or greater, in the presence of hypoglycaemia, is diagnostic of hyperinsulinaemia. In renal failure patients, the C-peptide concentration may be higher due to the fact that the kidney clears C-peptide. The main diagnostic test is a prolonged 72-hour fast. Other dynamic function tests are of little help. Localisation of the tumour is difficult; computer tomography and arteriography are only moderately successful. In difficult cases, transhepatic venous sampling and intra-operative ultrasound localisation are currently the best methods for localisation of the tumour.

The treatment of choice is surgical removal. When surgery is contraindicated, or when the tumour cannot be located, treatment with diazoxide, chlorothiazide or octreotide will relieve symptoms.

Case 10.3

A 45-year-old lorry driver went to see his GP complaining of dizzy spells. On questioning, he admitted that these were often in the morning or before meals and eating relieved the symptoms. His GP suspected that he may have insulinoma and referred him to the endocrine clinic. He was admitted and underwent a 72-hour fast. After 18 hours of fast, his glucose was 2.1 mmol/L and blood was taken for analysis of insulin and C-peptide. The results of these were:

Insulin	156 pmol/L (10–50)
C-peptide	1250 pmol/L (200–650)

This patient had inappropriate insulin and C-peptide at the time of hypoglycaemia. This is typical of increased endogenous insulin secretion and could be due to either insulinoma or sulphonylurea administration. A urinary sulphonylurea test was negative and imaging studies showed that he had a small nodule in his pancreas. This was later removed and he recovered fully.

Non-pancreatic neoplasms

Non-pancreatic neoplasms, particularly large mesenchymal tumours such as retro peritoneal fibrosarcomas, hepatocellular and adrenal carcinomas and carcinoid tumours may cause hypoglycaemia.

Manifestations of hypoglycaemia are similar to that of insulinoma except that the progression is rapid and relentless, and sometimes requires large amounts of intravenous glucose to relieve symptoms. There is fasting hypoglycaemia, hypoketonaemia and low plasma insulin and C-peptide concentrations. Hypoglycaemia is caused by overproduction of an abnormally large IGF2, which suppresses endogenous growth hormone. This in turn depresses the production of IGF binding protein by the liver. The low plasma IGFBP-3 results in a high concentration of free IGF2, leading to the inhibition of glucose release from the liver and an increased uptake of glucose by peripheral tissues. The diagnosis depends on the demonstration of low values for plasma insulin, C-peptide, and IGF1 together with a high IGF2 concentration (high IGF2 and IGF1 ratio).

Alcohol-induced hypoglycaemia

Alcohol-induced hypoglycaemia develops 6–36 hours after ingestion of moderate to large amounts of alcohol in a fasting or malnourished individual. The hypoglycaemia is due to suppression of gluconeogenesis by alcohol. The patient is usually stuporous or unconscious when first seen but may be aggressive or uncooperative; the symptoms are attributed to alcohol intoxication rather than hypoglycaemia. Hypothermia is more common than other causes of hypoglycaemia and this may be the first clue to the correct diagnosis. Plasma and urinary ketones are high while the plasma insulin and C-peptide concentrations are appropriately low.

Hypoglycaemia can also occur in alcoholic ketoacidosis. In these patients, alcohol may not have been taken for several days prior to admission. Clinically, alcoholic ketoacidosis resembles diabetic ketoacidosis except these patients have hypoglycaemia. Although alcoholic ketoacidosis is predominant in women, hypoglycaemia is more common in men.

Endocrine hypoglycaemia

Hypoglycaemia is an infrequent but important complication of endocrine disorders, of which Addison's disease is probably the most common. Hypoglycaemia is due to the lack of counterregulatory hormones.

Hepatic and renal disease

In the fasting state, glucose is produced from the liver by gluconeogenesis and glycogenolysis to maintain plasma glucose concentration. In severe liver disease, such as after severe paracetamol poisoning or other toxins, hypoglycaemia will be present due to the lack of an adequate amount of liver tissue. As the kidney is the other organ responsible for gluconeogenesis, hypoglycaemia may be a feature of terminal renal disease. Furthermore, the kidney is also an important organ in clearing insulin, therefore in renal failure, insulin degradation may be delayed.

Drug-induced hypoglycaemia

The sulphonylureas, apart from insulin, are the most important group of drugs that can cause hypoglycaemia. Hypoglycaemia may also be seen in the early stages of starting sulphonylureas such as glibenclamide. Sulphonylureas stimulate insulin from the β cells by a direct action as well as by sensitising the β cells to endogenous stimulants such as leucine. Sulphonylureas cause fasting hypoglycaemia, therefore this condition can be confused with insulinoma. As in insulinoma, plasma insulin, C-peptide and proinsulin concentrations are inappropriately high in the presence of hypoglycaemia. Plasma β-hydroxybutyrate and non-esterified fatty acids are low. The nature, frequency and severity of sulphonylurea-induced hypoglycaemia are difficult to assess. The diagnosis can be established by the demonstration of sulphonylurea in blood or urine.

Autoimmune hypoglycaemia

This is a rare condition caused by the interaction of endogenous antibodies with insulin and insulin receptors. Hypoglycaemia is most often postprandial but may be fasting and exacerbated by exercise. Diagnosis is suspected when insulin concentrations are high. The demonstration of antibodies will establish the diagnosis.

Inborn errors of metabolism

Inborn errors of carbohydrate metabolism, such as hereditary fructose intolerance, glucose-6-phosphatase deficiency and fructose-1,6-bisphosphatase deficiency, can cause hypoglycaemia and may first manifest in adult life either as fasting or more commonly as reactive hypoglycaemia (see Chapter 28).

Reactive postprandial hypoglycaemia

Some patients after gastric surgical procedures such as gastrectomy may have hypoglycaemic symptoms after a meal, typically 2 hours later. This is sometimes termed 'reactive hypoglycaemia'. Apart from this group, a group of subjects with vague symptoms, which come on after food ingestion have been described as 'idiopathic reactive hypoglycaemia' or 'idiopathic postprandial syndrome'. Investigations have shown that these symptoms are rarely due to low blood glucose concentration and these patients have postprandial autonomic symptoms without neuroglycopenia. Most experts are of the opinion that this is not a true hypoglycaemic disorder. The prolonged oral glucose tolerance test, which is sometimes used to diagnose 'idiopathic reactive hypoglycaemia' test has no diagnostic value.

Investigation of Patients with Hypoglycaemia

Demonstration of hypoglycaemia at the time of the patient's symptoms is vital. A normal blood glucose concentration obtained when the patient has symptoms eliminates the possibility of hypoglycaemia. If hypoglycaemia is established, further investigations are indicated. Careful drug history should be taken and associated conditions such as liver disease and renal failure etc. should be excluded by renal and liver function tests. A prolonged 72-hour fast is the classic diagnostic test. Patients should fast for 72 hours or until the symptoms appear, then a blood sample should be taken for the measurement of glucose, insulin, C-peptide and β-hydroxybutyrate. In patients with hypoglycaemia, low β-hydroxybutyrate and measurable insulin and C-peptide indicate endogenous hyperinsulinaemia. Sulphonylureas may need to be measured to exclude drug-induced hypoglycaemia. Serum IGF1/IGF2 ratio may be required if hypoglycaemia due to neoplasm is suspected.

Further Reading

1. Emanepator K. Laboratory diagnosis and monitoring of diabetes mellitus. *Am J Clin Pathol* 1999; 112:665–674.
2. Kjos SL, Buchanan TA. Gestational diabetes mellitus. *New Eng J Med* 1999; 341:1749–1756.
3. Melendez-Ramirez LY, Richards RJ, Cefalu WT. Complications of Type 1 Diabetes. *Endocrinol Metab Clin North Am* 2010; 39:625–640.
4. Smith J, Nattrass M. *Diabetes and Laboratory Medicine* 2004. ACB Venture publications: London.

Summary/Key Points

1. Blood glucose concentration is maintained within narrow limits by homeostatic mechanisms involving insulin and IGF1, which lowers blood glucose and cortisol, glucagon, growth hormone and adrenaline (the counterregulatory hormones), all of which raise blood glucose concentrations.
2. Diabetes mellitus is a common condition due to a deficiency of insulin or defect in insulin action or both. Diabetes mellitus is classified into type 1 (requiring insulin and at risk of developing diabetic ketoacidosis) and type 2 (no absolute requirement for insulin and usually do not develop diabetic ketoacidosis). Type 1 diabetes mellitus is an autoimmune disease and it accounts for 5–10% of all diabetics. Type 2 diabetes mellitus is due to a combination of genetic and environmental factors.

3. Diagnosis of diabetes mellitus is based on fasting plasma glucose ≥ 7.0 mmol/L, a random plasma glucose ≥ 11.1 mmol/L or hyperglycaemia after oral glucose tolerance test (≥ 11.1 mmol/L).

4. Gestational diabetes mellitus is defined as glucose intolerance that develops during pregnancy and is associated with several complications in the mother and foetus. In high risk mothers, an oral glucose tolerance test is recommended between 24 and 28 weeks.

5. Management of diabetes mellitus involves lifestyle changes and drug therapy (insulin or oral hypoglycaemic agents) to maintain normoglycaemia as far as possible. Good glycaemic control reduces long-term complications.

6. Diabetic subjects are usually monitored by frequent blood glucose measurements (self monitoring) or by means of HbA_{1c}, which gives an index of glycaemic control over the previous 8–10 weeks. Fructosamine (glycated serum proteins) is useful when an assessment of glycaemic control over a shorter time period (3–4 weeks) is required.

7. Acute complications of diabetes mellitus include diabetic ketoacidosis (most often in type 1), hyperosmolar non-ketotic coma (in type 2) and hypoglycaemia (in both types). Diabetic ketoacidosis is often precipitated by infection or other illness and it is a medical emergency.

8. Long-term complications of diabetes mellitus include diabetic nephropathy, retinopathy and neuropathy. Diabetic nephropathy is the leading cause of end-stage renal failure. Increased albumin excretion in urine (microalbuminuria or ACR) is the earliest indication of diabetic nephropathy. A high ACR (>2.5 mg/mol in males and >3.5 mg/mol in females) in diabetes mellitus, in at least two urine samples, indicates early diabetic nephropathy and treatment with ACE inhibitors is recommended. A high ACR in non-diabetics is also a risk factor for ischaemic heart disease.

9. Hypoglycaemia is a frequent complication of diabetes mellitus. In non-diabetic subjects, insulinoma, malignancy and administration of sulphonylureas or insulin are important causes of hypoglycaemia.

10. Investigation of hypoglycaemia requires a prolonged fast with measurements of glucose, insulin, C-peptide and ketones.

chapter 11

Lipids and Lipoproteins

Introduction

Lipids are a group of compounds, which are relatively insoluble in water but soluble in non-polar solvents such as chloroform and ether. Lipids are essential as energy stores, respiratory substrates, structural components of cells, vitamins, and hormones, and important for the protection of internal organs, heat conservation and digestion. The biologically important lipids are sterols, triglycerides, phospholipids and related compounds. Sterols include cholesterol, the various steroid hormones, bile acids and prostaglandins. Phospholipids are an important constituent of the cell especially in the nervous system. The major lipids in the circulation include cholesterol, triglycerides, phospholipids and non-esterified fatty acids. Fatty acids are monocarboxylic acids and sometimes classified into short chain, (2–6 carbon chain, C2–C6), medium chain (C8–C12) and long chain (C14–C24) fatty acids. Most of the fatty acids important in human metabolism contain an even number of carbon atoms. Fatty acids may be saturated, monosaturated or polyunsaturated. The numbering system for the fatty acids depends on the number of carbon atoms, the number of double bonds and the position of the bond. Fatty acids with double bonds are also classified into cis or trans depending on the plane in which these bonds occur. All fatty acids in mammals are of the cis variety.

The types of fats present in the diet varies. Majority of animal fats are saturated whereas plant fats such as olive oil are rich in monosaturated fatty acids. Diets rich in fish contain polyunsaturated fatty acids.

The human body can synthesise most fatty acids except linoleic and linolenic acids, which are therefore essential fatty acids. Triglycerides, which are made up of three fatty acids bound to glycerol, are the major storage form of energy in the adipose tissues. They are insoluble in water and are transported in the blood and lymph as complex lipoprotein particle — see below.

Phospholipids are compounds that are similar to triglycerides with the exception that the fatty acid on the third carbon atom of the glycerol is replaced by phosphate and an additional polar group such as alcohol. Phospholipids are the main lipid constituent of cell membranes. Cholesterol, a sterol, is an essential constituent of cell membranes. It is a precursor for bile acids and steroids including vitamin D. Cholesterol occurs either in the free or esterified form. Free cholesterol is a component of cell membrane, whereas cholesterol esters, which are insoluble, are found in serum.

Serum Lipids

Lipids are transported in the serum, in association with proteins. Free fatty acids are carried bound to albumin whereas triglycerides, cholesterol and phospholipids are transported in the form of lipoprotein complexes. Lipoproteins are macromolecular complexes carrying various lipids and proteins in plasma. Lipoprotein particles contain a hydrophobic core of cholesterol esters and triglycerides, which are surrounded by an outer layer of phospholipids, free cholesterol and proteins (apolipoproteins) (Figure 11.1). These apolipoproteins have important metabolic and structural functions. Some of the proteins such as apo B are deeply embedded into the particles while others, like apo C, are in looser contact on the surface and can be easily exchanged between different particles. Lipoproteins comprise a continuum of particles differing gradually in density and in the composition of lipid and apolipoproteins. However, they are usually classified into several major classes of lipoproteins based on their densities. These are chylomicrons, very low-density lipoprotein (VLDL), intermediate-density lipoproteins (IDL), low-density lipoprotein (LDL) and high-density lipoprotein (HDL). Properties of these major lipoproteins are

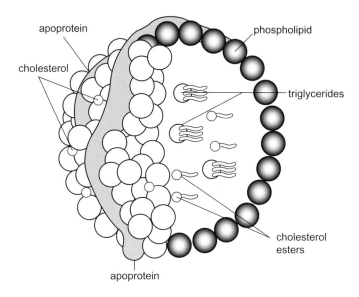

apoprotein

phospholipid

cholesterol

triglycerides

cholesterol esters

apoprotein

Figure 11.1   Structure of a lipoprotein particle.

given in Table 11.1. The largest and least dense particles are the chylomicrons, which are rich in triglycerides and have a relatively low protein content. With increasing densities of the lipoprotein particles, the triglyceride content and size decrease. High cholesterol content is found in LDL while HDL has the highest protein content and the lowest triglyceride content (Table 11.1). Chylomicrons are found in the circulation after a meal and they transport dietary triglycerides from the intestine to the peripheral tissues. HDL is further subclassified into HDL_2 and HDL_3. Lipoprotein (a) (Lp(a)) is an atypical lipoprotein whose function is unknown. It is denser and larger than LDL and its composition is similar to that of LDL except that it has a molecule of apo (a) for every molecule of apo B100.

Apolipoproteins

Protein components of lipoprotein particles are called 'apolipoproteins'. The major ones are listed in Table 11.2. Apolipoproteins provide structural stability, have a critical role in regulating lipoprotein

Table 11.1 Characteristics of the major lipoprotein classes

Lipoprotein	Density (kg/L)	Molecular weight (daltons)	Diameter (nm)	Lipid (%)			Electrophoretic mobility	Source	Principle function	Apolipoproteins
				TG	Chol	PL				
Chylomicrons	0.95	400×10^6	75–1200	80–95	2–7	3–9	Origin	Intestine	Transport of exogenous TG	B48, A, C, E
VLDL	0.95–1.006	10–80×10^6	30–80	55–80	5–15	10–20	Pre β	Liver, intestine	Transport of exogenous TG	B100, C, E
IDL	1.006–1.019	5–10×10^6	25–35	20–50	20–40	15–25	Broad β	VLDL	Precursors of LDL	B100, E
LDL	1.019–1.063	2.3×10^6	18–25	5–15	40–50	20–25	β	IDL	Cholesterol transport	B100
HDL	1.063–1.21	1.7–3.6×10^6	5–12	5–10	15–25	20–30	α	Liver, intestine VLDL Chylomicrons	Reverse cholesterol transport	A1, AII, C, E

Table 11.2 Characteristics of the major apolipoproteins

Apolipoprotein	Molecular weight	Lipoprotein	Metabolic function
A I	28,000	HDL, chylomicrons	Structural component of HDL LCAT activator
A II	17,000	HDL, chylomicrons	Unknown
A IV	46,500	HDL, chylomicrons	Unknown
B-48	264,000	Chylomicrons	Necessary for assembly and secretion of chylomicrons from the intestine
B-100	540,000	VLDL, IDL, LDL	Structural proteins of VLDL, IDL and LDL. Necessary for assembly and secretion of VLDL from the liver. Ligand for LDL receptor
C I	6600	Chylomicrons, VLDL, IDL, HDL	Msay inhibit hepatic uptake of chylomicrons and VLDL remnants
C II	8900	Chylomicrons, VLDL, IDL, HDL	Activator of lipoprotein lipase
C III	8800	Chylomicrons, VLDL, IDL, LDL	Inhibitor of lipoprotein lipase; may inhibit hepatic uptake of chylomicrons and VLDL remnants
E	34,000	Chylomicrons, VLDL, IDL, LDL	Ligand for binding of several lipoproteins to the LDL receptor
Apo (a)	250,000–800,000	Lp(a)	Not known. Independent predictor of CAD

metabolism, and some of them act as cofactors for plasma lipid-modifying enzymes.

Apo A1, a major component of HDL, is synthesised in the liver and small intestine and has a critical role in the reverse cholesterol transport. Apo A2 is the second most abundant apo protein in HDL. It is a cofactor for the enzyme lecithin cholesterol acyltransferase (LCAT)

Apo B100, which is a major apoprotein of VLDL, LDL and IDL, is synthesised in the liver and contains an LDL receptor binding domain. Unlike other apolipoproteins, this does not exchange between lipoproteins. Apo B100 is necessary for the initial assembly and secretion of VLDL by the liver. This protein has a crucial role in the catabolism of LDL and probably VLDL and LDL as well.

The same gene as apo B100 transcribes to apo B48. It is smaller than B100, is synthesised in the intestine and is necessary for intestinal assembly and secretion of chylomicrons.

Apo Cs are synthesised in the liver. Apo C1 is involved in the exchange of esterified cholesterol between lipoproteins and in the removal of cholesterol from tissues. Its main function is the inhibition of cholesteryl ester transfer protein (CETP). Apo C2 is an activator of lipoprotein lipase (LPL). Apo C3 is a major component of VLDL. It inhibits LPL and hepatic uptake of chylomicrons and VLDL remnants.

Apo E, synthesised in the liver, is found in all lipoproteins except LDL. It has a critical role in the removal of remnant lipoproteins from the plasma by interacting with several receptors in the liver, including the LDL receptor and the LDL receptor-related protein. Apo E containing lipoprotein is also recognised by the VLDL receptor, a recently identified receptor. Apo E has 3 major alleles: apo E3, the normal variant, is associated with normal VLDL and LDL metabolism; apo E2 cannot bind to LDL receptor and is associated with the accumulation of VLDL in the homozygous state; apo E4 binds normally to LDL receptor and is associated with high concentrations of LDL cholesterol. Apo E4 is associated with 10 to 30 times increased risk of Alzheimer's disease.

Apo (a) is found in Lp(a) and is present in a wide range of size and isoforms. The plasma concentration of Lp(a) varies widely in the population and is genetically determined to a great degree. Apo (a) has a high degree of homology to plasminogen, which is a key component of the fibronolytic pathway and might therefore interfere with fibrinolysis.

Apolipoprotein D is a glycoprotein and is a component of HDL. It is associated with the enzyme lecithin cholesterol acyltransferase.

Enzymes involved in Lipid Metabolism

Lipoprotein lipase (LPL)

LPL is synthesised and secreted by the adipose tissue and muscle. It is then transported to the capillary endothelial cell where it hydrolyses triglycerides into fatty acids and glycerol, thus allowing uptake of fatty acids by peripheral tissues. Several polymorphisms of LPL gene have been identified which predispose individuals to hypertriglyceridaemia.

Hepatic triglyceride lipase

Hepatic triglyceride lipase is synthesised in the liver and has homology to LPL. It functions as a triglyceride hydrolase and phospholipase. Its physiological function is to remove triglycerides and phospholipids from chylomicrons and VLDL remnants. Once secreted, it binds to the luminal surface of hepatic endothelial cells and hepatic sinusoids. It may also be involved in the conversion of larger HDL to small HDL species.

Cholesteryl ester transfer protein (CETP)

This is synthesised in the liver and mediates the exchange of triglycerides in the VLDL, chylomicrons and remnant particles for cholesterol esters in HDL and LDL.

Lecithin cholesterol acyl transferase (LCAT)

This converts cholesterol in HDL into cholesterol ester. Once formed, cholesterol esters move from the surface to the HDL core, allowing the particle to absorb more free cholesterol. Deficiency of this enzyme is associated with very low HDL concentration.

Metabolism of Lipoproteins

Exogenous Pathway

Chylomicrons (Figures 11.2 and 11.3)

Dietary lipids, which are absorbed as monoglycerides, fatty acids and free cholesterol, are re-esterified in the enterocytes where they

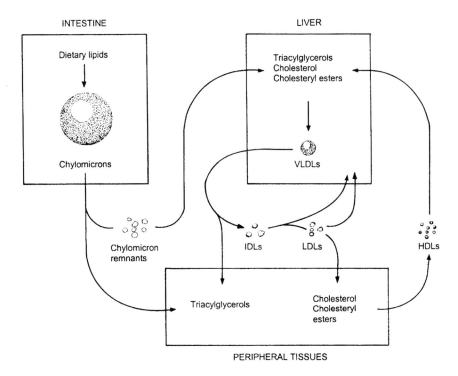

Figure 11.2   Overview of lipid metabolism.

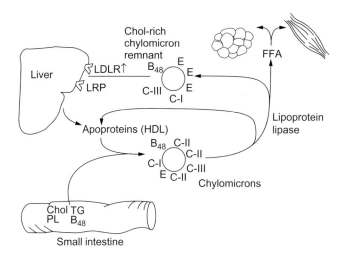

Figure 11.3 Metabolism of chylomicrons.

are reassembled with apo B48 to form chylomicrons that also contain apo (a). Chylomicrons enter the plasma via the lymph; once in the plasma, they acquire apo C1, apo C2, apo C3, and apo E from the surface of HDL. Hydrolysis of the triglyceride is initiated by LPL, which is activated by apo C2. The fatty acids released by this hydrolysis are taken up by muscle and adipose tissue and the glycerol is taken up by the liver for gluconeogenesis. The hydrolysis of triglycerides results in a reduction in the volume and surface area of the chylomicrons. This is accompanied by the transfer of phospholipid, free cholesterol, apo C2 and apo C3 back to HDL. This remnant particle, which is rich in cholesterol ester and apo E, interacts with receptors on hepatocytes and are removed from the circulation. The uptake of chylomicron remnants involves recognition of apo E by the receptors. The major function of chylomicrons is the transport of dietary triglycerides, cholesterol and fat-soluble vitamins to the liver. Chylomicrons are not normally detected in the plasma in the postprandial state. The main function of the exogenous pathway is to transport triglycerides to the adipose tissue and muscles, and cholesterol to the liver.

Endogenous Pathway

VLDL (Figures 11.2 and 11.4)

The liver secretes VLDL, which is involved in the transport of tri-glycerides from the liver to the peripheral tissues. Triglycerides synthesised in the liver, either from carbohydrates via acetyl CoA or from the re-esterification of fatty acids, are secreted into the circulation as VLDL, which contains apo B100 and phospholipids. Cholesterol in VLDL is either synthesised in the liver or is derived from the diet via the chylomicron remnants. After secretion, VLDL acquires apo C and apo E from HDL. The size of the VLDL particles secreted is determined by the availability of triglycerides. Large particles are secreted when excess triglycerides are synthesised, such as in obesity, untreated diabetes mellitus or in excess alcohol consumption. When triglycerides, not cholesterol availability, is limited, small VLDL particles are secreted, e.g. during weight loss. Insulin is thought to have an important but complex role in the regulation of VLDL secretion. The initial fate of VLDL is similar to that of chylomicrons. LPL in peripheral tissues hydrolyses the triglycerides and the fatty acids are

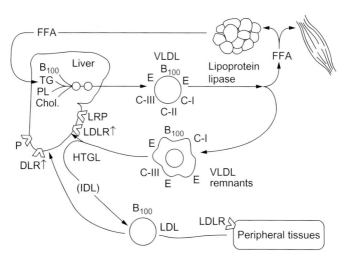

Figure 11.4   Metabolism of VLDL and LDL.

taken up by the peripheral tissues, mainly adipose tissue and muscles. VLDL shrinks in size and becomes IDL, which is removed by receptor-mediated pathways in the liver or converted to LDL. It has been suggested that the larger, more triglyceride enriched VLDL particles form IDL and are removed from plasma by receptor-mediated pathways whereas smaller, denser VLDL are converted to LDL. Conversion of IDL to LDL involves the removal of triglycerides by hepatic triglyceride lipase.

LDL (Figures 11.2 and 11.5)

LDL, which contains cholesterol, cholesterol esters and apo B100, is mainly formed from the catabolism of VLDL and IDL but there is some suggestion that the liver may directly secrete LDL as well. Clearance of LDL from plasma is determined by the availability and activity of LDL receptors. LDL receptor, which is a glycoprotein of molecular weight 160,000 kDa, is present in the cell surface of most tissues. In normal individuals, the LDL receptor-mediated pathways clear approximately 60–80% of LDL. The rest is cleared by non-receptor pathways, which include endocytosis. Uptake of the LDL by LDL receptors depend on the binding of apo B100 to the

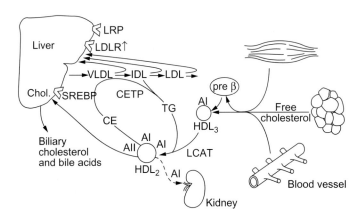

Figure 11.5   Metabolism of HDL.

receptor site. After binding, they are internalised and delivered to the lysosomes where apoproteins are degraded to amino acids and cholesterol ester is converted to free cholesterol via the action of an acid lipase. The free cholesterol interacts with endoplasmic reticulum membrane protein (SREBP), resulting in cleavage of this protein. The amino terminal portion enters the nucleus where it inhibits transcription of the genes for LDL receptor, HMG (hydroxymethylglutaryl) synthase and HMG CoA reductase, thus maintaining cholesterol homeostasis within the cell (Figure 11.5). Lack of LDL receptors results in unregulated cholesterol metabolism and markedly increased LDL cholesterol. The highest density of LDL receptors are found in organs such as liver, adrenals and gonads, which require cholesterol as precursors for synthesis of other steroids. If the LDL is oxidised, then the uptake of LDL by other tissues such as macrophages is increased. The uptake of LDL by macrophages in the arterial wall is an early event in the pathogenesis of atherosclerosis. Macrophages rich in cholesterol ester are converted to foam cells.

HDL (Figures 11.2 and 11.5)

HDL is involved in the transport of cholesterol from the periphery to the liver, so called 'reverse cholesterol transport system'. The liver and, to some extent, the intestine, secrete the precursor HDL, also called 'nascent HDL'. These are disc-shaped and contain phospholipids, apo E, apo A1 and apo A2 and are relatively poor in cholesterol. These HDL discs absorb free cholesterol from cell membranes and become nascent HDL3 particles. The HDL3 particles continue to absorb free cholesterol, which are then esterified by LCAT. The esterification increases the capacity of the HDL3 particles to accept more free cholesterol. Cholesterol esters move to the centre of the particle, which becomes spherical. The enlarged HDL3 accommodates apo C2 and apo C3 as well as phospholipids derived from chylomicrons and VLDL on their surface. This results in the formation of HDL2. Cholesterol ester in HDL2 is transferred to triglyceride rich lipoproteins (VLDL and chylomicrons), in exchange for triglycerides. This transfer is mediated by CETP. VLDL, which is enriched with cholesterol esters, is then removed by the liver as described earlier. The fate of HDL2, which is rich in triglycerides, is by one of two pathways: transfer of cholesterol components to other lipoproteins and cells or metabolism and removal of HDL2 apolipoproteins. Cholesterol transported this way from tissues to the liver can be excreted as bile acids.

Factors Affecting Plasma Lipid Concentration (Table 11.3)

Plasma cholesterol concentration is low at birth usually < 3.0 mmol/L and rapidly increases during the first year of life and reaches a value of < 4.1 mmol/L during childhood. In early adulthood, it increases further. In men, cholesterol concentration rises through the early and middle years and plateaus out in the elderly. In the female, the increase in cholesterol is slower until menopause when it increases rapidly and in postmenopausal elderly women, it is higher than men (Figure 11.6). Triglyceride concentrations also tend to increase with age. Serum HDL concentrations are higher and triglycerides are lower in females.

Exercise increases HDL cholesterol and lowers total cholesterol and triglycerides. Obesity decreases HDL cholesterol and increases triglycerides. Alcohol intake will increase HDL as well as triglycerides.

Table 11.3 Factors affecting serum lipid concentration

	Total cholesterol	HDL	LDL	TG
Age	I	I	I	I
Gender	M = F	F > M	M = F	M > F
Obesity	—	D	—	I
Alcohol	—	I	—	I
Exercise	D	I	D	D
Oestrogens	D	I	D	I
Androgens	I	?	?	?

*I = increased; D = decreased; M and F = males and females.

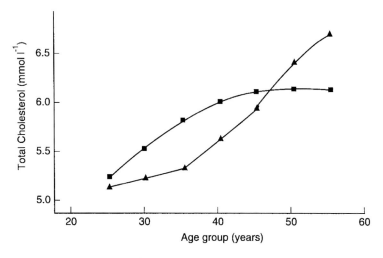

Figure 11.6   Changes in serum cholesterol with age.

Oestrogens tend to lower total cholesterol and LDL cholesterol; androgens raise total cholesterol.

Nutrition can influence serum lipids, e.g. a diet high in polyunsaturated fats will decrease LDL cholesterol.

In developing countries, the rise in cholesterol with age is not seen to the same extent as in developed countries. The risk of coronary heart disease (see Chapter 12) increases with increasing serum cholesterol concentration.

Disorders of Lipoprotein Metabolism

Hyperlipidaemia

Hyperlipidaemias were originally classified according to the type of lipoprotein involved, the WHO/Fredrickson's classification. This classification is no longer used as it gives little clue to the aetiology. The various phenotypes in this classification can be caused by primary or secondary disorders, and treatment with drugs or diet can change the classification. Hyperlipidaemia are now classified into primary, which are genetically determined, or secondary to other diseases.

Secondary hyperlipidaemia (Table 11.4)

Secondary hyperlipidaemias are common and its causes are listed in Table 11.4. The hyperlipidaemia will resolve when the primary condition is treated. Sometimes, secondary hyperlipidaemia may complicate an already existing primary hyperlipidaemia. Drugs causing secondary hyperlipidaemia are listed in Table 11.5. Antihypertensive drugs such as β-blockers and thiazide diuretics can cause hyperlipidaemia.

In diabetes, there is hypertriglyceridaemia, which may be associated with high total cholesterol and low HDL cholesterol.

Table 11.4   Causes of secondary hyperlipidaemia

	Cholesterol	TG	HDL
Diabetes mellitus	N or ↑	↑	↓
Obesity	N or ↑	↑	↓
Alcohol abuse	N	↑	↑
Hypothyroidism	↑	N or ↑	N
Chronic renal failure	N or ↑	↑	↓
Nephrotic syndrome	↑↑	↑↑	↓
Cholestasis	↑	N	

Table 11.5   Drugs causing dyslipidaemia

	Total cholesterol	Triglycerides	HDL cholesterol
Thiazide diuretics	↑	↑	—
Loop diuretics	↑	—	— (↓)
Potassium-sparing diuretics	—	—	↓
β-Blockers	—	↑	↓
α-Blockers	↓	↓	↑
Glucocorticoids	↑	↑	↑
Androgens	↑	—	↓
Oestrogens	↓	↑	↑
Tamoxifen	↓	(↑)	—
Raloxifene	↓	—	↓
Cyclosporin	↑	↑	↓
Retinoids	↑	↑	↓
Interferons	—	↑	—

Hypertriglyceridaemia is caused by decreased clearance as well as due to an increase in hepatic synthesis. In nephrotic syndrome, there is over-production of apo B100 as part of a generalised increase in protein synthesis. Renal failure is an important cause of secondary hyperlipidaemia and is due to increases in VLDL, LDL, and triglycerides. Lp(a) is markedly increased in renal disease even after transplantation. In hypothyroidism, LDL receptor function and LPL activity are reduced, giving risk to high LDL.

Case 11.1

A 65-year-old lady was referred to the lipid clinic for further investigation and management of her high cholesterol found on a routine examination. At the clinic, she admitted feeling tired most of the time and that she had put on nearly 10 kg in weight over the past 6 months. Results of investigations were as follows:

Serum		Reference Range
Cholesterol (mmol/L)	11.5	
Triglycerides (mmol/L)	2.1	0.6–1.7
TSH (U/L)	65	0.2–5.5
FT4 (pmol/L)	6	10.2–25.0

There was no protein in her urine. Blood glucose, liver and renal functions were all within the reference range.

High TSH with a low FT4 shows that she has hypothyroidism and her lipid abnormalities are likely to be secondary to the hypothyroidism. She was treated with thyroxine and her serum cholesterol and TSH decreased.

Primary hyperlipidaemias (Table 11.6)

1. *Familial hypercholesterolaemia*

   Familial hypercholesterolaemia (FH), an autosomal codominant hereditary disease is caused by a defect in the LDL receptor gene, which results in abnormal clearance of LDL particles from

Table 11.6 Classification of primary hyperlipidaemias

	Frequency	Defect	Serum lipids	Associated diseases
Familial hypercholesterolaemia (FH)	1 in 500 AD	LDL receptor defect	C, (LDL) ↑ TG (↑)	CHD +++
Familial defective apo B100	1 in 700 AD	Mutation of Apo B gene	C(LDL) ↑	CHD ++
Familial combined hyperlipidaemia (FCH)	1 in 20 AD	Overproduction of apo B100	TG(VLDL) ↑ C LDL↑	CHD +
Familial hypertriglyceridaemia	1 in 500 AD	Not known	TG (VLDL and chylomicron) ↑	Pancreatitis +
Polygenic hypercholesterolaemia	?	Interaction of genetics and dietary factors	↑C (LDL) ↑ TG (↑)	CHD +
Familial chylomicronaemia syndrome	1 in 1,000,0000	LPL deficiency Apo CII deficiency	TG (chylomicron +VLDL) ↑↑	Pancreatitis +++
Familial dysbetalipoproteinaemia	1 in 5000 AR	LDL Apo E_2	C (VLDL) ↑ TG↑	CHD ++ Peripheral vascular disease
Hyperalphalipoproteinaemia	AD	Mutation of CETP gene	C (HDL) ↑	None

*AD: autosomal dominant; AR: autosomal recessive; CHD: coronary heart disease; C: cholesterol; TRG: triglycerides.

plasma, leading to severe hypercholesterolaemia, xanthomatosis and premature coronary artery disease.

The frequency of heterozygotes is about 1:500 and that of homozygotes is about 1:1,000,000. In some areas of the world such as Finland, Israel, Quebec in Canada and South Africa, the frequency of heterozygotes can be as high in 1:80.

Five classes of mutations have been described. These results in the absence of LDL receptors, defective transport of the receptor from the endoplasmic reticulum to the Golgi apparatus, defective binding of LDL, defective internalisation of the receptor and defective recycling of the receptor.

Homozygous individuals develop severe atherosclerosis in childhood and do not survive beyond the age of 20 without aggressive treatment. They have severe xanthoma developing in early childhood and cardiac valve abnormalities especially aortic stenosis. Heterozygotes for FH have tendon xanthomas (over the extensor tendons of the fingers and Achilles tendon), premature atherosclerosis and a family history of premature coronary artery disease. Other manifestations include xanthelasmas, cutaneous cholesterol deposits around the eye and premature white rings in the cornea, corneal arcus.

Diagnostic criteria for FH is given in Table 11.7. Typically, plasma cholesterol concentrations are > 9 mmol/L with LDL

Table 11.7   Diagnostic criteria for familial hypercholesterolemia (Simon Broome criteria).

Definitive diagnosis of FH
- Serum cholesterol > 7.5 mmol/L (> 6.7 mmol/L in children < 16 years) or LDL cholesterol > 4.9 mmol/L in adults
- Tendon xanthomas present in patient or first or second degree relative

Possible diagnosis of FH
- Serum cholesterol > 7.5 mmol/L (> 6.7 mmol/L in children < 16 years) or LDL cholesterol > 4.9 mmol/L in adults
- A family history of myocardial infarction before the age of 60 in first-degree relative or serum cholesterol concentration > 7.5 mmol/L on a first or second degree relative

cholesterol > 6.4 mmol/L. HDL cholesterol and triglycerides are normal. In homozygotes, serum cholesterol concentration can be as high as 15–30 mmol/L.

Diagnosis of FH depends on the demonstration of very high serum cholesterol, family history, tendon xanthomata and history of premature coronary disease. Molecular diagnosis may be helpful when clinical manifestations and family history are equivocal in children or a condition mimicking FH is suspected or when the coexistence of two defects is suspected. Assays to determine the functional abnormality in the LDL receptor is available in some centres.

Management of heterozygotes with FH requires drug treatment to lower cholesterol; HMG CoA reductase inhibitors (statins) are the drug of choice. In homozygous FH, more aggressive treatments such as plasma exchange, LDL apharesis or heparin extracorporeal LDL precipitation (HELP) have been used. Liver transplantation may be required to provide LDL receptors to the patient.

Case 11.2

A 25-year-old man requested his GP to check his serum lipids as his father had died of myocardial infarction at the age of 48. His GP requested fasting glucose, lipids, renal and liver function tests and thyroid function tests. Results are as follows:

Serum		Reference Range
Cholesterol (mmol/L)	13.5	
Triglycerides (mmol/L)	1.0	0.6–1.7
TSH (U/L)	3.5	0.2–5.5
FT4 (pmol/L)	18	10.2–25.0

Blood glucose, renal and liver function tests were all normal excluding secondary causes of hyperlipdaemia. Based on the Simon Brook criteria, he was diagnosed as a possible case of familial hypercholesterolaemia and treated aggressively with statins.

2. *Familial defective apo B100*

Familial defective apo B100 is caused by a mutation of the apo B gene, resulting in impaired ligand interaction with the LDL receptor, causing delayed clearance of LDL. It is inherited as an autosomal dominant trait. Prevalence of this condition varies from 1:500 to 1:700. In some countries like Switzerland, prevalence may be higher, but it is absent in countries such as Turkey, Japan and Israel. Phenotypic expression of this disorder varies from no evidence of disease to clinical manifestations resembling that of heterozygote FH. Treatment of this condition is similar to that for heterozygote FH.

3. *Autosomal dominant hypercholesterolaemia*

This condition is indistinguishable from FH or familial defective apo B100. It is due to a mutation in proprotein convertase subtilisin kexin 9 (PCSK9) which increases the degradation of LDL receptors.

4. *Familial combined hyperlipidaemia (FCH)*

This is the most common form of primary hyperlipidaemia occurring in 1:50 to 1:30 Caucasians and accounts for 15–20% of patients with angiographically documented coronary artery disease before   the age of 60. FCH is inherited as an autosomal dominant trait with incomplete penetrance. The clinical manifestations of FCH are mild and non-specific. A mild to moderate elevation of plasma cholesterol (6–9 mmol/L), mild to moderate elevation of triglycerides (2–6 mmol/L) or both, lower HDL cholesterol and an increase in apo B are features in FCH. Subjects are frequently obese, have insulin resistance and have features of the metabolic syndrome. The aetiology of this condition remains unclear and is likely to be heterogeneous. Several metabolic abnormalities have been reported. These include hepatic over-production of apo B, prolonged postprandial elevation of plasma FFA, delayed clearance of postprandial chylomicron remnants and defective adipose tissue metabolism. At the moment, there is no specific or reliable marker of FCH. The lipid phenotype is affected by environmental factors. Diagnosis of FCH should be suspected if there is a family history of hyperlipidaemia, especially if family members show different lipoprotein phenotypes and if there is a family history of cardiovascular disease. Differential diagnoses include FH, and familial hypertriglyceridaemia. The treatment of FCH includes diet as well as lipid-lowering drugs.

5. *Familial hypertriglyceridaemia*

Familial hypertriglyceridaemia is rarely associated with clinical features other than corneal arcus and xanthelasmas. It is not thought to be associated with increased risk of CHD. In severe cases, there is increased risk of acute pancreatitis. Its prevalence is estimated to be 1:500. Triglycerides and VLDL are moderately raised, LDL and HDL are low. Typically triglyceride concentrations range from 2.3 to 5.7 mmol/L, which increase post prandially leading to postprandial lipaemia. In severe cases, the concentration may be higher than 20 mmol/L. This is often precipitated by obesity, high alcohol intake, diabetes mellitus or the use of oestrogens. The exact defect is not known but increased VLDL production with or without impaired catabolism of the VLDL has been demonstrated. Treatment involves dietary measures as well as the use of fibrates or omega-3 fatty acids.

Case 11.3

A 25-year-old woman was admitted with severe acute abdominal pain. She has had several previous episodes of abdominal pain. Investigations showed:

Serum		Reference Range
Cholesterol (mmol/L)	6.8	
Triglycerides (mmol/L)	34	0.6–1.7
Amylase (IU/L)	635	<235

> *The laboratory reported that the sample was lipaemic. Her high serum amylase suggested a diagnosis of acute pancreatitis. In view of the high triglycerides and a history of similar previous episodes a diagnosis of hyperchylomicronaemia causing acute pancreatitis was made.*

6. *Polygenic hypercholesterolaemia*

In most developed countries, hypercholesterolaemia is common; this is thought to be due to the interaction of genetic factors with environmental factors, mainly diet, hence the term polygenic (common) hypercholesterolaemia. The cholesterol concentration is not as high and the hypercholesterolaemia is seen later in life than in familial monogenic disorders. Tendon xanthomata is absent and there is overproduction of VLDL, which leads to high LDL, and there may be inability of non receptor-mediated catabolism. Obesity and high fat diets, particularly saturated fats, account for the wide variation in the prevalence of hypercholesterolaemia in different populations. The risk of coronary heart disease is increased. Treatment involves reduction in dietary fat and reduction in body weight to ideal weight. Occasionally lipid lowering drugs may be required.

7. *Familial chylomicronaemia syndrome*

Familial chylomicronaemia syndrome (FCS) is a rare disorder characterised by severe fasting hypertriglyceridaemia and accumulation of chylomicrons in plasma, which can lead to the development of eruptive xanthomas, lipaemia retinalis and pancreatitis. Genetic causes of the syndrome include familial deficiency of LPL, familial deficiency of apo C2 and familial inhibitor of LPL. In patients with familial forms of moderate hypertriglyceridaemia, chylomicronaemia may develop when other causes of hypertriglyceridaemia are present such as diabetes, drug treatment, or alcohol intake. Patients with FCS often present in childhood or adolescence with recurrent episodes of abdominal pain with or without pancreatitis precipitated by ingestion of

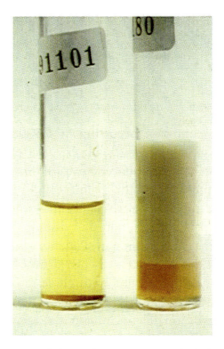

Figure 11.7 Appearance of plasma from a patient with chylomicronaemia. Plasma was left at 4°C overnight. A normal plasma sample is shown for comparison.

fatty meals. Serum triglycerides are increased to 20–70 mmol/L and may be as high as 200 mmol/L. Chylomicrons, which normally appear after a meal and disappear rapidly, persist up to 12 hours after a meal. VLDL concentrations are frequently increased, resulting in lipaemic plasma. Chylomicronaemia is easily detected as a discrete milky layer at the top when plasma samples are kept at 4°C (Figure 11.7).

Patients with FCS are not at increased risk of coronary heart disease. The diagnosis of FCS should be suspected in the presence of hypertriglyceridaemia, especially when they present at an early age in the absence of secondary causes. Confirmation of diagnosis may require the measurement of plasma LPL after intravenous heparin administration.

Treatment of FCS involves a low fat diet and avoidance of treatment or agents such as alcohol, diuretics, oestrogens, and β-adrenergic drugs which may increase triglycerides. Treatment with fibrates or nicotinic acid may also be necessary.

8. *Familial dysbetalipoproteinaemia (Type 3 hyperlipoproteinaemia, remnant or broad β hyperlipidaemia)*

This is a rare condition occurring in 1:5000 individuals. It is known as broad β hyperlipidaemia because of the characteristic serum lipoprotein electrophoretic pattern. Striated xanthomas or orange pigmentation of the palmar and plantar creases and orange yellow tuberoeruptive xanthomas on elbows, knees or buttocks are pathognomic of this disease. Xanthelasmas and corneal arcus may also be present. Patients with this condition are at increased risk of peripheral vascular disease as well as coronary artery disease. Biochemically, both cholesterol (9–10 mmol/L) and triglycerides are elevated. HDL is usually low and LDL also may be normal or low. This is inherited as a recessive trait and the phenotype is expressed when a second factor which raises triglyceride such as diabetes mellitus, hypothyroidism, obesity, hormones, or drugs, is present. The biochemical defect in this disorder is a reduced clearance of chylomicrons and VLDL remnants and is associated with apo E2. Apo E, which is important for the binding of remnant particles to the remnant receptor, is present as three variants (E2, E3 and E4). The presence of apo E2 seems to cause poor binding to the remnant receptor, leading to decreased clearance from the circulation. Treatment involves weight reduction if the patient is obese and drugs such as fibrates, or statins.

9. *Hyperalphaliproteinaemia*

In this disorder, serum HDL cholesterol is high and it is inherited as an autosomal dominant trait. Total serum cholesterol may be elevated with normal LDL cholesterol. It is not associated with increased risk of coronary heart disease. It does not require any treatment.

10. *Familial phytosterolaemia*

This is a rare condition presenting clinically with manifestations of FH such as xanthomatosis and premature atherosclerosis. The disease is characterised by increased intestinal absorption and reduced biliary secretion of plant sterols. The underlying genetic defect is unknown.

Further Reading

1. Crook M. The hyperlipidaemia and their management. *CPD bull Clin Biochem* 1999; 1:49–52.
2. Hoeg JM. Lipid disorders. *Endocrinol Metab Clin North Am* 1998; 21:503–738.
3. JBS2: The joint british societies guidelines on the prevention of cardiovascular disease in clinical practice. *Heart* 2005: Supplement V.
4. Knopp RH. Drug treatment of lipid disorders. *N Engl J Med* 1999; 341:498–511.
5. Tulenko TN, Sumner AE. The physiology of lipoproteins. *J Nucl Cardiol* 2002; 9:638–649.

Summary/Key Points

1. Lipids are essential energy sources and are structural components of cells. As lipids are insoluble in water, they are carried in association with proteins. These complexes, lipoproteins, are classified according to their density into chylomicrons, VLDL, LDL and HDL. Each lipoprotein particle has a lipid core surrounded by a layer of phospholipids, free cholesterol and proteins.
2. Proteins associated with lipoproteins are called apolipoproteins and they have specific functions.
3. Chylomicrons which contains apo B48 are the lightest of the particles and carry lipids absorbed from the diet to the liver and adipose tissue. VLDL is involved in the carriage of endogenous triglycerides from the liver and has apo B100. LDL is involved in the transport of lipids from the liver to peripheral tissues. HDL is

involved in reverse cholesterol transport, transport of cholesterol from peripheral tissue to the liver.

4. LDL binds to a receptor in the cell membrane and the complex is internalised and LDL is metabolised. Defects in the LDL receptor can give rise to familial hypercholesterolaemia.

5. Hyperlipidaemia may be secondary to other diseases, such as diabetes mellitus, nephrotic syndrome, renal failure and hypothyroidism; or due to primary defect in lipid metabolism, primary hyperlipidiaemis.

6. Of the primary lipid disorders, familial combined hyperlipdaemia is the most common disorder. The exact molecular defect is not well understood.

7. Familial hypercholesterolaemia (FH) is a well-characterised disorder due to a defect in the processing of LDL, or due to defect in the receptor or its function. The frequency of heterozygotes is about 1 in 500. FH causes premature coronary artery disease and xanthomatosis. Treatment with statins is necessary to reduce the cholesterol level. Homozygotes with FH die early in their 20s unless treated aggressively.

8. Other primary hyperlipidaemias includes familial hypertriglyceramia and familial chylomicronaemia. In these conditions, risk of acute pancreatitis is increased.

9. Polygenic hypercholesterolaemia is the most common form of hypercholesterolaemia and as the name implies, it is a result of interaction between genetic factors and environmental factors such as increased intake of saturated fat.

chapter 12

Hypertension and Cardiovascular Disease

Hypertension

Hypertension is not a disease but a risk factor for other diseases. There is a strong positive and continuous correlation between blood pressure and cardiovascular disease (CVD) (stroke, myocardial infarction, and heart failure), renal disease and mortality. Blood pressure is highly variable and has a normal distribution that is slightly skewed to the right. It increases with age — low at birth, rises rapidly to childhood values and then rises more slowly between the ages of 20 and 45 and then rapidly again. It tends to be higher in men than in women, especially under the age of 50. Blood pressure varies between different ethnic groups — African Americans tend to have higher blood pressure than Caucasians. In some communities (nomads in East Africa, Bushmen of South America and inhabitants of some pacific islands), blood pressure does not increase with age as in developed countries. Blood pressure increases during exercise, stress, and pain. It shows a circadian variation; lowest values are seen in the early hours of the morning and the highest values are seen on rising.

The definition of hypertension is arbitrary and is based on the association between blood pressure and increased risk of cardiovascular and other complications. The prevalence of hypertension varies, depending on the cut-off value used. Various international bodies have defined hypertension as a value of blood pressure of 140/90 mm of Hg or greater measured at least on two or more occasions. Based on this definition, the prevalence of hypertension is 33% in Caucasian men and 38% in African American men.

Aetiology of Hypertension

In most patients, there is no identifiable cause and this is described as essential hypertension. In a small minority (up to 10–15% of patients), a secondary cause can be found.

Pathogenesis of Essential Hypertension (Table 12.1)

Blood pressure is a result of the interaction between environmental and genetic factors. Genetic factors may account for up to 30% of the variation in blood pressure. Blood pressure among family members tends to be correlated and adoption studies show that the correlation in blood pressure between parents and natural children is higher than that between adoptive parents and children. However, this study also shows the importance of environmental influences, as there is a significant correlation in blood pressure between adopted siblings. Twin studies further confirm the importance of genetic factors. The correlation of blood pressure in monozygotic twins is greater than that in dizygotic twins. Genes responsible for hypertension have not yet been identified.

Table 12.1   Factors influencing blood pressure

Genetic
Age
Obesity
Alcohol intake
Nutrition
  • Sodium
  • Potassium
  • Calcium
  • Magnesium
Stress
Maternal nutrition

Environmental factors

1. *Nutrition*

   There is a strong body of evidence showing an association between salt intake and blood pressure and intervention studies show that reduction in salt intake will reduce blood pressure. The consensus is that moderate reduction in salt intake will be beneficial in reducing blood pressure. Intake of potassium, calcium, magnesium and possibly trace elements have a negative effect on blood pressure. The importance of nutrition in the pathogenesis of hypertension is shown by a recent study; the dietary approaches to stop hypertension (DASH) study which showed that a diet containing fruits, vegetables, low fat and adequate amounts of calcium, magnesium and potassium, but with low salt produced an anti-hypertensive effect comparable to that achieved with many pharmacological interventions. Recent studies also suggest that nutrition *in utero* may influence the development of hypertension in later life.

2. *Obesity*

   Obesity, especially central (visceral) obesity, correlates with blood pressure. The mechanism by which obesity raises blood pressure is not clear although there is some evidence that it may be related to sympathetic nervous system activity.

3. *Alcohol intake*

   High alcohol intake, i.e. more than 6 units a day (a unit being half pint of beer or one measure of spirit or a glass of wine), is associated with increased blood pressure. This relationship is independent of body mass index, age or social class and reduction in alcohol intake will lower the blood pressure. Some studies show that the association between alcohol intake and blood pressure is a J-shaped curve, i.e. absenteeism from alcohol may be associated with a slight rise in blood pressure compared to moderate intake of alcohol (1–2 drinks a day).

4. *Stress*

Acute stress increases the blood pressure. However, it is not known whether chronic stress will produce a sustained elevation of blood pressure.

Secondary Causes of Hypertension (Table 12.2)

About 10% of hypertensive subjects have an underlying disorder that could be treatable. The causes of secondary hypertension are given in Table 12.2.

Renal and renovascular hypertension

Renal disease leading to hypertension may account for 3–4% of hypertensive patients. In these patients, the hypertension is probably due to retention of salt and water. Abnormalities in the production of nitric

Table 12.2   Secondary causes of hypertension

Renal disease
Renovascular hypertension

Endocrine causes
- Primary hyperaldosteronism
- Phaeochromocytoma
- Cushing's syndrome
- Congenital adrenal hyperplasia
- Renin-secreting tumours

Coarctation of aorta

Drugs
- Carbenoxolone
- Liquorice
- Steroids
- Oral contraceptives

Rare genetic disorders
- Glucocorticoid suppressible hyperaldosteronism
- Liddle's syndrome
- Apparent mineralocorticoid excess

oxide due to the presence of circulating endogenous inhibitors may also contribute.

Renovascular hypertension, which accounts for 1–5% of the hypertensive population, is due to stenosis of one or both renal arteries. It is commonly due to atheromatous disease in older subjects and fibromuscular dysplasia in younger subjects. Hypertension in renovascular disease arises because of underperfusion of the kidney, leading to activation of the renin–angiotensin system.

In patients with the characteristics given in Table 12.3, renal and renovascular hypertension should be suspected and investigated further. Initial investigations should include urine analysis for protein, renal function tests (including electrolytes, urea and creatinine), fasting glucose, fasting lipids and chest radiography. Common renal diseases causing hypertension are given in Table 12.4.

Primary hyperaldosteronism

The incidence of primary hyperaldosteronism in an unselected population of hypertension is not known. Although many reports suggest that it is < 1%, recent reports suggest that it may be higher (up to

Table 12.3   Features suggestive of renal or renal vascular hypertension

History (past or family) or renal disease
Young age
Features of renal diseases — nocturia, polyuria, oedema, dysuria
Palpable kidney
Abdominal bruit
Reduced or absent leg pulses
Presence of protein or blood in urine

Table 12.4   Renal diseases causing hypertension

Diabetic nephropathy
Chronic glomerulonephritis
Chronic interstitial nephritis
Polycystic diseases of kidney

10%). Primary hyperaldosteronism is most commonly due to an adenoma of the adrenal gland (Conn's syndrome). Hypertensive patients with hypokalaemia should be investigated further for possible primary hyperaldosteronism (see Chapter 21 for further investigations).

Phaeochromocytoma

Phaeochromocytomas are catecholamine-secreting tumours and arise from chromaffin cells that are neuroectodermal in origin. Its origin may explain why phaeochromocytomas are sometimes associated with other tumours or conditions arising from neuroectodermal cells such as medullary carcinoma of the thyroid and parathyroid adenoma (multiple endocrine neoplasia syndrome (MEN) type 2), neurofibromatosis (Von Recklinghausen's disease), cerebellar haemogioblastoma and retinal angiomas (Von Hipple-Lindau disease). In adults, 90% of phaeochromocytomas occur in the adrenal medulla. Extra-adrenal forms of phaeochromocytomas are sometimes known as paragangliomas and these are more common in children (35%). The extra-adrenal sites include the sympathetic ganglia in the paraortic area of the abdomen, the bifurcation of the aorta, urinary bladder, the hilum of liver, kidney and very rarely, in the chest. Phaeochromocytomas are familial in about 10% of cases such as in MEN type 2 syndrome, neurofibromatosis and Von Hipple-Lindau syndrome. Most tumours are benign. Phaeochromocytomas account for a small proportion (< 1%) of hypertensive patients. However, it has been suggested that a high percentage of phaeochromocytomas are unsuspected during life and only identified at autopsy. About 10% of phaeochromocytomas are discovered incidentally.

1. *Catecholamine metabolism*

The synthesis and metabolism of catecholamine are shown in Figure 12.1. Tyrosine, the precursor of catecholamine, is converted to L-dopa by tyrosine hydroxylase, which is the rate-limiting step in the biosynthetic pathway. Decarboxylation of L-dopa by the enzyme dopa decarboxylase forms dopamine.

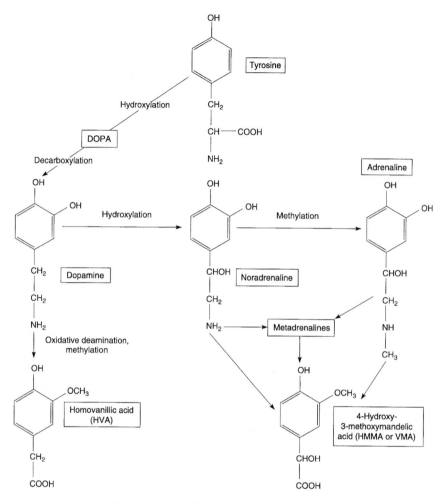

Figure 12.1   Catecholamines metabolism.

Dopamine is the principle catecholamine secreted by neuroblastomas, but only occasionally by phaeochromocytomas. Dopamine is further hydroxylated to noradrenaline, which is then methylated to adrenaline. As the enzyme catalysing this reaction, phenylethanolamine N-methyltransferase (PNMT), is relatively absent outside the adrenal gland, extra-adrenal phaeochromocytomas rarely produce adrenaline.

Catecholamines are metabolised by catechol-O-methyltransferase (Figure 12.1) to normetanephrine and metanephrine which are further metabolised to vanillylmandelic acid (VMA). Monoamine oxidase converts noradrenaline to dihydroxyphenylglycol, which is converted to hydroxymethoxyphenylglycol (HMPG).

2. *Clinical features*

Patients with phaeochromocytoma may have symptoms such as headaches, inappropriate and excessive sweating, palpitations, paroxysmal hypertension, nausea and pallor. In over 50% of patients, hypertension is sustained. In hypertensive patients, the triad of sweating, tachycardia and headache is highly suggestive of phaeochromocytoma.

3. *Diagnosis*

Biochemical diagnosis involves the demonstration of increased excretion of catecholamines or its metabolites. There is now consensus that measurement of metanephrines is better than catecholamines in the diagnosis of phaeochromocytoma. One of the reasons is that for a phaeochromocytoma to be detected biochemically, catecholamine secretion has to increase 15-fold whereas for metanephrines a four-fold increase is adequate. Furthermore, catecholamine secretion is usually episodic but there is a continuous metabolism of catecholamines to metanephrines. Although plasma metanephrines are more sensitive, they are less specific and it has been suggested that urinary fractionated metanephrines may be a better screening test, followed by plasma-free fractionated metanephrines (fractionated refers to quantitation of individual metabolites, namely metanephrine, normetanephrine and methoxytyramine and free metanephrines refers to unconjugated metanephrines). It is usual to collect three 24-hour urine samples before excluding phaechromocytoma. In children and when 24-hour urine is difficult to collect, a spot urine can be used and the results are expressed in relation to creatinine excretion.

Plasma chromogranin A is a secretory protein located in the secretory vesicle of neuroendocrine cells. It is co-secreted with catecholamines. Measurements of chromogranin A have been suggested as a test for phaeochromocytoma; however, this is non-specific. A combination of chromogranin A and fractionated metanephrines may be more sensitive. Chromogranin A is also useful as a tumour marker for follow ups of patients with phaeochromocytoma.

Factors which may increase noradrenaline excretion include stress, hypoglycaemia, heart failure, surgery, exposure to cold, exercise, sleep apnoea and myocardial infarction. Many drugs such as β-blockers and calcium channel blockers can influence the results of catecholamine and metanephrines (see Table 12.5).

Occasionally, suppression tests may be necessary when other tests are equivocal. Ganglion-blocking drug pentolinium or centrally active α_2-agonist clonidine are the most widely used drugs to suppress catecholamine secretion. Pentolinium is effective in suppressing noradrenaline and adrenaline from the adrenal medulla and it has a short half-life. Before a pentolinium suppression test is done, renal function should be assessed, as this drug is entirely excreted by the kidney. It should not be used if serum creatinine is > 150 μmol/L. The patient should rest supine for 30 minutes before the test. After

Table 12.5 Drugs causing interference in plasma and urinary catecholamines and metanephrines

	Catecholamines		Metanephrines	
Drug	Plasma	Urine	Plasma	Urine
Physiological Interference				
α-Blockers	?	Y	—	—
β-Blockers	—	Y	Y	Y
Calcium channel blockers	Y	Y	—	—
Tricyclic antidepressants	Y	Y	—	Y
Phenoxybenzamine	Y	Y	Y	Y
MAO inhibitors, phenothiazines	Y	Y	Y	Y
Ephedrine, amphetamines	Y	Y	Y	Y
Analytical Interference				
Buspirone	—	Y	—	Y
Paracetamol	Y	Y	—	Y

taking two blood samples, 5 minutes apart from an intravenous cannula, a bolus injection of pentolinium (2.5 mg) is given and further samples taken at 10 and 20 minutes. The normal response to pentolinium is a fall in both adrenaline and noradrenaline to the reference range or by 50% from the base line value. Clonidine is an α_2-adrenoceptor agonist and it suppresses the release of noradrenaline from neurones but not from the adrenal medulla or phaeochromocytoma. A 50% reduction in plasma noradrenaline or normetanephrine excludes phaeochromocytoma.

Localisation of the tumour is usually done by imaging techniques such as computer tomography (CT), magnetic resonance imaging (MRI) or metaiodobenzylguanidine (MIBG) isotopic scanning. MIBG scanning is useful in detecting extra-adrenal tumours. In rare instances, selective venous sampling and measurement of plasma catecholamines may be necessary to localise the tumour.

4. *Management*

Surgical removal of the tumour is the treatment of choice. Perioperatively, patients should be treated with α-blockers such as phenoxybenzamine; some patients may also need a β-blocker to reduce the effect of catecholamine surge as a result of handling the tumour.

Case 12.1

A 67-year-old man was referred by GP to the hypertension clinic for investigation and management of hypertension. He was on three types of anti hypertensive drugs and his blood pressure was still 175/100. On further questioning, he gave a history of several episodes of sweating and anxiety in the past. Urine was collected and sent for analysis.

Urine:

Metanephrine	3.5 μmol/24-hr	(< 1.9)
Normetanephrine	7.9 μmol/24-hr	(< 3.9)

When hypertension is difficult to control with several drugs, a secondary cause should always be excluded. With a history of episodes of anxiety and

---

> *sweating, phaeochromocytoma is likely and the urine metanephrines confirmed the diagnosis. Imaging studies showed a right adrenal mass and he underwent a right adrenalectomy. His blood pressure returned to normal after the operation.*

Exogenous corticosteroids or drugs related to mineralocorticoids

Liquorice, carbenoxolone and mineralocorticoid-containing nasal sprays may lead to hypertension. The active ingredient in liquorice and carbenoxolone, glycyrrhizic acid, inhibits 11β-hydroxysteroid dehydrogenase enzyme in the kidney, the enzyme responsible for inactivation of cortisol. Inhibition of the enzyme allows cortisol to activate the mineralocorticoid receptor and cause hypertension. Cases of hypertension have also been described with the use of nasal sprays containing mineralocorticoids or with the use of chewing tobacco-containing liquorice. Plasma renin activity and aldosterone concentration will be suppressed in these patients.

Cushing's syndrome

About 80% of patients with Cushing's syndrome have hypertension, which is usually mild (see Chapter 21).

Congenital adrenal hyperplasia

In the rare forms of congenital adrenal hyperplasia due to deficiencies of 11β-hydroxylase or 17α-hydroxylase, hypertension may be seen. In 11β-hydroxylase deficiency, deoxycorticosterone is increased causing hypokalaemia, hypertension and virilisation. In 17α-hydroxylase deficiency, there is no associated virilisation but there is hypokalaemia, hypertension and hypogonadism (see Chapter 21).

Coarctation of the aorta

Coarctation of the aorta is a narrowing of the aorta, most commonly at the region of the ductus. This is usually diagnosed in childhood and rarely in adult life. It is more common in males (male:female ratio 4:1) and may be associated with other congenital abnormalities. The diagnosis is usually suspected if there is delay or absence of femoral pulse or when enlarged collateral vessels are present. The blood pressure is increased in the arms but not in the legs. Arteriography will establish the diagnosis and treatment is by surgical correction.

Rare genetic causes of hypertension

Glucocorticoid suppressible hyperaldosteronism is a rare inherited (autosomal dominant) syndrome, where a chimeric gene is produced from unequal crossover at meiosis between genes encoding aldosterone synthase and adrenal 11β-hydroxylase. This results in aldosterone secretion, which is ACTH dependent.

Liddle's syndrome, a rare inherited syndrome where there is early onset of hypertension and hypokalaemia, is due to a mutation in the α- or β-subunit of the amiloride sensitive epithelial sodium channel, leading to increased activity, thereby causing increased sodium reabsorption and potassium secretion. Plasma renin activity and aldosterone concentration are both suppressed.

Apparent mineralocorticoid excess syndrome is an autosomal recessive disorder due to a mutation in the renal 11β-hydroxysteroid dehydrogenase. Features of this syndrome are similar to that seen after liquorice ingestion.

Investigation of Hypertensive Patients

Investigations of hypertensive patients are necessary to assess the effect of hypertension on other organs, to identify associated risk factors such as hyperlipidaemia, and to identify secondary causes.

Investigations necessary in all hypertensive patients are listed in Table 12.6. Urine analysis by dipstick is necessary to identify proteinuria, haematuria and glycosuria. If there is proteinuria, urinary tract infection should be excluded and the patient may need further investigation to exclude renal hypertension. In patients with glycosuria,

Table 12.6   Investigation of hypertensive patients

Initial investigations

- Urine analysis (dipstick)
  — Protein
  — Glucose
  — Blood
  — Nitrate

- Blood

  — Serum electrolytes, e.g. sodium, potassium, calcium
  — Urea/creatinine
  — Uric acid
  — Glucose
  — Gamma-glutamyl transpeptidase (GGT)
  — Lipids — cholesterol, triglyceride and HDL cholesterol

- Full blood count
- Chest X-ray
- ECG

Further investigations

- 24-hour urine protein
- Urinary catecholamines
- Plasma renin:aldosterone ratio
- Imaging studies, e.g. ultrasound of kidney

---

diabetes mellitus, a common cause of renal disease leading to hypertension, should be excluded.

Serum sodium concentration may alert to the possibility of primary hyperaldosteronism if it is raised or near the upper end of the reference range. Low sodium suggests secondary hyperaldosteronism such as malignant hypertension, renal or renovascular disease. Sodium concentration may also be lowered by the use of diuretics.

Serum potassium concentration is important to exclude hyperaldosteronism. Patients with low or low normal potassium concentration should be further investigated. Usually, there is accompanying metabolic alkalosis. High serum potassium may be seen in patients with renal impairment or in patients taking potassium-sparing diuretics.

Creatinine is important to exclude renal disease or renovascular disease. A high serum calcium suggests primary hyperparathyroidism in which hypertension is a common finding. Uric acid will be raised in patients with renal disease and in 40% of patients with essential hypertension. There is some suggestion that high uric acid itself is an independent risk factor for cardiovascular disease. If uric acid is elevated, the cause of this should be investigated. The measurement of GGT may help to identify those hypertensive subjects in whom high alcohol intake may be a contributing factor. Blood glucose estimation is done to exclude diabetes (see Chapter 10). Hyperlipidaemia in hypertensive patients disproportionately increases the risk of cardiovascular disease; cholesterol, triglycerides and HDL should be measured in all hypertensive patients.

A second line of investigation, which is necessary to identify secondary causes of hypertension, will depend on initial findings and is discussed under the appropriate sections.

Management

The management of hypertensive patients depend on the cause. Patients without a treatable cause should be advised to take active exercise, reduce their alcohol intake, reduce weight if obese and eat a low fat, low salt diet rich in vegetables and fruits. Drugs used in the treatment of hypertension include diuretics, β-blockers, calcium channel blockers, ACE inhibitors, α-blockers, and angiotensin receptor blockers. The regular monitoring of renal function and electrolytes during treatment is required to avoid complications and side effects.

Cardiovascular System

Ischaemic Heart Disease (IHD)

Ischaemic heart disease (IHD) is the most common cause of death in most industrialised countries and is rapidly increasing in developing countries. IHD is the manifestation of atherosclerosis in the coronary arteries. Atherosclerosis, which affects medium and large vessels, is an

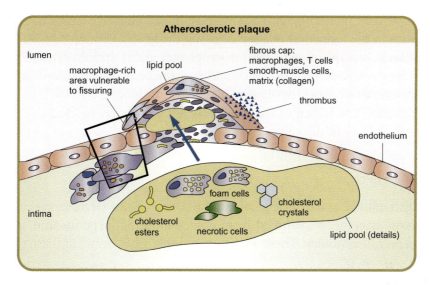

Figure 12.2 Diagram showing an atherosclerotic plaque.

inflammatory process that begins early in life and results in the deposition of lipid, fibrin and calcium in the intimal layers of the arteries, leading to the hardening, narrowing and eventual occlusion of the vessel. Atherogenesis is a complex process due to the interaction of several factors (Figure 12.2). The exact mechanism of atherosclerosis is not fully understood. The most commonly accepted theory is the response to injury theory. According to this theory, an important early event in this process is the damage to the endothelium of the blood vessel. Oxidized lipoprotein, infectious agents, hyperglycaemia, hyperhomocystenemia, smoking or hypertension may cause this damage. The damage to the endothelium allows for the migration of LDL into the vessels, which gets deposited in the intima. LDL may undergo oxidation and oxidized LDL stimulates the production of adhesion molecules, VCAM-1 and MCP-1. These attract monocytes into the vessel wall where they become macrophages. These macrophages take up oxidized LDL via the scavenger receptor and become 'foam' cells. When a large number of foam cells accumulates, fatty streaks can be seen on the intimal surface of the vessel. Various growth factors and cytokines produced by the endothelium and smooth muscles stimulate the proliferation of smooth muscle cells, which migrate towards the lumen and secrete collagen. This results in the formation of a cap covering the lipids. This cap consists of collagen, smooth muscle cells, macrophages and T lymphocytes. This plaque protrudes into the lumen and over the years it grows slowly and may eventually block the lumen. Reduction in blood flow will lead to ischaemia. The lumen of the vessel may get completely blocked due to the formation of thrombus on this plaque.

Risk factors

Although the exact cause(s) of atherosclerosis is not clear, several risk factors have been identified from epidemiological, experimental and intervention studies. These are listed in Table 12.7. Some of these risk

Table 12.7   Risk factors for the development of IHD

Primary risk factor
- Genetic predisposition
- Hypertension
- Smoking
- Hyperlipidaemia

Secondary risk factors
- Age
- Male sex
- Lack of exercise
- Obesity
- Diabetes
- Stress
- Alcohol
- Ethnicity
- Socioeconomic factors
- Homocysteine
- C-reactive protein

Life style factors
- Nutrition
  — Fat
  — Antioxidants
  — Salt intake

Environmental influence in early life

factors are inherent biological traits such as age, sex and genetic predisposition, while others are physiological characteristics such as blood pressure, adiposity and serum lipids and the third group of factors are related to life style factors such as smoking, alcohol consumption and nutrition. Lipid metabolism and the causes of hyperlipidaemia are discussed in Chapter 11.

Numerous studies have shown that the risk of CHD increases with increasing total cholesterol concentration (Figure 12.3). This association is mainly due to the LDL cholesterol fraction and more particularly to small dense LDL particles. HDL cholesterol is inversely associated with CHD risk and this is more so in women. In addition, lipoprotein (a) is independently associated with increased risk of CHD.

- Homocysteine

  Homocysteine is a sulphur-containing non-essential amino acid formed from methionine. Homocysteine is metabolised either by transsulphuration or remethylation (Figure 12.4). In transsulphuration,

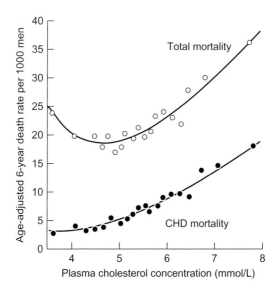

Figure 12.3 Relationship between serum cholesterol and risk of coronary heart disease.

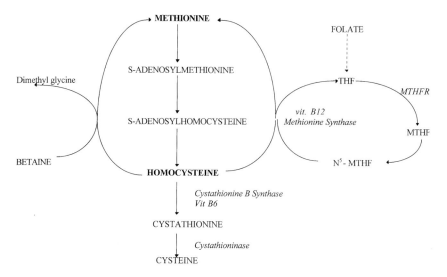

Figure 12.4   Metabolism of homocysteine.

homocysteine condenses with serine to form cystathionine in a reaction catalysed by a vitamin B_6-dependent enzyme. In the remethylation reaction, homocysteine is converted to methionine by an enzyme system that requires vitamin B_{12} and folate. Deficiency of vitamin B_{12}, folate or vitamin B_6 will result in high serum homocysteine. Other factors, which cause high serum homocysteine, are renal failure and inherited disorders. Severe hyperhomocysteinemia (homocystinuria) is due to deficiency of the enzyme cystathionine β-synthase. This is a rare inborn error of metabolism, associated with thromboembolic complications. Minor degrees of hyperhomocysteinemia may arise from mutations in the gene coding for the enzyme methyl tetrahydrofolate reductase (MTHFR); up to 5% of general population may have a mutation in this gene. Hyperhomocysteinemia is associated with increased risk of thromboembolic complications and cardiovascular disease. It is now being recognised as a modifiable risk factor for IHD. Recent intervention trials in which homocysteien concentration was lowered by folate, vitamins B_{12} and B_6, however, failed to show significant improvements in IHD.

Myocardial Infarction

Acute myocardial infarction (MI) occurs when circulation to a part of the heart is interrupted and necrosis follows. Acute MI is characterised by severe chest pain. When the blood supply is inadequate, myocardial ischaemia follows and this usually manifests as angina pectoris. Acute MI is the leading cause of death in many countries.

Acute coronary syndrome (ACS) is the term applied to signs and symptoms related to myocardial ischemia and includes MI with typical ECG changes of ST segment elevation (STEMI), MI without ST segment elevation (NSTEMI) and unstable angina. About 10% of patients attending emergency departments present with chest pain and of these, only 5–10% are diagnosed with MI. It is important to diagnose MI early to start thrombolytic treatment and to exclude those without MI for early discharge.

WHO criteria for the diagnosis of MI include two of the following three criteria: typical history of chest pain, ECG changes and elevated cardiac enzymes. More recently, the rise in troponins has been included as an essential criteria for the diagnosis of MI.

Diagnosis of acute MI

Biochemical diagnosis of acute MI is based on the release of myocardial contents such as enzymes and proteins into the circulation. Some characteristics of biochemical cardiac markers are given in Table 12.8.

Table 12.8 Characteristics of cardiac markers

	M. wt.	Rise (hour)	Peak (hour)	Time to return to normal (day)
CK	86,000	3–8	10–24	3–4
CKMB	86,000	3–8	10–24	2–3
LD1	135,000	8–12	72–144	8–14
Myoglobin	18,000	1–3	6–9	1
Troponin I	23,000	3–8	24–48	4–10
Troponin T	42,000	3–8	24–48	4–10

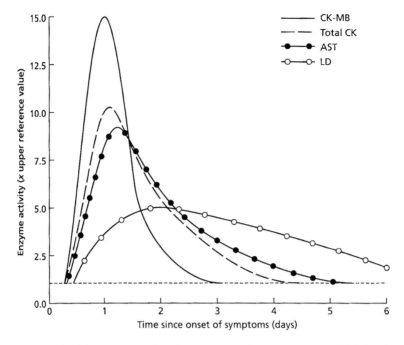

Figure 12.5   Time course of cardiac enzymes after acute myocardial infarction.

CK, AST and LD are the common enzymes measured in the diagnosis of MI. The pattern of release of these enzymes is shown in Figure 12.5.

CKMB activity starts to rise 3 to 8 hours after MI to values above the reference range and reaches a peak at approximately 24 hours and returns to baseline after 48–72 hours. This basic pattern may be affected by several factors including size of the infarct, concomitant skeletal muscle injury, composition of the myocardium and reperfusion (spontaneous or following thrombolytic treatment). Skeletal muscle contains < 2% of CKMB and injury to skeletal muscle may increase the absolute amount of CKMB but the relative concentration is usually < 5%. Measurement of CKMB mass is more sensitive and specific for MI than measurement of activity by immunoinhibition assays. Sensitivity of CKMB at admission varies from 17–62% and rises to 92–100% at 3 hours after presentation. Furthermore, serial measurements of CKMB at 0, 3, 6 and 9 hours after presentation will be more sensitive and allow earlier detection of MI.

1. *Troponins*

Troponin is a regulatory complex of the muscle and consists of three proteins T, C and I. Troponin T and I are very specific for cardiac muscle therefore, they are more useful than enzymes. Troponins are rapidly released and the rate of release is similar to that of CKMB. Both reach a peak at about 24 hours but troponins remain high for 4–10 days probably due to continuous release from the bound myofibrillar pool. The sensitivity of troponin I and troponin T for the diagnosis of acute MI approaches 100% by 12 hours after the onset of chest pain. Troponin T is cleared by the kidney and may be increased in renal failure. Although troponins are very specific for myocardium, they are not good in the early detection of MI. Recently, more sensitive troponin assays capable of detecting values below 0.01 μg/L have been developed and early studies show that these may be useful in the diagnosis of MI as early as 3 hours after chest pain.

Several studies have shown that increases in troponin, even without evidence of MI, are a poor prognostic marker of future cardiac events. As troponins are sensitive markers of myocardial damage, elevated troponins have been found in many conditions including sepsis, pulmonary embolism, hypothyroidism, cardiac failure, pericarditis and myocarditis. In all these conditions, this is a poor prognostic sign.

2. *Myoglobin*

Myoglobin is an oxygen-carrying haem protein present in skeletal and cardiac muscle. It is present in the cytoplasm and is of relatively low molecular weight (17.8 kDa). Myoglobin is released earlier than other proteins due to its small size. It becomes abnormal within 2 hours and peaks at 6 to 9 hours. As myoglobin is rapidly cleared from the circulation, it returns to normal in 24 hours and is a useful biochemical marker of early diagnosis of MI. Although myoglobin is highly sensitive for early diagnosis of MI, it has poor specificity (60–95%). It is increased in skeletal muscle injury and renal failure. Therefore, a negative serum myoglobin result is helpful in excluding MI, but a positive result needs confirmation with other cardiac-specific markers. The specificity of myoglobin can be increased by the simultaneous measurement of carbonic anhydrase III (CAIII) that is present in skeletal muscle but not in cardiac muscle. The ratio of myoglobin/CAIII will increase in MI but not in skeletal muscle injury. However, assays for CAIII are not widely available.

Other biochemical markers, which have been examined, include ischemia-modified albumin, heart fatty acid binding protein and glycogen phosphorylase BB. Further evaluation is awaited on the usefulness of these markers.

Case 12.2

A 68-year-old man was brought to the emergency department at 11 p.m. with a history of severe chest pain which started after his evening meal at 9 p.m. On admission, he was in severe pain and ECG failed to show typical changes seen in myocardial infarction. A blood sample was taken for troponin T and the results was 0.01 μg/L (< 0.1).

Although serum troponin T was not raised, myocardial infarction cannot be excluded as the sample was taken within 3 hours of his chest pain. Troponins start to rise 4 to 6 hours after a myocardial infraction and remain elevated for up to 10 days. A repeat sample taken 3 hours later showed an elevated troponin and he was treated with thrombolytic agents.

---

3. *Clinical uses of biochemical markers of myocardial damage*

As CK-MB and troponins do not start to rise until 3 to 8 hours after infarction, these are not useful for the exclusion of MI on admission. A negative myoglobin test will exclude MI at this stage. By 12 hours, troponins are 100% sensitive and they are also useful in late diagnosis of MI.

Acute MI presenting within 12 hours of onset of symptoms is treated with thrombolytic treatment. If this treatment fails, other forms of treatment such as angioplasty may be necessary. The success of thrombolytic treatment can be assessed by the accelerated released of CK and other markers (such as troponin and myoglobin) — wash out phenomenon. Of these, myoglobin is the best marker of reperfusion — a four-fold increase in myoglobin 90 minutes after start of therapy indicates complete reperfusion.

Troponins are particularly useful in the diagnosis of myocardial ischaemia during cardiac and non-cardiac surgery as CK and CKMB are released from damaged skeletal muscle.

In patients with unstable angina, troponins are found to be useful to identify those patients with poor prognosis. A high troponin on admission in patients with unstable angina predicts higher mortality.

Heart Failure

Heart failure is defined as the inability of the heart to meet the need of the tissues. Primary abnormality in heart failure is a decrease in the left ventricular function, leading to abnormalities in other systems such as skeletal muscle and kidney. A decrease in cardiac output as a result of left ventricular dysfunction leads to compensatory mechanisms, which include alteration in renin–angiotensin system, sympathetic nervous system, natriuretic peptides, ADH and endothelin. Stimulation of the renin–angiotensin system leads to vasoconstriction and increased aldosterone, which causes retention of sodium and water. Sympathetic system is activated via baroreceptors and this helps to maintain cardiac output. Chronic sympathetic stimulation leads to further activation of renin angiotensin system.

Natriuretic peptides, ANP and BNP, are also released in heart failure as a compensatory response to volume expansion. ADH is increased in heart failure; in severe heart failure, it may contribute to the development of hyponatraemia. Endothelin secreted by vascular endothelium is also increased, causing vasoconstriction and sodium retention. In addition, there is an increase in prostaglandins E_2 and I_2, activation of kallikrein system and an increase in TNF-α, which may contribute to the cachexia of heart failure.

In early asymptomatic stages of heart failure, plasma noradrenaline, renin and BNP and ANP are increased. Concentration of noradrenaline and BNP and ANP appear to be predictors of development of symptomatic heart failure. In established heart failure, concentration of adrenaline, ANP and BNP are good predictors of prognosis.

In the investigation of patients suspected of heart failure, in addition to chest X-rays, ECGs and echocardiography, laboratory investigations are also necessary. Full blood count is done to exclude anaemia. Serum electrolytes and renal function are necessary to detect abnormalities such as hyponatraemia, hypo- or hyperkalaemia and to assess renal function, which may be impaired. Electrolyte abnormalities such as hypo- or hyperkalaemia increase the risk of cardiac arrhythmias. Serum magnesium measurement is necessary to detect hypomagnesaemia that develops as a result of long-term diuretic treatment. Heart failure may lead to abnormalities in liver function tests as a result of liver congestion. It is usual to do thyroid function tests to detect and correct thyroid disease that contributes to the heart failure.

Naturetic peptides, ANP, BNP are elevated in heart failure. BNP is secreted as a large molecule, proBNP, which is then cleaved to produce BNP and a large N terminal fragment, NT-proBNP. In heart failure, both BNP and NT-proBNP are elevated. Many laboratories measure NT-proBNP as this is more stable *in vitro* after collection of blood. Measurements of BNP or NT-proBNP are useful to exclude heart failure. Many studies have shown that if the peptides are normal in serum, the patient is unlikely to have heart failure and costly echocardiography is unnecessary. This test is now routinely used in patients suspected of heart failure. In addition, this test is useful as a prognostic marker of outcome in heart failure.

Case 12.3

A 72-year-old lady went to see her GP feeling unwell, tired and getting out of breath on exertion. On examination, she had no clinical signs and her GP sent a sample to the laboratory for renal function tests thyroid function tests and NT-proBNP. Her renal and thyroid function results were all normal:

Serum NT-proBNP 750 ng/L (< 350)

A raised NT-proBNP suggested heart failure and her GP sent her for an echocardiogram, which confirmed the diagnosis.

Further Reading

1. Barron J. Phaeochromocytoma: Diagnostic challenge for biochemical screening and diagnosis. *J Clin Pathol* 2010; 63:669–674.
2. Christenson RH. Biochemical marker of myocardial infarction — cardiac troponin. *Clin Chem* 2007; 53:545–546.
3. Clark GH. Cardiac markers. CPD *Bull Clin Biochem* 1999; 1:108–111.
4. Davies MK, Gibbs GR, Lip GYH. ABC of heart failure: Investigation. *Br Med J* 2000; 320:297–300.
5. Hammer F, Stewart PM. Investigating hypertension in a young person. *B Med J* 2009; 338:b1043.
6. Tang WH, Francis GS, Marrow DA *et al*. NACB laboratory practice guidelines: Clinical investigation of cardiac biomarker testing in heart failure. *Circulation* 2007; 116:e99-e109.
7. Wu AHB, Apple FS, Gibler WB *et al*. Recommendations for the use of cardiac markers in coronary artery disease. *Clin Chem* 1999; 45:1104–1121.

Summary/Key Points

1. Hypertension is a common condition and is a major risk factor for cardiovascular disease — myocardial infraction, stroke and renal failure.
2. About 10% of patients with hypertension have treatable secondary causes of hypertension. These include renal disease, renal artery stenosis, primary hyperaldosteronism, phaeochromocytoma and Cushing's syndrome.
3. Investigation of hypertensive patient is important to exclude secondary causes, as well as to assess organ damage as a result of hypertension and to assess cardiovascular risk.

4. Investigations include, full blood count, urine examination, blood glucose, lipids, electrolytes, GGT, uric acid, renal function tests and calcium.

5. Atherosclerosis is a chronic inflammatory disease of the arteries and leads to ischemic heart disease. There are many risk factors for atherosclerosis and laboratory investigations are helpful in identifying these risk factors.

6. Myocardial infarction is the leading cause of death and early diagnosis is important to start thrombolytic treatment.

7. Biochemical markers are essential for the diagnosis of myocardial infarction and these include CKMB, myoglobin and troponins. Of these, troponins are highly specific and used widely. Myoglobin is useful to exclude MI within 4 hours of presentation.

8. Heart failure is a common condition in the elderly and these patients have non-specific symptoms. Naturetic peptides are produced in the heart and are elevated in heart failure.

9. Measurement of BNP or NTproBNP (product released with BNP) is useful in excluding heart failure.

chapter 13

Clinical Enzymology

Introduction

Enzymes are proteins that catalyse specific reactions. Many enzymes are synthesised within the cell and function intracellularly while others, a small number, are secreted and function in the extracellular environment. The latter enzymes are usually secreted in an inactive form, which have to be activated before they can function. Examples of this type of enzymes are those found in the gastrointestinal tract and those in the blood clotting and fibrinolytic cascades. Enzymes, which function intracellularly, find their way into the circulation in small amounts; probably released as a result of cell turnover or leakage. Increased activities of these enzymes are indicative of damage or pathology of tissues. As the concentration of enzymes intracellularly are several thousand folds greater than that in the plasma, an increase in plasma enzyme activity is a sensitive index of even minor cellular damage.

Factors Affecting Serum Enzyme Activity (Table 13.1)

The activity of a given enzyme in serum at a given time is a balance between entry into circulation and the rate of removal of the enzyme (Table 13.1). Decreased production of intracellular enzymes due to genetic factors (e.g. alkaline phosphatase (ALP) in hypophosphatasia) or disease (e.g. cholinesterase in liver disease) will result in reduced entry of enzymes into circulation. An increase in the number or activity of cells may cause increased entry. For instance, increased osteoblast

Table 13.1   Factors affecting serum enzyme activity

Factors affecting enzyme entry into serum
- Production
  — Genetic variation
  — Number and activity of cells
- Leakage of enzymes into serum
  — Size of the enzyme molecule
  — Distribution within the cell
  — Mode of transfer
  — Vascularity of organ
  — Permeability of capillaries

Factors affecting clearance
- Reticuloendothelial cells
- Renal clearance

number and activity cause increased ALP. Increased synthesis of ALP enzyme occurs in cholestasis as a result of enzyme induction.

Enzymes are large molecules and are retained within the cell by the cell membrane, the integrity of which depends on ATP production. Any interference in the production of ATP, for example hypoxia, will lead to increased cell membrane permeability, hence leakage of intracellular enzymes. Cell membranes may also be damaged by direct toxic effects of chemicals (alcohol, drugs, etc.), and immune mechanisms (cytotoxicity, analphylaxis) leading to increased entry of enzymes into circulation.

The rate of leakage of enzymes into the circulation from 'damaged' cells depends on several factors: (i) intracellular location: enzymes in subcellular organelles such as mitochondria appear later than cytoplasmic enzymes; (ii) size of the enzyme molecule: smaller ones appearing earlier; (iii) mode of transfer to circulation: in tissues such as the liver where the capillaries are permeable and very vascular, enzymes directly enter the circulation from the cells. In tissues such as the muscles where the capillaries are relatively impermeable, enzymes enter the circulation via the lymph.

Most enzymes entering the circulation are removed by the reticuloendothelial system via a receptor-mediated endocytosis. This is followed by fusion with lysosomes and finally digestion of the enzyme protein. Renal clearance does not play an important role in the clearance of most enzymes except amylase, which is small enough to be cleared by glomerular filtration. The half-life of enzymes in circulation varies from 3 hours (e.g. creatine kinase) to 170 hours (e.g. placental alkaline phosphatase). The rate of clearance of an enzyme may affect the serum activity.

Specificity of Enzyme Tests

Enzymes are present in more than one tissue and therefore are not very specific for a given organ. Table 13.2 lists the enzymes, which are diagnostically useful, their tissue origin and their potential clinical

Table 13.2 Distribution of some enzymes used diagnostically

Enzyme	Origin	Main application
Acid phosphatase	Prostate, erythrocytes, bone	Prostatic carcinoma
Alkaline phosphatase	Liver, bone, placenta, GI tract, kidney	Bone disease Liver disease
Alanine aminotransferase	Liver, skeletal muscle, heart	Hepatocellular disease
Amylase	Pancreas, saliva	Acute pancreatitis
Aspartate aminotransferase	Liver, heart, skeletal muscle, erythrocytes	Hepatocellular disease Muscle disease Myocardial infarction
Cholinesterase	Liver	Organophosphorous poisoning Suxamethonium sensitivity Hepatocellular disease
Creatine kinase	Skeletal muscle, heart, brain	Myocardial damage Skeletal muscle disease
Gamma-glutamyl transferase	Liver, pancreas, kidney	Cholestasis Alcohol abuse
Lactate dehydrogenase	Skeletal muscle, heart, liver, erythrocytes, platelets	Myocardial infarction Haemolysis Hepatocellular disease

value. Creatine kinase (CK) is found in the heart and skeletal muscles, thus serum CK would be increased in myocardial infarction as well as skeletal muscle disease. Although some enzymes are specific for a given tissue or organ, they are not used in clinical practice. Isocitrate dehyrodgenase is high in heart muscles but it is not used in the diagnosis of acute myocardial infarction, as it is rapidly inactivated in circulation. Ornithine carbamoyltransferase is liver specific but it is not used clinically as it is difficult to measure.

Specificity of enzyme tests can be improved either by measuring multiple enzymes or by measuring isoenzymes or isoforms (see below). As the distribution of enzymes varies between different organs, measurement of multiple enzymes may help to localise the organ of interest. For example, alkaline phosphatase (ALP) is present in the liver and bone while gamma-glutamyl transferase (GGT) is present in liver but not in bone. Thus, elevation of ALP and GGT will indicate liver disease whereas an increase in ALP without GGT will indicate bone disease.

Multiple Forms of Enzymes — Isoenzymes

Enzymes with the same catalytic activity exist in multiple forms. Multiple forms may be present in different organs, in different cells of the same organ or even within the same cell. Multiple forms arise as a result of genetic or non-genetic variation. The term 'isoenzyme' is strictly applied to those that arise due to variation at the gene level giving rise to differences in structure. Differences may be due to multiple gene loci and these isoenzymes are present in all individuals, e.g. amylase and ALP. Another form of variation is due to allelic genes at a particular locus giving rise to inter-individual variation. For example, 150 distinct forms of glucose-6-phosphate dehydrogenase have been identified. Another form of genetic variation is where enzymes, that are made up of multiple subunits and are encoded by different genes, combine to give different isoenzymes, e.g. lactate dehydrogenase (LD) is a tetramer made up of two different subunits (H and M) and five isoenzymes are described.

Non-genetic variation in enzymes arises due to posttranslational modification. These variations are called isoforms and may be due to (i) variation in carbohydrate content, e.g. ALP from bone and liver differ in their sialic acid content; (ii) by alteration of the peptide chain, e.g. CK released into the circulation undergoes removal of a lysine residue giving rise to further isoforms (CK-MM$_1$, CK-MM$_2$ and CK-MM$_3$); (iii) aggregation of enzyme molecules with one another (e.g. cholinesterase) or with other proteins (e.g. amylase with immunoglobulins giving rise to macroamylase) to give isoforms. Other forms of posttranslational modifications include acylation, deamidation, sulphahydryl oxidation, phosphorylation and association with other proteins. When enzymes form complexes with other proteins such as immunoglobulins, the clearance of the enzyme may be prolonged, leading to increased activity in serum. These are called 'macroenzymes' and have been described for amylase, CK and LD.

Distribution of isoenzymes and isoforms make it possible to improve the specificity of enzyme tests in diagnoses. Differences in properties between the various isoenzymes and isoforms such as electrophoretic mobility, resistance to inactivation and differences in substrate specificity, can be utilised to identify these. Structural differences in isoenzymes also make it possible to measure the enzyme using specific antibodies.

Enzymes of Diagnostic Value

Alkaline Phosphatase (ALP) (EC 3.1.3.1)

This group of enzymes catalyses a large number of substrates at alkaline pH to release phosphate. The natural substrate for this enzyme and the function of this enzyme *in vivo* are unknown. In individuals with an inherited absence of this enzyme, ethanolamine phosphate is excreted in large quantities suggesting that this may be a natural substrate. ALP is present in practically all tissues at or in the cell membrane. High concentrations are found in bone (osteoblasts), liver, placenta and intestine. Four genes code for four isoenzymes, namely intestinal, placental, germ cell and tissue non-specific forms. The latter gene encodes for the enzyme in the liver, bone and kidney. Germ cell enzyme is present in testes and thymus. Liver and bone iso-forms of ALP show posttranslational modifications in the sialic acid content of the enzyme.

In adults, approximately half the serum ALP activity comes from liver and the other half from bones with a small fraction coming from the intestine. During pregnancy, total ALP increases due to placental contribution. Total ALP activity in serum shows marked variation with age due to increased osteoblastic activity during growth. It is high at birth and falls to two or three times the upper limit of normal values for adult until adolescence, when it rises again (Figure 13.1). This increase is earlier in females than in males. The activity falls to adult values when the growth spurt ceases. A slight increase in ALP is seen in elderly women compared to men.

ALP is raised in bone disease and in hepatobiliary disease (Table 13.3). In response to biliary tract obstruction, there is

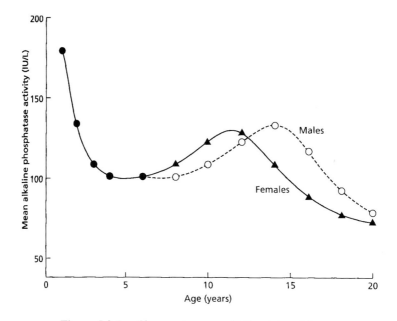

Figure 13.1   Changes in serum ALP activity with age.

Table 13.3   Causes of increased serum ALP activity

Physiological
- Growth periods
- Pregnancy
  — Infancy
  — Puberty

Pathological
- Bone disease
  — Paget's disease of the bone
  — Hyperparathyroidism
  — Vitamin D deficiency
  — Osteomyelitis
  — Healing fracture

- Hepato biliary disease
  — Cholestasis
  — Hepatitis
  — Cirrhosis
  — Space-occupying lesion

- Others
  — Carcinoma of the bronchus — Regan isoenzyme

increased synthesis of ALP in the hepatocytes adjacent to the biliary canals and some of this newly formed enzyme enters the circulation. The increase in ALP in cholestasis, especially in extrahepatic obstruction, usually reaches five times the upper limit of normal (ULN). In parenchymatous liver disease, such as infectious hepatitis, ALP is increased due to biliary stasis but values are usually not greater than three times the ULN. The degree of elevation of ALP is useful in distinguishing jaundice due to parenchymal damage from jaundice due to cholestasis. ALP is increased in metabolic bone disease and highest values are seen in Paget's disease of bone where serum activity may reach 10 to 25 times ULN. Moderate elevations are seen in vitamin D deficiency (osteomalacia), during healing of bone fractures and in hyperparathyroidism. The value of ALP in osteoporosis is discussed under 'Metabolic Bone Diseases' (Chapter 9).

High ALP, seen in malignancy is due to secondaries in bone (bone isoenzyme) or liver (liver isoenzyme). Some tumours (e.g. bronchial carcinoma) secrete an ALP with characteristics similar to placental isoenzyme and this is called Regan isoenzyme. This isoenzyme is due to derepression of the placental ALP gene. When serum ALP is increased and the clinical picture does not indicate the source, the origin of ALP can be established by isoenzyme analysis or by measurement of another enzyme, e.g. GGT.

Low serum ALP values are usually of little diagnostic significance except in hypophosphatasia. This is a rare autosomal dominant metabolic bone disease due to a mutation and these patients can present with a wide spectrum of features including early death and multiple fractures. This is associated with increased excretion of phosphoethanolamine in urine.

Case 13.1

A 68-year-old man went to see his GP with some upper abdominal discomfort and poor appetite. His GP performed liver function tests and the results are as follows:

Serum		Reference Range
Bilirubin (μmol/L)	13	< 20
ALT (IU/L)	25	< 40
Albumin (g/L)	42	35–45
ALP (IU/L)	425	< 120

These results show an isolated increase in ALP. His GP asked the laboratory for advice and the duty biochemist suggested doing serum GGT. Serum GGT was found to be elevated. Increase in GGT and ALP strongly suggests a liver pathology and the patient was sent for an ultrasound which showed a small mass in the right lobe of his liver. The mass was causing obstruction of bile flow in that side. As the rest of the biliary tree was normal, bilirubin was not elevated. He was referred urgently to a hepatologist who confirmed the diagnosis of a small hepatocellular carcinoma, after imaging studies and a biopsy.

Acid Phosphatase (ACP, EC3.1.3.2)

This refers to a group of enzymes which have phosphatase activity at pH below 7.0. Acid phosphatase (ACP) is found in high concentrations in prostate, osteoclasts, erythrocytes and platelets. At least five isoenzymes have been recognised. Four gene loci encoding for these isoenzymes have been identified. ACP of clinical interest comes from prostate and osteoclasts. Most ACP present in the normal serum comes from osteoclasts and this isoenzyme is resistant to inhibition by tartrate. ACP activity therefore varies with bone turnover, high during growth spurts. Tartrate-resistant ACP is useful in monitoring bone turnover (see Chapter 9).

The prostatic isoenzyme is inhibited by tartrate but not by formaldehyde; this isoenzyme has been used in the management of prostatic carcinoma. However, this is now superseded by prostate-specific antigen (PSA). Tartrate-resistant ACP measurement is useful in the diagnosis of Gaucher's disease, a lysosomal storage disorder, in which abnormal macrophages in the spleen and other tissues overexpress ACP, giving rise to high serum ACP. The hairy cells of hairy cell leukaemia also express tartrate-resistant ACP but this does not enter plasma.

ACP is present in high concentrations in the semen. Its measurement is useful in the field of forensic medicine, rape investigations and similar offences.

Transaminases/Aminotransferases

Alanine aminotransferase (EC 2.6.1.3, ALT) and aspartate aminotransferase (EC 2.6.1.1, AST) are enzymes involved in the interconversion of amino acids and their 2-oxoacids by the transfer of amino groups. These enzymes are widely distributed in tissues. There are two isoenzymes of AST, one in the cytoplasm and other in the mitochondria. In mild injuries, the cytoplasmic form is found in the serum while in severe injuries, mitochondrial AST is also found. However, isoenzymes of AST are not routinely measured in many clinical laboratories.

Table 13.4   Causes of elevation of aspartate amino transaminase

---

Physiological
  • Neonates

Pathological
  • Liver disease
    — Hepatitis and hepatic necrosis
    — Cholestasis
    — Chronic hepatitis
  • Cardiac disease
    — Myocardial infarction
  • Skeletal muscle
    — Crush injuries
    — Trauma/surgery
    — Myopathy

Erythrocytes
  • Haemolysis (*in vivo* and *in vitro*).

AST is present in high concentrations in heart, liver, skeletal muscle, kidneys and erythocytes. High concentrations of ALT are found in liver and kidneys.

Causes of elevation of AST in serum are given in Table 13.4. Artefactual elevation of AST can occur due to haemolysis. In any disease of the liver where there is damage to hepatocytes, elevation of serum AST and ALT will be seen. In acute hepatitis, the levels are usually 10 to 20 times the upper limit of normal (ULN) and the peak values are seen between 7th and 12th days. Occasionally, the values may reach 100 times the ULN. Sometimes, this increase is in the prodromal stage before the patient is jaundiced. Some laboratories use the ratio of ALT/AST to differentiate infectious hepatitis and other inflammatory conditions of the liver from other causes of hepatic necrosis. This ratio is < 1 in normal subjects and in non-infective hepatic necrosis, but is > 1 in hepatitis. In hepatitis, cytoplasmic AST is released whereas in toxic hepatic necrosis, mitochondrial AST is also released, decreasing the ALT/AST ratio. A ratio > 2 suggests alcoholic liver disease. In cholestasis, moderate elevation in serum transaminase is seen. In cirrhosis, the degree of elevation of serum transaminases varies with the stage of the cirrhotic process. A moderate increase in serum transaminases may be seen after alcohol intake or following drugs such as anticonvulsants, opiates and ampicillin. Over 300 drugs are reported to cause elevation of serum AST. Serial measurements of serum AST are valuable in detecting graft rejection after liver transplantation.

Serum AST activity is increased after MI (see Chapter 12). Serum ALT is not usually increased in uncomplicated MI. An increase in ALT after MI suggests liver congestion.

In diseases of or injuries to muscle, serum AST will be increased and in crush injuries, the serum activity can be 10 to 50 times the ULN. In progressive muscular dystrophy and dermatomyositis, serum AST is increased. Mild to moderate elevation of serum AST can be seen after pulmonary embolism, acute pancreatitis and haemolytic disease.

Gamma-glutamyl Transferase (GGT, EC 2.3.2.2)

This enzyme transfers γ-glutamyl group from peptides to an acceptor, which may be a peptide or an amino acid. It is present in most tissues except skeletal muscle. High concentrations are found in the kidney and significant amounts are found in the liver and pancreas. Most of the enzyme is located in the cell membrane with some in cytoplasm.

Although the highest concentration of GGT is found in the kidney, serum GGT is primarily from the liver and is increased in all forms of liver disease. The highest values are seen in biliary obstruction with values reaching 5–30 times the ULN (Table 13.5). It is more sensitive than ALP in detecting cholestasis. Moderate elevations are seen in all forms of hepatitis, cirrhosis and primary and secondary malignancy of liver. In pancreatitis and in some pancreatic malignancies, serum GGT may be moderately elevated.

Serum GGT is high in heavy alcohol drinkers without evidence of liver disease and in patients taking drugs such as phenytoin, phenobarbitone and rifampicin. All these agents increase serum GGT due to enzyme induction. In up to 70% of heavy alcohol drinkers, GGT will be high and it will remain elevated up to 2 to 3 weeks after stopping alcohol intake. The sensitivity and specificity of GGT in detecting

Table 13.5   Causes of increased GGT

Hepatobiliary disease
  • Cholestasis
  • Hepatitis
  • Neoplasm
  • Cirrhosis
  • Alcoholism
      — alcoholic liver disease
      — chronic alcoholism
Drugs
  • Alcohol
  • Anticonvulsants
  • Antidepressants

alcoholism, however, are not high. Some heavy drinkers may have normal GGT values while there may be other causes of increases in GGT.

Creatine Kinase (CK, EC 2.7.3.2)

This enzyme catalyses the reversible phosphorylation of creatine. High concentrations of CK are found in the skeletal muscle, heart and brain. CK is a dimer made up of two subunits of molecular weight 40,000 each, B and M. Three different isoenzymes are seen CKBB (CK1), CKMB (CK2) and CKMM (CK3). All three of these isoenzymes are found in cytosol and a fourth isoenzyme is present in the mitochondria (CK-Mt). CKBB is predominantly found in the brain, prostate, gut, lung, bladder, uterus, placenta and thyroid; CKMM is found predominantly in the skeletal and cardiac muscles. CKMB is found in varying degrees in the heart (25–46%) and to a smaller extent in skeletal muscles (< 5%). The proportion of CKMB in skeletal muscle varies in different muscle groups and is affected by training. Causes of elevation of CK are in Table 13.6.

As serum CK is predominantly of muscle origin, it is higher in those with high muscle mass. In healthy African American subjects, serum CK values are 2–3 times the ULN as that in Caucasian subjects and these values are greater than expected for muscle mass. Serum CK is higher in males than females probably due to the higher muscle

Table 13.6   Causes of increased serum CK activity

Physiological
- Muscle mass
- Race — higher in blacks
- Exercise

Pathological
- Cardiac
  - Acute myocardial infarction
- Skeletal muscle
  - Rhabdomyolysis
  - Malignant hyperpyrexia
  - Muscular dystrophy
  - Myositis
  - Hypothyroidism

Drugs causing increased serum CK activity (predominantly CKMM)
- Aminocapric acid
- Amphotericin B
- Carbenoxolone
- Fibrates
  - Clofibrate, fenofibrate and beclofibrate, benzafibrate
- β-blockers
  - Pindolol
- Tricyclic antidepressants
  - Amphetamine
  - Amitryptyline
- Phenothiazines
- HMG CoA reductase inhibitors
  - 'Statins' — simvastatin, lovastatin, pravastatin and fluvastatin
- Drugs of abuse
  - Heroin
  - Opiates
  - Methylenedioxymetamphetamine (MDMA, 'Ectasy')

mass. The clinical value of CK and its isoenzymes in the diagnosis of MI is discussed in Chapter 12.

In all types of muscular dystrophy, serum CK is elevated sometime during the course of the disease. Serum CK exceeding 200 times the ULN may be found in acute rhabdomyolysis. Other causes of elevated CK include muscle injury, sepsis, hypothermia, seizures and neuroleptic malignant syndrome. A large number of drugs can cause elevation and they are listed in Table 13.6.

Serum CK is increased in hypothyroidism and decreased in hyperthyroidism probably due to alteration in clearance. In severe hypothyroidism, it is usually five times the ULN but may reach 50 times the ULN.

During parturition, serum CK can increase six fold probably due to enzymes from the placenta and uterus. In neonates, especially in premature infants and infants with brain damage, serum CK is high.

In cerebrovascular disease, serum CK may increase and this is mainly due to CKMM (CK-3) isoenzyme. In head injuries, the brain isoenzyme CKBB (CK1) can be elevated and the degree of elevation is correlated to the severity of the injury and to the prognosis.

Amylase (EC 3.2.1.1)

These enzymes hydrolyse complex carbohydrates. They are small molecules of 55,000 to 60,000 molecular weight and are small enough to be filtered and excreted in urine. Although they are present in a number of tissues, high concentrations of amylase are found in pancreas and salivary gland. These enzymes are normally secreted into the gastrointestinal tract and take part in the digestion of food. Serum amylase is derived from the pancreas and saliva in normal subjects. Pancreatic and salivary amylases are products of related genes and there are allelic variants of each gene. Twelve salivary and six pancreatic phenotypes have been described.

Amylase in serum is sometimes found complexed to high molecular weight proteins such as immunoglobulins. Clearance of this complex is reduced and serum amylase will increase to 6–8 times the ULN. This phenomenon is called macroamylasemia and is of no clinical consequence.

Causes of increased amylase are listed in Table 13.7. Serum amylase measurement is used largely in the diagnosis of acute pancreatitis. In acute pancreatitis, serum amylase increases 2–12 hours after the onset and reaches a peak at 12–74 hours and returns to normal in 3–4 days. The greater the elevation, the greater the probability of

Table 13.7   Causes of increased serum amylase activity

Pancreatic disease
- Acute pancreatitis
- Pancreatic trauma
- Pancreatic carcinoma

Biliary tract disease
Intra-abdominal disease
- Perforated peptic ulcer
- Intestinal obstruction
- Peritonitis
- Ruptured 'ectopic' pregnancy

Diabetic ketoacidosis
Renal failure
Macroamylasaemia
Drugs
- Opiates

Salivary gland disorders
- Mumps
- Calculus disease

Neoplastic hyperamylasaemia
- Carcinoma of bronchus or ovary

acute pancreatitis. Values greater than 10 times the ULN are virtually diagnostic. Serum amylase however, may be within the reference range in up to 20% of patients with acute pancreatitis.

In chronic pancreatitis, serum amylase is often normal but in complications of acute pancreatitis, such as pancreatic pseudocyst, amylase is high. Diseases of the biliary tract such as cholecystitis can cause an increase in serum amylase up to five times the ULN. In diabetic ketoacidosis, serum amylase is often high in up to 80% of patients.

Serum amylase is high usually up to 5–10 times the ULN, in other acute abdominal conditions such as perforated peptic ulcer. In renal impairment, serum amylase is high (2–3 times the ULN) due to reduced clearance and is proportional to the decrease in renal function. Serum amylase may be elevated in tumours of the lung and ovary. Salivary isoenzyme is also high in diseases of the salivary gland such as mumps.

Urine amylase and clearance of amylase are sometimes used in the diagnosis of acute pancreatitis. Clearance of amylase is calculated by the formula:

$$\text{Amylase clearance} = \frac{\text{U amy}}{\text{S amy}} \times \frac{\text{S Cr}}{\text{U Cr}} \times 100,$$

where U amy and S amy are urine and serum amylase concentrations respectively, S Cr and U Cr are serum and urine creatinine concentrations respectively. In acute pancreatitis, the tubular reabsorption of amylase along with other proteins is reduced and amylase clearance is high. Some studies suggest that this is more sensitive than serum amylase.

Lipase (EC 3.1.1.3)

Lipase, which hydrolyses triglycerides, is present in high concentrations in the pancreas. Serum lipase is useful in the diagnosis of acute pancreatitis. As serum amylase may be normal in 20% of patients with acute pancreatitis, serum lipase serves as an additional investigation. Furthermore, serum lipase is not often increased in other abdominal conditions such as perforated peptic ulcer. Serum lipase is increased in renal disease but is normal in diseases of the salivary gland.

Cholinesterase (EC 3.1.1.7)

Two types of cholinesterase are present: (1) Acetylcholinesterase (or true cholinesterase or cholinesterase I) which is present in erythrocytes, nerve endings, grey matter of the brain, spleen and lungs hydrolyses acetylcholine released at nerve endings. (2) Acylcholinesterase (cholinesterase II or pseudocholinesterase) which is found in liver, pancreas, heart, white matter of the brain and serum. The serum enzyme originates from the liver. The function of serum cholinesterase is not known. The two enzymes have different substrate specificity.

There are several genetic variants of cholinesterase II. The gene controlling the synthesis of serum cholinesterase exists in many allelic forms. The most common forms are given the symbols E_1^u, E_1^a, E_1^f and E_1^s. The most common phenotype is $E_1^u E_1^u$ or UU. The allelic form E_1^a is called the atypical gene and people homozygous for this gene have the phenotype $E_1^a E_1^a$ or AA. This enzyme does not hydrolyse most substrates and is resistant to inhibition by dibucaine. In the phenotype $E_1^f E_1^f$ or FF, the enzyme is weakly active and is resistant to fluoride while the E_1^s gene is associated with absence of enzyme activity. The homozygous forms of the atypical variation of AA or FF are found in 0.5% of Caucasians.

The main clinical use of serum cholinesterase is in the diagnosis of scoline apnoea. Scoline (suxamethonium), a short-acting muscle relaxant given during induction of anaesthesia, is rapidly hydrolysed by serum cholinesterase. In subjects with atypical or reduced activity of serum cholinesterase, the drug is not hydrolysed or hydrolysed slowly and apnoea is prolonged (scoline apnoea). Family studies are necessary to prevent the use of scoline in other affected members of the family. The activity of the enzyme in serum is measured together with the inhibitory effect of fluoride and dibucaine (dibucaine and fluoride numbers). From these measurements, the phenotype is derived. Phenotypes most susceptible to scoline apnoea are AA, AS, FF, FS, SS, AF and to some extent UA. Molecular biology techniques are now being used to identify the genotype.

Another important use of this enzyme is in the diagnosis of poisoning by organophosphorous compounds, which inhibit cholinesterase activity. These compounds, which are used extensively in agriculture as pesticides, can cause poisoning due to accidental or intentional (suicidal) exposure. Poisoning may result from inhalation, skin exposure and occasionally oral ingestion. Chronic poisoning can occur in agricultural workers exposed over a period of time. In acute poisoning, both serum and red cell cholinesterase are reduced and symptoms occur when 40% of the enzyme is inhibited; in severe poisoning, 80% of the enzyme is inhibited. In chronic poisoning, red cell cholinesterase may be a better index.

As serum cholinesterase is synthesised by the liver, low values indicate the synthetic capacity of the liver and it has been used as a liver function test.

Lactate Dehydrogenase (LD,EC 1.1.1.27)

LD which catalyses the reversible oxidation of L-lactate to pyruvate is a tetramer composed of four peptide chains of two subunits, M and H. Five isoenzymes are recognised: LD_1 (H_4), LD_2 (H_3M), LD_3 (H_2M_2), LD_4 (HM_3) and LD_5 (M_4). LD is present in all cells of the body with the highest concentration found in liver, heart, kidney, skeletal muscle and erythrocytes. In the heart, erythrocytes and kidney, LD_1 and LD_2 predominate while in the liver and skeletal muscle, LD_4 and LD_5 are the main forms.

Serum LD is elevated in diseases of the liver, heart and skeletal muscle. Total LD is very non-specific and many laboratories do not offer this test any more. As LD has a long half-life, it remains elevated for 7–10 days after myocardial infarction and is sometimes useful in the late diagnosis of AMI. However, this is now superseded by troponin (see Chapter 12). LD1 isoenzyme has greater specificity towards α-hydroxybutyrate and hence, is sometimes called hydroxybutyrate dehydrogenase (HBD) and has been used in the diagnosis of MI.

Serum LD is elevated in several malignancies. Very high values are seen in patients with Hodgkin's disease, abdominal and lung cancer, germ cell tumours, such as teratoma and seminoma of the testis. However, the increase is not specific and of poor clinical value.

The measurement of LD in pleural fluid has been suggested to be valuable in differentiating malignant from non-malignant effusion. Pleural fluid LD is often greater than serum LD in malignant effusions whereas they are lower than serum in non-malignant effusions.

Serum Enzymes in Disease

Myocardial Infarction

This is discussed in detail in Chapter 12.

Liver Disease

The use of enzyme measurement in liver disease is discussed in Chapter 15. In patients with jaundice, a transaminase value > 10

times the ULN suggests that the jaundice is due to hepatocellular damage such as acute hepatitis where ALP is usually moderately elevated. In cholestasis, ALP is elevated to > 5 times the ULN. Serum transaminases are often used to detect hepatotoxicity of drugs.

Muscle Disease

Skeletal muscle contains high concentrations of CK, AST, aldolase and LD. Of these, serum CK is the most widely used in the diagnosis of muscles disease. Serum CK will be high in any form of skeletal muscle trauma, muscular dystrophies, polymyositis and rhabdomyolysis, but normal in neurogenic myopathies such as poliomyelitis and motor neurone disease. CK values are often high after exercise and values may reach 10–20 times the ULN in subjects undertaking severe exercise such as marathon running especially if the subjects are poorly trained. Some sports physiologists use the rise in serum CK after exercise as a guide to monitor the training of athletes.

Duchenne's muscular dystrophy is an X-linked recessive disorder caused by abnormal dystrophin gene. It causes progressive weakness starting at the age of five. Serum CK is elevated before symptoms appear and it is a useful diagnostic test before symptoms develop. In this muscular dystrophy, serum CK falls with age as the muscle mass gets lower. Serum CK is useful to detect carriers for genetic counselling as it is raised in up to 75% of female carriers.

Becker muscular dystrophy which presents at a later age runs a more benign course. The pattern of change in serum CK is similar to the CK in Duchenne's muscular dystrophy. In other forms of muscular dystrophy, changes in CK are not so marked.

Many drugs (Table 13.6) can cause an increase in CK, some by causing rhabdomyolysis, others by disturbing the membrane function or by a toxic effect on the muscle cell. HMG-CoA reductase inhibitors ('statins') cause elevated CK and sometimes severe rhabdomyolysis can occur. These drugs decrease the serum concentration of coenzyme Q, which is part of the oxidative respiratory pathway.

A decrease in this would lead to myocyte dysfunction and a rise in CK. Patients on these drugs are monitored for evidence of muscle damage.

Malignant Hyperpyrexia

This is a rare, potentially fatal complication of general anaesthesia using halothane, methoxyflurane and succinylcholine. It is characterised by a rapid rise in temperature, muscle stiffness, tachycardia, shock and fits. Serum CK values are very high during the attack. It is an autosomal dominant inherited disorder and the defect is in the gene encoding the skeletal muscle ryanodine receptor. In suspected individuals, preoperative CK is measured but a normal CK does not exclude this condition. Diagnosis is based on a functional test on a skeletal muscle biopsy.

Rhabdomyolysis

Rhabdomyolysis is a condition when there is rapid destruction of the skeletal muscles; this can be caused by a wide variety of disorders (Table 13.8). Myoglobin and other muscle proteins such as CK are released into the circulation. Myoglobin being a small molecular weight protein (17,000 kDa) is filtered by the glomerulus and excreted. Severe rhabdomyolysis causes hyperphosphataemia due to the release of phosphate from muscle cells, hypocalcaemia as a consequence of hyperphosphataemia, hyperuricaemia due to the breakdown of purines from muscle, metabolic acidosis due to release of organic acids from muscle and acute renal failure. The underlying mechanism is thought to be hypovolaemia, renal vasoconstriction, intraluminal cast formation and direct haem-protein–induced cytotoxicity.

Case 13.2

A 52-year-old man was started on simvastatin (a HMG CoA reductase inhibitor) by his GP to treat his hypercholesterolaemia. One week later,

Table 13.8   Causes of rhabdomyolysis

Muscle injury
- Crush injury
- Pressure necrosis
- Electric shock
- Freezing
- Burns

Severe exercise
Ischaemic necrosis
- Vascular occlusion/external compression

Metabolic disorders
- Severe hypokalaemia
- Severe hypophosphataemia
- Hypothyroidism
- Diabetic coma
  — Ketoacidosis
  — Non-ketotic hyperosmolar coma
- Water intoxication

Sepsis
- Bacterial
  — Typhoid
  — Shigellosis
  — Haemolytic streptococcus
- Viral
  — Influenza
  — Coxsackie

Inflammatory myopathies
- Polymyositis, infectious myopathies

Inherited disorders
- Deficiencies in glycogenolytic enzymes
  — McArdle's syndrome (glycogen phosphorylase deficiency)
- Disorders of lipid metabolism
  — Carnitine palmityl acyl transferase deficiency
- Others
  — Malignant hyperpyrexia

Drugs
- Cocaine, amphetamine, opiates, fibrates, statins, neuroleptics
Toxins
- Snake/insect toxin

he returned with severe muscle aches. Blood tests done at this time showed these results:

Serum		Reference Range
CK (IU/L)	12,500	> 250
Bilirubin (μmol/L)	15	< 20
ALT (IU/L)	25	< 42
AST (IU/L)	87	< 45
ALP(IU/L)	152	< 250
Albumin (g/L)	42	35–45

Grossly elevated CK and elevated AST suggest severe muscle damage, rhabdomyolysis. Urine examination showed the presence of myoglobin, confirming the muscle damage. Rhabdomyolysis is a known complication of treatment with 'statins'.

Further Reading

1. Lane R, Philips M. Rhabdomyolysis. *Br Med J* 2003; 327:115–116.
2. Moss DW, Rosalki SB, *Enzyme Tests in Diagnosis* 1996. Edward Arnold: London.
3. Whitfield, JB. Gammaglutamyl transferase. *Crit Rev Clin Lab Sci* 2001; 38:263–355.
4. Xavier B, Poch E, Grau JM. Rhabdomyolysis and acute kidney injury. *N Engl J Med* 2009; 361:62–72.

Summary/Key Points

1. Measurement of enzymes in serum and urine are diagnostically useful. However, they lack specificity.
2. Enzymes with the same catalytic activity may exist in different forms, isoenzymes. Diagnostically useful enzymes such as ALP and CK exist in several isoenzymes. Measurement of isoenzymes may help to increase the diagnostic value of enzymes.
3. There are four isoenzymes of alkaline phosphatase (ALP). Total ALP is elevated in bone as well as liver disease. Measurement of

isoenzymes of ALP or measurement of GGT may help in identifying the origin of ALP.

4. Transaminases (ALT and AST) and GGT are used as part of liver function tests.

5. Creatine kinase is present in skeletal muscle, heart muscle and brain. This enzyme is mainly measured in the diagnosis of skeletal muscle disorders. CKMB, which is heart specific, has been replaced by troponins to diagnose myocardial infarction.

6. Amylase, which is a small molecular weight enzyme, is commonly used in the diagnosis of acute pancreatitis. Lipase is another enzyme useful in the diagnosis of acute pancreatitis.

7. Cholinesterase measurement is useful in detecting those people who are liable to get scoline apnoea (prolonged muscle relaxation after administration of this short-acting muscle relaxant) and in detecting poisoning with organophosphorus pesticides.

8. Rhabdomyolysis is a condition where there is rapid destruction of muscle. This can lead to acute renal failure, hyperkalaemia, hypocalcaemia, and hyperphosphataemia. Serum CK will be grossly elevated in this condition.

chapter 14

Proteins

Introduction

Several hundred proteins with a wide range of functions are found in plasma but only a few of these are measured in clinical practice. Major functions of the plasma proteins are listed in Table 14.1.

The concentration of any protein in plasma is the net balance between its synthesis, catabolism and distribution. Most proteins in the plasma are synthesised by hepatocytes at a rate of 25 g per day. Proteins are mainly found in the vascular compartment. Proteins are also found in the interstitial fluid at low concentration. However, as the total volume of interstitial fluid is about 3 to 4 times that of the vascular compartment, a significant amount of protein is found in the interstitial compartment. The amount of protein entering the interstitial compartment is dependent on the vascular permeability, which is increased in inflammation.

Most proteins are taken up by capillary endothelial cells or mononuclear phagocytes and catabolised. The amino acids released are then reutilised. Small molecular weight proteins are filtered by the glomerulus and reabsorbed by the renal tubular cells and catabolised in these cells. A small amount of protein may leak into the intestinal tract, where they are digested and the amino acids reabsorbed.

Assessment of Plasma Proteins

Plasma proteins can be assessed quantitatively by measuring total protein or a specific protein such as albumin. The semi-quantitative method of

Table 14.1   Functions of plasma proteins

1. Transport function
   - Albumin — T4 and T3
   - Thyroxin-binding globulin — T4
   - Cortisol-binding globulin — cortisol

2. Coagulation and fibrinolysis
3. Maintenance of colloid oncotic pressure
4. Defence function/inflammatory response
   - Immunoglobins
   - Complement system
   - Acute phase proteins

5. Buffering
6. Protease inhibitors e.g. α_1-antitrypsin

assessment is usually by electrophoresis. Electrophoresis is a technique by which plasma proteins are separated on the basis of their electrical charge. A small amount of serum is applied to a support medium, usually cellulose acetate or agarose gel. A current is applied for a period of time and the proteins are visualised by staining with a dye. In normal individuals, such separation will show five or six bands: albumin, α-, α_1-, β- and γ-globulins (Figure 14.1). Albumin, which forms the most obvious band, is closest to the anode. α_1 Globulin consists mainly of α_1-antitrypsin (AAT), α_2 mainly α_2 macroglobulin and haptoglobin, while β globulin consists mainly of transferrin and complement (Table 14.2). Although abnormal electrophoretic patterns are seen in many diseases such as liver disease, and nephrotic syndrome, these are not diagnostic. Important abnormalities detected by electrophoresis are the presence of paraproteins (see later) and AAT deficiency.

Plasma Total Protein

Measurements of total protein concentration have limited clinical value, as a change in one protein or a group of proteins may be masked by an opposite change in another. Total protein concentration

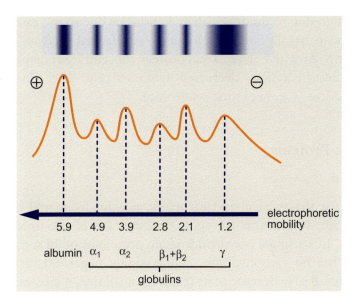

Figure 14.1 Normal serum electrophoretic pattern.

Table 14.2 Some properties of some plasma proteins

Protein	Electrophoresis	Molecular weight	Concentration g/L
Albumin		66,000	38–48
Prealbumin		55,000	0.1–0.4
α_1-Antitrypsin	α_1	54,000	1.4–4.0
α_1-Acid glycoprotein	α_1	44,000	0.6–1.4
Retinol binding protein	α_1	21,000	0.03–0.1
Caeruloplasmin	α_2	132,000	0.2–0.6
Haptoglobin	α_2	100,000	1.2–2.6
α_2 Macroglobulin	α_2	800,000	1.5–3.5
C-reactive protein	β	105,000	< 0.003
β_2 Microglobulin	β	11,800	< 0.002
Transferrin	β	77,000	1.8–2.7
Fibrinogen	γ	341,000	2.0–4.5
Immunoglobulins	δ	160,000–900,000	

will be high due to fluid loss (dehydration) or due to increase in specific protein usually immunoglobulins. Artefactual increase is often seen due to excessive stasis. Total protein concentration may decrease due to overhydration, decreased synthesis (e.g. liver disease) or increased loss (e.g. nephrotic syndrome).

Specific Proteins

Albumin

Albumin is the most abundant plasma protein from mid-gestation until death. It accounts for approximately 50% of the total plasma proteins. It is a single polypeptide chain protein of 66 kDa molecular weight, has no carbohydrates, and is highly negatively charged at physiological pH. Because of its relatively small size, it crosses the vascular and glomerular basement membrane. Because of this, albumin is the most important protein in extravascular fluids such as interstitial fluid, CSF, urine and amniotic fluid. Approximately 50% of the total body albumin is present in the interstitial compartment. Albumin is synthesised primarily by the liver except during foetal life when it is synthesised by the yolk sac. Liver has a large synthetic reserve and in protein-losing states, such as nephrotic syndrome synthesis can increase three fold. The synthetic rate is thought to be controlled primarily by the colloid oncotic pressure and secondarily by protein intake. Catabolism of albumin occurs in all tissues by pinocytosis and the catabolic rate depends on local tissue metabolism. Small amounts are also lost in the intestinal tract. Albumin is filtered at the glomerulus, most of which is reabsorbed and catabolised in the tubules leaving a very small amount to be excreted in the urine. The amount of albumin filtered is less than what you would expect from the molecular size because of the high negative charge. The half-life of albumin is between 15 and 19 days.

Albumin contributes to 80% of the plasma colloidal oncotic pressure, which is important in maintaining the fluid distribution between vascular and interstitial space. Albumin binds to many substances including metal ions, free fatty acids, amino acids, drugs, hormones,

Table 14.3   Causes of hypoalbuminaemia

Artefactual
- Sample from drip arm

Physiological
- Pregnancy
- Recumbency

Pathological
- Decreased synthesis
  — Malnutrition
  — Malabsorption
  — Liver disease
- Increased distribution/redistribution
  — Overhydration
- Increased capillary permeability
  — Septicaemia
  — Burns
- Increased loss
  — Renal loss — nephrotic syndrome
  — GI loss — protein-losing enteropathy
  — Skin — burns
- Increased catabolism
  — Severe sepsis
  — Malignancy

and bilirubin. Albumin is a source of amino acids for peripheral tissues, has antioxidant activity and acts as a buffer.

Albumin concentration can be high due to venous stasis and in dehydration. Albumin concentration however, is not a useful index of dehydration. Albumin concentration may decrease as a result of redistribution, decreased synthesis, increased loss or catabolism or a combination of these (Table 14.3). One of the most common causes of hypoalbuminaemia in hospital patients is redistribution from the vascular to the interstitial compartment as a result of increased permeability of the vascular endothelial membrane caused by cytokines and other inflammatory mediators. The degree of decrease in albumin is proportional to the severity of the inflammation or injury. Decreased synthesis may arise as a result of genetic causes or acquired causes. In the inherited disorder of analbuminaemia, plasma albumin concentration can be less than 0.5 g/L. In hepatic diseases such as cirrhosis, albumin synthesis is low. In malabsorption and starvation, albumin concentration decreases because of the lack of amino acid supply to the liver. Increased loss of albumin in the urine in renal diseases, such as nephrotic syndrome, can also lead to hypoalbuminaemia (see Chapter 5). Inflammatory diseases of the gastrointestinal tract, especially protein-losing enteropathy, may cause increased loss of albumin. Hypoalbuminaemia is a contributing factor in the accumulation of fluid in the extravascular space, i.e. oedema and ascites.

α_1-Acid Glycoprotein

α_1-Acid glycoprotein (AAG), also called 'orosomucoid', is a glycoprotein of 40 kDa, containing 45% carbohydrate. It is synthesised primarily by the parenchymal cells of the liver, with contributions from granulocytes and monocytes. The function of this protein is not clear. This protein is now classified as one of the lipocalins, which are a group of proteins that bind to lipophilic substances. AAG may also have a function in the downregulation of immune response. A 3- to 4-fold increase in concentration of AAG is seen as an acute phase response (see below) between 3–5 days after the initial insult. It is less sensitive than other proteins, such as C-reactive protein (CRP) and α_1-antichymotrypsin (ACT), as a marker of acute phase response. Serum AAG may be useful as a marker of chronicity of inflammation. It has been used as a marker of disease activity in Crohn's disease and ulcerative colitis in combination with CRP and ACT.

α_1-Antitrypsin

α_1-Antitrypsin (AAT), a glycoprotein of molecular weight 54 kDa accounts for majority of the α_1 globulin fraction in the serum and majority of the plasma tryptic inhibitory activity. Because of its small size, AAT can pass into body fluids and inhibit serine proteases, especially leucocyte elastase, which is released during phagocytosis. Elastase can react with elastin in the vascular endothelium and the tracheo-bronchial tree, leading to loss of elastic recoil. In the absence of this enzyme, uninhibited activity of elastase on the bronchial tree can lead to emphysema. AAT is predominantly synthesised in the liver although monocytes and macrophages also contribute. AAT is an acute phase protein increasing within 24 hours after an inflammatory process and reaching a peak in 3–4 days. Serum AAT is increased during pregnancy especially in the late stages due to stimulation by oestrogens. Decreased concentration of AAT is seen as a genetic variation.

AAT deficiency (see below) is associated with increased risk of developing pulmonary emphysema due to the degradation of elastin by the unopposed action of leucocyte elastase. This process is accelerated by air pollution and cigarette smoking, which leads to recruitment, activation and lysis of leucocytes. In addition, smoking causes oxidation of methionine in the active site of AAT, rendering it incapable of inhibiting elastase. The resulting unrestricted activity of elastase eventually leads to emphysema. The age of onset of symptoms in AAT-deficient subjects is significantly earlier in those who smoke. AAT deficiency is also associated with diseases of the liver including neonatal cholestasis, hepatitis, cirrhosis and hepato-cellular carcinoma. Hepatic damage is believed to be due to the accumulation of AAT. As a result of a mutation, AAT is not processed within the hepatocytes where it accumulates and causes damage. AAT deficiency may account for 6% of cases of emphysema and 20% of neonatal cholestasis.

Genetic variation of AAT

The gene locus for AAT is highly polymorphic and nearly 75 distinct genetic variants or isotypes have been described. These variants are classified by the Pi system of nomenclature. Each allele is designated by a letter B–Z corresponding to the electrophoretic mobility and this is followed by a number or a subscript to indicate the subtypes. PiM is the most common phenotype in all populations. It is found in about 84% of the Caucasian and 95% of the Indian population. Reduced plasma AAT concentration of 25%, 60%, 80%, 15% and 0% are associated with variants PiP, PiS, PiW, PiZ and PiO respectively. The prevalence of these genetic variants varies in different populations, e.g. PiZ is the most common in Northern Europe whereas PiS is common in South Western Europe. Variant proteins are only partially secreted resulting in accumulation within the endoplasmic reticulum causing hepatocellular damage. Alleles giving rise to functional deficiency are S, P, W, Z, M-malton and M-durate and null, which causes complete deficiency. M-durate is a functionally deficient protein although the mass concentration is normal. Patients who are homozygous for one or other of these alleles or who are heterozygotes for any two will have severe reduction in serum AAT concentration. A combination of these alleles with many other alleles will result in the heterozygote state with intermediate AAT concentrations. Incidence of severe genetic deficiency of AAT in the UK is approximately 1 in 6000 whereas in Denmark it is 1 in 1600.

Caeruloplasmin

Caeruloplasmin is an α_2 globulin, which contains 95% of the total serum copper, in a tightly bound complex. It is synthesised in the liver, has ferroxidase and superoxide dismutase activity and plays an important role in the control of membrane lipid oxidation.

Decreased caeruloplasmin concentrations are seen in liver disease, in dietary copper deficiency and genetic disorders such as Menke's disease and Wilson's disease (see Chapter 18).

Increased caeruloplasmin concentrations are seen as acute phase response, the increase occurring relatively slowly, peaking at 4–20 days after the acute insult. Caeruloplasmin is increased in oestrogen treatment, pregnancy, biliary obstruction and in reticuloendothelial neoplasia.

β_2 *Microglobulin*

β_2 Microglobulin is a low molecular weight protein of 11 kDa found on the surface of all cells as part of the polypeptide chain of the HLA antigen complex. The function of β_2 microglobulin is not known. Being a small molecular weight protein, it is freely filtered in the kidney, being almost completely reabsorbed in the proximal tubule, leaving a small amount to be excreted in the urine.

Serum β_2 microglobulin has been used as a marker of GFR. However, it has not found widespread application. Urinary β_2 microglobulin is useful in detecting renal tubular damage. As almost all filtered β_2 microglobulin is reabsorbed in the proximal tubules, any damage to the proximal tubules will cause a decrease in tubular reabsorption, leading to an increase in the excretion of β_2 microglobulin in the urine. However, β_2 in the urine is unstable at acidic pH therefore, the value of this measurement in renal tubular disorders is limited. The uses of β_2 microglobulin in malignancy and myeloma are discussed in Chapter 24. Serum β_2 microglobulin concentration is an accurate predictor of disease progression in AIDS and AIDS related complexes.

Retinal Binding Protein and Transthyrin (prealbumin)

Retinol binding protein (RBP) is a small molecular weight protein, which transports retinol. In serum, it is present as a complex with transthyrin (prealbumin), which is also a low molecular weight protein. These proteins have short half-lives and they are useful in the assessment of nutritional status (see Chapter 17). Being a small molecular protein, RBP is filtered and reabsorbed in the tubules and it has been used in the assessment of renal tubular function.

Transferrin

Transferrin, a β glycoprotein, is the principle transport protein of iron. It is a negative acute phase protein and has a molecular weight of 80,000. Transferrin is synthesised in the liver and its half-life is approximately 7 days. Serum concentration of transferrin is increased in iron deficiency.

Transferrin measurements, together with serum iron, are useful in the differential diagnosis of anaemia. Transferrin concentrations are also low in protein calorie malnutrition and are useful in nutritional assessment (see Chapter 17). Increased concentration is seen in pregnancy and during oestrogen treatment. Urine transferrin measurement is useful to calculate protein selectivity index in minimal change renal disease (see Chapter 5).

Carbohydrate deficient transferrin

Transferrin normally contains about 6% carbohydrate. In chronic excess alcohol ingestion, transferrin with less sialic acid, carbohydrate-deficient transferrin (CDT), is produced. The concentration of CDT is a useful index of alcohol abuse. CDT, which moves towards the anode during electrophoresis, has a sensitivity and specificity of 90% and 99% respectively for chronic alcohol abuse. However, the sensitivity of CDT in alcohol abuse is lower in women as hormonal factors and iron status can also affect transferrin glycosylation. Raised CDT concentrations are also seen in patients with chronic liver disease, premenopausal women, and in women receiving hormone replacement therapy.

CDT is also associated with the rare metabolic disorder, disialotransferrin development deficiency syndrome that is associated with mental deficiency and characteristic external features.

An asialylated form of transferrin is present in CSF. This protein is useful in the detection of CSF leak, e.g. CSF rhinorrhoea (Chapter 29).

Haptoglobin

Haptoglobin is a α_2 glycoprotein made up of two subunits, α and β. Its main function is to conserve iron by binding to free haemoglobin liberated during intravascular haemolysis. The haptoglobin/haemoglobin complex is too large to be lost in the urine and it has a short half-life being cleared by the reticuloendothelial system from where the iron is returned for haemoglobin synthesis. Haptoglobin concentrations are low in any situation where there is intravascular haemolysis. Haptoglobin is also an acute phase protein. Increased concentrations are seen in acute inflammatory conditions, trauma and neoplasia. Concentration of haptoglobin increases in hypoalbuminaemic states such as nephrotic syndrome. Haptoglobin shows considerable polymorphism in the α-subunit and there are three major phenotypes. Haptoglobin polymorphism is associated variation in prevalence of many diseases including infection, atherosclerosis and autoimmune disorders.

Acute Phase Response

During inflammation, a large number of systemic changes occur at distant sites from the site of inflammation. This response, referred to as acute phase response (APR), results in changes in concentration of many plasma proteins. These proteins are therefore referred to as acute phase proteins. An acute phase protein is defined as one whose plasma concentration changes by at least 25% during inflammatory disorders; proteins which increase are called 'positive' acute phase proteins, and those that decrease are called 'negative' acute phase proteins (Table 14.4). The changes in acute phase proteins are due to changes in synthesis in the liver and the magnitude of the increase varies from 50% in the case of caeruloplasmin to 1000-fold or more in the case of C-reactive protein and amyloid A. Conditions that give rise to changes in acute phase proteins include infection, trauma, surgery, burns, infarction, inflammatory conditions mediated by immunological mediators, crystals and cancer. These changes also may be seen but to a milder degree after strenuous exercise, heat stroke and childbirth. In response to injury, inflammatory cells (neutrophils and macrophages) produce cytokines which are intracellular signalling polypeptides. Cytokines involved in inflammation include interleukin-6, tumour necrosis factor-α, interferon-α, transforming growth factor-β and interleukin-8. These cytokines stimulate the liver to produce acute phase proteins.

Table 14.4   Acute phase proteins

Positive acute phase proteins
— Complement system
— Coagulation and fibrinolytic system
— Fibrinogen
— Plasminogen
— Tissue plasminogen activator
— Urokinase
— Proteins
— Plasminogen — activator inhibitor

- Antiproteases
— α_1-Antichymotrypsin

- Transport proteins
— Caeruloplasmin
— Haptoglobin

- Others
— C-reactive protein
— Serum amyloid
— α_1-Acid glycoprotein
— Fibronectin
— Ferritin

Negative acute phase protein
— Albumin
— Transferrin
— Transthyretin
— Thyroxine-binding globulin
— α-Fetoprotein
— Retinol-binding protein (RBP)
— Cortisol-binding globulin (CBG)
— Antithrombin

Acute Phase Proteins

Acute phase proteins are divided into two groups — Type I which is induced by IL-1 group of cytokines and includes CRP, serum amyloid A (SAA) and α_1-acid glycoprotein (AGP); Type II induced by IL-6 and include α_1-antichymotrypsin (ACT), α_1-antitrypsin (AAT) and haptoglobin. Acute phase proteins can also be classified into three groups according to the speed of response and magnitude of response. Group I shows approximately 50% increase and include complement component C3 and C4, caeruloplasmin and lipoprotein Lp (a); Group 2 proteins increase by 200–400%, respond within 10–24 hours and include ACT, AGP, AAT, haptoglobin and fibrinogen and Group 3 proteins which increase up to 1000 fold or greater starting within 10 hours and reach a maximum by 48–72 hours, and include SAA and CRP.

C-reactive protein

C-reactive protein (CRP), which belongs to the family of proteins known as pentraxins, has five non-covalently linked identical non-glycosylated polypeptide subunits. The main role of CRP is thought to be the recognition and removal of potentially toxic autogenous substances released from damaged tissues. It binds to damaged cell membranes, forming a complex which activates the complement pathway, leading to opsonisation, phagocytosis and lysis of invading cells. CRP is synthesised in the liver and it is a sensitive acute phase protein. The synthesis of CRP in the liver increases within 6 hours following inflammation and the concentration may reach 2000 times normal. It is the most useful measure of acute phase response and is more sensitive than ESR in detecting organic disease. Increased CRP concentration is an indication of inflammation and is used to detect and assess organic disease.

Diagnostic use of CRP

1. CRP measurements are useful in assessing the extent or activity of inflammation. The concentration of CRP gives a guide to the amount of tissue involved in inflammation and to the severity of inflammatory response.
2. It is useful in assessing the response to treatment in inflammatory conditions, such as rheumatoid arthritis, giant cell arthritis and inflammatory bowel disease.
3. It is useful in the detection of bacterial infection, which is the most potent stimulus for the acute phase response whereas viral

infection does not cause an increase. A rapid increase in CRP is an indicator of intercurrent sepsis, for example after surgery. A decrease in CRP in these situations is a useful indication of response to antibacterial treatment. Persistently increased concentration of CRP also indicates a poor prognosis in inflammatory and malignant diseases.

4. Cardiovascular risk marker: Recent studies also show that mild elevation of CRP is associated with increased risk of coronary artery disease.

Procalcitonin

Procalcitonin (PCT) is a 116 amino acid peptide with a sequence identical to that of the prohormone of calcitonin. In normal circumstances, PCT is only found in the C cells of the thyroid gland. However, pro-inflammatory cytokines stimulate the synthesis of PCT by many cell types and organs. Serum PCT rises within 3 hours of bacterial infection and reaches a peak after 6–12 hours, making it a good early marker of bacterial infection. Serum PCT is not raised in viral infections, chronic inflammatory disorders and autoimmune conditions. However, serum PCT can be elevated without bacterial infection in neonates within 48 hours of birth, the first day after major trauma or surgery and in patients with invasive fungal infections. It may also remain low in subacute bacterial endocarditis and if the infection is localised. Several recent studies have shown that serum PCT may be a better at detecting bacterial infection than serum CRP.

Serum amyloid A (SAA)

Serum amyloid A belongs to a group of polymorphic apolipoproteins synthesised by the liver. Three genes encode for SAA and two of these are responsible for the acute phase increase in SAA. The acute phase response of SAA is similar to that of CRP, peak values are reached at 24–48 hours and the serum concentration may increase by 1000-fold. It may be a more sensitive marker than CRP. SAA like CRP shows higher values in bacterial infection compared to viral infection. SAA is also increased in malignant diseases especially when there are second-aries. Combination of CRP and SAA may be useful in differentiating acute allograft rejection from infection. An increase in both CRP and SAA indicates infection, while an increase in SAA without an elevation in CRP suggests acute rejection. As SAA is an apolipoprotein associated with HDL, it may be a predictor of cardiovascular disease.

Serum α_1-antichymotrypsin

α_1-Antichymotrypsin (ACT), a 68-kDa protein with 23% carbohydrate is an inhibitor of serum proteases like AAT. ACT forms complexes with prostate specific antigen. It is important in limiting proteolytic damage to tissues from enzymes released during inflammation. High concentrations are found in bronchial secretions. It rises rapidly within 8 hours of injury. However, the rise is not as high as CRP or SAA. In combination with CPR, it is useful in the management of inflammatory bowel disease.

Immunoglobulins

Immunoglobulins are large molecular weight proteins, which are important in recognising foreign antigens and destroying them. An immunoglobulin molecule is produced by a single plasma cell or by a clone of identical plasma cells. There are six basic types of immunoglobulins — IgG, IgA, IgM, IgD and IgE.

All immunoglobulin molecules have a basic structure that consists of two identical heavy chains and two identical light chains (Figure 14.2), each with a variable and a constant region. The variable region is involved in antigen recognition and binding. The antigenic specificity of the particular antibody is determined by the amino acid sequence of the variable region. The constant region is the same for every immunoglobulin molecule for a given subclass. There are several heavy chains and the nature of the heavy chain determines the class of the immuno globulin. IgG, A, M, D and E are characterised by γ, α, μ, δ, and ε heavy chains, respectively. There are two types of light chains, κ and

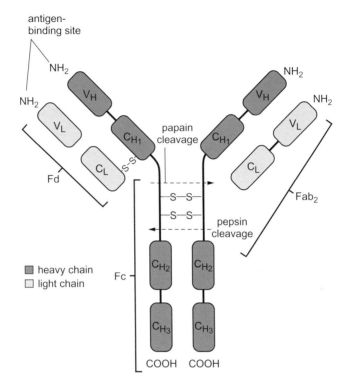

Figure 14.2   Structure of immunoglobulin.

λ, and both types may be found in any one class of immunoglobulin but not in the same molecule (Table 14.5).

Immunoglobulin G (IgG)

IgG the major circulating form of immunoglobulin, has a molecular weight of 160,000 accounts for 75% of serum immunoglobulin in adults with a half-life of 22 days. Sixty-five percent of the total immunoglobulin in the body is in the extravascular space, the remainder is in the circulation. IgG has 4 subclasses; IgG_1, IgG_2, IgG_3 and IgG_4. These subclasses show age, sex and genetic variation. IgG_1 is the major IgG (60–70%) in adult serum. From 18–20 weeks of gestation, IgG principally IgG_1 is actively transported across the placenta

Table 14.5   Characteristics of immunoglobulin classes

	Molecular weight	Number of immuno-globulin units	Heavy chain	Light chain	Serum con-centration	% in serum	Time to reach adult value(yrs)
IgG	160,000	1	χ	κ, λ	7–16 g/L	75	3–5
IgA	160,000 (320,000)	1,2	α	κ, λ	0.7–4 g/L	10–15	15
IgM	970,000	5	μ	κ, λ	04–2 g/L	5–10	0.7
IgD	180,000	1	ρ	κ, λ	< 0.04 g/L	< 1	15
IgE	200,000	1	ε	κ, λ	< 0.5 mg/L	< 1	15

and provides humoral immunity to the foetus and then to the neonate. After birth, the concentration of IgG_1 decreases initially before rising again as IgG synthesis by the infant increases. Specific deficiencies of the subclasses of IgG have been described, some patients exhibiting clinical symptoms, whilst others show no evidence of disease.

Immunoglobin M (IgM)

Immunoglobin M, a high molecular mass immunoglobulin of 970 kDa, normally circulates as a pentomer and accounts for 5–10% of plasma immunoglobulins. It has a half-life of about 5 days. Because of the high molecular weight, most IgM is found in the intravascular space.

Immunoglobin A (IgA)

IgA forms approximately 10–15% of circulating immunoglobulins, has a molecular weight of 160 kDa with a half-life of 6 days. About 10–15% is present in a polymeric form in the serum. An important form of IgA is the secretory form of IgA, which is found in tears, sweat, saliva, milk, colostrum, and gastrointestinal and bronchial secretions. Secretory IgA has a molecular weight of 380 kDa and consists of two molecules of IgA. It is synthesised primarily by plasma cells in mucous membrane of the gut, bronchi and ductules of lactating breast. The secretory IgA is resistant to enzymes and gives protection to mucosa from bacteria and viruses. There are two subclasses of IgA_1 and IgA_2. IgA_1 is found in serum and IgA_2 in mucosal secretions.

Immunoglobin D (IgD)

IgD, which accounts for < 1% of immunoglobulins in serum, has a molecular weight of 180 kDa. It is a surface receptor for antigen in B lymphocytes. Its primary function however is not known. It is probably involved in processing antigens in immature B cells.

Immunoglobin E (IgE)

IgE, which has a molecular weight of 200 kDa, is found only in trace amounts in the serum normally. Many IgE molecules are attached to surfaces of mast cells. When an allergen (antigen) crosslinks two of the attached IgE molecules, the mast cell is stimulated with the release of histamine and other vasoactive compounds. These compounds are responsible for increased vascular permeability, smooth muscle contraction, etc. and account for the allergic reactions seen in hay fever, asthma, urticaria and eczema. IgE is mainly synthesised at mucosal surfaces and sensitises mast cells locally and this suggest that IgE plays a protective role in mucosal defence.

Increased IgE concentration is strongly suggestive of an atopic predisposition. Increased concentrations are also found in parasitic disorders. IgE does not cross the placenta and the finding of an increased concentration in the neonate or child less than 2 years of age is very suggestive of development of atopic disease. However, in an adult, a normal IgE does not exclude atopic disorders as there is considerable overlap between normal and atopic individuals.

Allergen specific IgE is of value to determine the causative allergen.

Hypoimmunoglobulinaemia

Transient hypogammaglobulinaemia

The changes of immunoglobulin concentration during development are shown in Figure 14.3. IgG is transferred from the mother to the foetus across the placenta during the last 3 months of pregnancy. Maternal IgG concentration in the neonate slowly declines with time and as the infant starts to produce its own, IgG concentration slowly starts to rise from the age of about 3 months to reach adult values by 1 year. Concentrations of IgA and IgM are low at birth and slowly start to rise and reach the adult value by the end of the first decade.

Premature infants are at risk of hypogammaglobulinaemia because they do not have the full complement of maternal IgG. Hypogammaglobulinaemia may also be seen in infants in whom initiation of IgG synthesis is transiently delayed, transient hypogammaglobulinaemia of infancy. These infants are susceptible to infection. The exact frequency of transient immunoglobulin deficiency is not known. These infants present with frequent or severe infection. Recovery is full and spontaneous by the age of 18–36 weeks.

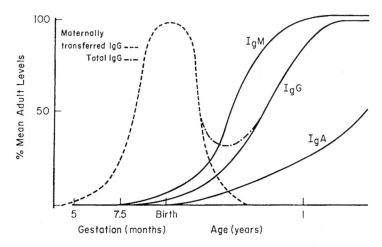

Figure 14.3 Serum concentration of immunoglobulins (expressed as a percentage of adult values) before birth and during the first few years of life.

Table 14.6   Causes of hypogammaglobulinaemia

Primary
- Selective IgA deficiency
- X-linked or autosomal hyper IgM syndrome
- Common variable immune deficiency
- X-linked agammaglobulinaemia (Burton's disease)

Combined deficiency of cellular and humoral immunity
- Severe combined immunodeficiency

Secondary
- Defective synthesis
  - Lymphoid malignancy
  - Toxic reaction e.g. renal failure
  - Drugs — e.g. phenytoin
  - Malnutrition
  - Infection
  - Radiotherapy
- Abnormal loss of protein
  - e.g. nephrotic syndrome

Pathological hypogammaglobulinaemia

Hypogammaglobulinaemia can be primary or secondary (Table 14.6).

1. *Secondary hypogammaglobulinaemia*

   Secondary causes include haematological malignancies (such as multiple myeloma, chronic lymphocytic leukaemia and Hodgkin's disease) causing bone marrow suppression, bone marrow disorders (such as hypoplasia, metastases of the bone and myelosclerosis), increased loss (such as in nephrotic syndrome and protein losing nephropathy), and toxic factors (such as cytotoxic drugs, severe infection, gluten sensitive enteropathy and prolonged renal failure).

2. *Primary hypogammaglobulinaemia*

   Primary hypogammaglobulinaemias are inherited disorders of immunoglobulin synthesis. Four types are described. Of these, the most common is IgA deficiency. Frequency of IgA deficiency varies from 1 in 150 in Arabs to 1 in 15,000 in Japanese. As most patients are asymptomatic, the frequency may be higher. Some patients present with increased frequency of infections especially those with associated IgG_2 subclass deficiency.

Hyper IgM syndrome is a rare disorder where there is an inability to switch production of IgM antibodies to IgG, A or E. These patients have low levels of serum IgG, IgA and high levels of IgM. This syndrome can be caused by a variety of gene defects, commonest is due to a mutation in the gene coding for CD40 ligand.

Common variable immune deficiency is a heterogeneous group of disorders of unknown aetiology where there is defective antibody production. These patients present with frequent infections.

Burton's disease or X-linked agammaglobulinaemia is rare disorder where there is a defect in B cell development, leading to severe hypogammaglobulinaemia. These patients present with recurrent infections after the age of 4–6 months.

Hypergammaglobulinaemia

Increased immunoglobulin concentration may result from stimulation of many clones of B cells, polyclonal hypergammaglobulinaemia or due to proliferation of single clone, monoclonal hypergammaglobulinaemia. Polyclonal hypergammaglobulinaemia is seen in chronic bacterial infections, autoimmune diseases and chronic liver disease.

Monoclonal immunoglobulins (paraproteins)

Proliferation of a single clone of B cells results in immunoglobulin molecules of identical structure. Such immunoglobulin molecules will represent a homogeneous protein migrating as a compact band in electrophoretic separation. This is called a paraprotein or monoclonal immunoglobulin. Clinical conditions associated with paraproteinaemia are given in Table 14.7. These monoclonal immunoglobulins or

Table 14.7   Conditions associated with paraproteinaemia

	Relative incidence
Malignant	
• Multiple myeloma	60%
• Lymphoma	5%
• Waldenstrom's macroglobulinaemia	4%
• Soft tissue plasmocytoma	3%
• Chronic lymphocytic leukaemia	1%
• Heavy chain disease	<1%
Non-malignant	
• Benign (asymptomatic)	18%
• Primary cold agglutinin disease	1%
• Amyloidosis	1%
• Others	1–2%

paraproteins may be polymers, monomers or fragments of immunoglobulin molecules, which are most commonly light chains (Bence-Jones proteins) and rarely heavy chains.

- *Multiple myeloma*

  Multiple myeloma, a malignant neoplasm of a single clone of plasma cells usually presents as a disseminated malignancy where plasma cells proliferate diffusely throughout the marrow. Occasionally, they may form an isolated solitary tumour called the plasmacytoma. Multiple myeloma is more common in African Americans and the incidence increases with age. Proliferation of plasma cells produces osteolytic lesions in the bone. At the same time, other bone marrow cells are reduced so that there is thrombocytopenia, anaemia and leucopenia. The development of normal clones of plasma cells is reduced, causing a reduction in other immunoglobulins. Patients with multiple myeloma may present with a wide variety of symptoms; it may be local due to the bone lesion or general non-specific symptoms such as weight loss, etc. (Table 14.8). Diagnosis of multiple

Table 14.8   Features of multiple myeloma

Clinical

• Bone pain	— Pathological fracture
	— Osteoporosis
	— Local lytic lesion
• Anaemia	— Normocytic normochromic
	— Marrow invasion
	— Renal failure
• Recurrent infection	— Reduced normal immunoglobulins
• Renal failure	— Proteinuria
	— Hypercalcaemia
	— Hyperuricaemia
	— Amyloid
• Hyperviscosity syndrome	— Hyperproteinaemia

Biochemical

• High serum protein	— Paraprotein
• Low immunoglobulins	— Immune paresis
• Proteinuria	— Bence-Jones Proteins
• Fanconi syndrome	— Renal tubular damage due to Bence-Jones proteins
• Hypercalcaemia	— Osteolytic lesion
• Hyperuricaemia	
• Uraemia	
• Hyponatraemia	— Pseudohyponatraemia
• Raised ESR	
• Coagulopathy	

myeloma depends on the demonstration of two of the following three characteristics:

(a) detection of paraproteins in urine or serum,
(b) the presence of increased plasma cells in the bone marrow,
(c) destruction of bone.

In multiple myeloma, any type of immunoglobulin may be seen. The most common is IgG followed by IgA. In 20% of patients, only light chains (Bence-Jones protein) are seen and in these patients diagnosis may be missed if only serum is examined. It is therefore important to examine the urine for Bence-Jones proteins. The amount of paraprotein is related to the tumour mass and quantitation of paraproteins is important as a tumour marker (see Chapter 24). Patients with myeloma have hypercalcaemia, due to increased protein binding of calcium and an increase in ionised calcium due to the osteolytic lesions. They often have renal failure due to the effect of hypercalcaemia and due to obstruction of nephrons by proteins. There is increased uric acid due to increased turnover of cells and renal failure. Patients might be hyponatraemic due to hyperproteinaemia (pseudohyponatraemia). Despite the osteolytic lesions in myeloma, bone formation markers such as alkaline phosphatase and osteocalcin are normal. This is thought to be due to the production of a factor by the tumour, which may be one of the cytokines, inhibiting bone formation. Serum β_2 microglobulin is a useful prognostic indicator in myeloma. Based on the concentrations of serum β_2 microglobulin and albumin, myeloma is classified into three stages. When a paraprotein is detected, it is important to identify the type of immunoglobulins as they have prognostic significance. Features that suggest poor prognosis include anaemia, renal failure, hypercalcaemia, hypoalbuminaemia, high β_2 microglobulin and high amount of paraprotein.

Myeloma is treated with cytotoxic drugs and sometimes local radiotherapy may be required to treat localised tumours (plasmacytoma).

Case 14.1

A 68-year-old man went to see his GP because of repeated upper respiratory tract infection and backache of 4 months duration. He had lost some weight and felt tired most of the time. His GP did some basic investigations and the results were as follows:

Serum		Reference Range
Sodium (mmol/L)	132	135–145
Ura (mmol/L)	13.2	2.7–7.0
Creatinine (μmol/L)	165	66–110
Corrected calcium (mmol/L)	2.85	2.25–2.55
Total protein (g/L)	98	60–75
Albumin (g/L)	32	35–45
Hb (g/L)	85	125–145
ESR (mm/h)	85	< 10

> *These results show hypercalcaemia, high globulins (66 g/L), anaemia with raised ESR. Multiple myeloma was suspected and serum and urine electrophoresis were done and they showed a paraprotein band in the serum and urine. This was identified as IgG kappa paraprotein. Diagnosis of multiple myeloma was confirmed by demonstrating increased plasma cells in his bone marrow.*

- *Waldenstrom's macroglobulinaemia*
  Waldenstrom's macroglobulinaemia is a B cell tumour producing IgM. These patients do not have bone pain or hypercalcaemia but have lymphadenopathy and hepatosplenomegaly. It is a slow growing tumour such that paraprotein concentration may increase to 30 g/L before causing symptoms, which are due to hyperviscosity (hyperviscosity syndrome). Hyperviscosity leads to impaired circulation of blood through capillaries causing lassitude, confusion, coma, blindness, fatigue, anaemia, renal failure, congestive heart failure and increased tendency to thrombosis. In a small proportion of cases, the IgM paraproteins are cryoglobulins (see below).

  Treatment of Waldenstrom's macroglobulinaemia consists of chemo-therapy and plasmaphoresis to remove paraproteins, which relieves the symptoms. The prognosis is usually better than in multiple myeloma.

- *Heavy chain disease*
  Some lymphoma and chronic lymphocytic leukaemia are associated with synthesis of heavy chains without light chains. These are rare disorders and the heavy chains can be α, δ or μ and of these α is the most common. Alpha chain disease is common in people of Middle East origin and they present with severe malabsorption syndrome due to the involvement of the small intestine.

- *Benign paraproteinaemia — Monoclononal gammopathy of unknown significance (MGUS)*
  This term is applied when a monoclonal protein has been identified with no other feature of malignancy. Nearly 20% of paraproteins

found on electrophoresis are benign. It is a diagnosis of exclusion. Criteria for the diagnosis of MGUS include no clinical features of myeloma or associated disorders, paraprotein concentration < 10 g/L, no Bence-Jones protein, no suppression of normal immunoglobulins, normal bone marrow, no lytic lesions in the bone, no change in paraprotein concentration with age and no evidence of malignancy on follow up after 5 years.

Case 14.2

A 66-year-old man was started on simvastatin for hypercholesterolaemia. Two weeks later, his GP requested blood tests (CK and liver function tests) as part of the follow up. Results were as follows:

Serum		Reference Range
CK (IU/L)	120	>250
Bilirubin (μmol/L)	15	<20
ALT (IU/L)	30	<42
AST (IU/L)	33	<45
ALP (IU/L)	125	<250
Albumin (g/L)	37	35–45
Total protein	82	60–75

The only abnormality was a raised total protein due to raised globulins. The laboratory suggested serum electrophoresis and this showed a paraprotein band. Further investigations showed that there was no paraprotein in the urine and the concentration of paraprotein in serum was 8 g/L. As he had no other features of myeloma, a diagnosis of monoclonal gammopathy of unknown significance (MUGS) was made and he was followed regularly at the haematology clinic.

---

2. *Cryoproteins*

Cryoproteins are proteins which form a precipitate on cooling the serum or plasma and re-dissolve on warming to 37°C. In general, proteins that precipitate at 21°C are more likely to cause symptoms. There are many proteins, which aggregate at temperatures lower than 16°C and these are usually laboratory artefacts and have no clinical features. Patients with cryoproteins may present with cold intolerance, with pain in exposed areas, Raynaud's phenomenon, which is an episodic ischaemic pain in the extremities precipitated by cold weather, skin lesions including purpura, urticaria and ulcers. A more serious complication of cryoprotein is renal failure. The type of proteins causing cryo-precipitation includes cryoglobulins, which may be polyclonal or monoclonal, the latter are usually IgA and these are very often malignant. Rarely cryofibrinogen and cold agglutinins may also be present. In the investigation of cryoproteins, it is important to take the blood sample into a warm syringe and keep the sample at 37°C and separate the sample at 37°C. Failure to do this will cause loss of cryoproteins and false-negative results.

Cytokines

The term cytokine includes a large number of regulatory peptides produced by a variety of cells. They have either a paracrine (acting on adjacent cells) or autocrine (acting on the same cell that secretes the cytokines). Cytokines include interleukins, lymphokines, and cell signalling molecules such as tumour necrosis factor or interferon and growth factors. They act via cell surface receptors and can be stimulatory or inhibitory. Many of the cytokines have diverse functions on different cell types. Cytokines have important role in regulating the inflammatory response and the immune response. Although cytokines can be measured in serum, at present there are no clear indications for measuring those in routine practice.

Further Reading

1. Milfrord-Ward A, Sheldon J, Rowbottom A, Wild DG. *PRU Handbook of Clinical Immunochemistry* 2007. Sheffield: PRU Publishing 9th Edition.
2. Silverman EK, Sandhaus RA. Alpha1-Antitrypsin Deficiency. *N Engl J Med* 2009; 360:2749–2757.

Summary/Key Points

1. There are several hundred proteins in circulation and of these, albumin is the most abundant protein. Albumin and other proteins are also present in the interstitial fluid. The total amount of albumin in interstitial fluid is nearly 50% of the total amount of albumin in the body because of the large volume of the interstitial fluid.

2. The most common cause of a low serum albumin concentration in hospital patients is acute illness where there is redistribution from plasma to interstitial fluid as a result of increased permeability of the capillary membrane. Other important causes are decreased synthesis (e.g. liver disease) or increased loss (e.g. nephrotic syndrome).

3. α_1-Antitrypsin (AAT) is an important protein, which protects against elastase released during phagocytosis. Deficiency of AAT can be caused by a large number of genetic variants of AAT and this can lead to emphysema and liver disease.

4. Acute phase proteins are those, which change by at least 25% during inflammatory disorders. Proteins like C-reactive protein increase (positive acute phase proteins) while albumin and transferrin decrease (negative acute phase proteins).

5. Serum CRP and procalcitonin are two proteins which are useful in distinguishing bacterial infection from viral infections.

6. Immunoglobulins are large molecular weight proteins involved in the recognition and destruction of foreign proteins. There are five classes of immunoglobulins (IgG, IgA, IgM, IgD, IgE).

7. Hypogammaglobulinaemia can be secondary to haematological malignancies and bone marrow disorders or can be primary due to genetic defects.

8. Hypergammaglobulinaemia can be polyclonal due to chronic infection or autoimmune diseases. One of the important causes of monoclonal increase in gamma globulins is multiple myeloma, a malignant tumour of plasma cells. Diagnosis of multiple myeloma requires demonstration of paraproteins in serum or urine, presence of increased plasma cells and/or

evidence of destruction of bone. Biochemical changes seen in multiple myeloma include hypercalcaemia, renal impairment, and hyperuricaemia.

9. Monoclonal gammopathy of unknown significance (MGUS) is a benign condition where there is a monoclonal paraprotein without any evidence of malignancy. These patients need to be followed up regularly as some may go on to develop multiple myeloma.

10. Cytokines include a large number of regulatory proteins which play an essential role in regulating inflammation and the immune response.

chapter 15

The Liver

Introduction

The liver is a vital organ which plays a central role in the metabolism of carbohydrates, proteins and lipids and in the detoxification of endogenous and exogenous substances. The liver is made up of two lobes and the lobes consist of lobules. The liver consists of two types of cells: (1) parenchymal cells or hepatocytes which constitute about 80% of the mass of the liver, and (2) Kupffer cells which belong to the reticuloendothelial system. Hepatocytes are arranged in single-cell thick cords or sheets radiating between the portal tract and the small branches of the hepatic vein (Figure 15.1). The functional unit of the liver is the acinus, which represents a group of hepatocytes supplied by a terminal branch of the portal vein and the hepatic artery and drained by a terminal branch of the bile duct. The portal triad, which consists of the portal vein, hepatic artery and bile duct, lies at the centre of the acinus. Hepatocytes are separated by sinusoids, which are lined by endothelial cells and Kupffer cells. Blood flows from the portal vein and the hepatic artery, through the sinusoids and enters the hepatic vein. Bile canaliculi are microvilli covered grooves in the lateral surface of the hepatocytes. These canaliculi eventually drain into bile ducts lined by biliary cells. The flow of bile in the canaliculi is in the opposite direction to the blood flow (Figure 15.1).

Hepatocytes in the acinus are sometimes divided into three zones — zone 1 (periportal) being nearest to the portal tract and zone 3 (perivascular) being closest to the hepatic vein. Due to the nature of the blood flow, zone 1 has the highest while zone 3 has the

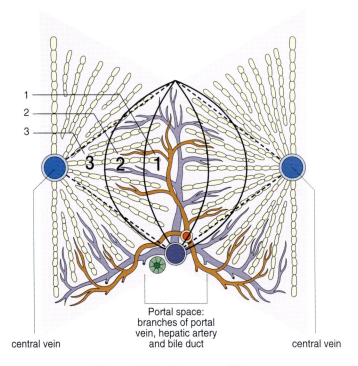

1
2
3

3 2 1

Portal space:
branches of portal
vein, hepatic artery
and bile duct

central vein   central vein

Figure 15.1   Structure of liver.

least oxygen and nutrient content. Cells in zone 1 are more active in oxidative function, gluconeogenesis and bile formation. They have greater lysosomes and mitochondria, ALP and transaminases than other areas. Cells in zone 3 are enriched with endoplasmic reticulum and have the highest concentration of drug metabolising enzymes. Because of this functional heterogeneity, the effect of disease processes are different in these different zones. For example, zone 3 is more susceptible to viral, toxic and anoxic damage.

Functions of the Liver

The liver has many important functions and these are listed in Table 15.1. Functions of the liver in relation to general metabolic functions (carbohydrates, fat, proteins) are described in relevant chapters.

Table 15.1   Functions of the liver

Intermediary metabolism
- Carbohydrate metabolism
  — Gluconeogenesis
  — Glycogen synthesis and breakdown
  — Metabolism of lactate
  — Galactose metabolism
- Protein metabolism
  — Urea synthesis
- Lipid metabolism
  — Lipoprotein synthesis
  — Fatty acid synthesis
  — Cholesterol synthesis and excretion
  — Ketogenesis
  — Bile acid synthesis
- Synthetic function
  — Synthesis of plasma proteins, coagulation factors
  — Bile acid synthesis
- Detoxification and excretion
  — Bilirubin metabolism
  — Foreign compounds — e.g. drugs
- Metabolism of hormones
  — Hydroxylation of vitamin D to 25-OH vitamin D
  — Metabolism of peptide hormones
  — Metabolism of steroid hormones
- Storage function
  — Storage of A, B_{12}, iron, glycogen, etc.

Synthetic functions

Large numbers of proteins are synthesised by the liver. All plasma proteins except immunoglobulins, all clotting factors except factor VIII, inhibitors of coagulation (protein C, protein S, antithrombin III), α_2-antiplasmin (inhibitor of plasmin) and complement component 3 are synthesised by the liver. Proteins synthesised by the liver are listed in Table 15.2. In liver disease, synthesis of proteins are reduced only if the disease is severe as the liver has a large reserve capacity. Concentration of proteins with short half-life such as clotting factors are the first to decrease in severe liver disease.

Table 15.2   Proteins synthesised by liver

---

1. Plasma proteins except immunoglobins
2. Proteins involved in haemostasis and fibrinogen
    a. All clotting factors except factor VIII
    b. Inhibitors of coagulation — proteins C and S, antithrombin III
    c. Fibronolysis factors — plasminogen
    d. Antifibrinolysis — $\alpha 2$-antiplasmin
3. Hormones:
    IGF1, Thrombopoietin (a hormone regulating platelet production)
4. Apolipoproteins — all except apoB48
5. Prohormones — angiotensinogen
6. Carrier plasma proteins — sex hormone-binding globulin (SHBG), cortisol-binding globulin (CBG), etc.

Bilirubin metabolism

Bilirubin is the catabolic product of haem, which is present in haemo-globin, myoglobin and other haem-containing proteins or enzymes. Approximately $500\ \mu mol$ (275 mg) of bilirubin is produced daily. Majority (80%) of this is formed from the breakdown of erythrocytes while the remaining comes from myoglobin, other haem-containing proteins and from premature destruction of newly formed red cells and its precursors in the bone marrow (ineffective erythropoiesis). Formation of bilirubin takes place in the reticuloendothelial system. The first step is the conversion of haem to biliverdin by haemoxygenase, a microsomal enzyme, and then to bilirubin by biliverdin reductase. In this process, an equimolar amount of carbon monoxide is released. The bilirubin produced is unconjugated, lipid soluble and is transported to the liver bound to albumin. Binding of albumin prevents the uptake of this lipid-soluble bilirubin by other tissues such as the brain where it can be toxic (see neonatal jaundice). A variety of compounds including sulphonamides, furosemide, salicylate, fatty acids and radiographic con-trast media may displace bilirubin from its albumin-binding site. Bilirubin binds reversibly to albumin. However, in prolonged jaundice, a small fraction binds irreversibly and this is termed δ-bilirubin. The half-life of δ-bilirubin is long and it is not excreted in the bile.

In the liver, the bilirubin–albumin complex dissociates by an unknown mechanism and is transported across the hepatocyte membrane by a carrier-mediated process via an organic anion transport protein called OATP-2. This process can be competitively inhibited by other organic anions such as indocyanine green, bromosulphthalein and contrast media. Inside the liver cell, bilirubin binds to cytoplasmic proteins, mainly ligandin (glutathione-S-transferase B) and Z protein, which shuttle bilirubin to the rough endoplasmic reticulum where it is conjugated with glucuronic acid by bilirubin uridine diphosphate (UDP) glucuronyltransferase (UGT1A1) to form bilirubin mono- and diglucuronides. Hepatic uptake of bilirubin is dependent on the concentration of unconjugated bilirubin in the blood as well as the rate of the bilirubin conjugation within the liver. The activity of the enzyme UGT1A1 is low at birth and it increases during the first 10 days of life. Drugs such as phenobarbitone, fibrates, diazepam and spironolactone can induce this enzyme. The liver cell actively secretes conjugated bilirubin by a carrier-mediated process into the bile canaliculi involving multi-drug-resistant protein 2 (MDRP2). The secretion of bilirubin into the bile canaliculi is a rate-limiting step in the metabolism of bilirubin.

Some of the bilirubin entering the large bowel is deconjugated, which makes the bilirubin lipid soluble. It is then absorbed and re-excreted by the liver via the enterohepatic circulation. Most of the bilirubin reaching the intestine is oxidised to urobilinogen, a colourless product, which is then oxidised to coloured stercobilin and urobilin and excreted in the faeces. About 20% of the urobilinogen produced is reabsorbed by the colonic mucosa. Most of this is taken up by the liver and re-excreted into the bile. As urobilinogen is water-soluble, a small fraction appears in the urine.

During cholestasis, conjugated bilirubin is transported back to the blood by multi-drug resistant protein 3 (MRP3) which is located in the lateral membrane of the hepatocyte. In the circulation, 95% of the bilirubin is unconjugated therefore, they do not appear in the urine. An increase in conjugated bilirubin will result in bilirubinuria. Conjugated bilirubin is also bound to albumin but with lower affinity.

Bile acids

Bile is made up of bile acids, bile pigments, cholesterol, and phospholipids dissolved in an alkaline electrolyte solution. Each day about 500 mls of bile is secreted.

Bile acids synthesised from cholesterol in the liver are the primary bile acids, cholic acid and chenodeoxycholic acid. These are conjugated with glycine or taurine to increase their solubility and facilitate their function.

Bile salts are stored in the gall bladder in between meals and released into the intestine during digestion. Bacterial enzymes deconjugate most of the bile salts and bile acids are reabsorbed in the terminal ileum, returned to the liver, reconjugated and re-excreted — an enterohepatic circulation. The small proportion reaching the colon is converted to secondary bile acids, deoxycholic and lithocholic acids, which are absorbed and returned to the liver. Deoxycholic acid is conjugated, excreted and reabsorbed like the primary bile acids. Lithocholic acid is conjugated and excreted but not reabsorbed.

Reabsorption of bile acids from the terminal ileum occurs by an efficient active transport process involving the bile acid transport protein. The synthetic rate of bile salt is 0.2–0.4 g/day and the total bile salt pool is approximately 3.5 g. It has been estimated that the entire bile acid recycles twice per meal and 6–8 times per day. The synthesis of bile acid is under negative feedback control by reabsorbed bile salts.

Bile acids help in the absorption of fat by forming micelles. They also play an important role in activating lipases in the intestine and modulate cholesterol metabolism via binding to nuclear receptors called farnesoid X receptor (FRX). Through these receptors, bile acids regulate their own enterohepatic circulation, as well as homeostasis of glucose, cholesterol and triglycerides.

Conjugation and detoxification or biotransformation and excretion

The liver plays an important part in detoxifying or transforming endogenous as well as exogenous compounds. This biotransformation occurs in two stages. In phase I, the polarity of the compound is increased by oxidation or hydroxylation, a reaction mediated by microsomal cytochrome P450 oxidase system in the endoplasmic reticulum. In phase II, the hydroxylated and carboxylated compounds are conjugated with glucuronic, acetyl or methyl groups making them water-soluble. These conjugated compounds are then excreted in the bile or in the urine. An alternative pathway for detoxification is via glutathione (see 'Paracetamol' in Chapter 25). While most biotransformation reactions result in detoxification, sometimes it may give rise to toxic or active metabolites. There are many cytochrome P450 isoenzymes. Each gene locus for these enzymes has multiple alleles and these allelic variations can lead to variations in drug effects (see Chapter 25). In addition to genetic polymorphism, other drugs and chemicals can also induce these enzymes. For example, the induction of enzymes by phenytoin increases the metabolism of endogenous substances such as vitamin D as well as exogenous chemicals and drugs (e.g. amiodarone).

Jaundice

Jaundice is the yellowish discolouration of tissues caused by the deposition of bilirubin. The discolouration occurs when serum concentration exceeds 35–40 μmol/L. Jaundice is usually classified into (i) prehepatic where the production of bilirubin is increased, (ii) hepatic, where there is an abnormality or reduction in conjugation or the excretory function of the liver, and (iii) obstruction to the bile flow, either intrahepatic or extrahepatic. The causes of jaundice are given in Table 15.3.

Pre-hepatic Jaundice

Pre-hepatic jaundice is caused by increased production of bilirubin which can either be due to destruction of mature cells, e.g. in haemolytic anaemias or due to ineffective erythropoiesis, e.g. in pernicious anaemia. Bilirubin in the serum is unconjugated and the concentration rarely exceeds 100 μmol/L. Serum bilirubin may exceed 100 μmol/L in prehepatic jaundice if there is associated liver

Table 15.3   Causes of jaundice

Pre-hepatic
- Ineffective erythropoiesis
  — Pernicious anaemia
  — Thalassaemia
- Increased erythrocyte breakdown
  — Haemolysis — e.g. congenital spherocytosis,
     autoimmune haemolysis
  — Internal haemorrhage

Hepatic
- Immature conjugating enzymes
  — Neonatal jaundice
- Inherited defects in bilirubin metabolism
  — Gilbert's syndrome
  — Crigler-Najjar syndrome
  — Rotar syndrome
  — Dubin-Johnson syndrome
- Hepatic dysfunction
  — Hepatitis
  — Cirrhosis
- Drug-induced
  — Paracetamol
  — Isoniazide

Cholestatic (Post-hepatic)
- Intrahepatic
  — Hepatitis
  — Biliary cirrhosis
  — Drugs — anabolic steroids, phenothiazine
  — Hepatic malignancy
- Extrahepatic
  — Gallstones
  — Bile duct tumours
  — Compression of bile duct
  — Carcinoma of head of pancreas

disease (e.g. alcoholic hepatitis) or if there is haemolytic crisis when the rate of formation of bilirubin exceeds the rate of hepatic clearance. In prehepatic jaundice, as a result of more bilirubin production, urine will show increased urobilinogen but no bilirubin.

Hepatic Jaundice

Hepatic jaundice commonly occurs due to a decreased capacity of the liver to conjugate and/or secrete bile pigments. The most common cause of hepatic jaundice is hepatitis where there is destruction of hepatocytes. Jaundice is due to reduced hepatic uptake, decreased conjugation and impaired transport of bilirubin. The serum bilirubin is a mixture of conjugated and unconjugated forms and bilirubin as well as excess urobilinogen may be found in the urine.

See below for inherited disorders of bilirubin metabolism.

Cholestasis/Post-hepatic Jaundice

Cholestasis is when there is interference to the bile flow, which can be intrahepatic or extrahepatic. Intrahepatic cholestasis may result from diseases of the liver such as hepatitis, primary biliary cirrhosis and hepatic malignancy or may be caused by drugs such as phenothiazines (chlorpromazine). Extrahepatic obstruction may be caused by obstruction to bile flow in the biliary tree by gallstones, enlarged lymph nodes in the porta hepatis or carcinoma of the head of pancreas. In this type of jaundice, serum bilirubin is conjugated and bilirubin can be detected in the urine, but urobilinogen will be absent if the obstruction is complete.

The Inherited Hyperbilirubinaemias

Gilbert's syndrome

Gilbert's syndrome, a benign condition occurring in approximately 5–7% of the population, is characterised by recurrent episodes of mild jaundice, which is usually associated with intercurrent illness and starvation. Other liver function tests and histology of the liver are all normal. It is inherited in an autosomal dominant fashion and is due to a mutation in the UDP-glucuronyl transferase gene leading to reduced activity of the UDP-glucuronyl transferase enzyme (UGT1A1). The bilirubin in the blood is unconjugated. It is usually detected accidentally during routine blood test. Recognition of this syndrome is important to avoid unnecessary investigations. Diagnosis can be confirmed either by nicotinic acid or caloric deprivation test. When 500 mg of nicotinic acid is given or when the patient is fasted for 48 hours, there is a 2–3-fold increase in unconjugated bilirubin in patients with Gilbert's syndrome. However, this test is not used often, as the specificity of the test is not high. As UGT1A1 is also responsible for the detoxification of some drugs, subjects with this syndrome may have side effects when these drugs are given. For example, irinotecan (an anticancer drug) can cause diarrhoea and neutropenia.

The Crigler-Najjar syndrome

This is a rare autosomal recessive condition of which there are two types. In type 1, there is a complete absence of UDP glucuronyl transferase activity leading to failure of conjugation of bilirubin. The child becomes deeply jaundiced during the first few days of life. The unconjugated, lipid soluble bilirubin crosses the blood brain barrier and gets deposited in parts of the brain causing kernictus and death occurs within the first year of life. In a much less severe form, type 2, where the enzyme is not completely absent, jaundice may appear later in life. Phenobarbitone may be given to induce the enzyme and increase conjugation.

Dubin–Johnson syndrome

This is an uncommon disease inherited in an autosomal recessive form. It is more frequently found in some people, e.g. people of Arabian origin. It is characterised by mild fluctuating jaundice and the jaundice is mixed, both conjugated and unconjugated. It is due to a mutation in the gene controlling the multidrug resistance related protein 2 (MRP2) which is an anion transporter. It may present at any age from birth to adult life, often precipitated by pregnancy or the contraceptive pill. Other liver function tests are normal but a bromosulphthalein (BSP) excretion test is abnormal due to failure of excretion of this anion. The liver becomes black due to the accumulation of brownish pigments.

Rotor's syndrome

This is a benign autosomal recessive condition characterised by fluctuating jaundice due to failure to secrete bilirubin. The liver is not pigmented and MRP2 is normal. Its exact pathogenesis is not fully understood.

Neonatal jaundice — see Chapter 28.

Assessment of Hepatic Function: Liver Function Tests

Due to the multiple functions of the liver, it is impossible to assess all these functions using only one or two investigations. Liver function tests are conveniently grouped into three classes. The first group consists of commonly used standard liver function tests. The second group consists of tests that quantitatively assess the functions of the liver, and the third group are tests which are used to diagnose specific diseases of the liver (Table 15.4).

Liver function tests are useful to (i) establish whether there is any evidence of liver dysfunction, (ii) establish a specific diagnosis, (iii) assess the severity of liver dysfunction, (iv) monitor the progression of disease, and (v) to assess response to treatment.

The standard liver function tests are tests of liver cell damage or liver cell dysfunction rather than function tests in the true sense of the word. As the liver has considerable reserve capacity, some of these tests are insensitive to liver damage and may be normal even when there is considerable liver damage.

Standard Liver Function Tests

Serum bilirubin

Serum bilirubin is not always increased in liver disease and it may be high in other diseases, e.g. in haemolytic disease. Metabolism of bilirubin and causes of jaundice have been discussed above. As the liver has a large reserve capacity to excrete bilirubin, serum bilirubin is not a sensitive test for liver disease. It is a useful test to monitor the progress of liver disease (e.g. primary biliary cirrhosis, neonatal

Table 15.4   Liver function tests

Test	Function
Commonly used tests	
Transaminases	Hepatocellular damage
Alkaline phosphatase	Cholestasis
Gamma-glutamyl transferase	Cholestasis/enzyme induction
Albumin	Synthetic capacity
Prothrombin time	Synthetic capacity
Bilirubin	Hepatocellular damage/cholestasis
Bile acids	
Quantitative tests/Less commonly used tests	
Galactose elimination capacity	Hepatocyte mass
Aminopyrine breath test	Microsomal metabolism
MEGX formation rate	Hepatic blood flow/microsomal metabolism
Bromosulphthalein clearance	Hepatic blood flow/biliary excretion
Indocyanine green clearance	Hepatic blood flow
Caffeine clearance	Microsomal metabolism
Arterial ketone body ratio	
Tests to establish a specific diagnosis	
α-Fetoprotein	
α_1-Antitrypsin	
Tests for haemochromatosis	
Tests for viral hepatitis	
Tests for hepatic fibrosis	

jaundice) and to assess response to treatment (e.g. after relieving obstruction). Serum bilirubin can be measured as unconjugated (indirect) and conjugated (direct) fractions. Identification of the type of bilirubin is sometimes helpful in the differential diagnosis of jaundice, e.g. if serum bilirubin is predominantly unconjugated, prehepatic jaundice is likely and if it is conjugated, cholestasis is likely.

Aminotransferases

Aspartate and alanine aminotransferases (AST and ALT) are intracellular enzymes. These enzymes are released into the circulation when there is liver cell damage, hepatic necrosis or increased permeability of hepatocyte plasma membrane (see Chapter 13). Aminotransferases are sensitive indicators of hepatocellular damage or dysfunction. Increases in transaminases activity in the serum, however, does not always correlate with the degree of necrosis or inflammatory activity as seen on liver biopsy. Very high values (> 10 upper limit of normal (ULN)) are usually due to primary liver disease such as viral hepatitis. Some suggest that the ratio of AST/ALT is useful in distinguishing alcoholic hepatic disease from other causes (a value > 2 is very suggestive of alcohol-related liver pathology).

Markers of cholestasis: Alkaline phosphatase (ALP) and gamma-glutamyl transpeptidase (GGT)

Alkaline phosphatase is found in the canalicular plasma membrane of hepatocytes and its synthesis is increased within hours of biliary obstruction. This is thought to be triggered by an increase in tissue bile acid concentration. GGT is found in the membranes of the smooth endoplasmic reticulum, in the canalicular plasma membrane of hepatocytes and biliary epithelial cells. GGT is a non-specific but sensitive indicator of hepatocellular disease and it is markedly elevated in cholestatic conditions. Certain drugs like antiepileptics and chronic alcohol intake may also increase GGT. When serum GGT and ALP are both elevated, hepatobiliary disease is very likely.

Tests of synthetic capacity: Serum albumin and prothrombin time

Synthetic function of the liver can be assessed by measurement of proteins synthesised by the liver. Of these, albumin and clotting-related proteins are commonly measured. Serum albumin concentration falls in liver disease. However, because of its long half-life, (about 20 days) a low serum albumin is indicative of chronic rather than acute liver disease. Furthermore, serum concentration of albumin can be affected by many other factors (see Chapter 14).

Prothrombin time which measures the rate at which prothrombin is converted to thrombin is an indication of the activity of vitamin K dependent clotting factors — prothrombin and factors VII, IX and X which are synthesised by the liver. Of these, factor VII has the shortest half-life of 4–6 hours. Prolongation of prothrombin time is an early indication of impaired liver synthetic capacity. However, prothrombin time may be abnormal due to vitamin K deficiency from other causes such as malabsorption syndromes.

Bile acids

Concentration of serum bile acids depends on the absorption of bile acids from the gut as well as on the uptake of reabsorbed bile acids by the liver. Increased fasting serum bile acid concentrations are seen when there is impaired hepatic uptake, impaired secretion of bile acids or when there is shunting of blood from the portal to systemic circulation. Serum bile acid can be used as an endogenous clearance test. However, the sensitivity and specificity of bile acid are no better than conventional standard liver function tests and thus it is not routinely used except in the investigation of liver disease during pregnancy (intrahepatic cholestasis of pregnancy). In this condition, which occurs in the third trimester, there is intense itching, jaundice and other features of cholestasis. Serum bile acids are often markedly elevated and the condition is associated with increased risks of premature delivery and stillbirth.

Quantitative Liver Function Tests

These tests are mainly used in specialised liver units. In the galactose elimination test, the rate of disappearance of galactose after an intravenous dose is measured. This has been used as a prognostic marker in liver disease. In the aminopyrine breath test, ^{14}C-labelled aminopyrine is given. Aminopyrine is demethylated by microsomal enzymes of the liver and the methyl group is converted to carbon dioxide. By measuring ^{14}C carbon dioxide in the breath after giving ^{14}C-labelled aminopyrine, a quantitative assessment of hepatic detoxifying function can be made.

The caffeine clearance test assesses the capacity of liver to metabolise caffeine. Other clearance measurements used are bromo-sulphthalein (BSP) clearance and indocyanine green clearance which are measures of hepatic secretory function. The BSP test has now been abandoned because of its side effects.

In the monoethylglycinexylidide (MEGX) formation test, the rate of MEGX appearance, a metabolite of lignocaine, produced by the cytochrome P450 enzyme system is measured after administration of lignocaine. This is a reflection of uptake, metabolism and clearance by the liver. This test is simple to perform and has been used to assess liver function in the liver donors, in the staging of liver damage in liver recipients and as a prognostic marker.

Other Liver Function Tests

Glutathionine S-transferase α isoenzyme is found in highest concentrations in the liver. Plasma GST-α activity is almost entirely from the liver. It is a more sensitive marker than amino transferase in detecting liver cell damage. However, its use is currently limited to research studies.

Markers of fibrosis

The ability to detect fibrosis, which ultimately leads to cirrhosis, is useful. A large number of biochemical tests as well as imaging techniques have been examined. Collagen deposited in the liver during fibrosis is type I and III. Collagen is synthesised as procollagen and a peptide (procollagen) is then cleaved to give the mature molecule. Procollagen type III has been used as a marker of liver fibrosis. It was found to be useful in monitoring patients who at risk of developing liver fibrosis, e.g. those on methotrexate treatment. Hyaluronate, a glycosaminoglycan synthesised in the liver during fibrosis has also been used. Others have developed combinations of tests to detect fibrosis. These include APRI (AST:platelet ratio) and Fib 4 (which include platelet, AST and gamma globulin). Other combinations such as Fibrotest requires tests such as α_2 macroglobulin, haptoglobin and apo A1 (in addition to bilirubin, GGT and ALT). As these tests are not routinely available, it has been suggested that tests such as APRI or Fib 4 should be done first and if one of these is abnormal, other more costly investigations could be done.

Breath tests such as the methacetin breath test, which assesses the liver microsomal function, or aminopyrine test have also been investigated as potential markers.

Tests for Diagnosis of Specific Liver Disease

These include tests such as α-fetoprotein (AFP) for the diagnosis of hepatoma, iron and iron binding capacity for haemochromatosis, urine copper and caeruloplasmin concentration for Wilson's disease, serological tests for viral hepatitis, and auto-antibodies for autoimmune liver disease.

Liver Disease

Acute Hepatitis

Hepatitis is commonly caused by viral and other infectious agents, but can also be caused by drugs and toxins. Acute hepatitis can be detected by the standard liver function tests. Typically, the aminotransferases are elevated often to > 10 ULN with a modest increase (less than twice the ULN) in ALP and varying degrees of bilirubinaemia. In general, the degree of increase in aminotransferase is a reflection of the severity of the disease.

Viral hepatitis can be caused by hepatitis viruses A, B, C, D, E and by other hepatotropic viruses. Hepatitis A is transmitted by the oro-faecal route while hepatitis B, D, C and E are transmitted by blood products or body fluids. Patients infected with hepatitis virus develop a flu-like illness after a variable incubation period (2–6 weeks for hepatitis A, 40–180 days for hepatitis B, and 2–26 weeks for hepatitis C) when serum transaminases will be elevated. Up to 90% of patients with hepatitis A infection may go unrecognised, as they do not develop jaundice. When symptoms develop, they usually coincide with the maximal increase in aminotransferases. Symptoms are often non-specific with malaise, anorexia, nausea and fever. The urine may become dark due to the presence of bilirubin and the stools may be pale as a result of cholestasis and jaundice becomes evident a few days later. Impaired excretion of bilirubin by hepato-cytes is the major cause of hyperbilirubinaemia, which is therefore mainly conjugated. Urobilinogen may be increased due to decreased reuptake of urobilinogen absorbed from the gut. In hep-atitis A infection, jaundice subsides after a few days but some patients may go into a cholestatic phase when there will be increases in GGT and ALP that may persist for weeks and in these patients, urobilinogen will be absent in urine.

Most cases of acute hepatitis with hepatitis A or E completely recover. Infection with hepatitis B, C and D viruses can lead to chronic liver disease; up to 10% of patients with hepatitis B may progress to chronic liver disease. A small number of cases (< 1%) may progress to acute liver failure. Some patients with hepatitis B may become asymptomatic carriers of the virus.

Case 15.1

A 20-year-old university student saw his GP a week after returning from a trip to Africa. He was feeling unwell, nauseated and had some pain in the right side of abdomen. The GP thought he looked jaundiced and sent a sample for liver function tests. Results of these tests were:

Serum		Reference range
Bilirubin (μmol/L)	112	<20
ALT (IU/L)	2152	<40
Albumin (g/L)	39	35–50
ALP (IU/L)	165	30–120

The liver function tests show there is predominant liver cell damage (ALT more than 50 times the upper limit of normal) with a moderate rise in ALP. Further serology tests showed he was positive for hepatitis A. On conservative management, he made an uneventful recovery.

Case 15.2

A 72-year-old man presented with a loss of weight and loss of appetite. He had lost nearly 10 kg in weight during the preceding 3 months. He also noticed that his urine was dark all the time. On examination, he was found to be jaundiced and results of his liver function tests were as follows:

Serum		Reference range
Bilirubin (μmol/L)	115	<20
ALT (IU/L)	65	<40
Albumin (g/L)	35	35–50
ALP (IU/L)	695	30–120

Examination of his urine showed bilirubin but urobilinogen was absent. These results are typical of a cholestatic jaundice: very high ALP with a moderate elevation of ALT. With a history of weight loss, malignancy was suspected. Further investigations (ultrasound and CT scan) revealed tumour of the head of the pancreas obstructing the common bile duct.

Acute Liver Failure

Acute liver failure is a severe abnormality in liver function characterised by hepatocellular jaundice, hepatic encephalopathy and an increase in prothrombin time. It is commonly caused by paracetamol overdose and viral hepatitis. Aminotransferase are grossly elevated and there is deep jaundice. Prothrombin time is prolonged and clotting factors II, V, VII, IX, X are decreased. Hypoglycaemia due to impaired gluconeogenesis and impaired glycogen synthesis may be present. A wide range of electrolyte and acid–base disturbances may be seen. Hyponatraemia due to impaired free water clearance and/or administration of fluids may be seen. Respiratory alkalosis due to hyperventilation and hypophosphataemia is common. Metabolic acidosis (lactic acidosis) may be present if there is hypoxia. Hypocalcaemia and hypoalbuminaemia may be present and serum urea may be low due to reduced synthesis.

Chronic Hepatitis

Chronic hepatitis is usually diagnosed when clinical or biochemical features of liver disease (chronic inflammation) persist for more than 6 months. Chronic hepatitis used to be divided into chronic persistent hepatitis and chronic active hepatitis based on histology. This classification is no longer in use.

Chronic viral hepatitis

Infections with hepatitis B and C can lead to chronic hepatitis. About 90% of neonatal hepatitis cases due to hepatitis B, 5% of adults with acute hepatitis B infection and almost 85% of patients with acute hepatitis C infection show evidence of chronic hepatitis. Of the hepatitis B patients, 20% go on to develop cirrhosis and 5% develop hepatocellular carcinoma (HCC). Treatment with antiviral agents can slow further progression.

Autoimmune Liver Disease

Chronic liver diseases due to autoimmunity are autoimmune hepatitis, primary biliary cirrhosis and primary sclerosing cholangitis.

Autoimmune hepatitis

This is a chronic liver disease of unknown cause, that predominantly affects females and is characterised by continuing chronic inflammation and necrosis. The disease tend to progress to cirrhosis. This is now recognised as a heterogeneous disease and has been classified into three types; depending on the type of autoantibodies present. Type 1 can occur at any age and is characterised by anti-smooth muscle antibodies, anti-nuclear antibodies and anti-actin antibodies. Type 2 is predominantly a paediatric disease with anti-liver/kidney microsomal antibodies (LKM). Type 3 is seen in adults with antibodies against cytokeratines. Liver function tests show a 'hepatitic' picture, elevated AST, ALT and bilirubin but normal or slightly elevated ALP. The disease runs a fluctuating course and eventually leads to cirrhosis. Serum globulins are increased due to increases in γ globulin, mainly IgG.

Primary sclerosing cholangitis

This is a rare chronic disorder, affecting the intrahepatic and extrahepatic bile duct causing strictures that can occur at any age and is associated with inflammatory bowel disease. This can ultimately lead to cirrhosis, liver failure and liver cancer. Live function tests show a 'cholestatic' picture, serum IgG is increased and anti-nuclear antibodies are present. Diagnosis depends on demonstrating characteristic cholangiographic appearance. Clinical features include chronic fatigue, jaundice and evidence of malabsorption (due to the stricture of the bile duct).

Primary biliary cirrhosis

This is a chronic, autoimmune liver disease affecting the bile duct, most commonly seen in middle-aged women. Many patients are asymptomatic and diagnosis is suspected after incidental finding of raised ALP. Patients may present with pruritis, jaundice or non-specific symptoms such as tiredness. Liver function tests show characteristic cholestatic picture with high ALP, GGT, and high bilirubin concentration.

Hypergammaglobulinaemia due to high IgM is seen and the disease is characterised by anti-mitochondrial antibodies directed against E2 subunit of the pyruvate dehydrogenase complex. Plasma bilirubin is a prognostic factor in primary biliary cirrhosis.

Case 15.3

A 50-year-old lady was referred to the hospital for investigation of jaundice and itching. Her serum results were as follows:

Serum		Reference range
Bilirubin (μmol/L)	83	< 20
ALT (IU/L)	76	< 40
Albumin (g/L)	36	35–50
ALP (IU/L)	480	30–120
GGT (IU/L)	865	< 55

> *The liver function tests showed a predominant cholestatic picture (high ALP with a moderate elevation of ALT). In a female with this picture and a history of itching, autoimmune liver disease should be excluded. Serum anti-mitochondrial antibodies were strongly positive. A provisional diagnosis of primary biliary cirrhosis was made and this was later confirmed by liver biopsy and histological examination.*

Non-Alcoholic Fatty Liver Disease (NAFLD)

This is the most common liver disease in many countries. Accumulation of fat in the liver not due to alcohol is termed NAFLD and there are two stages. In the early stage, there is fatty liver while the late stage is called non-alcoholic steatohepatitis (NASH). In the first stage, there is fat deposition in the hepatocyte which is reversible and benign. In NASH, there is inflammation (steatohepatitis) and scarring. This can progress to cirrhosis, liver failure and cancer.

NAFLD is associated with insulin insensitivity. Type 2 diabetics, obese subjects and those with increased triglycerides are at increased risk of NAFLD. NAFLD is also associated with the use of several drugs including amiodarone, antiviral drugs (nucleoside analogues), methotrexate, tamoxifen, tetracycline, aspirin and corticosteroids. South Asian men tend to show high prevalence. Polymorphism in apo C3 is also associated with increased risk of NAFLD.

Up to 20–35% of adults in the western world have NAFLD; of these, 20–50% have NASH. Of the subjects with NASH, 20–50% will develop cirrhosis. It is the leading cause of non-alcoholic cirrhosis.

NAFLD is closely linked to metabolic syndrome. As a result of insulin resistance, free fatty acids increase and these are taken up by liver cells, causing a fatty change. Further damage results from oxidative stress and the upregulation of inflammatory mediators such as TNF-β and TNF-α.

Most patients with NASFLD are asymptomatic and are identified incidentally when liver function tests or imaging studies are done. There are no specific biochemical tests for the diagnosis of this disorder. Transaminases are elevated, usually not more than four times the upper limit of normal and AST/ASLT ratio is < 2. Unexplained abnormal liver function tests should prompt further investigations. Non-invasive methods such as ultrasound will support the diagnosis, but histological examination of liver biopsy may be required to confirm the diagnosis and to establish the degree of abnormality.

Some suggest tests such as Fibrotest (which is a combination of α_2 macroglobulin, apo A1, haptoglobin, bilirubin, GGT and ALT), AST to platelet ratio or such similar indices should be done at intervals to detect progress to cirrhosis.

Current methods of management rely on reduction or treatment of risk factors (weight reduction and exercise) and drugs such as metformin to reduce insulin resistance.

Cirrhosis of the Liver

Cirrhosis is a chronic liver disease characterised by diffuse hepatic fibrosis, nodular regeneration and disturbance of the normal hepatic architecture. Causes of cirrhosis of the liver are given in Table 15.5;

Table 15.5   Causes of hepatic cirrhosis

Alcohol abuse
Chronic viral hepatitis
Non-alcoholic steatohepatitis (NASH)
Venous obstruction
  • Budd-Chiari syndrome
Lupoid hepatitis
Cryptogenic cirrhosis

Autoimmune hepatitis
  • Primary biliary cirrhosis
  • Primary sclerosing cholestasis
Metabolic disorders
  • Haemochromatosis
  • Wilson's disease
  • α_1-Antitrypsin deficiency
  • Tyrosinaemia
  • Galactosaemia
  • Type IV glycogen storage disease
Toxins and drugs

chronic excessive alcohol abuse, hepatitis B and C and non-alcoholic steatohepatitis are common causes (Table 15.5). Cirrhosis is usually, but not always, associated with recurrent episodes of hepatocellular necrosis and attempts at liver cell regeneration. Standard liver function tests are often normal due to the large reserve capacity of the liver. In the early stages, if liver function tests are abnormal it is usually due to the initiating agent. Liver function tests become abnormal in the late stages of the disease. Until the stage when liver function tests become abnormal, the cirrhosis is said to be 'compensated'. The standard liver function tests are not reliable to diagnose early compensated cirrhosis. Some of the dynamic function tests described earlier are of value but they are not available in most laboratories. Fibrosis of the liver could be monitored by markers of collagen such as amino terminal procollagen III peptide (PIIINP) or other markers of fibrosis.

In decompensated cirrhosis, there is jaundice accompanied by increased urobilinogen and bilirubin in the urine. There is hypoalbuminaemia due to decreased synthesis and increase in transaminases. Serum immunoglobulins are often increased due to antigens from the gut bypassing the liver and reaching the systemic circulation.

Electrolyte abnormalities are common in cirrhosis. Hyponatraemia is common and usually results from excessive fluid administration and/or diuretic therapy. Hypokalaemia may result from a combination of diuretics, poor diet and vomiting. Hypomagnesaemia is often seen in alcoholic cirrhosis. Serum urea may be lower in late decompensated cirrhosis. Insulin resistance is common.

Consequences of cirrhosis include hepatic encephalopathy, ascites, hepatorenal syndrome, portal hypertension, development of hepatic malignancy and endocrine dysfunction.

Hepatic encephalopathy

Hepatic encephalopathy is a syndrome of impaired mental status and abnormal neuromuscular function due to impaired hepatocellular function. It is characterised by the inversion of sleep rhythm, confusion, disorientation and impairment of higher functions often precipitated by a large protein meal, gastrointestinal bleeding, infection, hypoglycaemia, electrolyte imbalance or drugs. The diagnosis of hepatic encephalopathy is mainly clinical. EEG changes are sometimes helpful. Blood ammonia concentration is usually raised but is not diagnostic and does not correlate with the degree of encephalopathy. Manifestation of hepatic encephalopathy is attributable to a net increase in substances that cause neuronal inhibition. The exact pathogenesis is not fully understood but is probably multifactorial. An important factor in the development of hepatic encephalopathy is thought to be toxic products from the gastrointestinal tract (which are normally removed by the liver) reaching the systemic circulation either because there are not enough hepatocytes or due to portal blood by passing the liver via collateral circulation. Toxic products thought to be involved are accumulation of ammonia, manganese (substances normally removed by the liver), increased levels of natural benzodiazepines, and accumulation of tryptophan metabolites and other neurotoxins.

Treatment of hepatic encephalopathy includes removing/treating any precipitating factors; reducing the absorption of nitrogenous substances from the gut by restriction of dietary protein intake and sterilising the gut by using a non-absorbable antibiotic such as neomycin, giving enemas or laxatives such as lactulose; adequate energy intake in the form of carbohydrates, and maintenance of fluid and electrolyte balance.

Ascites

Ascites, an accumulation of extracellular fluid in the peritoneal cavity, is a late complication of cirrhosis. Ascites can also occur due to non-hepatic causes (Table 15.6). The exact mechanism of ascites is not clear but several factors are involved: retention of sodium and water by the kidney, decreased colloid oncotic pressure due to hypoalbuminaemia, portal hypertension and increased hepatic lymph production. The exact mechanism of increased retention of

Table 15.6 Causes of ascites

Venous hypertension
- Cirrhosis of liver
- Congestive heart failure
- Constrictive pericarditis

Hypoalbuminaemia
- Nephrotic syndrome
- Malnutrition (a contributing factor in cirrhosis)

Malignant disease
- Secondary deposits
- Primary mesothelioma

Infection
- Tuberculous peritonitis

Miscellaneous
- Chylous ascites

sodium is not well understood but involves the activation of renin–angiotensin–aldosterone and sympathetic systems. Analysis of ascitic fluid is sometimes useful in establishing the cause of ascites. In cirrhosis, the ascitic fluid is clear, the protein concentration is < 25 g/L, serum:ascitic fluid albumin ratio is > 1.1 and there are fewer than 500 cells per cu mm. In other conditions such as tuberculosis or malignancy, the protein concentration is higher and the serum:ascetic fluid albumin ratio is < 1.0 and LD activity is high. In ascites, due to tuberculosis, adenosine deaminase activity in the fluid is high.

Management of ascites involves a combination of sodium restriction, diuretic treatment, withdrawal of ascitic fluid and albumin infusion. However, ascitic fluid should not be removed rapidly as this may precipitate renal failure.

Case 15.4

A 45-year-old man was admitted to hospital with haematemesis. He admitted to drinking large amount of alcohol for nearly

25 years. On examination, he was found to have ascites and his blood results were as follows:

Serum		Reference range
Bilirubin (μmol/L)	58	< 20
ALT (IU/L)	66	< 40
Albumin (g/L)	20	35–50
ALP (IU/L)	175	30–120
GGT (IU/L)	365	< 55

Hypoalbuminaemia together with ascites and a history of chronic alcohol abuse are indicative of cirrhosis of liver secondary to alcohol abuse. The low serum albumin is due to decreased synthetic capacity of the liver. Poor nutrition may have also contributed.

Hepatorenal syndrome

Hepatorenal syndrome is a functional reduction in GFR and is usually associated with advanced liver disease, ascites and encephalopathy. The kidneys are normal under histological examination and appearances of acute tubular necrosis are absent. The reduction in GFR is not reversed by fluid replacement, showing that it is not a prerenal volume depletion syndrome. The exact pathogenesis of this syndrome is unknown. There is intense renovascular constriction, leading to decreased renal blood flow and reduced glomerular filtration rate. One theory (the underfill theory) is that there is vasodilatation in the splanchnic area due to the release of vasodilator substances such as nitric oxide and prostaglandins from the liver. This causes a reduction in effective circulating blood volume, triggering the activity of the renin–angiotensin–aldosterone system, thus leading to the retention of sodium and vasoconstriction of vessels in the systemic circulation especially renal vessels. Hepatorenal syndrome is characterised by a dramatic degree of sodium retention, with urine sodium concentration < 10 mmol/L, and there is no response to volume expansion.

Treatment involves infusion of albumin and medications such as teripressin, a vasopressin analogue.

Endocrine disturbances

Disturbances in endocrine function are common in patients with cirrhosis. Men with cirrhosis are usually impotent, infertile and features of feminisation such as testicular atrophy, gynaecomastia, female distribution of body hair are seen. Testosterone concentrations are low and SHBG concentrations are elevated giving rise to a decrease in free testosterone. LH concentration may be increased but there may also be primary testicular failure. Oestrogen concentrations are usually increased due to enhanced peripheral conversion.

Portal hypertension

Portal hypertension is another feature of cirrhosis and can lead to oesophageal varices, which may lead to severe gastrointestinal bleeding and haematemesis. Portal hypertension in cirrhosis is due to fibrosis in the liver that causes distortion of the liver architecture, leading to obstruction in the portal circulation.

Management of cirrhosis

Treatment of cirrhosis is symptomatic as it is an irreversible process. Progression of cirrhosis could be delayed by treating the primary cause of the cirrhosis. The curative treatment of cirrhosis is by liver transplantation. This is especially useful in inherited or other metabolic disorders. Conservative treatment of cirrhosis includes treating the underlying cause (e.g. withdrawal of alcohol), treating ascites and adequate nutrition. Prognostic indices such as the MELD score (Model for End-stage Liver Disease) which incorporates serum creatinine, bilirubin and INR has been used in an attempt to predict the need for transplantation.

Alcoholic Liver Disease

Alcoholic liver disease is the commonest cause of liver disease in many parts of the world. Alcoholism causes a wide spectrum of social, psychological and pathological effects. Development of liver disease in alcoholism depends on the amount of alcohol consumption, genetic susceptibility, gender (F > M), nutrition, and associated infection with hepatitis B and C. Estimates of genetic susceptibility to alcoholism range between 50% and 60%. Serotonin 1B receptor, dopamine D2 receptor, CRF receptor and neuropeptide Y are possible candidate targets of genetic susceptibility to alcohol. Alcohol is metabolised to acetaldehyde and then to acetate by alcohol and acetaldehyde dehydrogenases. In heavy drinkers, alcohol is also metabolised by cytochrome P450 enzyme systems. These enzymes systems can be induced by alcohol. The intoxicating and metabolic effects of alcohol are due to the direct effect of alcohol whereas liver damage is caused by direct and indirect effects of alcohol. These indirect mechanisms include acetaldehyde-mediated oxidative stress, immune mechanism and activation of pro-inflammatory factors.

Three types of liver disease can be produced by alcohol: fatty liver, alcoholic hepatitis and cirrhosis. Fatty liver is present in all heavy drinkers and is associated with mild elevation of aminotransferases, elevation in GGT and slight increase in bilirubin concentration. This seldom progresses to chronic liver disease and resolves when alcohol is withdrawn.

Alcoholic hepatitis

Alcoholic hepatitis presents in a wide spectrum of features ranging from asymptomatic to hepatic failure. It usually follows very heavy bouts of drinking and in subjects who have been drinking heavily for several years. The pathogenesis is not fully understood and patients have symptoms and signs resembling viral hepatitis but the AST/ALT ratio is >2 and the transaminases are usually < 10 ULN. There may be evidence of fibrosis. Patients with alcoholic hepatitis are at greater risk of developing cirrhosis.

Excessive alcohol consumption appears to unmask haemachromatosis in patients with this genetic disease.

The detection of liver disease in subjects who are known to drink excessively is difficult, as raised aminotransferases and GGT do not reflect histological damage. Liver biopsy is the only means of assessing the degree of liver damage in excessive alcohol users. In patients who are known to have liver disease, elevated GGT, which falls on admission to hospital and an ALT/AST ratio >2 are very suggestive of alcohol as the aetiology of liver disease. Other features, which may suggest alcoholic liver disease, are macrocytosis and the presence of disialylated transferrin in serum. In apparent healthy subjects, GGT, MCV and disialyated transferrin are potential markers of excessive alcohol intake; of these, desialylated (or carbohydrate-deficient) transferrin is most sensitive.

Malignancy of the Liver and Biliary Tract

Tumours of the liver can be primary or secondary. Liver is a common site for secondaries from primary tumours especially gut. Primary tumours of the liver are hepatocellular carcinoma, (HCC), cholangiocarcinoma and angiosarcoma. Of these, HCC is the most common and is associated with infection with hepatitis B and C, cirrhosis and carcinogens such as aflotoxins. Serum α-fetoprotein is increased in 70–80% of patients with primary hepatocellular carcinoma and is a useful tumour marker (see Chapter 24). However, AFP can also be elevated in other conditions where there is regeneration of liver cells.

The presence of primary tumours in the liver or infiltration of the liver may not produce any abnormality in liver function tests, or may cause isolated elevation of plasma ALP due to the obstruction of one of the bile ducts.

Drugs and Liver

The central role of liver in the detoxification and excretion of drugs and toxins makes it vulnerable to liver damage. Some drugs are toxic in therapeutic doses; others only in toxic doses, and some others cause liver damage due to the idiosyncratic reaction. The type of liver damage includes hepatocellular damage, cirrhosis, cholestasis, and tumours. Examples of these drugs and the liver damage they cause are given in Table 15.7. Paracetamol and methotrexate are examples

Table 15.7   Drugs causing liver damage

Hepatocellular damage

- Paracetamol
- Alcohol
- Halothane
- Methotrexate

Cholestasis

- Anabolic steroids
- Naproxen
- Oral contraceptives
- Chlordiazepoxide

Cirrhosis

- Alcohol
- Methotrexate
- Methyldopa

Chronic cholestasis

- Cloxacillin
- Phenothiazines (chlorpromazine)
- Tricyclic antidepressants (amitriptyline)

Liver tumours

- Oral contraceptives
- Anabolic steroids

Chronic hepatitis
Hepatic granulomas

- Amidarone, cimetidine
- Nitrofurantoin
- Amoxicillin
- Chlorpromazine
- Phenylbutazone

where there is hepatocellular damage which is dose related. Halothane causes hepatocellular damage by an idiosyncratic reaction in some individuals. Phenothiazines are examples of drugs that cause hepatic cholestasis due to idiosyncratic reaction. Anabolic steroids cause cholestasis in a dose-dependent manner.

Gallstones

Gallstones are common and affect 10–20% of the population. Gallstones are classified as cholesterol or pigment stones. More than 75% of stones are cholesterol stones with varying amounts of bilirubin and calcium salt. Pathogenesis of gallstones is complex. One of the predisposing factors for stone formation is the supersaturation of bile with cholesterol. As cholesterol is insoluble in water, it is solubilised by aggregating with lecithin and bile salts to form mixed micelles. Whenever there is an increase in cholesterol or a decrease in bile acids or lecithin in bile, the risk of stone formation is increased. Obesity, ageing, drugs (such as clofibrate, and nicotin) and hormones (such as oestrogens) predispose to increased cholesterol secretion. Decreased bile acid pool is found in stone formers and this increases lithogenicity. The cholesterol content of bile increases with age and treatment with oestrogens and fibric acid analogues reduce bile acid secretion resulting in more lithogenic bile. Incidence of gallstones is higher in women, especially multiparous women. This is thought to be due to incomplete emptying of gallbladder, which occurs in late pregnancy. Genetic factors may also be important as it is common in some groups, e.g. Pima Indians have an incidence of 70%.

Hard pigment stones occur in chronic haemolytic diseases and in chronic biliary obstruction. Soft pigment stones are associated with infections of the biliary tract.

Gallstone does not always cause symptoms. Symptoms, when present, are due to obstruction causing biliary colic, pancreatitis, cholangitis or cholecystitis.

Management of gallstone disease is by surgical removal or by dissolution either by chemical or physical methods. Chemical dissolution involves oral treatment with bile acids.

Further Reading

1. Angulo P. Non-alcoholic fatty liver disease. *N Engl J Med* 2002; 346:1221–1231.

2. Bernal W, Auzinger G, Dhawan A, Wendon J. Acute liver failure. *Lancet* 2010; 376:190–201.

3. Brunt EM. Pathology of non-alcoholic fatty liver disease. *Nat Rev Gastroenterol Hepatol* 2010; 7:195–203.

4. Cobbold JFL, Anstee QM, Thomas HC. Investigating mildly abnormal serum aminotransferase values. *Br Med J* 2010; 341:c4039.

5. Denzer UW, Luth S. Non-Invasive diagnosis and monitoring of liver fibrosis and cirrhosis. *Best Pract Res Clin Gastroenterol* 2009; 23:453–460.

6. Grren RM, Flamm MBA. AGA technical review on the evaluation of liver chemistry tests. *Gastroenterol* 2002; 123:1367–1384.

7. Hohenester S, Oude-Elferink RPJ, Beuers U. Primary Biliary Cirrhosis. *Semin Immunopathol.* 2009; 31:283–307.

8. Lewis JR, Mohanty SR. Non-alcoholic fatty liver disease: A review and update. *Dig Dis Sci* 2010; 55:560–578.

9. Macfarlane I, Bomford A, Sherwood RA. *Liver Disease and Laboratory Medicine* 2000. ACB venture publications: London.

10. Myers R. Non-invasive diagnosis of non-alcoholic fatty liver disease. *Ann Hepatol* 2009; 8:suppl S25–S33.

11. Navarro VJ, Senior JR. Drug-Related Hepatotoxicity. *N Engl J Med* 2006; 354:731–739.

12. Renner EL. Liver function tests. *Ballieres Clin Gastroenterol* 1995; 9:661–772.

13. Tredger JM, Sherwood RA. The liver: New functional, prognostic and diagnostic tests. *Ann Clin Biochem* 1997; 34:121–141.

Summary/Key Points

1. Liver plays an important role in the metabolism of carbohydrates, fats and proteins. It is vital for detoxification of endogenous and exogenous substances and, for the synthesis of large number of proteins. Liver is also a storage organ.

2. Commonly used liver function tests are indicators of liver cell damage (aminotransferases, AST and ALT), synthetic function (albumin and clotting factors), cholestasis (ALP and GGT) and excretory function (bilirubin).

3. These tests, especially AST and ALT, are sensitive indicators of liver cell damage. Patterns of changes in liver function tests may indicate the type of liver disease but are not diagnostic. Predominant elevation of ALT and AST are indicative of liver cell necrosis (e.g. hepatitis) and predominant elevation of ALP and GGT indicates cholestasis.

4. Jaundice (elevation of serum bilirubin) is not an uncommon finding in liver disease. Elevation of conjugated bilirubin indicates obstructive type of liver disease and elevation of unconjugated bilirubin suggests increased production of bilirubin, e.g. haemolytic disease.

5. Quantitative test such as galactose elimination test, or aminopyrine breath test are available but not routinely used.

6. Gilbert's syndrome is a common benign inherited condition due to mutation in the enzyme which conjugates bilirubin. Isolated mild elevation of serum bilirubin which is unconjugated is seen in this syndrome.

7. Alcoholic liver disease is common and is a leading cause of cirrhosis. Markers of alcoholism include GGT, mean corpuscular volume (MCV) of erythrocytes and desialyated transferrin, of which the last is more specific and sensitive.

8. Non-alcoholic steatohepatitis (NASH) is now recognised as a common condition and is associated with insulin resistance and metabolic syndrome. Mild elevation of liver enzymes, AST and ALT are early indicators of NASH, which can progress to cirrhosis.

9. Biochemical test are not diagnostic of cirrhosis and routine liver function tests may not become abnormal until late stages of the disease.

chapter 16

The Gastrointestinal Tract

Introduction

The major function of the gastrointestinal tract is the digestion and absorption of nutrients. In addition, it is a major endocrine organ and a major target organ for many hormones — both locally and systemically. Endocrine cells are found throughout the mucosa of the intestine and are so numerous that the gut is the largest endocrine organ in the body. Gastrointestinal hormones influence all aspects of gastrointestinal function — motility, secretion, digestion and absorption. Many of the gastrointestinal hormones are also present in the central nervous system where they act as neurotransmitters. Some of the gastrointestinal hormones also affect the secretion of other hormones, e.g. gastric inhibitory polypeptide (GIP) influences the secretion of insulin.

Stomach

In the stomach, food is mixed, stored for some time and then finally discharged into the duodenum. The stomach secretes enzymes, intrinsic factor, hormones, mucin and hydrochloric acid. The secretion of gastric juice is controlled by neural mechanisms involving the vagus nerve and by gastrin, the stomach hormone.

Gastrin is a polypeptide hormone produced and stored by the G cells found in the antrum of the stomach and in the proximal duodenum. Gastrin secretion is stimulated by distension of the stomach, the presence of proteins and polypeptides in the stomach, vagal stimulation, plasma calcium concentration and by circulating catecholamines. Gastrin secretion is inhibited by acid in the antrum and by blood-borne factors such as secretin, GIP, vasoactive intestinal peptide (VIP), glucagon and calcitonin. The effect of acid on gastrin secretion forms the basis of a negative feedback. Gastrin increases acid secretion which then feeds back to inhibit further gastrin secretion. Gastrin circulates in many forms; the principal circulating form is a 34 amino acid peptide. Other forms include little gastrin (17 amino acids), mini gastrin (14 amino acids) as well as sulphated and non-sulphated forms of these peptides. The significance of these different forms is not clear. All the different forms have the same C-terminal end, which is required for the physiological effects of gastrin. Gastrin stimulates secretion of acid, pepsinogen and intrinsic factor, increases gastric motility and stimulates the growth of gastric mucosa. Gastrin stimulates acid secretion by acting directly on parietal cells as well as indirectly via the release of histamine from the enterochromaffin cells. Histamine binds to H2 receptors in the parietal cells, stimulating acid production. H2 receptor antagonists, cimetidine and ranitidine reduce acid production in the stomach. These drugs are now less commonly used to control acid secretion as proton pump inhibitors such as omeprazole and lansoprazole are in use now. These drugs block the hydrogen–potassium–ATPase (proton pump) of the parietal cells irreversibly.

Ghrelin is a 28 amino acid peptide produced by the P/D1 cells of the fundus of the stomach. Secretion of ghrelin is inhibited by food intake and the concentration is highest just before meals. Ghrelin stimulates growth hormone secretion and is an appetite-regulating factor. This peptide is also produced in the hypothalamus.

Investigation of Gastric Function

Hypersecretion of gastric fluid occurs in duodenal ulceration and hyposecretion is found in gastric ulcers and pernicious anaemia (an autoimmune destruction of the parietal cells). Zollinger-Ellison syndrome is a rare disorder of gastric hypersecretion caused by gastrin-secreting tumours, gastrinomas, the majority of which (90%) are found in the pancreas. Occasionally, gastrinomas are part of the multiple endocrine neoplasia syndrome I (MEN1). Increased gastrin secretion from these tumours causes hypersecretion of hydrochloric acid and ulceration of the mucosa.

Investigation of gastric function is mainly endoscopic and radiological. Biochemical tests are seldom used. A frequently used laboratory test is the test for infection with *Helicobacter pylori* (*H. pylori*) which is the main cause of gastric and duodenal ulcers. These can be functional tests such as breath tests or the detection of specific antibodies. The breath test is based on the production of labelled CO_2 from labelled urea by *H. pylori*. Measurement of gastric secretion either under basal conditions or after stimulation with pentagastrin, (a synthetic peptide), is rarely done now. Serum gastrin is measured in suspected cases of Zollinger-Ellison syndrome. Serum gastrin, along with other gastrointestinal hormones, is very labile due to proteolytic enzymes in blood. Blood should be collected in bottles containing a protease inhibitor such as aprotinin. Typically, gastrin concentrations are increased in Zollinger-Ellison syndrome, achlorhydria, pernicious anaemia and in patients after vagotomy.

The Pancreas

The major exocrine function of the pancreas is the production and secretion of pancreatic juice, which contains bicarbonate and enzymes — trypsin, chymotrypsin, carboxypeptidase, lipase and amylase. These enzymes are secreted in inactive or proenzyme forms. The exocrine secretion of pancreas is controlled by two gut hormones, secretin and cholecystokinin/pancreozymin (CCK-PZ). Secretin, which is a 27 amino acid peptide, secreted mainly by the S cells of the duodenum causes a watery alkaline pancreatic secretion. Secretin is stimulated by acid bathing the mucosa of the duodenum. CCK-PZ is secreted by the I-cells of the upper small intestinal mucosa mainly duodenum. It is present in multiple forms (33, 39, 59 and 8 amino acids). Secretion of CCK-PZ is stimulated by a protein or fat-rich chyme entering the duodenum. CCK regulates the contraction of the gall bladder, increases the motility of the duodenum and small intestine, stimulates pancreatic enzymes and pancreatic growth and reduces gastric emptying. It is also a satiety hormone.

Pancreatic secretion neutralises the acidity of the gastric contents to a pH of approximately 6.5, which is the optimum pH for many of the pancreatic and small intestinal enzymes. Pancreatic enzymes break down nutrients to oligosaccarides, disaccharides, small peptides, amino acids and fatty acids.

Disorders of Pancreatic Function

Acute pancreatitis

Acute pancreatitis develops when pancreatic enzymes, which are normally activated in the duodenal lumen, become activated within the gland itself. Pancreatic enzymes escape into the tissues and the peritoneal cavity, causing necrosis of the pancreas and adjacent organs. Necrosis of fat may lead to liberation of fatty acids, which forms calcium salts; hypocalcaemia may follow. Acute pancreatitis presents as an abdominal emergency with acute abdominal pain and variable degrees of shock. The causes of acute pancreatitis are listed in Table 16.1.

Table 16.1   Causes of acute pancreatitis

Major causes
- Biliary disease
- Alcohol abuse

Minor causes
- Trauma
  — Blunt abdominal injury
  — Endoscopic retrograde cholangiopancreatography (ERCP)
- Post-abdominal surgery
- Metabolic
  — Severe hypertriglyceridaemia
  — Hypercalcaemia/hyperparathyroidism
  — Renal failure
- Infection/infestations
  — Mumps
  — Coxackie B virus
  — Intestinal parasites — worms
- Autoimmune disease — SLE, polyarteritis nodosa

Drugs
- Valproic acid
- Azathioprine
- Corticosteroids

The most frequent causes are gallstones and excessive alcohol intake. In severe pancreatitis, there is haemorrhage and methaemalbumin may be detectable in plasma. Hypoxia is a characteristic finding in acute pancreatitis and can reflect its severity. The exact mechanism of hypoxia is not clear but it may be due to shunting of blood in the pulmonary vascular tree. Activation of phospholipase A2 which hydrolyses phospholipids leads to increased vascular permeability and ischemia. It also destroys pulmonary surfactant and leads to acute respiratory distress syndrome (ARDS). Several cytokines are released and these account for the systemic inflammatory response. Activation of the kinins, kallikrein and complement pathways play a role in disseminated intravascular coagulation, renal failure and shock. Sepsis may result due to gram-negative bacteria entering from the gut. Table 16.2 gives laboratory indicators of severity. Presence of hypocalcaemia, low pO_2, high glucose, high urea and low albumin are poor prognostic features. In acute pancreatitis caused by gallstones, there may be elevation of aspartate transaminase. The diagnostic value of amylase and lipase is discussed in Chapter 13.

Patients with acute pancreatitis are usually managed conservatively by nasogastric aspiration, fluid and electrolyte balance and parenteral nutrition if necessary.

Table 16.2 Laboratory indicators of severe acute pancreatitis

On admission	48 hours later
Glucose > 11 mmol/L	Decreasing haematocrit by > 10%
LD > 350 IU/L	Increasing serum urea > 10 mmol/L
AST > 250 IU/L	Serum calcium < 2.00 mmol/L
WBC > 16 000/cu mm	PO_2 < 8.0 kPa
	Base excess < −4 mmol/L
	Serum albumin < 32 g/L

Case 16.1

A 47-year-old man was brought to the emergency department with a history of severe abdominal pain, which radiated to the back. There was no previous history of any ill health. On examination, his blood pressure was 110/70, his pulse was 110/min and there was tenderness and guarding of the epigastric region. Results of investigation done on admission were as follows:

Serum		Reference range
Sodium (mmol/L)	138	135–145
Potassium (mmol/L)	4.2	3.5–5.0
Ura (mmol/l)	11.2	2.7–7.0
Creatinine (μmol/L)	125	66–110
Corrected calcium (mmol/L)	1.95	2.25–2.55
Glucose (mmol/L)	14.5	
Amylase (IU/L)	4252	< 200

The very high amylase strongly suggested acute pancreatitis as the diagnosis and CT scanning confirmed this. Two of the most common causes of acute pancreatitis are alcohol and gallstones. This patient denied excessive alcohol intake and the CT scan also showed the presence of gallstones. High urea and creatinine are due to the hypotension and consequent poor renal perfusion. Low serum calcium and high glucose are features of acute pancreatitis.

Chronic pancreatitis

Chronic pancreatitis, is an uncommon condition, and is most often caused by alcohol abuse. Acute pancreatitis may also lead to chronic pancreatitis. In chronic pancreatitis, inflammation with subsequent fibrosis leads to exocrine deficiency. Patients may present with repeated episodes of abdominal pain or malabsorption, which does not occur until majority of the gland is destroyed. In some patients, there may be impaired glucose tolerance. Investigations in patients suspected of chronic pancreatitis include tests to establish malabsorption and to demonstrate pancreatic insufficiency. The diagnosis is established by endoscopic retrograde cholangiopancreatography (ERCP), which will show characteristic anatomical changes.

Treatment of chronic pancreatitis involves treating the underlying cause and replacement with pancreatic extracts if required.

Case 16.2

A 52-year-old man was referred to the gastroenterology clinic for investigation of chronic abdominal pain, weight loss and diarrhoea. His abdominal pain was exacerbated by food and the stools were foul smelling and bulky. He admitted drinking 7–8 pints of beer every night as well as whiskey or other sprits at weekends for the past 32 years. Blood investigation at the clinic showed:

Serum		Reference range
Bilirubin (μmol/L)	16	< 20
ALT (IU/L)	28	< 40
Albumin (g/L)	28	35–50
ALP (IU/L)	325	30–120
Corrected calcium (mmol/L)	2.1	2.25–2.55
Phosphate (mmol/L)	0.6	0.8–1.2
Glucose (mmol/L)	12	
Amylase (IU/L)	185	< 200

His low serum calcium associated with low phosphate and high ALP suggested vitamin D deficiency, which was confirmed by measurement of serum calcidiol (< 20 nmol/L). His history suggested malabsorption and this was confirmed by demonstration of low faecal elastase (65 μg/g faeces; reference range > 200 μg/g faeces). A provisional diagnosis of chronic pancreatitis was made and this was confirmed by an ultrasound examination, which showed dilatation and calcification of the pancreatic duct.

Carcinoma of the pancreas

Pancreatic carcinomas are notoriously difficult to diagnose and may present with abdominal pain, weight loss or jaundice. Jaundice is caused when the tumour is in the head of the pancreas causing obstruction of the bile duct. Patients may present with features of metastases rather than due to the direct effect of the tumour.

Cystic fibrosis

Cystic fibrosis is the most common inherited metabolic disorder in Caucasians (see Chapter 25). In addition to pulmonary dysfunction, pancreatic deficiency is present in about 85% of patients.

Tests of Pancreatic Function

Pancreatic function can be evaluated by the measurement of enzymes in serum or other body fluids and by dynamic function tests. Dynamic function tests can be direct where pancreatic juice is sampled before and after stimulation or indirect where assessment is made without intubation. Direct tests are where bicarbonate and enzyme activity are measured before and after a test meal (Lundh meal test) or after injection of secretin–CCK. Direct tests are rarely used now. The most frequently used methods for investigating pancreatic disease are non-biochemical imaging methods such as CT scanning and ERCP. Biochemical tests other than serum and urine enzymes play only a minor role in the investigation of exocrine pancreatic function. The measurement of enzymes (amylase and lipase) is discussed in Chapter 13.

Indirect tests

1. *PABA test*

    In this test a synthetic tripeptide, N-benzoyl-L-tyrosyl, p-amino benzoic acid is administered orally together with a test meal in order to stimulate pancreatic secretion. The peptide is hydrolysed by chymotrypsin releasing p-aminobenzoic acid (PABA), which is excreted in the urine as PABA-glucuronide and PABA-acetylate. The amount of PABA in the urine is an indirect estimate of the activity of chymotrypsin in the duodenum. This test is affected by various drugs especially antibiotics, analgesics, sulphonamides and diuretics. Foods such as prunes and cranberries, which contain hippurate precursors, interfere with the determination of the chromogen in the urine and these foods should be discontinued for several days before the test. Other factors affecting this test include liver and renal diseases. In order to compensate for these errors, radiolabelled PABA can be administered simultaneously. The recovery of radiolabelled PABA indicates non-pancreatic abnormalities such as intestinal disease or decreased renal excretion. The diagnostic sensitivity has been reported to vary from 60–97%.

2. *Fluorescein dilaurate test (Figure 16.1)*

The basis of this test is similar to the PABA test. Fluorescein dilaurate (FDL) is given orally in the middle of a standard breakfast. FDL is hydrolysed by pancreatic esterase and the fluorescein released is absorbed, conjugated in the liver and excreted in the urine. This

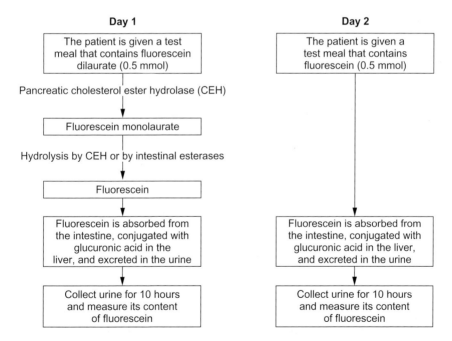

Figure 16.1 Principles of non-invasive pancreatic function test — fluroscein dilaurate test.

test is repeated a few days later using free fluorescein to correct for individual variability, intestinal absorption, hepatic conjugation and urinary excretion. False positive results are obtained in liver disease as bile salts are necessary for the digestion and absorption of FDL. Bacterial overgrowth gives false negative results due to the hydrolysis of FDL by bacteria.

These indirect tests are simple to perform and have similar sensitivity and specificity but less specific than the direct tests.

3. *Breath test (Figure 16.2)*

^{14}C-triolein breath test is an alternative indirect test where the excretion of labelled carbon dioxide in expired air following the

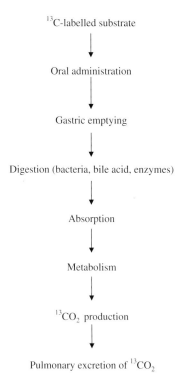

^{13}C-labelled substrate

↓

Oral administration

↓

Gastric emptying

↓

Digestion (bacteria, bile acid, enzymes)

↓

Absorption

↓

Metabolism

↓

$^{13}CO_2$ production

↓

Pulmonary excretion of $^{13}CO_2$

Figure 16.2   Principles of ^{13}C-labelled breath test.

administration of ^{14}C-triolein is measured. The major disadvantage of this test is the radiation hazard, which can be avoided by the use of stable isotope ^{13}C-labelled triolein. The breath test will be abnormal in pancreatic as well as intestinal diseases. These can be differentiated by simultaneous administration of labelled triglycerides and non-esterified fatty acids. In pancreatic disease, the absorption of fatty acids should be normal whereas in intestinal disease, the absorption of both fatty acids and triglycerides will be abnormal. However, these tests are not reliable and the breath test is mainly used as a test of fat malabsorption.

4. *Enzymes in stools*

Enzymes such as trypsin, chymotrypsin and elastase can be measured in stools and have been used for the diagnosis of pancreatic insufficiency. Detection of trypsin in stools has been used as a screening test for cystic fibrosis in neonates. Chymotrypsin is more stable and more sensitive than trypsin in the detection of pancreatic insufficiency. Elastase in stool can be measured by an immunoassay and it appears to be more sensitive than chymotrypsin.

Intestinal Function

In the small intestine, the final stages of digestion and absorption take place. Oligosaccharidases such as maltase and lactase present in the intestinal brush border, hydrolyse oligosaccharides to monosaccharides. Glucose and galactose are absorbed by a common active transport mechanism and fructose is transported by an independent facilitated diffusion process. Digestion of proteins by gastric and pancreatic enzymes results in peptides which are hydrolysed by peptidases in the brush border of the intestine. Amino acids and dipeptides are transported across the intestine by specific transporters, the dipeptides are hydrolysed in the enterocytes and the amino acids then enter the circulation.

Dietary fat is emulsified by a mixture of bile salts and lipids, particularly phospholipids, to form polymolecular aggregates, micelles. Pancreatic lipase hydrolyses triglycerides to monoglycerides and fatty acids, while pancreatic esterase hydrolyses cholesterol esters to free cholesterol. Monoglycerides, cholesterol and fatty acids are then absorbed by passive diffusion. Fatty acids containing less than 10–12 carbon atoms enter the portal blood while longer chain fatty acids are re-esterified in the enterocytes to triglycerides. Some of the absorbed cholesterol is also esterified. The triglyccrides and cholesterol esters are then coated with a layer of lipoprotein, and phospholipids to form chylomicrons, which enter the lymphatics.

Intestinal Disorders

The term malabsorption in clinical practice refers to the failure to assimilate nutrients either due to impaired digestion, or defective absorption. The causes of malabsorption syndrome are given in Table 16.3. In some conditions, more than one mechanism may lead to malabsorption (see later).

Malabsorption

Pancreatic diseases causing maldigestion are described in the section above. Excessive secretion of gastric acid in diseases such as Zollinger-Ellison syndrome may lead to malabsorption by inactivating pancreatic enzymes. Malabsorption is a complication of partial gastrectomy due to several contributing factors such as reduced pancreatic enzyme secretion caused by a decrease in the release of secretin and CCK-PZ, inadequate mixing of enzymes and bowel contents, reduced intestinal transit time and stasis in the afferent loop of the duodenum. Malabsorption may occur from reduced secretion of bile acids in diseases of liver and biliary tract disease (Chapter 15). The small intestine is normally free of significant amounts of bacteria, which are present mainly in the large intestine. Bacterial overgrowth in the small intestine may occur due to stasis caused by anatomical factors, motor abnormalities or contamination of small bowel by large bowel content. Bacterial overgrowth causes deconjugation of bile salts, thus interfering with the digestion and absorption of fat. Deficiency of vitamin B_{12} is common in bacterial overgrowth due to utilization of B_{12} by bacteria.

Table 16.3   Causes of malabsorption

Disturbances in digestion
- Pancreatic insufficiency
  — Chronic pancreatitis
  — Cystic fibrosis
  — Pancreatic carcinoma
- Bile salt insufficiency
  — Liver disease
  — Biliary tract obstruction
  — Bacterial overgrowth
  — Interruption of enterohepatic circulation
  — Ileal resection
  — Regional enteritis
- Drugs
  — Cholestyramine
  — Broad spectrum antibiotics
- Reduced transient time
  — Postgastrectomy
- Inactivation of pancreatic enzymes
  — Zollinger-Ellison syndrome

Impaired absorption
- Reduction in small intestinal length
  — Resection
  — Ileojejunal bypass surgery
- Primary disease of intestinal mucosa
  — Villous atrophy
  — Coeliac disease
  — Tropical sprue
  — AIDS
- Infiltrative disease
  — Crohn's disease
  — Amyloidosis
  — Small bowel lymphoma
- Vascular disease
  — Mesenteric vascular insufficiency
- Metabolic disorders
  — Diabetes mellitus
  — Carcinoid syndrome
  — Abetalipoproteinaemia

1. *Short bowel syndrome*

When a large part of the ileum is resected, short bowel syndrome follows. When a part of the intestine is removed, the absorptive capacity of the remaining gut increases. However, malabsorption will result if more than 50% of the small intestine is removed or if the proximal intestine and/or terminal ileum are removed. Removal of the ileocaecal valve can lead to bacterial overgrowth. Jejunal ileal bypass, a procedure used in the treatment of morbid obesity, causes malabsorption.

The short-term complication of short bowel syndrome is the loss of fluids and minerals; replacement is necessary until adaptation occurs in the small intestine. Adaptation in the small intestine depends on the presence of nutrients in the lumen of the gut. Intravenous nutrition may be required initially but this should be supplemented by small amounts of enteral feeding to stimulate the gut. Enteral feeding is then gradually increased and parenteral nutrition reduced.

Long-term complications of short bowel syndrome include persistent diarrhoea, nutritional deficiencies, gallstones and renal calculi. Specific deficiencies of B_{12} or fat-soluble vitamins may occur. Malabsorption of vitamin D, calcium and magnesium can cause metabolic bone disease.

2. *Mucosal atrophy*

Coeliac disease is an immunological reaction to gliadin, a fraction of the wheat protein gluten, causing atrophy of the intestinal villi. Villous atrophy is also found in patients with AIDS and in tropical sprue. Malabsorption in these conditions is due to reduced surface area and deficiencies of small intestinal enzymes.

3. *Inflammatory disease*

Crohn's disease, a chronic inflammatory disease of uncertain aetiology, causes malabsorption as a result of several contributing factors. The terminal ileum is often affected, causing interruption of enterohepatic circulation of bile acids and impaired B_{12}

absorption. Inflammation of the bowel may cause adhesions, fistula and strictures. This may lead to blind loops and bacterial overgrowth syndrome, which causes bile salt deconjugation. Inflammation of the small bowel can lead to reduced absorptive capacity and if there is extensive inflammation, there may be loss of protein into the gut (protein losing enteropathy).

Malabsorption can occur in diabetes mellitus as a result of autonomic neuropathy. Malabsorption is also present in the rare inherited metabolic disorder of abetalipoproteinaemia, where the enterocytes cannot synthesise the apolipoprotein B (see Chapter 11).

4. *Defects in specific absorptive mechanisms*

In addition to generalised malabsorption syndromes, there are specific defects of absorptive processes, such as deficiency of oligosaccharidases. The most common form of oligosaccharidase deficiency is lactase deficiency (lactose intolerance), of which there are two types:

(1) Congenital lactase deficiency
This is a very rare disorder in which lactase levels in the mucosa are low or absent at birth and symptoms occur as soon as milk is given.

(2) Acquired lactase deficiency/Hypolactasia
Normal lactase activity declines as the child ages and the age at which this decline begins is genetically determined in an autosomal recessive fashion and differs among ethnic groups. For example, in Thai children, lactase activity decreases by the age of 2 whereas in American children they do not decrease until after the age of 5. Prevalence of lactose intolerance also varies widely among various ethnic groups. In the nomadic, milk-dependent populations of Africa, Arabia as well as populations of European ancestry, the prevalence is 5–10% whereas in native populations of Australia, South East Asia and parts of Africa, the prevalence is between 45% and 95%.

The symptoms of hypolactasia are due to the presence of undigested sugar, which is fermented in the colon by bacteria producing many small molecules including carbon dioxide, methane and hydrogen and organic acids such as acetic and propionic acid. These lower the pH of the stools. As a result of the osmotic effect of the small molecules, water enters the lumen from the extracellular fluid leading to diarrhoea. The gases produced are absorbed and metabolised or excreted.

(3) Sucrase-isomaltase and trehalase deficiencies

Sucrase-isomaltase deficiency is inherited as an autosomal recessive trait. The enzymes occur together in the mucosa as a complex. The deficiency is rare in North Americans but common in Eskimo tribes. Malabsorption of monosaccharides can cause intestinal symptoms such as bloating, flatulence, abdominal pain and occasionally diarrhoea. Disaccharide deficiency may also occur secondary to small intestinal diseases such as coeliac disease.

Disaccharide deficiency can be diagnosed by measuring the enzyme activity in intestinal biopsy specimens or by measuring breath hydrogen following ingestion of the relevant sugar.

Investigation of Intestinal Function

Assessment of fat absorption

Fat malabsorption is often a prominent feature of generalised malabsorption syndromes. Two commonly used methods to assess fat malabsorption are faecal fat determination and triolein breath test.

1. *Faecal fat test*

In this test, the fat content of the stool over a given time period is determined when the patient is on a normal diet for at least 48 hours. However, this test is rarely used as it is difficult to carry out. This has been replaced by other tests such as triolein breath test.

2. *Triolein breath test*

This is described in the section above.

Xylose absorption test

Xylose is a pentose not normally present in blood. When given orally, it is passively absorbed in the proximal small intestine and excreted by the kidneys. Some xylose is metabolised or excreted in bile. Renal clearance is about 90% of the filtered load indicating some tubular reabsorption. The amount of xylose recovered in the urine or the concentration in blood at a specified time after administration of a given dose of xylose can be used to evaluate the absorptive capacity of the mucosa. As reduced excretion of xylose can occur in the presence of renal impairment, blood xylose concentration is considered a better method. In bacterial overgrowth syndrome, where xylose may be metabolised by bacteria, or in oedematous states where xylose may accumulate in the oedema fluid, false negative results may occur. Delayed gastric emptying and rapid intestinal transit times are other situations where misleading results may be obtained. Xylose absorption is abnormal in severe coeliac disease and disorders of the proximal small intestine. Xylose absorption is normal in pancreatic disease and is therefore useful in differentiating pancreatic from intestinal disease. The diagnostic performance of the xylose test could be improved by simultaneous administration of 3-O-methyl-D-glucose and comparing the absorption of the two sugars. A reduction in the ratio of xylose to 3-O-methyl-D-glucose indicates mucosal disease.

Permeability and the absorptive capacity of small intestinal mucosa can be assessed by giving a mixture of sugars (d-xylose, 3-O-methyl-D-glucose, L-rhamnose and lactulose) and measuring their urinary excretion. An increased ratio of lactulose/rhamnose excretion in urine is seen in diseases of the small intestine where permeability is increased, e.g. coeliac disease and Chron's disease. The ratio helps to correct for any variation in renal clearance. This test, however, is not often used in many hospitals.

Faecal calprotectin

Faecal calprotectin is a useful test to differentiate inflammatory bowel disease such as Crohn's disease and ulcerative colitis from irritable bowel syndrome. This is a protein present in the cytoplasm of neutrophils and is released when leucocytes are activated. Faecal calprotectin is very stable in stools and the concentration can be measured easily by an immunoassay. Increased concentration is found in inflammatory bowel disease but not in irritable bowel syndrome. This test is very useful in selecting patients for endoscopic evaluation; those with normal values can be safely reassured.

Breath hydrogen test

Disaccharidase deficiency can be diagnosed by measuring hydrogen excretion after the administration of the appropriate disaccharide.

Tests for transit time

Transit time can be measured by giving a disaccharide such as lactulose, which is not hydrolysed by small intestinal enzymes. Lactulose is broken down by colonic bacteria producing hydrogen. The time from ingestion to the appearance of breath hydrogen is an indication of mouth to caecal transit time.

Tests for bacterial overgrowth

Bacterial overgrowth can occur in a variety of situations. Bacterial overgrowth leads to deconjugation of bile acids, which results in malabsorption of fat. Bacterial overgrowth can be diagnosed by culture of duodenal contents. Urinary indicans, which are products of bacterial metabolism of tryptophan, are raised in bacterial overgrowth but this test has poor sensitivity.

Bacterial overgrowth can also be assessed by a breath test using ^{14}C-labelled xylose or ^{14}C-glycocholate. Xylose is metabolised by the bacteria producing ^{14}C-labelled carbon dioxide, which is measured in the expired air. ^{14}C-glycocholate acid is deconjugated by bacteria releasing ^{14}C-labelled glycine, which is metabolised to ^{14}C-carbon dioxide. This latter test is less reliable than ^{14}C-labelled xylose test.

Faecal occult blood

Bleeding into the gastrointestinal tract will lead to the appearance of blood in the stools. However, if the bleeding is in the upper region of the gastrointestinal tract, blood will not be visible as it will be digested and discoloured. Detection of such blood is called faecal occult blood test. The presence of blood can be detected either by detecting the haem by guaiac-based test or by detecting the globin by an antibody-based test. Haem-based assays are cheap but are prone to interference from dietary haem. Detection of blood in the stools implies bleeding in the gastrointestinal tract and should prompt further investigation by endoscopy. Faecal occult blood test is now used in many countries as a screening test to detect colorectal cancer, which is the second most common cancer. Several studies have shown that by screening asymptomatic subjects with faecal occult blood test mortality from colorectal cancer can be reduced. In the UK, all adults are invited to participate in the screening test and those who show a positive test are sent for further investigations.

Further Reading

1. Abdehshaheed NN, Goldbery DM. Biochemical tests in diseases of the intestinal tract. *Crit Rev Clin Lab Sci* 1997; 34:141–163.
2. Braden B. Methods and functions: Breath tests. *Best Pract Res Clin Gastroenterol* 2009; 23:3337–352.
3. Keller J, Aghsassi AA, Lerch MM, Mayerle JV. Tests of pancreatic exocrine function — clinical significance in pancreatic and non-pancreatic disorders. *Best Pract Res Clin Gastroenterol* 2009; 23:425–439.
4. Lankisch PG. Secretion and absorption (methods and function). *Best Pract Res Clin Gastroenterol* 2009; 23:325–335.
5. Schulzke JD, Troger H, Amasheh M. Disorders of intestinal secretion and absorption. *Best Pract Res Clin Gastroenterol* 2009; 23: 395–406.
6. Whitcomb DC. Acute pancreatitis. *N Eng J Med* 2006; 354:2142–2150.

Summary/Key Points

1. Stomach produces hydrochloric acid, intrinsic factors, enzymes, gastrin and ghrelin. Assessment of acid production is no longer used. Infection with *Helicobacter pylori* can be detected by labelled urea test. Labelled urea is administered and is split by urease present in the bacteria. The labelled CO_2 released is measured in breath.

2. Gallstones or alcohol commonly causes acute pancreatitis. It causes severe abdominal pain and serum amylase and lipase are elevated. In severe acute pancreatitis hypocalcaemia, acute respiratory distress syndrome and hypoxia are poor prognostic indices.

3. Chronic pancreatitis is a rare condition often due to alcohol abuse and presenting features are abdominal pain and malabsorption.

4. Function of the pancreas can be assessed by the PABA test where a peptide linked to a marker (PABA) is given and pancreatic enzymes digest this peptide and the marker excreted in urine. Reduced excretion of the marker indicates poor pancreatic exocrine function.

5. Intestines are the final place for absorption of nutrients. Failure of digestion or absorption will result in malabsorption syndrome. Malabsorption can be diagnosed by trolein breath test. After administration of labelled trolein, the amount of labelled CO_2 in the expired air is measured. If there is malabsorption, less label will appear in the breath.

6. Measurement of enzymes such as elastase in the stool is an alternative way of detecting malabsorption.

7. Calprotectin, a protein present in leucocyte appear in the stools when there is inflammation. Measurement of stool calprotectin is useful in differentiating inflammatory bowel disease from irritable bowel syndrome.

8. Faecal blood test is used as a screening test for early detection of colon cancer.

chapter 17

Nutrition

Introduction

Proper (adequate, good) nutrition is important during all stages of life — during growth and development and for the maintenance of health during adult life. In developing countries, generalised malnutrition such as protein-energy malnutrition is common. There may also be deficiency of specific nutrients, e.g. vitamin A deficiency. In developed countries, and in the rapidly expanding 'middle classes' of developing countries, obesity is common and has reached epidemic proportions. Malnutrition is not uncommon in hospital patients and the elderly. This may contribute to and exacerbate existing diseases.

It is also becoming increasingly recognised that nutrition plays an important role in many chronic diseases such as atherosclerosis, hypertension, some forms of malignancy, osteoporosis, diabetes mellitus, dental caries, gallstones and urinary stones.

Protein-energy Malnutrition (PEM)

Protein-energy malnutrition (PEM) is a spectrum of disorders in which the intake of energy and protein are inadequate, often accompanied by deficiency of vitamins and minerals. The clinical spectrum depends on the severity of various deficiencies. PEM is very common in developing countries, especially in children, and is mainly due to inadequate food availability and increased metabolic demand may contribute. In developing countries, up to 17% of children may be affected by mild to moderate PEM, while 2% may suffer severe malnutrition. Marasmus and Kwashiorkor are two recognised syndromes in PEM.

Marasmus

Nutritional marasmus is the most common severe form of PEM and is caused by lack of protein and energy. It affects young children (weaning age) and is commonly due to early weaning, followed by consuming diluted feed because of poverty and/or ignorance. Gastroenteritis and poor appetite further exacerbate the condition; intestinal atrophy develops, increasing the susceptibility to diarrhoea. Classical features of marasmus include very low body weight, very little body fat, gross muscle wasting, without oedema and normal hair.

Kwashiorkor

Kwashiorkor is an acute form of PEM due to protein deficiency with relatively adequate carbohydrate intake. Features include oedema, enlarged liver, moon face, pale and thinned hair, patches of pigmentation and desquamation of skin; child looks miserable, apathetic and refuses to eat. The pathogenesis appears to be a very low protein intake with relatively adequate carbohydrate intake, which maintains insulin secretion. Insulin spares muscle protein but liver proteins are not. Synthesis of albumin and LDL by the liver is reduced leading to oedema and the accumulation of fat in the liver.

Marasmic kwashiorkor has some features of both conditions.

The reason why some children develop marasmus while others develop kwashiorkor is not fully understood. In addition to differences in protein and energy intake, deficiencies in micronutrient and antioxidant are thought to contribute.

PEM is usually associated with deficiencies of other nutrients including zinc, vitamin A, iron, folate, magnesium, potassium, thiamine, riboflavin, niacin and iodine. Some of these deficiencies are specific for certain areas and depend on local dietary/food habits. For example, thiamin and riboflavin deficiency are common in Thailand. It is important to recognise these associated deficiencies as they may cause specific problems (e.g. vitamin A deficiency may lead to blindness) or require additional treatment supplements.

Management of PEM is in three stages. In the resuscitation stage, dehydration, electrolyte disturbances, hypoglycaemia and infection are treated. The second stage is when feeding is introduced gradually and then increased. These children cannot accept normal food due to deficiencies of digestive enzymes. Vitamin and mineral supplementation is often required. The third phase is nutritional rehabilitation where normal nutrition is resumed and catch up growth is completed.

Malnutrition in Hospital Patients

Malnutrition is common in hospital patients in developed countries. Several studies have shown that up to 50% of hospitalised patients may be malnourished and up to 70% of these patients are undiagnosed.

Malnutrition in ill patients develop due to several factors, which include anorexia, the inability to eat and the action of cytokines released in response to illness or injury (Table 17.1). Malnutrition in hospital

Table 17.1   Factors contributing to malnutrition in hospital patients

Reduced intake
Anorexia, nausea
- Depression
- Chronic disease

Inability to eat
- Confused or reduced consciousness
- Neurological disorder
- Oesophageal disease — dysphagia
- Oropharyngeal disease

Food related factors
- Poor quality of food in hospital
- Less appetising food

Increased requirement
- Inflammatory response
- Cytokines

Reduced absorption or digestion
- Inflammatory bowel disease
- Radiation enteritis
- Gluten enteropathy
- Short bowel syndrome

Increased loss of nutrients
- GI losses — vomiting, diarrhoea, fistula
- Skin — exudates

patients contributes and exacerbates existing disease states and may have other adverse effects (Table 17.2). Loss of weight in excess of 30% is associated with increased mortality. Malnutrition will cause delayed wound healing, increase susceptibility to infection, and prolong hospital stay. Malnutrition increases the susceptibility to disease which in turn leads to worsening of malnutrition causing a vicious circle. Hospital patients at increased risk of developing malnutrition include elderly patients, patients with AIDS, chronic renal failure patients, and those with intestinal disease and gut resection.

Diagnosis of Malnutrition

Awareness that malnutrition may develop in hospital patients is the first step in recognising malnutrition. Clinical assessment is the most important way of diagnosing malnutrition. Features suggestive of malnutrition include recent weight loss, reduced body mass index, reduced mid-arm muscle circumference and skin fold thickness. However, the state of hydration may affect these measurements. A simple screening system to identify patients at risk of malnutrition has been developed. Laboratory assessment includes haemoglobin, serum

Table 17.2   Possible effects of malnutrition in hospital patients

Mental function
- Apathy
- Fatigue
- Inability to cooperate with treatment

Muscle function
- Respiratory failure
- Delayed mobilisation

Immune function
- Increased susceptibility to infection

Miscellaenous
- Impaired wound healing
- Reduced digestive capacity
- Impaired thermogenic response

proteins such as albumin, retinol binding protein (RBP) and transferrin (see below) and 24-hour creatinine excretion as an index of muscle mass (see below for details of nutritional assessment).

Nutritional Support

Nutritional support is essential for those who are unable to eat or have prolonged intestinal failure. Patients who are malnourished will benefit from nutritional support as studies have shown that it will reduce morbidity and hospital stay. Nutritional support should be considered in those who have lost at least 10% of their body weight and in those who cannot maintain adequate food intake for 10 days or more as a result of illness.

Nutritional support may be given orally, by enteral tube feeding or parenterally (intravenous nutrition).

Enteral Nutrition

Oral supplements

When possible, nutrition supplements in addition to food may be given to augment the diet. Supplements include energy in the form of high calorie drinks (such as High Cal) and mineral and vitamin supplements. These supplements are available in a range of flavours and are used between meals.

Enteral tube feeding

This form of feeding is cheaper, safer and more physiological than parenteral nutrition. It stimulates intestinal and biliary motility, provides a wider range of nutrients including glutamine and short chain fatty acids which are important substrates for intestinal epithelial cells. Enteral nutrition may also provide a mucosal barrier in ill patients.

Enteral tube feeding can be delivered by many methods and these include nasogastric tube, percutaneous gastrostomy or jejunostomy. A commonly used enteral feeding preparation is polymeric whole-protein feed, which is cheaper and avoids the use of hypertonic solutions. In some patients, e.g. those with intestinal failure, special preparations containing peptides and medium-chain triglycerides may be used. Feed preparation supplemented with glutamine, arginine, nucleotides and fish oil are sometimes used in critically ill patients.

Complications of enteral feeding

These include metabolic complications (see "Complications of parenteral nutrition"), gastrointestinal complications, such as diarrhoea, and complications related to the delivery of nutrients, such as aspiration pneumonia, peritonitis and infection of the stomach.

Parenteral Nutrition

Parenteral nutrition is required when the intestinal tract is unavailable, inadequate or cannot be used. Indications for parenteral nutrition are given in Table 17.3.

Parenteral nutrition solutions are usually chemically defined preparations and contain glucose, lipids, amino acids, electrolytes, minerals, vitamins and trace elements. Typically, it provides non-protein energy of 17–40 kcal per kg and nitrogen of 0.2–0.3 g per kg in 2 to 3 litres of fluid. Non-protein energy is provided as glucose and lipids with approximately 50% coming from each of these sources. This provides essential fatty acids and avoids metabolic complications associated with excess glucose or fat. These solutions contain all essential amino acids but not all non-essential amino acids, due to potential problems with

Table 17.3   Indications for parenteral nutrition

Severe inflammatory bowel disease
Severe acute pancreatitis
Mucositis following chemotherapy
Patients with multi-organ failure
Short bowel syndrome
Radiation enteritis
Scleroderma

stability. Under some circumstances, some of these 'non-essential' amino acids such as glutamine become 'conditionally essential' in stressed patients. Some solutions provide glutamine in the form of dipeptides and this is thought to improve immune and gut barrier functions. Parenteral nutrition is usually given via a central venous catheter, as the feeds are hypertonic. If the osmolality of the solution is kept isotonic and/or if the feeding is for a short period of time, it can be given via a peripheral vein.

Complications of parenteral nutrition

Complications include metabolic and catheter-related complications (Table 17.4). Fluid depletion and overhydration are not uncommon during parenteral nutrition. Refeeding syndrome is the most important acute complication of enteral and parenteral nutrition. In ill patients, hyperglycaemia can occur due to relative insulin resistance and insulin

Table 17.4   Complications of parenteral nutrition

Metabolic complications
- Short-term
  — Refeeding syndrome
  — Hyperglycaemia
  — Hypo- or hyperkalaemia
  — Hypo- or hypernatraemia
  — Hypophosphataemia
  — Abnormal liver function tests
  — Metabolic acidosis
  — Hypoglycaemia
  — Hyperlipidaemia
  — Dehydration/fluid overload
- Long-term
  — Metabolic bone disease
  — Deficiency of micronutrients
  — Biliary disease

Catheter-related
- Infection
- Occlusion
- Central vein thrombosis
- Fracture

infusion may be necessary to improve glucose tolerance. If feeding is abruptly interrupted, hypoglycaemia may develop. During parenteral feeding, the nutrients reach the systemic circulation directly without first going through the liver, which plays an important role in metabolic regulation. Hypophosphataemia used to be a frequent complication as a result of phosphate utilisation during glucose metabolism. Provision of phosphate in the fluids has reduced the incidence of hypophosphataemia (see Chapter 8). Provision of amino acids as chloride can lead to hyperchloraemic metabolic acidosis.

Long-term complications of parenteral nutrition include metabolic bone disease and osteoporosis. The aetiology of these is not fully understood.

Refeeding Syndrome

This is an acute life-threatening complication that may develop when malnourished patients are refed either enterally or parenterally. This was first described during the Second World War and may have caused deaths when malnourished prisoners of war were refed quickly. Features of this syndrome include cardiac problems, hypokalaemia, hypomagnesaemia and hypophosphataemia (Table 17.5). Refeeding syndrome develops within 4 days of starting the feed. It is caused by the rapid shift of electrolytes and fluids stimulated by an increase in insulin secretion, which increases the

Table 17.5   Features of refeeding syndrome

Cardiac effects
- Cardiac failure
- Pulmonary oedema
- Dysarrthymias

Electrolyte disturbances
- Hypophosphataemia
- Hypokalemia
- Hypomagnesemia
- Hypocalcaemia

Hyperglycaemia

uptake of phosphate, potassium and magnesium by cells. Those at risk of developing this syndrome include those with body mass index < 16 kg/cm², unintentional weight loss of 15% within the previous 3–6 months, very little or no food intake for > 10 days and low levels of potassium, phosphate and magnesium prior to refeeding. Patients at risk of this syndrome should be started at low rates of energy and protein with adequate vitamins.

Wernicke-Korsakoff syndrome is another syndrome seen during refeeding due to acute thiamine deficiency (see Chapter 18).

Laboratory monitoring of parenteral nutrition (Table 17.6)

Patients receiving parenteral nutrition should be carefully monitored in order to avoid complications and to prevent deficiencies. Monitoring should include regular clinical examinations, fluid balance chart and daily body weight measurements when possible. Serum glucose and potassium should be monitored daily and at shorter intervals (say every 6 hours) if there is hyperglycaemia or hypoglycaemia.

Serum sodium is usually monitored daily as hypo- and hypernatraemia can develop. Hyponatraemia can be artificial caused by hyperlipidaemia as a result of impaired lipid clearance. Hyponatraemia can also arise as result of hyperglycaemia, volume depletion or over-hydration (see Chapter 2). Hypernatraemia is commonly caused by inadequate water intake.

Renal function tests, serum urea and creatinine should be measured daily, especially in the initial stages. A rise in serum urea without a rise in serum creatinine is indicative of excess amino acid infusion. Serum creatinine should be monitored regularly if the renal function is poor.

Liver function tests should be monitored twice weekly. Abnormalities in liver function tests frequently occur in patients on parenteral nutrition. Overprovision of energy may lead to fatty infiltration and subsequent increases in ALT, usually in the early stages of parenteral nutrition. Subsequently, a cholestatic picture with elevated ALP, GGT and occasionally high bilirubin may occur. This is thought to be due to biliary sludging as a result of decreased bile secretion due to the absence of stimulation from the gut (gut hormones and other signals).

Table 17.6   Laboratory monitoring of patients on parental nutrition

Body weight (when possible)	Daily
Fluid balance	Daily
Sodium, potassium, urea & creatinine	Baseline, daily until stable and then one or two times weekly
Glucose	Baseline, once to twice a day until stable then weekly
Magnesium & phosphate	Baseline, daily and then three times a week until stable, weekly thereafter
Liver function tests	Baseline, twice weekly until stable and then weekly
Calcium and albumin	Baseline and then weekly
CRP	Baseline, two or three times a week until stable
Zinc and copper	Baseline, two or four times weekly
Selenium	Baseline and monthly
Full blood count	Baseline, 1–2 times a week until stable and then weekly
Iron and ferritin	Baseline, every 3–6 months
Vitamin B_{12} and folate	Baseline, 2–4 times weekly
Manganese	Every 3–6 months
25-OHD	Every 6 months
Bone mineral density	Baseline and every 2 years (those on long-term parentral nutrition)

These abnormalities in liver function tests are reversible in adults but may occasionally lead to progressive liver damage in children.

The measurement of short half-life proteins such as transferrin, retinol binding protein and prealbumin may help to assess the adequacy of nutritional support. Measurement of serum albumin has very little value.

Weekly measurements of magnesium and zinc are required to detect depletion of these minerals. Patients on parenteral nutrition for more than a few weeks require measurements of trace elements such as selenium, copper, and manganese at intervals.

Urine urea excretion (as a surrogate for nitrogen estimation) may be useful to monitor nitrogen balance.

Nutritional Assessment

Assessment of nutritional status is important to decide nutritional support as well as to monitor the effects of such support. The methods used for assessing nutritional status are listed in Table 17.7. These tests can be broadly classified into tests which determine the whole body status, and tests which assess the adequacy of recent intake. Of these, anthropometric measurements such as height, weight, body mass index (weight divided by square of height) (BMI), skin fold thickness and arm muscle circumference are commonly used. A BMI between 17 and 18.4 indicates mild malnutrition while a BMI < 17 indicates chronic severe malnutrition. Skin fold thickness measures subcutaneous fat and arm muscle circumference gives an indication of the amount of muscle. There is a large variability between individuals in these measurements. These measurements are useful in assessing progress within an individual over a period of time. Of the other methods listed in Table 17.7, bioelectrical impedance method is the most practicable. This is based on the principle that fat is a very poor conductor of electricity whereas fat-free tissues which contain water and electrolytes are good conductors. Simple apparatus are now available for measuring body composition by this method. Other methods of measuring body composition are not routinely used in clinical practice. Handgrip strength is sometimes used at the bedside. The grip strength is dependent on the mass and function of the forearm muscles and thus, grip strength is related to whole body muscle mass.

Adequacy of recent intake can be assessed by simple clinical assessment or by assessment of dietary intake. Methods such as calorimetry are beyond the scope of this book. Nitrogen balance can be useful to assess recent intake, but is difficult to perform in clinical practice. A simple method is to measure 24-hour urine urea as urea accounts for nearly 90% of total nitrogen loss. However, this proportion can vary from 60–90%, depending on the metabolic state of the patient. Nitrogen balance gives an assessment of the adequacy of recent changes in body composition. The serum concentration of transthyrin/prealbumin, retinol-binding protein and transferrin are also useful as these

Table 17.7   Assessment of nutritional status

Tests of whole body status
- Body composition
    — Anthropometry
        o Weight, body mass index
        o Skinfold thickness
        o Arm muscle circumference
    — Total body potassium
    — Densitometry
    — Total body water
    — *In vivo* neutron activation analysis
    — Bioelectrical impedance
    — Imaging method
        o Computerised axial tomography (CAT)
        o Nuclear magnetic resonance (NMR)
        o Dual energy X-ray (DEXA)
- Handgrip strength
- Delayed hypersensitivity states
- Biochemical tests
    o Albumin
    o Urine creatinine

Tests of adequacy of recent intake
- Clinical assessment
- Assessment of dietary intake — e.g. 24-hour recall method
- Indirect calorimetry
- Nitrogen balance
- Serum prealbumin, transthyretin
- Serum retinol-binding protein
- Serum transferrin
- Serum amino acids
- Serum IGF-1

proteins have short half-lives and their concentration will decrease in malnutrition (see Chapter 14). These proteins may be useful to assess the progress of a patient who is on nutritional supplementation.

Insulin-like growth factor 1 (IGF 1) is influenced by growth hormone status as well as by protein and energy intake. It is useful in assessing nutritional status, as the change in concentration of IGF 1 is greater than albumin, transferrin or anthropometric measurements. It is also useful in following the progress of a patient on nutritional support.

Prognostic indices have been used to predict those who are more likely to benefit from nutritional intervention. Various combinations of measurements including anthropometric, immunological as well as biochemical tests have been used but their value has not been established.

Obesity

Obesity has become an increasing health problem and has reached epidemic proportions in developed countries and in the 'middle class' of developing countries. In many industrialised countries, up to 17–25% of the adult population are obese.

Obesity can be defined as an accumulation of adipose tissue high enough to cause a health hazard. However, measurements of adipose tissue in clinical practice is difficult. Body mass index is the most common method of assessing body fat. According to WHO, a BMI between 25 and 29.9 is classified as overweight, that between 30 and 39.9 as obese and that above > 40 as gross obesity. These cutoff values are based on studies in Caucasians and may not apply to all populations. In certain ethnic groups, e.g. South Asians, WHO's definition is less applicable as studies have shown that a BMI of 22 or greater in South Asians is associated with increased complications. Measurements of body fat using electrical impedance are now gaining popularity, as this is a simple and easy method to use.

Although this definition is based on BMI, it is now recognised that fat distribution may be more important in determining long-term complications of obesity. Waist circumference and waist/hip (W/H) ratio may be better index, a W/H ratio of 0.90 in men and 0.85 in women or a waist circumference of 102 cm in males and 88 cm in females are indicative of increased risks of complications.

Obesity can be primary, where there is no obvious predisposing condition or secondary, where there is an underlying disorder such as myxodema, Cushing's syndrome or Prader-Willi syndrome (a hypothalamic disorder). Secondary causes account for only a small proportion of obesity.

The amount of fat in a person is strongly influenced by genetic factors as shown by family and twin studies. Although genetic factors are important, the exact genetic mechanism is not clear. Rare single gene disorders may cause obesity. In the vast majority of patients, there is no single factor responsible for obesity. Probably a combination of little exercise and overeating for various psychological and social reasons may result in obesity.

The pathophysiological basis of obesity is complex and is under intense research. It is now recognised that energy balance is maintained tightly by a complex interplay of neuronal and humoral mechanisms. These include factors released by the adipose tissue such as leptin and those from the gastrointestinal tract including ghrelin, peptide YY, CCK, GLP and hypothalamic neuropeptides including neuropeptide Y, agouti-related peptide, melanocortin, cocaine and amphetamine-regulated transcript (CART) and serotonin.

Complications of Obesity

Risks of cardiovascular complications and diabetes are greater in people with abdominal obesity. Other complications of obesity include sleep apnoea, cerebrovascular disease coronary heart disease, respiratory disease, gallstones, hernia, arthritis, varicose veins, hypertension, diabetes mellitus and cancer. In addition to medical complications, there are many social complications.

Management of Obesity

The mainstay of treatment for obesity involves a reduction in food intake and an increase of energy expenditure by exercise. Occasionally, drug treatment to reduce appetite has been tried. The appetite suppressant sibutramine, a serotonin–norepinephrine reuptake inhibitor, has been previously used but was withdrawn due to cardiac valve disease and pulmonary hypertension. Orlistat is a lipase inhibitor that reduces absorption of fat by 30%. A third drug, Rimonabant, which reduces appetite by blocking endocannabinoid pathway, has also been withdrawn due to its side effects.

In severe obesity, bariatric surgery is indicated. This include gastric banding to reduce the volume of stomach, gastric bypass surgery (the most common is Roux-en-Y gastric bypass). Weight loss with gastric bypass surgery is sustained. Surgical treatment also reduces the risk of diabetes mellitus, cardiovascular disease and cancer. The weight loss associated with gastric bypass surgery is primarily due to a reduction in appetite brought about by the reduction in gut hormones like ghrelin.

Complications of bariatric surgery include dumping syndrome due to the rapid entry of food into the small bowel, diarrhoea, bacterial overgrowth, blind loop syndrome and gall stones. Patients with dumping syndrome have nausea and diarrhoea shortly after eating.

Gallstones develop in up to 40% of patients and are thought to be due to rapid weight loss. Long-term complications include nutritional deficiencies due to a combination of reduced intake and malabsorption. These include deficiencies of vitamin B_{12}, folate, fat-soluble vitamins and thiamine. Deficiencies of calcium and vitamin D increase the risk of osteoporosis and osteomalacia. Patients after bariatric surgery require regular monitoring to detect and treat such deficiencies.

Eating Disorders

These are a group of disorders characterised by abnormal eating habits that lead to impaired physical and mental function. Anorexia nervosa and bulimia nervosa are the most common forms. Both are more common in females but an increasing number of males are being recognised as having these syndromes.

Anorexia is characterised by the refusal to maintain normal body weight. These subjects have a BMI < 17.5 kg/m^2 and secondary hypogonodotrophic hypogonadism. They tend to have high plasma and urine free cortisol and low T3 and T4 levels.

Bulimia nervosa is characterised by recurrent binge eating followed by compensatory actions like vomiting, excessive use of diuretics, laxatives and exercise.

There are no laboratory tests to diagnose these disorders. However, measurements of potassium and phosphate may be important to recognise severe hypokalaemia and hypophosphataemia.

Metabolic Response to Trauma and Sepsis

Following trauma, surgery, sepsis or burns, a series of changes in the body is seen. These metabolic changes are conveniently divided into an early or ebb phase and a late or flow phase.

The ebb phase is characterised by hypovolaemia, shock and the priority in this stage is survival. There is a reduction in cardiac output, oxygen consumption, blood pressure, body temperature and tissue perfusion. Hyperglycaemia is a feature of this phase and it is caused by a combination of a reduction in insulin and glucose oxidation and an increase in the mobilisation of glucose due to increased catecholamines, growth hormone and cortisol. The duration of this phase is usually 1–2 days, depending on the severity of the trauma.

The flow phase is characterised by increased energy expenditure and increased protein breakdown. This phase lasts 3–7 days, depending on the severity of the trauma. There is increased secretion of cortisol, growth hormone, glucagon and catecholamines. There is increased production of acute phase proteins (see Chapter 14).

Factors mediating the metabolic response include cytokines such as TNF-α, IL-1, IL-6 and IL-10, afferent nerves, stress hormones and bacterial toxins. During the early phase, there is reduced urine volume in response to hypovolaemia and ADH release, and low urine sodium due to increased aldosterone. As a result of increased catabolism, there is increased urine nitrogen, which may reach 25 g/day in extreme sepsis. Fat becomes the principal source of energy and there is increased catabolism of muscle proteins to provide gluconeogenic substrates and for the increased synthesis of acute phase proteins. Factors that modify the response include severity of trauma, genetic factors, nutrition, ambient temperature and drugs.

This phase is followed by a period of anabolism when the skeletal muscle mass is restored and the body returns to normal.

Further Reading

1. Ayling R, Marshall W. *Nutrition and Laboratory Medicine* 2007. ACB Venture publications: London.
2. Bauer JM, Kaiser MJ, Sieber CC. Evaluation of nutritional status in older persons: Nutritional screening and assessment. *Curr Opin Clin Nutr Metab Care* 2010; 13:8–13.
3. Mehanna HM, Moledina J, Travis J. Refeeding syndrome: What it is, and how to prevent and treat it. *Br Med J* 2008; 336:1495–1498.
4. NICE Guidelines: http://guidance.nice.org.uk/CG32
5. Smith T, Elia M. Artificial nutrition support in hospital: Indications and complications. *Clin Med* 2006; 6:457–460.
6. Truswell AS. *ABC of Nutrition* 2003. BMJ Publication, 4th edition.

Summary/Key Points

1. Protein energy malnutrition is a global problem and the spectrum includes marasmus and kwashiorkor.
2. Malnutrition is common among hospital patients (up to 50%). Recognition of malnourished patients is important as malnutrition will extend hospital stay and increase the risk of complications.
3. Nutritional support either enterally or parenterally is essential in those who are unable to eat or who have prolonged intestinal failure.
4. Metabolic complications of enteral and parentral feeding include hypo- or hyperglycaemia, hypo- or hyperkalaemia, hypophosphataemia and abnormal liver function tests.
5. Refeeding syndrome is a potentially fatal complication of nutritional support. When starving, subjects are refed quickly, severe electrolyte disturbances may develop. These include severe hypokalaemia, hypophosphataemia and hypomagnesaemia.
6. Patients given nutritional support should have regular monitoring to avoid these complications.
7. Methods for assessing nutritional status of an individual include anthropometric measurements and biochemical measurements (serum proteins and IGF1).

8. Obesity is an increasing problem throughout the world. In gross obesity, bariatric surgery is increasingly used. This surgical procedure may cause complications such as dumping syndrome, diarrhoea, bacterial overgrowth, blind loop syndrome and gall stones.

9. In response to injury, sepsis or surgery a series of metabolic changes take place in the body and these changes are conveniently divided into and early ebb phase and a late flow phase. In the ebb phase, there is reduced energy expenditure and hyperglycaemia; this phase lasts 1–2 days. In the flow phase, there is increased catabolism, muscle protein breakdown, synthesis of acute phase proteins and metabolic rate.

chapter 18

Vitamins and Trace Elements

Vitamins

Vitamins are compounds that take part in specific biochemical functions. They are required in small amounts, are not usually made in the body or not made in sufficient quantities. Deficiencies of vitamins can occur as part of general malnutrition or as specific deficiencies. Vitamin deficiency may arise in developing countries mainly as a result of inadequate intake. In developed countries, vitamin deficiency is due to disease (increased requirement, or decreased absorption) or as a result of drugs (impaired or altered metabolism). Many people in developed countries take vitamins in doses higher than that required for physiological function, as they are believed to cure/'relieve' many diseases/symptoms. As a result of this 'megavitamin' therapy, an overdose may result. Table 18.1 summarises the important vitamins and their requirements.

Fat-Soluble Vitamins

Vitamin A

Vitamin A is a term used to describe retinol (an alcohol) and retinal (an aldehyde). Retinal can be converted to retinoic acid in the body. Retinol, retinal, retinoic acid and related compounds are called retinoids. Retinol is not found in foods, but retinol palmitate, the storage form, is found in many animal products — liver, fish, liver oils (rich sources), dairy products, eggs and in fortified margarine. It is present as the precursor β-carotene, in many plant products such as

Table 18.1   Summary of vitamins, their function, requirements, effects of deficiency and methods of assessment

Vitamins	Main function	Daily requirement	Effects of deficiency	Assessment
Fat-soluble vitamins				
Vitamin A	Vision, Epithelial function	700–900 μg	Night blindness, Keratomalacia	Serum vitamin A
Vitamin D	Calcium homeostasis	400 U	Rickets, Osteomalacia	Serum 25-hydroxyvitamin D
Vitamin E	Antioxidant, Membrane stability	15 mg	Haemolytic anaemia	Serum tocopherol
Vitamin K	γ carboxylation	150 μg	Clotting defects	Prothrombin time, PIVKA*
Water-soluble vitamins				
Thiamine (B$_1$)	Coenzymes	1.1–1.2 mg	Beriberi, Cardiac myopathy	Activation of red cell tansketolase by thiamine
Riboflavin (B$_2$)	Coenzymes	1.1–1.63 mg	Angular stomatitis, Glossitis, Dermatitis	Activation of red cell gluathione reductase by FAD
Niacin	Coenzymes	14–16 mg	Pellagra	Urinary niacin or its metabolites
Pyridoxine (B$_6$)	Coenzymes	1.5–1.7 mg	Dermatitis, Stomatitis	Red cell AST activation

(*Continued*)

Table 18.1 (*Continued*)

Vitamins	Main function	Daily requirement	Effects of deficiency	Assessment
Biotin	Coenzymes	30 μg	Dermatitis, Depression	
Folic acid	Coenzymes	400 μg	Megaloblastic anaemia	Serum or red cell folate
Vitamin B_{12}	Coenzymes	2–3 μg	Megaloblastic anaemia	Serum B_{12}, Methylmalonic acid excretion or holotranscobalamin
Vitamin C	Collagen formation	65–75 mg	Scurvy, Anaemia	Plasma or leucocyte vitamin C

*PIVKA: Protein induced by vitamin K absence

carrots, red palm oil, apricots, melon and dark green leafy vegetables. β-carotene is converted to retinol by an intestinal enzyme. After absorption, vitamin A is transported by a specific protein, retinol binding protein (RBP), and is stored in the liver. The best known effect of vitamin A is its role in night vision. Retinol in circulation is taken up by retina and stored as retinyl ester. When required, retinyl ester is hydrolysed and isomerised to give 11-cis retinol, which is then oxidised to 11-cis-retinal. This is taken up by rod cells, where it combines with the protein opsin to form rhodopsin. When light falls on the rod cells, 11-cis-retinal isomerizes to all-trans-retinal and this process releases energy in the form of electrical impulses, which are carried by the optic nerve (Figure 18.1). All-trans-retinal is converted to all-trans-retinol and are transported to the retinal epithelial cells.

Retinol is also taken up by many cells and oxidised to retinoic acid or 9-cis-retinoic acid. These are transported to the nucleus where they bind to specific receptors and initiate genetic transcription and regulate differentiation. Retinol is essential for the maintenance and integrity of skin and mucosal cells. Retinoic acid also plays a role in the development and differentiation of lymphocytes, thereby influencing immunity. Retinoic acid and retinol are essential for embryo development.

In vitamin A deficiency, there is metaplasia of many epithelial cells. In the conjunctiva, there is loss of mucous production and

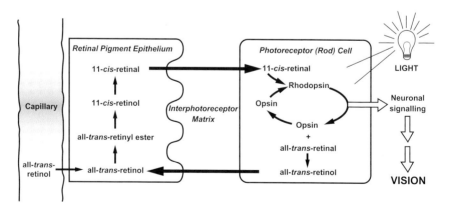

Figure 18.1 Vitamin A cycle.

metaplasia, leading to xerophthalmia (dry eyes). The respiratory epithelium is similarly affected and resistance to infection is lowered. In vitamin A deficiency, cell-mediated immunity is also impaired. The mortality rate in children with vitamin A deficiency is significantly higher. Xerophthalmia is the most common cause of blindness in the world and WHO has estimated that 170 million children may be suffering from vitamin A deficiency worldwide. The first sign of vitamin A deficiency is night blindness. This is followed by increased keratinisation in the skin, blocked sebaceous glands causing follicular keratosis, keratinisation and squamous metaplasia of the conjunctiva (conjunctival keratosis) and this is followed by the appearance of Bitots' spots, which are white plaques of desquamated epithelium. Corneal ulceration or keratomalacia (which is a punched out ulcer) develop and may lead to blindness. In developed countries, vitamin A deficiency is rarely seen and is usually associated with malabsorption syndromes. Vitamin A is stored in the liver and the stores can last up to 2 years. The vitamin A status can be assessed by measuring serum vitamin A concentration. Serum concentration, however, may be low due to low binding protein. Serum concentration only falls when stores are severely depleted.

In pharmacological doses, vitamin A reduces keratinisation and sebum production. Derivatives of retinol such as 13-cis-retinoic acid are used topically in the treatment of acne.

Vitamin A is teratogenic and should not be given in large doses to pregnant women. Occasionally, vitamin A toxicity may occur due to the ingestion of large amounts of vitamin A containing tablets. Vitamin A toxicity may cause hypercalcaemia due to a direct effect of vitamin on bone resorption. Excess β-carotene on the other hand is not toxic as its conversion to vitamin A is regulated, but the patient may have yellowish skin and plasma. Chronic ingestion of large amounts of vitamin A can lead to osteoporosis.

Vitamin D

See Chapter 6.

Vitamin E

There are eight compounds in this vitamin group — four tocopherols, α, β, γ and δ, of which α is the most potent, and four tocotrienols, which have double bond side chains. Tocotrienols are less potent than tocopherols. α-Tocopherol is the main form present in the human body. Tocopherol is found in the cell membrane and is an important antioxidant. It is believed to reduce peroxidation of unsaturated fatty acids by free radicals and prevent the oxidation of lipoproteins. Rich sources of vitamin E are vegetable oils (wheat germ oil being the richest), nuts, whole grain and leafy vegetables.

Vitamin E deficiency is seen in patients with malabsorption syndromes such as cystic fibrosis and in the rare inherited disorder of abetalipoproteinaemia. Features of vitamin E deficiency include mild anaemia, ataxia (due to spinocerebellar degeneration), loss of tendon reflexes and pigmentary retinopathy.

Because of its antioxidant activity, vitamin E is believed to reduce atherogenesis and many people take supplementation. Recent trials, however, have not confirmed any benefit of vitamin E supplementation on coronary heart disease, diabetes mellitus, dementia or cancer. No toxic effect of high doses of vitamin E has been reported.

Vitamin E status can be assessed by measuring α-tocopherol in plasma. It is recommended that the results be expressed as tocopherol/cholesterol ratio as an increase in lipoproteins can cause an increase in tocopherol.

Vitamin K

Vitamin K is present in two forms: K_1 (phylloquinone) found mainly in vegetables, and K_2 (menaquinones) produced by bacteria in the gut. Major dietary sources of vitamin K are green leafy vegetables (greens, broccoli, cabbage, and lettuce) and some vegetable oils (soyabean, olive and cotton seed oil). Although gut bacteria can synthesise vitamin K_2, their contribution to body requirement is very small. Vitamin K is absorbed in the presence of bile and enters the circulation with the chylomicrons and is then transported to the liver.

Vitamin K is a cofactor for γ carboxylation reaction in proteins. Proteins requiring γ carboxylation are prothrombin, factors VII, IX and X and osteocalcin. This action of vitamin K is blocked by warfarin. Vitamin K deficiency leads to hypoprothrombinaemia and bleeding. It has been suggested that vitamin K deficiency contributes to bone loss and osteoporosis.

Neonates are prone to vitamin K deficiency as there is very little placental transfer and breast milk has very little vitamin K. It is the policy in many countries to give vitamin K to all newborns at birth or those at increased risk (low birth weight babies or difficult delivery). In adults, vitamin K deficiency develops in malabsorption syndrome and in liver disease.

Vitamin K status can be assessed by measuring prothrombin time. Direct measurement of vitamin K is possible but seldom used clinically. Measurement of a protein, induced by vitamin K absence (P1VKA), is sometimes used to assess vitamin K status.

Water-Soluble Vitamins

Thiamine (vitamin B1)

Thiamine in the form of thiamine pyrophosphate plays an important role in the metabolism of carbohydrates, alcohol and branched chain amino acids. Thiamine pyrophosphate is a cofactor for the enzymes pyruvate dehydrogenase, α-ketoglutarate dehydrogenase and branched chain ketoacid dehydrogenase. These enzymes are involved in the decarboxylation of pyruvate, α-ketoglutarte and branched chain amino acids to form acetyl CoA, succinyl CoA, and branched chain amino acid derivatives respectively. It is also involved in a pentose phosphate pathway reaction catalysed by transketolase. Wheat germ, which is found in wholemeal bread, and wheat germ bran, fortified cereals, oatmeal, peas and marmite, are good sources. The requirements of thiamine are proportional to the intake of non-fat energy. The amount of thiamine in the body (30 mg) is about 30 times the daily requirement therefore, deficiency will develop in a month on a thiamine-free diet.

Thiamine deficiency may develop in chronic alcoholics due to poor nutrition, reduced absorption caused by alcohol and increased requirement. Patients on regular haemodialysis may have increased losses leading to deficiency.

Thiamine deficiency can lead to beriberi and Wernicke-Korsakoff syndrome. Beriberi used to be common in countries such as Japan, Malaysia and Indonesia where large amounts of milled (polished) rice are eaten. Consumption of raw fish, which contain thiaminase, can also cause its deficiency. Beriberi is now occasionally seen in alcoholics. Features of beriberi include high output cardiac failure, peripheral vasodilatation and oedema.

Wernicke-Korsakoff syndrome is seen in alcoholics, those who are starving (hunger strikers) and in those with prolonged vomiting, such as in hyperemesis gravidarum. Features include encephalopathy (Wernicke's syndrome with ataxia, nystagmus, ophthalmoplegia), peripheral neuropathy and psychosis (Korsakoff syndrome).

Thiamine deficiency can be diagnosed by measuring the activity of red cell transketolase before and after the addition of thiamine pyrophosphate *in vitro*. Activation of the enzyme by thiamine pyrophosphate is high in thiamine deficiency, even when the basal transketolase activity is normal.

If thiamine deficiency is suspected, treatment should be started immediately as delays in treatment may cause permanent memory loss. Thiamine should be given prophylactically to those on regular haemodialysis, those with persistent vomiting or prolonged gastric aspiration and those who are on prolonged fast. Toxicity of thiamine is very low.

Riboflavin (Vitamin B$_2$)

Riboflavin is an essential component of flavine mononucleotide (FMN) and flavine adenine dinucleotide (FAD), which have vital roles in oxidative metabolism. Antioxidant enzymes glutathione reductase, glutathione peroxidise and xanthine oxidase are also FAD-dependent enzymes. Sources of riboflavin include liver, kidney (richest sources), milk, yoghurt, cheese and marmite. Deficiency is rare and may be seen in alcoholics. Pregnant women, thyrotoxic patients and those taking chlorpromazine, imipramine and amitripty-line may have increased requirements and are at risk of developing deficiency. Clinical features of riboflavin deficiency include angular stomatitis, glossitis, seborrhoeic dermatitis and anaemia.

Riboflavin status is assessed by measuring the activation of red cell glutathione reductase by FAD. An increase in activity of 30% or more is indicative of deficiency.

Nicotinic acid (Niacin)

Niacin is an essential part of the coenzymes, nicotinamide adenine dinucleotide (NAD) and nicotinamide adenine dinucleotide phosphate (NADP) which are required for many oxidation-reduction reactions. Sources of niacin include liver, kidneys, meat, poultry, fish and marmite. Nicotinamide is also produced endogenously from the amino acid tryptophan via kynurenine, but this is insufficient to meet all the requirements.

Deficiency leads to pellagra, which is common in parts of Africa where maize is the staple diet. In maize, niacin is in a bound form which is unavailable and tryptophan becomes a limiting amino acid. Niacin deficiency may develop in chronic renal failure patients on low protein diet or on dialysis. It may also develop in disorders of tryptophan metabolism such as Hartnup disease and carcinoid syndrome. In Hartnup disease, there is decreased absorption of tryptophan along with other amino acids. In carcinoid syndrome, tryptophan is metabolised to 5-hydroxytryptamine. Features of pellagra include weight loss, anaemia, photosensitive dermatitis, dementia and diarrhoea. Diagnosis is established by measuring urinary excretion of niacin or its metabolite (N-methylnicotinamide).

High doses of niacin can cause flushing, gastric irritation, hyperuricaemia, impaired glucose tolerance, abnormal liver function tests and occasionally cholestatic jaundice.

Vitamin B_6

Vitamin B_6 includes pyridoxal, pyridoxamine, their 5′-phosphates and pyridoxine. Pyridoxal phosphate is coenzyme for over 100 reactions involving amino acids, especially transamination and decarboxylation. It is also involved in the synthesis of delta aminolevulinic acid (see Chapter 23). Vitamin B_6 is involved in the metabolism of homocysteine. Low levels of B_6 can lead to hyperhomocysteinaemia which is thought to be a risk factor for ischaemic heart disease. Foods containing B_6 include wheat germ, bran, potatoes, nuts, seeds, peanut, butter and fortified cereals. Dietary deficiency is rare. Several drugs such as isoniazide, hydralazine, penicillamine and possibly oestrogens may interfere with vitamin B_6. Polyneuropathy of isoniazide treatment can be prevented by high doses of B_6. Deficiency can be diagnosed by the activation of red cell aspartate aminotransferase by pyridoxal phosphate or by measurement of plasma concentration of pyridoxal phosphate. The measurement of urinary xanthuric acid is an indirect measure of pyridoxal status as it is involved in the metabolism of tryptophan to nicotinic acid.

High doses of B_6, which is used in the treatment of conditions such as homocystinuria and hyperoxaluria, can cause a neuropathy.

Biotin

Biotin is the prosthetic group for carboxylic enzymes such as pyruvate carboxylase and acetyl CoA carboxylase. It is found in egg whites where it is bound to a protein avidin. Cooking releases the bound biotin. If raw eggs are ingested, deficiency may result. Biotin is also synthesised by intestinal bacteria. Deficiency may occur during total parenteral nutrition. The clinical features of deficiency include dermatitis, anorexia, nausea and depression.

Pnatothenic acid

Pnatothenic acid, also known as vitamin B_5, is a constituent of coenzyme A which is a vital enzyme for fat and carbohydrate metabolism. This is widely distributed in food stuffs and deficiency is rare.

Vitamin B_{12}

The term vitamin B_{12} refers to a group of active substances classified as cobalamins. These compounds are composed of tetrapyrrole rings surrounding a central cobalt atom and nucleotide side chains attached to the cobalt. The different cobalamins differ in the nature of additional side groups bound to cobalt. These include methyl (methylcobalamin), 5-deoxyadenosine (deoxyadenosylcobalamin or coenzyme B_{12}), hydroxyl (hydroxycobalamin or vitamin B_{12}), water (aquocobalamin, vitamin B_{12b}) and cyanide (cyanocobalamin). The predominant physiological form of cobalamin in serum is methyl-cobalamin, whereas in cytosol it is 5-deoxyadenosylcobalamin. Cyanocobalamin is the most commonly used form for supplements and it is converted to 5-deoxyadenosyl methylcobalamin in the body.

Vitamin B_{12} is the coenzyme for two important functions: synthesis of methionine from homocysteine and conversion of methylmalonic acid to succinic acid (Figure 18.2). Methylcobalamin is required for the enzyme methionine synthase, a folate-dependent enzyme. Methionine synthase converts homocysteine to methionine, which in turn is required for the synthesis of s-adenosylmethionine, a methyl group donor for many reactions including methylation of DNA and RNA (Figure 18.2).

5-Deoxyadenosylcobalamin is required for L-methyl-malonyl-CoA mutase, an enzyme which converts L-methymalonyl-CoA to succinyl-CoA, a reaction important in the production of energy from fat. Succinyl-CoA is also important for the synthesis of haem.

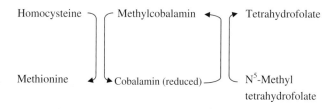

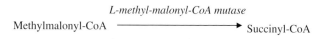

Figure 18.2 Reactions involving vitamin B_{12}.

1. *Absorption of B₁₂*

Small amounts of crystalline B_{12} can be absorbed through oral mucosa. B_{12} in food is bound to proteins. In the stomach B_{12} is released by proteolytic enzyme in the gastric juice at an acidic pH. Free B_{12} binds to R proteins, haptocorrins and cobalophilins. Gastric parietal cells secrete an intrinsic factor, a glycoprotein that is essential for the absorption of B_{12}. In the duodenum, B_{12} is freed from R proteins by protease and B_{12} binds to the intrinsic factor (IF) to form a complex (B_{12}–IF complex) that is resistant to bacterial degradation. This complex binds to receptors in the terminal ileum called cubilin and the complex is absorbed. B_{12} is liberated within the intestinal cell, binds to the transport protein and enters the circulation. The absorbed B_{12} reaches the liver where it is stored and released into the systemic circulation to meet the physiological demands. In a normal individual, approximately 2 mg of B_{12} is stored in the liver. Vitamin B_{12} is transported in the plasma, bound to a transcobalamin II, which is a β globulin, synthesised by the liver. The transcobalamin bound B_{12} complex is the biologically active form and is also called holo-transcobalamine. This complex binds to receptors in cell membrane and enters the cell by pinocytosis where lysosomal proteinolysis degrades the transcobalamin, releasing B_{12}.

The total amount of B_{12} in the body is about 2–5 mg in adults and about 0.1% of this is lost daily by secretions into the gut. Therefore, the requirement is low, 2–3 μg per day. B_{12} is mainly present in animal foods, especially liver. It is also present in fish and milk products and in some fortified breakfast cereals. Vegetable food has no B_{12} and vegans who do not eat any animal products are at risk of developing B_{12} deficiency. As the stores are large, B_{12} deficiency does not develop for 5 years from the time the dietary supply is removed.

2. *Vitamin B₁₂ deficiency*

As B_{12} is required for DNA synthesis, deficiency of B_{12} affects almost all organs especially where the cell turnover is rapid, i.e.

bone marrow and gastrointestinal tract. It also causes a serious neurological disorder called subacute combined degeneration of the spinal cord. This neurological effect may be due to the accumulation of abnormal lipids, as well as the requirement of B_{12} for myelin synthesis. B_{12} deficiency causes hypersegmentation of neutrophils, macrocytic anaemia, leucopoenia, thrombocytopenia and megablastic changes in bone marrow. Causes of B_{12} deficiency are listed in Table 18.2. B_{12} deficiency is common in the elderly and the prevalence has been reported to be as high as 21%. Pernicious anaemia, where there is destruction of gastric parietal cells by autoantibodies against gastric hydrogen–potassium–ATPase leading to a lack of intrinsic factor, is seen commonly among Caucasian women over the age of 60. In patients who had partial gastrectomy, B_{12} deficiency will develop due to removal of parietal cells.

Abnormal bacterial colonisation may occur in diverticulosis and blind loop syndrome. The resulting bacterial overload leads to increased utilisation of B_{12} leading to deficiency. Rarely infestation of the small intestine by the fish tapeworm *Diphyllobothrium latum* causes B_{12} deficiency because of utilisation by this parasite.

In malabsorption syndromes where the absorptive capacity of the terminal ileum is reduced (surgical resection or Crohn's disease), B_{12} deficiency may develop.

Table 18.2   Causes of vitamin B_{12} deficiency

Pernicious anaemia — autoimmune disease
Inadequate intake, e.g. vegans
Post-gastrectomy
Abnormal intestinal flora
Small intestinal disease

- Surgical resection
- Crohn's disease

Tapeworm infestation
Drugs

- Nitrous oxide
- Phenytoin
- Dihydrofolate reductase inhibitors

As B_{12} is predominantly of animal origin, vegetarians and vegans can develop B_{12} deficiency. Drugs, which interfere with B_{12} absorption or metabolism, include nitrous oxide, phenytoin, and dihydrofolate reductase inhibitors. Patients with HIV may develop B_{12} deficiency due to malabsorption.

- Diagnosis of B_{12} deficiency
  Presence of macrocytic anaemia will raise the possibility of B_{12} deficiency. This can be confirmed by measurement of serum B_{12} concentration. However, there is evidence that the current lower limit of reference interval may be too low and some patients with apparently normal level may be deficient. An alternative approach is to measure the active form holotranscobalamin or serum methylmalonic acid, which is increased in B_{12} deficiency. Recent studies suggest that serum methylmalonic acid is affected by other factors. However, a normal methylmalonic acid value excludes B_{12} deficiency. This test is not widely used. Diagnosis of pernicious anaemia depends on the demonstration of autoantibodies to parietal cells. B_{12} absorption test (Schilling test), which is used to determine B_{12} malabsorption caused by intrinsic factor deficiency, is no longer used in most countries.

- Treatment
  Treatment of the disorder is by injection of B_{12} in pernicious anaemia and by oral hydroxycobalamin supplementation in others. B_{12} has very little toxic effects in high doses.

Folate

Folate is a general term for a family of compounds related to pteroic acid. Pteroylglutamic acid (folic acid) is the pharmaceutical form of the compound. The biological form of the compound is tetrahydrofolate. Folate coenzymes are essential for the transfer of single carbon units. Some of the important reactions involving folate coenzymes are the conversion of serine to glycine, catabolism of histidine, synthesis of thymidylate, methionine and purine. Some of these reactions are important in DNA and RNA synthesis. Folate is present in many vegetables. However, prolonged heating and processing of food may cause loss of folate. Folate in food is present in conjugated form (polyglutamate) and these are converted to monoglutamate before absorption as methyltetrahydrofolate. The daily requirements of folate are approximately 50 μg and deficiency can quickly develop, as the stores are not large.

Folate deficiency may develop due to poor diet (e.g. elderly), in those with malabsorption particularly in conditions affecting the upper small intestine (e.g. coeliac disease and tropical sprue), due to increased requirements (e.g. pregnancy), in situations where there is increased cell proliferation (e.g. haemopoiesis and lymphoproliferative disorders) and due to drugs affecting folate metabolism. Some drugs (e.g. anticonvulsants) interfere with folate absorption, some are folate antagonists (e.g. methotrexate) and others inhibit the reduction of folate to the active form, tetrahydrofolate (e.g. cotrimoxazole). Deficiency may develop in alcoholics due to poor intake and due to the antagonistic action of alcohol on the reduction of folate.

The major effect of folate deficiency is megablastic anaemia, but subacute combined degeneration of spinal cord does not occur in folate deficiency. Folate deficiency during pregnancy is associated with increased incidence of neural tube defects. Folate deficiency can lead to increased homocysteine, which is associated with an increased risk of vascular disease.

Folate deficiency is diagnosed by a measurement of serum or red cell folate. Serum folate is affected by short-term fluctuations in dietary intake, whereas red cell folate reflects tissue stores more accurately.

Vitamin C/Ascorbic acid

Ascorbic acid is the major antioxidant in the aqueous phase of the body. It is a cofactor for protocollagen hydroxylase, which hydroxylates proline and lysine residues in newly formed collagen. Therefore, vitamin C is important for the normal maintenance of connective tissue and for normal wound healing. Vitamin C increases the absorption of non-haem iron when taken at the same time. It is important for the synthesis of noradrenaline and carnitine. Vitamin C may also be important in the synthesis of corticosteroids as very high concentrations are found in the adrenal gland.

Vitamin C can be synthesised *in vivo* by most animals except guinea pigs and primates who depend on fruits and vegetables as the main sources of vitamin C.

Vitamin C is easily destroyed by cooking especially in alkaline conditions. Fresh food, juice and salads are important sources.

Vitamin C deficiency can arise in children aged 6–12 months who receive processed milk without ascorbic acid supplements. Dietary vitamin C deficiency can occur in the elderly especially those who live alone. Vitamin C deficiency in hospital patients is due to increased demand and/or decreased absorption. Trauma and surgery increase the requirement of vitamin C as more collagen is synthesised. Drugs such as steroids, aspirin, indomethacin, phenylbutazone and tetracycline antagonise vitamin C.

Clinical manifestations of vitamin C deficiency (scurvy) include skin papules, purpura, haemorrhages, poor wound healing, gum disease and anaemia. Osteoporosis may develop as a result of defective collagen synthesis. Because of its antioxidant capacity, large doses of vitamin C have been recommended to reduce/prevent common colds, to reduce the incidence of cardiovascular disease, to delay regenerative diseases such as cataract and to prevent stomach cancer. However, there is no evidence that vitamin C in large doses is effective. However, high intake of ascorbate is associated with increased excretion of oxalate, a metabolite of ascorbate. This theoretically increases the risk of renal stones.

Plasma vitamin C concentrations are poor indicators of vitamin C deficiency and measurement of leucocyte ascorbic acid is a better index.

Trace Elements

Trace elements are those that are present in $< 0.01\%$ of dry weight of the body. Sometimes the term 'ultra trace elements' is used to describe those that occur in μg/kg amount ($< 0.001\%$). Some of the

Table 18.3   Biochemical functions of trace elements

Enzyme action
    as cofactors
    as prosthetic groups, e.g. zinc
Transport of oxygen — iron
Activity of vitamins — cobalt and vitamin B_{12}
Organisation and structure of macromolecules, e.g. silicon and connective tissue
Hormonal activity — iodine and thyroid hormone

trace elements are essential for normal health and function whereas others, like lead, are non-essential. The biochemical functions of trace elements are listed in Table 18.3.

Zinc

Zinc, the second most abundant trace element after iron, forms an integral part of nearly 300 enzymes. High concentration of zinc is found in tissues such as the prostate, semen, liver, kidney, parts of brain and muscle. Zinc plays a major role in protein synthesis, gene expression, stabilising the structure of proteins and nucleic acids, pre-serving the integrity of subcellular organelles, transport processes, immune responses and in wound healing.

Zinc is present in foods such as meat, fish and dairy products and 30–40% of ingested zinc is absorbed. Absorption takes place mainly in the duodenum and proximal jejunum by an active energy-dependent process. Other dietary substances like cellulose and dietary fibres can reduce zinc absorption. Zinc is transported in the blood bound to albumin (60–70%) and α_2 microglobulin (30–40%) with a very small proportion being associated with transferrin and free amino acids. Zinc is mainly excreted in faeces and urine.

Zinc deficiency due to poor diet is not uncommon in certain parts of the world, e.g. Turkey, Portugal, Morocco and Yugoslavia. The features of severe zinc deficiency include retardation of growth and skeletal maturation, testicular atrophy, delayed wound healing, impaired immune response, alopecia, weight loss and neuro-psychiatric disorders.

Zinc deficiency can occur in patients with malabsorption syndromes and in patients on parenteral nutrition, if adequate amounts of trace metals are not included. Requirement for zinc is increased during pregnancy; and pregnant women are at increased risk of zinc deficiency, which can lead to foetal abnormalities and complications during pregnancy. An inherited metabolic disorder of zinc metabolism is acrodermatitis enteropathica. This is an autosomal recessive condition in which there is failure to absorb zinc when infants are weaned from breast milk. Breast milk contains an unidentified ligand, which promotes the gastrointestinal absorption of zinc. After weaning, the child develops symptoms and signs of zinc deficiency. Untreated babies may die.

Zinc deficiency has also been associated with sickle cell anaemia.

Laboratory assessment of zinc status

A large number of tests are available for assessing zinc but none of them seems to be satisfactory. Plasma or serum zinc, which is the commonly done test, can be affected by alteration in albumin and other binding proteins. For example, in acute illnesses, the plasma zinc concentration will decrease due to decreased albumin concentration. Other tests used to assess zinc status include analysis of zinc in urine, hair and leucocytes and analysis of zinc-containing enzymes. A decreased urinary excretion of zinc is usually associated with zinc deficiency. However, in liver disease, post-surgical patients and patients on parenteral nutrition, zinc excretion may be increased even in the presence of zinc deficiency. Leucocyte zinc is a good index; however, it is time-consuming and difficult in practice. Zinc-containing metallo-enzymes, such as alkaline phosphatase and carbonic anhydrase, are useful indicators of zinc. Recent studies indicate that metallothionein concentration in plasma or red cells may be a useful index of assessing zinc status.

Copper

Copper is an important trace element associated with a number of metalloproteins and the major function of copper–containing metalloproteins is in oxidation–reduction reactions. Copper is an integral component of many metal enzymes such as caeruloplasmin and cytochrome C oxidase. Copper also plays an important role by oxidising ferrous to ferric ion in iron metabolism.

Copper is found in a variety of foods, and is most plentiful in meats, shellfish, nuts and seeds. Dietary copper is absorbed in the upper intestine being maximal in the duodenum. Copper is transported as copper–albumin or copper–histidine complexes to the liver where it is stored mainly as metallothionein-like copper proteins. From the liver, copper is released mainly as caeruloplasmin, which accounts for nearly 90% of plasma copper. The liver is the main storage site for copper; relatively high amounts of copper are also found in heart, brain and kidneys. Copper is excreted primarily in the bile.

Copper deficiency

Copper deficiency is rare and can occur in premature infants, in general malnutrition, malabsorption syndromes and in those on total parenteral nutrition. Features of severe deficiency include neutropenia, fractures, and microcytic hypochromic anaemia. Menkes disease, an inherited disorder of copper metabolism, is an X-linked disorder occurring with a frequency of 1:100,000 live births. It is a disorder of intestinal absorption of copper and possibly also an abnormality of intracellular copper transport. Clinical features develop early in life and include kinky, brittle hair, depigmentation of the skin and hair, hypothermia, fits, cerebral degeneration, vascular defects, and bone changes including osteoporosis. In affected infants, copper accumulates in the intestinal mucosa.

Wilson's disease

Wilson's disease is a genetic disorder of copper accumulation and presents between the ages of 6 and 40. This is a rare autosomal recessive disorder affecting 30 per million population. The metabolic basis is a defect in the protein ATP7B, a copper transport ATPase, leading to a reduced excretion of copper in the bile. Features of the disease are caused by the deposition of copper in the liver, brain, kidney and cornea. Copper excretion is increased in the urine while the synthesis of the copper-containing caeruloplasmin is decreased. Although free and albumin-bound copper are increased, total serum copper is low. Hepatic symptoms are more frequent in children and it is the most usual cause of chronic liver disease in this age group. Neurological presentations are usually seen between the ages of 20 and 40. Features include poor coordination, tremor, dysarthria, disorders of muscle tone, posture, balance and the characteristic Kayser-Fleisher rings in the cornea. Deposition of copper in the kidney may lead to renal tubular dysfunction such as Fanconi syndrome.

Measurement of serum copper, caeruloplasmin and urinary copper excretion are important in the diagnosis of Wilson's disease. Calculation of non-caeruloplasmin-bound copper has been thought to be of value. Caeruloplasmin-bound copper is calculated by multiplying caeruloplasmin (in g/L) by 50.4. Subtracting this from total copper gives an indication of free copper (normal between 3–5 μmol/L). In difficult cases measurements of liver copper may be indicated followed by a penicillamine challenge test. Molecular biological techniques are now available for family studies. Wilson's disease can be treated with penicillamine, which chelates and removes copper. Compliance and efficiency of treatment have to be monitored by measuring urine copper.

Indian childhood cirrhosis is a condition characterised by jaundice, enlarged liver and spleen and signs of liver failure. The aetiology of this condition is uncertain but it is associated with increased ingestion of copper from contaminated food cooked in copper-containing vessels. Histological features of the liver resemble that of Wilson's disease. This disease is seen in children from a few weeks to 10 years. Absence of Keyser-Fleisher rings and a normal concentration of copper and caeruloplasmin distinguish this from Wilson's disease.

Serum concentration of copper is increased by oral contraceptives as well as in rheumatoid arthritis.

Assessment of copper status involves the measurement of serum copper and caeruloplasmin and/or urine copper. Occasionally, penicillamine challenge test is indicated.

Case 18.1

A 7-year-old boy was referred to the paediatric neurology clinic with slurring of speech, tremors and deterioration in school performance. On examination, his liver was enlarged and the registrar noticed a ring around the cornea. Wilson's disease was suspected and the following investigations were done:

Serum		Reference Range
Bilirubin (μmol/L)	13	< 20
ALT (IU/L)	88	< 40
Albumin (g/L)	42	35–50
ALP (IU/L)	155	30–120
GGT (U/L)	225	< 55
Caeruloplasmin (g/L)	0.08	0.2–0.6
Copper (μmol/L)	10	11 to 20

Wilson's disease was confirmed on the basis of low serum copper and low caeruloplasmin, together with elevated urine copper. He was treated with D-penicillamine.

Selenium

Selenium is an essential element for the function of many selenoproteins, including glutathione peroxidase and iodothyronine dehydrogenases. Selenium in tissues is present in two forms, selenocysteine and selenomethionine. Selenocysteine, the biologically active form of selenium, is used in the synthesis of ribosome-mediated protein synthesis, and it is tightly regulated.

Selenium is present in the diet in the inorganic form and as selenoamino acids such as selenomethionine (plants) and selenocysteine (animal source). Selenium is absorbed well from the gastrointestinal tract and homeostasis is maintained by urinary excretion. Selenium content of food varies depending on the selenium content of the soil. Thus, dietary intake of selenium varies widely in different countries.

Two diseases have been described in relation to selenium. Keshan disease is an endemic cardiomyopathy affecting children and women in certain parts of China. The most common symptoms are dizziness, malaise, loss of appetite, cardiogenic shock, cardiac enlargement and congestive heart failure. It is due to selenium deficiency in the diet. Kashin-Peck disease is endemic osteoarthritis that occurs during adolescence and pre-adolescent years in some parts of the China and is associated with low selenium. However, other factors such as mineral imbalance and mycotoxins may be involved.

Epidemiological studies have shown an association between low selenium intake and increased incidences of cancer and cardio-myopathies. Selenium deficiency can occur in patients on total long term parenteral nutrition. Chronic selenium poisoning can occur in areas where there are high concentrations of selenium in the soil. This is associated with loss of hair and nails, skin lesions, tooth decay and CNS abnormalities. Acute selenium toxicity can occur as an occupational hazard. It can cause circulatory collapse and systemic failure.

Selenium status can be assessed by measurement of selenium in plasma, serum or whole blood. Measurements of glutathione peroxidase activity and/or other cellular proteins such as selenoprotein P have also been used in assessing selenium status.

Manganese

Manganese is an essential constituent of some metalloenzymes and is an enzyme activator. Manganese deficiency is associated with clotting defects, hypocholesterolaemia and dermatitis. Low tissue manganese is seen in children with maple syrup urine disease and phenylketonuria and in those on long-term parenteral nutrition. Manganese deficiency can be detected by measurement of serum or whole blood manganese.

Lead

Lead is a heavy metal found in the environment and can lead to acute or chronic toxicity. High concentrations of lead are found in paints manufactured before 1970. Significant amounts of lead can be found in ceramic products used in the home such as dishes and bowls and lead can be leached out from these by weak acids such as vinegar or fruit juices. Lead is also present in leaded fuels and this can contribute to increased lead in the soil. When water is transported through lead-containing pipes, lead can leach out. Lead is absorbed and distributed throughout the body and significant amounts can accumulate in the bone and red cells. Lead is also found in some cosmetic products used by South Asians, and in traditional Chinese and Ayurvedic herbal medicines.

Lead inhibits ALA dehydrase, causing increased excretion of ALA (Figure 23.2). Lead also avidly binds with sulphide groups of cysteine in proteins and thus protein in all tissues exposed to lead will have lead bound to them. Keratin in hair, which contains high fractions of cysteine, binds to lead and hair analysis has been used as a marker of lead exposure. Lead toxicity progresses from mild symptoms to severe encephalopathy and death. Children are particularly prone to the effects of lead.

Exposure to lead can occur through ingestion, inhalation or via the skin. Tetraethyl lead, used as a fuel additive, can be absorbed through skin. Occupational exposure is the main cause of lead poisoning in adults: workers in battery factory, smelting, ship-breakers, etc.

Whole blood lead measurement is the accepted method for assessing lead exposure. Whole blood lead concentration of < 100 $\mu g/L$ are considered normal in children. WHO has defined that whole blood lead concentration > 300 $\mu g/L$ is indicative of significant exposure and concentrations > 600 $\mu g/L$ require chelating treatment. Red cell protoporphyrin concentrations can also be used as an index of exposure but it is not a specific indicator.

Treatment of lead poisoning is with chelating agents, such as dimercaprol (BAL) or dimercaptosuccinic acid (succimer, DMSA) given intravenously.

Further Reading

1. Mak CM, Lam CW. Diagnosis of Wilson's disease: A comprehensive review. *Crit Rev Clin Lab Sci* 2008; 45:263–290.

2. Taylor A. Detection and monitoring of disorders of essential trace elements. *Ann Clin Biochem* 1966; 33:486–510.
3. Truswell AS. *ABC of Nutrition* 2003. BMJ Publication, 4th edition.

Summary/Key Points

1. Vitamins are essential compounds that are required in small quantities.
2. Vitamin A is essential for vision and epithelial cell surfaces. Deficiency can lead to night blindness and keratomalacia. Vitamin A deficiency is common worldwide and can lead to blindness.
3. Vitamin K is involved in γ carboxylation of proteins including prothrombin and osteocalcin. Deficiency can lead to bleeding disorders.
4. Thiamine or vitamin B_1 is essential for the metabolism of carbohydrates and deficiency can lead to beriberi.
5. Vitamin B_{12} and folate are essential for nucleic acid synthesis. Deficiency can lead to megaloblastic anaemia. B_{12} deficiency is common in vegetarians and in the elderly (due to deficiency of intrinsic factor, pernicious anaemia).
6. Trace elements are those present in small quantities in the body and many of them are essential for normal body functions.
7. Zinc is required for the function of several hundred enzymes. Deficiency can give rise to retardation of growth. Acrodermatitis enteropathica is an inherited disorder of zinc metabolism. Zinc status can be assessed by plasma or white cell zinc.
8. Copper is an essential element for many enzymes. An inherited disorder of copper is Menke's disease where there is a defect in the intestinal absorption of copper. Wilson's disease is an inherited disorder, leading to an accumulation of copper in the brain, liver and the cornea.
9. There are many selenium-containing enzymes in the body and deficiency can lead to either Keshan disease or Kashin-Peck disease.

chapter 19

The Hypothalamus and the Pituitary

Introduction

The hypothalamus forms part of the floor of the third ventricle and receives neural input from many parts of the brain. The hypothalamus is a major centre for the regulation of many functions such as sleep, temperature, thirst, appetite, behaviour, etc. It secretes many hormones, and through them it regulates the synthesis and secretion of peptide hormones from the pituitary. The pituitary gland, which consists of two parts, anterior and posterior, lies within the pituitary fossa beneath the hypothalamus (Figure 19.1). The pituitary gland is connected by means of the pituitary stalk, which carries both neural and blood vessel connections between the hypothalamus and pituitary.

The two parts of the pituitary gland are of different embryological origins, have different functions, and different relationships with the hypothalamus. The anterior pituitary or adenohypophysis is ectodermal in origin, and contains different cell types, which secrete six hormones. The anterior pituitary hormones are in turn regulated by factors secreted by the hypothalamus and carried to the anterior pituitary by the hypophyseal/portal system. The posterior pituitary gland is of neural origin, and the two hormones secreted by the posterior pituitary are synthesised in the hypothalamus.

The hormones secreted by the anterior pituitary gland, their regulation by the hypothalamus, and their effects/target organs are listed in Table 19.1.

475

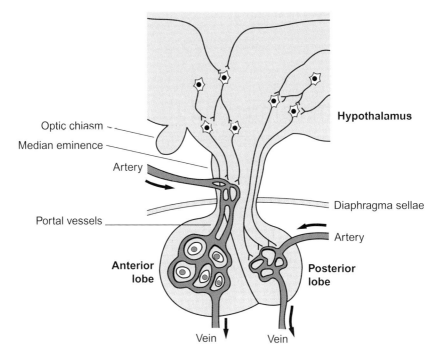

Figure 19.1   Relationship between the pituitary and the hypothalamus.

Table 19.1   Anterior pituitary hormones, their target organs and their regulation

Hypothalamic factors	Effect	Anterior pituitary hormone	Target gland secretion or effect
CRH	+	ACTH	Cortisol
GnRH	+	Gonadotrophins (FSH and LH)	Testosterone/ oestradiol/ progesterone
GHRH	+	GH	IGF-I
TRH	+	TSH	T4 and T3
Somatostatin (SRIF)	–	GH	IGF-I
Dopamine	–	Prolactin	Milk secretion

* '+' indicates stimulation and '–' indicates inhibition

Anterior Pituitary Hormones

Adrenocorticotrophin Hormone (ACTH) (Corticotrophin)

ACTH, a 39 amino acid peptide, secreted by the corticotrophs of the anterior pituitary, is controlled by corticotrophin-releasing hormone (CRH) from the hypothalamus. CRH, a 41 amino acid peptide, stimulates ACTH and other biologically active peptides. ACTH is derived from a larger precursor molecule, pro-opiomelanocortin (POMC). POMC undergoes a series of post-translational modification before it is cleaved by protelytic enzyme to produce several physiologically active peptides including ACTH, β- and γ-lipotrophins, melanocyte-stimulating hormone (MSH) and β endorphin. The secretion of ACTH is pulsatile with a diurnal rhythm — highest during the morning. ACTH in turn stimulates adrenal steroids, mainly glucocorticoids with minor effects on mineralocorticoids and adrenal androgens. The secretion of CRH and ACTH are in turn inhibited by cortisol; a negative feedback loop. Stress, including hypoglycaemia, is a powerful stimulus for the secretion of CRH and ACTH. The placenta also produces CRH, which is thought to play a role in parturition.

Low ACTH concentration are seen in hypopituitarism and Cushing's syndrome due to adrenal adenoma and exogenous glucocorticoid administration. High ACTH levels are seen in adrenal failure, Cushing's disease and in ectopic production by some tumours. Very high ACTH concentrations are seen in adrenal failure or after bilateral adrenalectomy (Nelson's syndrome). These patients have increased pigmentation due to the increased MSH that is co-released with ACTH.

Thyroid-Stimulating Hormone (TSH)

TSH, a glycoprotein of molecular weight 28,000, stimulates the secretion and synthesis of thyroid hormones. TSH is stimulated by thyrotrophin releasing hormone (TRH), a tripeptide from the hypothalamus. TSH consists of two subunits, α and β. The α-subunit is similar to that in other glycoprotein hormones, gonadotrophins and human chorionic gonadotrophin (hCG). Secretion of TSH and TRH are inhibited by thyroid hormones via a negative feedback loop. TSH acts on the thyroid gland via the TSH receptor.

Growth Hormone

Growth hormone (GH) is a 191 amino acid peptide (molecular weight 21,500) secreted by the somatotrophs of the anterior pituitary. GH secretion is stimulated by GH-releasing hormone (GHRH), a 44 amino acid peptide, released from the hypothalamus and by ghrelin, a hormone released by the fundus of the stomach. GH secretion is inhibited by somatostatin, a 16 amino acid peptide, secreted from the hypothalamus. Somatostatin, in addition, inhibits the TSH response to TRH. Drugs such as clonidine and L-dopa increase GH secretion by stimulating GHRH release. Human GH bears marked structural resemblance to prolactin and to the placental hormone — human chorionic somatomammotropin (hCS) also known as human placental lactogen (hPL). GH is secreted in bursts every 3–4 hours, with the secretory activity being highest at night. Other factors which stimulate GH secretion are hypoglycaemia, exercise, fasting, increase in some amino acids such as arginine, ornithine and lysine in blood (after a meal), and stress. GH secretion is inhibited by glucose, free fatty acids, glucocorticoids, insulin-like growth factor-1 (IGF-1) and GH itself. Secretory bursts and therefore integrated 24-hour secretion of GH is highest in children and young adults. The half-life of GH in circulation is about 20 minutes. In the circulation, about 60% of GH is bound to a binding protein (29 kDa), which is identical to the extracellular domain of the GH receptor.

GH has many actions, some by direct effect and others via the production of IGF-I. The direct actions of GH include effects on metabolism of carbohydrates and lipids. These are synergistic to cortisol and opposite to that of insulin and IGF-I, namely increased gluconeogenesis, decreased uptake of glucose by peripheral tissues and increased lipolysis. These actions are brought about by the interaction of GH with its receptor on the cell surface (GHR), which belongs to the haematopoietic cytokine family of receptors. GH is an anabolic hormone, stimulating the uptake of amino acids, increasing protein synthesis and decreasing the oxidation of proteins.

GH, as the name implies, is important for growth. GH appears to stimulate growth by a direct, as well as an indirect effect. GH acts directly on prochondrocytes to stimulate differentiation to chondrocytes, which respond to GH by forming IGF-1 and IGF-1 receptors. GH also acts indirectly via stimulation of IGF-1 synthesis and its binding proteins by the liver.

Insulin-like growth factor 1 (IGF-1) is a 70 amino acid peptide produced by the liver. It has considerable homology with proinsulin. Secretion of IGF-1 is regulated by GH, insulin, as well as by nutritional status. A reduction in energy or protein intake will reduce IGF-1. IGF-1 in extracellular fluid is bound to proteins, mainly IGF binding protein-3 (IGFBP-3), whose synthesis by the liver is also regulated by GH. IGF-1 binds to a receptor (IGFR), which is structurally similar to the insulin receptor. IGF-1 stimulates glucose uptake and oxidation in adipose tissue and muscle, increases uptake of amino acids in the muscle and increases the synthesis of collagen and proteoglycans. Serum concentration of IGF-1 decreases with age, peak values are seen at puberty. In addition to hepatic synthesis, IGF-1 is also synthesised by many tissues where they act as an autocrine or paracrine factor. IGF-1 stimulates DNA synthesis and is important in the differentiation and growth of cartilage.

Prolactin

Prolactin, a 199 amino acid polypeptide hormone, is secreted by the lactotrophs of the pituitary and has some homology with GH. The secretion of PRL is mainly regulated by dopamine from the hypothalamus. Dopamine inhibits PRL secretion, a tonic inhibition such that when the secretion of dopamine is abolished, PRL secretion increases. PRL secretion can be stimulated by TRH, other hypothalamic factors such as vasoactive intestinal peptide (VIP) and oxytocin, steroids (oestrogen) and growth factors such as epidermal growth factor. Stress is another stimulus for PRL secretion.

Release of PRL is episodic (every 2–3 hours) and varies during the day, lowest at midday and the highest levels during REM sleep. An important physiological stimulus for the secretion of PRL is suckling. Effects of PRL include the development of the mammary gland, synthesis of milk and maintenance of milk secretion. PRL also has a role in leuteal function and in reproductive and parental behaviour. In addition to these, PRL has been suggested to have a role in immunity and homeostasis. PRL secretion increases steadily during pregnancy in response to rising oestrogen concentration. During late pregnancy, PRL concentrations are 10 times the non-pregnant value. Breast feeding is accompanied by episodic secretion of PRL reaching 5–8 times non-pregnant values. Sexual orgasm may also induce high PRL concentrations. In non-breast feeding mothers, PRL concentrations fall rapidly after childbirth, and reach non-pregnant values within 7–10 days. PRL concentrations decrease after 3 months, even if breastfeeding is continued beyond this point. It is now recognised that many tissues also secrete PRL, where it may function as a paracrine or autocrine factor.

PRL acts via PRL receptor, which is widely distributed in the body in almost all cell types. The significance of this wide distribution is not clear.

Gonadotrophins

Gonadotrophins, follicle-stimulating hormone (FSH) and luteinising hormone (LH) are glycoprotein hormones of molecular weight approximately 28 kDa. Like other glycoprotein hormones, they have α- and β-subunits; the α-subunit is identical to that in other glycoprotein hormones. The β-subunit of both FSH and LH are composed of 115 amino acids. In both males and females, FSH and LH are regulated by the hypothalamic hormone, gonadotrophin-releasing hormone (GnRH), which is a decapeptide. Secretion of GnRH is pulsatile and in the male, the frequency of pulses is constant whereas in the female, the frequency varies according to the stage of the menstrual cycle and there is a surge just before ovulation. Low-frequency GnRH pulses cause the release of FSH and high frequency pulses release LH. Kisspeptin, a peptide released in the hypothalamus,

is an important regulator of GnRH release. GnRH pulsality is important for reproductive function. This pulsality can be disturbed by
hypothalmic-pituitary disease (trauma, tumour). Hyperprolactinaemia
suppresses GnRH pulsality and hyperinsulinaemia (as in polycystic
ovarian syndrome) increases the pulsality. In Kallmann syndrome,
GnRH pulsality is absent. GnRH binds to its receptor at the
gonadotrophic cell surface and stimulates FSH and LH. In addition
to GnRH, sex steroids (oestrogens and androgens) and inhibin, a
peptide hormone secreted by the gonads affect gonadotrophin secretion probably by modulating GnRH. Oestrogens increase and
androgens decrease GnRH receptors. Feedback by oestrogens can be
positive or negative, depending on the stage of the menstrual cycle.

In men, LH acts on Leydig cells to increase the synthesis and
secretion of testosterone, while FSH together with testosterone, acts
on Sertoli cells to stimulate spermatogenesis and increase the synthesis of other proteins, including androgen-binding protein and inhibin.
Androgen-binding protein is secreted into the seminiferous tubular
lumen, where it binds to testosterone and helps to maintain a high
androgen concentration. Inhibin has a negative effect on the secretion
of gonadotrophins.

In women, during the first few days after the start of a menstrual
period (initial stages of the menstrual cycle), FSH is the predominant
hormone that stimulates oestradiol secretion and the development of
the ovarian follicle. As oestradiol rises, FSH secretion falls; this
combination is responsible for the selection and development of a dominant follicle. Oestrogen concentration continues to rise (Figure 20.1),
and at mid-cycle, the surge in oestrogen has a positive feedback effect
causing a surge of LH which, together with steroid hormones, causes
the rupture of the follicle, ovulation. Ovulation is accompanied by a
sharp fall in oestrogen followed by a fall in LH. The ruptured follicle
now becomes the corpus luteum, which secretes progesterone and
oestradiol. Progesterone and oestradiol, together with inhibin secreted
by the oocyte, prepare the uterine endometrium to accept the fertilised
ovum. The rise in steroid hormones inhibits FSH and LH. In the
absence of fertilisation, the secretion of progesterone and oestradiol by
the corpus luteum decreases the endometrium undergoes vascular

changes, and menstruation occurs. The fall in oestrogen secretion stimulates FSH secretion to start a new cycle (see Chapter 20).

High levels of gonadotrophins are seen in ovarian failure and at menopause. High LH and low FSH are seen in polycystic ovarian syndrome. Low FSH and LH are seen in hypogonotrophic hypogonadism, Kallmann syndrome, eating disorders and in pan-hypopituitarism. In the male, azoospermia causes high FSH and low testosterone causes a rise in LH.

Disorders of the Anterior Pituitary Gland

Hypopituitarism

Deficiency of one or more pituitary hormones is defined as hypopituitarism, and when all hormones are deficient it is referred to as panhypopituitarism. Isolated pituitary hormone deficiency is usually due to a hypothalamic disorder. Causes of hypopituitarism are listed in Table 19.2.

A chromophobe adenoma in the adult and craniopharyngioma in children are the most common tumours causing hypopituitarism. Hypopituitarism following surgical or radiation treatment of the pituitary is a frequent cause. Sheehan's syndrome or postpartum pituitary necrosis used to be a common cause of hypopituitarism; it is now rarely seen in developed countries due to improved obstetric care. In developing countries, it is still an important cause. During pregnancy, the pituitary gland undergoes hyperplasia. In Sheehan's syndrome, hypotension following postpartum haemorrhage causes spasm of the hypophyseal arteries and eventual necrosis of the anterior pituitary.

The clinical features of hypopituitarism (Table 19.3) depend on the degree of pituitary failure, the pattern of failure of the hormones and the cause of hypopituitarism, e.g. pressure from space-occupying lesion. Earliest hormones to be affected are LH and GH, and this is followed by FSH, ACTH and TSH. Early manifestations are disorders of reproduction (menstrual irregularities, secondary amenorrhoea, infertility and loss of libido, etc). Children may present with growth retardation

Table 19.2   Causes of hypopituitarism

Tumours
- Pituitary adenoma
- Craniopharyngioma
- Cerebral tumour
  — Primary and secondary

Vascular disease
- Sheehan's syndrome
- Vascular malformation

Trauma
Iatrogenic
- Surgery
- Radiation

Infiltration
- Sarcoidosis
- Haemochromatosis
- Lymphocytic hypophysitis

Infection
- Tuberculosis, meningitis, syphilis

Hypothalamic disease
Miscellaneous
- Genetic/embryological disorders

Table 19.3   Features of hypopituitarism

GH — growth retardation in children due to GH and TSH deficiency

Gonadotrophins
- Amenorrhoea or infertility
- Atrophy of secondary sexual characteristics — loss of axillary and pubic hair
- Loss of libido
- Impotence
- Delayed puberty in children

ACTH
- Features of glucocorticoid deficiency — fatigue, weight loss, hyponatraemia and hypoglycaemia

TSH — secondary hypothyroidism

due to GH failure. Prolactin concentration may be high in pituitary tumours due to the release of tonic inhibition from the hypothalamus.

Posterior pituitary dysfunction may be present especially in patients with large tumours. However, it may become apparent when ACTH deficiency is corrected with glucocorticoid administration as cortisol is essential for the excretion of free water.

1. *Treatment of hypopituitarism*

Treatment depends on management of the specific cause and the replacement of deficient hormones. ACTH deficiency is corrected by glucocorticoid treatment and TSH deficiency by thyroxine replacement. Gonadotropin deficiency requires testosterone replacement in the male and oestrogen in the female. GH deficiency in children requires GH replacement and in adults only if GH deficiency is severe enough to cause symptoms.

2. *Assessment of anterior pituitary function*

Initial assessment of anterior pituitary function consists of measurements of basal hormone concentrations, including FT4, TSH, LH, FSH, testosterone or oestradiol, nine am cortisol and prolactin. Hypopituitarism is suggested by a low FT4 accompanied by low or normal TSH, a low testosterone with normal or low LH, and in post-menopausal women absence of a rise in gonadotrophins. In pre-menopausal women a regular menstrual cycle excludes gonadotrophin deficiency.

Dynamic function tests are done to assess individual hormones or to assess the pituitary reserve. The pituitary–adrenal axis is assessed by short and long ACTH (Synacthen) tests (described in Chapter 21). Failure of cortisol to rise after a single dose (short Synacthen test) and a rise after a depot injection (long Synacthen test) indicate failure of pituitary ACTH secretion. When deficiencies of TSH and gonadotrophins are suspected, TRH and GnRH tests (combined or separately) may help to distinguish between hypothalamic and pituitary dysfunction. Failure of TSH to show any rise after TRH and gonadotrophins after GnRH usually indicates pituitary failure, while a rise indicates hypothalamic dysfunction. However, the TRH and GnRH tests are seldom used at present, as they do not give any additional information to that gained by basal hormones.

As GH secretion is episodic, basal GH concentrations are undetectable in many normal individuals. An undetectable basal GH concentration does not indicate GH deficiency, whereas a basal GH concentration > 20 mU/L excludes it. GH deficiency is assessed by the insulin hypoglycaemia test.

The insulin hypoglycaemia test is done as part of the investigation of anterior pituitary function to assess pituitary reserve of GH and ACTH. Hypoglycaemia induced by insulin is a powerful stimulus for the secretion of GH and ACTH. Insulin 0.15 u/kg body weight is given intravenously, and blood samples are taken before and at 30, 45, 60 and 90 minutes after injection for GH, cortisol, and PRL measurements. The dose of insulin needs to be high (0.30 u/kg body weight) in obese subjects who are insulin resistant and low in patients who are suspected of hypopituitarism and therefore sensitive to insulin. The test is potentially dangerous, and is contraindicated in patients with ischaemic heart disease and in patients with low basal cortisol concentration. During the test, the patient should be closely monitored, and intravenous glucose should be available if severe hypoglycaemia develops. Adequate hypoglycaemia, as judged by a blood glucose < 2.2 mmol/L and sweating is necessary to interpret the results. In the absence of adequate hypoglycaemia, a repeat injection of insulin may be given. An adequate cortisol response is a rise in cortisol by at least 200 nmol/L and a maximum of > 550 nmol/L. Rise in GH concentration to > 20 mU/L is considered an adequate response.

When insulin hypoglycaemia test is contraindicated, a glucagon stimulation test is sometimes done to assess pituitary function. After an overnight fast, basal blood samples are taken and 1 mg glucagon is given subcutaneously. Further samples are taken at 90, 120, 150, 180, 210 and 240 minutes. An adequate response is a rise in cortisol by at least 200 nmol/L to > 550 nmol/L and GH to > 40 mU/L.

Case 19.1

A 32-year-old man was referred by his GP for investigation of loss of libido and erectile dysfunction. On questioning at the clinic, he admitted that he was shaving less frequently and was feeling very tired.

Investigation showed:

Serum		Reference Range
Sodium (mmol/L)	130	135–145
Potassium (mmol/L)	4.5	3.5–5.0
Testosterone (nmol/L)	3.5	10–20
FSH (U/L)	< 0.5	1–8
LH (U/L)	70	1–7
Cortisol (nmol/L)	300	
TSH (mU/L)	1.5	0.2–5.6
FT4 (pmol/L)	8	12.0–25.0
Prolactin (mU/L)	< 100	< 450

An insulin hypoglycaemia test was done and the results were:

Time (min)	Glucose (mmol/L)	Cortisol (nmol/L)	GH (mU/L)
0	4.5	235	< 2
15	2.0	275	< 2
30	2.2	315	< 2
60	3.2	275	< 2
90	4.0	290	< 2
120	4.5	220	< 2

His basal hormone results show that he has secondary hypothyroidism (low FT4 with an inappropriately low TSH). Hypogonadotrophic hypogonadism and the failure of GH and cortisol to increase after stimulation test are characteristic of hypopituitarism.

He was treated with hydrocortisone, thyroxine and testosterone. His serum sodium returned to normal and he felt a lot more energetic and was

> *due to return to work. Imaging of the pituitary showed a tumour as the cause of the hypopituitarism. This was resected and he was given lifelong replacement therapy.*

Growth hormone deficiency

In children, growth hormone deficiency is an important but uncommon cause of short stature. The causes of short stature are described in Chapter 28. In adults, growth hormone deficiency can cause reduced lean body mass, poor bone density, tiredness, dyslipidaemia, increased cardiovascular disease and psychological symptoms like poor memory, social withdrawal, and depression. The cause of growth hormone deficiency is usually due to pituitary disease as described above. Isolated growth hormone deficiency is more often seen in children.

Diagnosis of GH deficiency depends on the demonstration of low growth hormone levels during sleep or after exercise. Stimulation tests such as glucagon stimulation test, insulin hypoglycaemia test or, rarely, clonidine stimulation are alternative tests. In children, insulin hypoglycaemia test is seldom used because of the risks involved. Serum IGF-1 concentration will be low in GH deficiency, however, the sensitivity and specificity are poor to be used as a diagnostic test of GH deficiency.

Children with GH deficiency are treated with recombinant GH. Guidelines for treatment of adults with GH deficiency are published by the National Institute of Clinical Excellence (NICE).

Excess GH secretion — Gigantism and acromegaly

Excess growth hormone before puberty leads to gigantism and after puberty leads to acromegaly. In the majority of cases, a pituitary tumour is the cause. Prevalence of growth hormone secreting adenoma is 50–80 cases per million. A small proportion of cases of acromegaly are caused by hypothalamic or ectopic tumour (bronchial carcinoid tumour) producing growth hormone releasing hormone. The clinical features of acromegaly develop slowly and progressively over many years. The average delay between onset of symptoms and diagnosis is about 6 years. Acromegaly is characterised by local overgrowth of bone and increased growth of soft tissues, hands, feet, jaw and internal organs. In children and adolescents, excess growth hormone leads to gigantism because the associated secondary hypogonadism delays epiphyseal closure, which allows continued growth. Other features of acromegaly include fatty hyperhiderosis, goitre, osteoarthritis, carpal tunnel syndrome, menstrual disorders and decreased libido or impotence due to anterior pituitary insufficiency. In 30–40% of patients, there is co-secretion of prolactin. Other features include features of deficiency of other pituitary hormones and visual field defects caused by the pressure on the optic chiasm by the tumour.

Once clinically suspected, diagnosis of acromegaly or gigantism can be confirmed by measurement of growth hormone secretion. As growth hormone secretion is pulsatile and may be increased by stress, single measurements are not reliable. In most patients, growth hormone concentrations are > 20 mU/L. Diagnosis depends on suppression tests and the simplest and most specific test is the oral glucose tolerance test. In normal subjects, an oral glucose load suppresses growth hormone concentration to < 2 mU/L after 2 hours. In patients with acromegaly, there is failure of such suppression and sometimes, a paradoxical rise in growth hormone is also seen. Measurement of serum IGF-1 has been suggested to be a simpler test as IGF-1 concentrations are stable and reflect an integrated 24-hour growth hormone concentration. However, IGF-1 concentration can be affected by other factors such as nutrition. Measurement of IGFBP-3 does not have any greater diagnostic sensitivity than measurement of IGF-1 alone. Further investigation includes imaging techniques such as an MRI scan to identify pituitary tumour.

Management is by transphenoidal surgery. In those patients who have contraindications for surgery or those in whom surgery has failed, radiotherapy is required. Dopamine agonists, somatostatin analogues (octreotide or lanreotide) or GH receptor antagonists (such as pegvisomant) can be used alone or in combination if surgery or radiotherapy is contraindicated. Drug therapy is especially useful in elderly people.

Case 19.2

A 50-year-old lady went to see her GP with headaches. On questioning, she admitted that her ring had become tighter and she had to buy new shoes recently, which were one size bigger. Her GP thought she had large hands and referred her to the endocrinologist for investigation. She under went an oral glucose tolerance test and the results were:

Time	Glucose (mmol/L)	GH (mU/L)
0	6.5	27.0
30	10.2	26.5
60	11.5	22.4
90	9.8	20.3
120	6.9	26.9

There was failure to suppress GH during the oral glucose tolerance test. This together with the clinical features is diagnostic of acromegaly. MRI of the pituitary showed a tumour and she underwent surgical removal of the tumour.

Hyperprolactinaemia

Hyperprolactinaemia is an important cause of infertility and is seen in 10–40% of patients with secondary amenorrhoea. Causes of hyperprolactinaemia are listed in Table 19.4. Hyperprolactinaemia causes infertility in both men and women, impotence and gynaecomastia in men and menstrual irregularities and galactorrhoea in females. Prolactin inhibits the pulsatile secretion of GnRH leading to these abnormalities. As dopamine secretion from the hypothalamus inhibits prolactin secretion, any process interfering with dopamine synthesis, transport or its action can lead to hyperprolactinaemia. Microprolactinomas (adenomas < 10 mm) or macroprolactinomas (adenomas > 10 mm) are important pathological causes of hyperprolactinaemia. The prevalence of prolactinomas is not known. Autopsy studies suggest that microadenomas may be present in up to 23–27% of

Table 19.4   Causes of hyperprolactinaemia

Stress
Hypothyroidism
Chronic renal failure
Cirrhosis
Polycystic ovarian syndrome
Neurogenic
  • Breast stimulation
  • Chest wall lesions
  • Spinal cord lesions

Drugs — dopamine receptor blockers, e.g. phenothiazines,
        metoclopramide
  • Antihypertensives
    — Methyldopa
  • Oestrogens
  • Opioids
  • Antipsychotic agents
  • Others
    — Verapamil
    — Cimetidine

Hypothalamic disease
  • Tumours
    — e.g. Craniopharyngioma
  • Infiltration
    — e.g. Sarcoidosis, histiocytosis

Pituitary disease
  • Prolactinomas
    — Micro- or macroadenomas
  • Acromegaly
  • Cushing's disease
Physiological
  • Pregnancy
  • Lactation

Macroprolactin

individuals. Macroadenomas are less common and occur more often in men than women. This may reflect the duration of the disease and a gradual development of symptoms, which involves sexual and gonadal dysfunction. Most microadenomas (up to 90%) remain small and only a small proportion become macroadenomas.

Macroprolactin is an increase in prolactin concentration without symptoms. This is due to prolactin in the circulation, binding to immunoglobulins. This form of prolactin does not bind to the prolactin receptor and the clearance of the molecule is slower than the normal prolactin hence, higher serum concentration. This condition does not require treatment but should be excluded so as to prevent further unnecessary investigations or treatments.

1. *Investigations*

   In patients with symptoms suggestive of hypogonadism or galactorrhoea, serum prolactin should be measured. Samples should be taken preferably from a venous cannula after the patient has rested for 30 minutes to avoid stress-induced hyperprolactinaemia. If hyperprolactinaemia is found, secondary causes such as hypothyroidism and renal failure should be excluded and a careful drug history should be taken. In rare cases, hyperprolactinaemia may be due to a high molecular weight prolactin, macroprolactinaemia, and this can be excluded by reanalysis of the sample after removing immunoglobulins by treatment with polyethylene glycol. Functional dynamic tests have no place in the diagnosis of the cause of hyperprolactinaemia. Imaging studies such as CT or MRI scanning are required to identify a pituitary adenoma.

2. *Management*

   Management of patients with hyperprolactinaemia is usually medical with dopamine agonists such as bromocriptine or carbergoline. These drugs shrink tumours and can lead to tumour disappearance. Transphenoidal surgery or radiotherapy are alternative treatments in macroplactinomas.

Posterior Pituitary

The posterior pituitary is mainly a collection of axonal projections from the hypothalamus. The hormones released from the posterior pituitary, vasopressin (ADH) and oxytocin are synthesised in the hypothalamus, travel down the axonal projections, and are stored in the posterior pituitary from where they are released. Both hormones are synthesised as large precursor molecules together with large specific polypeptides called neurophysins, which are binding proteins and serve as carriers of the hormones during axonal transport and storage.

Antidiuretic Hormone (Arginine Vasopressin (AVP))

Antidiuretic hormone (ADH) is a cyclic peptide of molecular weight 1080. The major physiological function of ADH is the regulation of osmolarity of body fluids, and hence cell volume. Secretion of ADH is primarily regulated by osmolarity of ECF via osmoreceptors located in the hypothalamus. As little as 1–2% change in osmolarity can influence ADH secretion. A plasma osmolarity of 280 mOsm/kg is considered as the osmotic threshold for ADH release; this is the osmolarity below which there is no change in ADH secretion. This set point, however, varies between individuals. Apart from osmolality, ECF volume via baroreceptors is also a powerful stimulus for ADH release (Figure 2.1). ADH acts on the basolateral membrane of the collecting tubules of the kidney to increase water reabsorption. Binding of ADH to its receptor (V2 receptor) activates an intracellular messenger system, which stimulates the relocation of water channel, aquaporin 2, to the apical membrane. This increases the permeability of the cells to water, resulting in net water reabsorption. The role of ADH in water metabolism and its interrelationship to thirst, etc. are discussed in Chapter 2. In addition to its effect on water metabolism, ADH is a potent vasoconstrictor via ADH receptor 1B. ADH may be of some importance in constricting regional blood vessels (e.g. splanchnic, renal and hepatic vessels), thus helping to maintain blood pressure in volume depletion states.

Diabetes insipidus

Diabetes insipidus is a condition caused by a deficiency of ADH or failure of the renal tubules to respond to ADH, causing the passage of large amounts of dilute urine (polyuria). The resulting increase in plasma osmolality stimulates thirst and the subject is able to maintain water balance. Patients with diabetes insipidus present with polyuria and thirst. The causes and investigation of patients presenting with polyuria is described in Chapter 2.

- Management

    Patients with cranial diabetes insipidus are treated with intranasal desmopressin. Whenever possible, the underlying disease should be treated. To avoid dehydration, fluid intake should always be adequate. Patients with nephrogenic diabetes insipidus should have an adequate fluid intake. Desmopressin is not effective in this condition. Thiazide diuretics may be of some benefit in nephrogenic diabetes insipidus.

Case 19.3

A 40-year-old female presented with fatigue, amenorrhoea, polyuria and thirst for about 10 months.
    Investigations showed:

TSH   0.28 mU/L (0.2–5.6)
FT4   5.8 pmol/L (12.0–25.0)

A water deprivation test was done and the results were:

Time	plasma osmolarity	urine osmolality
Initial	298	120
After 6 hours of water deprivation	305	260
After desmopressin	304	750

Failure to increase urine osmolarity on water deprivation test followed by an increase after desmopressin established the diagnosis as neurogenic diabetic insipidus. An inappropriate TSH with low FT4 shows secondary hypothyroidism. Further investigations showed that she had an empty sell and a diagnosis of neurosarcoidosis was made. She was treated with desmopressin and glucocorticoids. Her menstrual periods returned to normal 12 months later.

Oxytocin

The structure and function of oxytocin overlap with those of ADH. The main function of oxytocin is the control of uterine contractility and ejection of milk from the lactating breast. Oxytocin release is stimulated by the mechanical distension of the vagina and suckling of the nipples. Oxytocin stimulates uterine contraction during parturition and smooth muscle contraction in the breast during suckling. Although oxytocin concentrations are high during pregnancy, it is balanced by a rise in oxytocinase and progesterone. The function of oxytocin in men is not known. Oxytocin has similar effects on water homeostasis to ADH. Infusion of oxytocin at a high rate during the induction of labour has been reported to cause severe water intoxication (see Chapter 2).

Further Reading

1. Barth J, Butler G, Hammond P. *Biochemical Investigations in Laboratory Medicine 2001*. ACB Venture publications: London.
2. Bonert V. Diagnostic challenges in acromegaly: A case-based review. *Best Pract Res Clin Endocrinol Metab* 2009; 23 Suppl 1:S23–30.
3. Melmed S. Acromegaly pathogenesis and treatment. *J Clin Invest* 2010; 119:3189–3202.
4. Reddy R, Hope S, Wass J. Acromegaly. *BMJ* 2010; 16:341:c4189.
5. Schneider HJ, Aimaretti G, Kreitschmann-Andermahr I, Stalla G, Ghigo E. Hypopituitarism. *The Lancet* 2007; 369:1461–1470.

Summary/Key Points

1. The anterior pituitary gland secretes ACTH, TSH, LH, FSH, GH, and prolactin. All these hormones are regulated by hypothalamic factors and there is feedback regulation from the target hormones of these trophic hormones.
2. Hypopituitarism results form destruction of pituitary cells by tumour, infection or infiltration and the manifestations are due to the deficiencies of the hormones secreted. Gonadotrophins and GH are the first to go followed by ACTH and TSH.

3. Tumours of the pituitary can cause increased secretion of one of the hormones or may cause destruction of the rest of the pituitary.
4. Pituitary function is assessed by measurement of basal pituitary hormones and the hormones secreted by their target organs (e.g. ACTH and cortisol). Insulin tolerance test is required to diagnose GH deficiency or ACTH deficiency.
5. Excess of GH from an adenoma causes acromegaly in adults and giagantisim in children. It is diagnosed by failure to suppress GH during an oral glucose tolerance test.
6. Hyperprolactinaemia is a common condition and most cases are caused by an adenoma (micro or macro). It can also be secondary to drugs, hypothyroidism and renal failure.
7. Posterior pituitary gland secretes ADH, which is important in the regulation of water balance, and oxytoxin, which is involved in parturition and lactation. Deficiency of ADH causes polyuria and polydipsia. Water deprivation test may be required to differentiate this form other causes of polyuria. Diabetes insipidus can also be due to insensitivity of renal tubules to ADH (nephrogenic diabetes insipidus).

chapter 20

The Reproductive System

Male Reproductive System

The testes consist of a network of tubules that produce sperm and the interstitial, or Leydig, cells which secrete testosterone. In the spermatogenic tubules, in addition to the germ cells, there are Sertoli cells which secrete androgen-binding protein (ABP), inhibin- and Mullerian-inhibiting factor. Secretion of testosterone by the Leydig cells is under the influence of the anterior pituitary via LH (see Chapter 19). Other androgens secreted by the testes include androstenedione and dehydroepiandrosterone (DHEA). Testosterone is metabolised in target organs such as the skin and prostate to dihydrotestosterone (DHT) by 5α-reductase. Testosterone and DHT bind to receptors in target organs and initiate actions, which include gonadotrophin regulation, spermatogenesis, sexual differentiation and at puberty sexual maturation. Testosterone is also metabolised by aromatase to oestradiol in the adipose tissue. Therefore, obese males have higher serum oestrogen concentrations. The main excretory route of testosterone is urine via formation of metabolites such as 17-ketosteroids and other conjugates.

Testosterone in serum is primarily bound to sex hormone-binding globulin (SHBG) and albumin. About 60% of testosterone is bound to SHBG, and about 40% is non-specifically bound to albumin with a small (2–3%) free fraction. It is now believed that the non-SHBG–bound testosterone is available for uptake by cells. Serum testosterone concentrations exhibit a diurnal variation, with

Table 20.1   Factors affecting SHBG concentration

Increase
- Oestrogen — pregnancy, contraceptive pill
- Hyperthyroidism
- Drugs — antiepileptic drugs

Decrease
- Androgens
- Hypothyroidism
- Obesity

highest values between 4 a.m. and 8 a.m. and lowest values between 4 p.m. and 8 p.m.

SHBG also binds to DHT and oestradiol. Affinity of SHBG for these steroids is as follows: DHT > testosterone > oestradiol. The concentration of SHBG is increased by oestrogens, hyperthyroidism and liver diseases (Table 20.1) while androgens, hypothyroidism, glucocorticoids and obesity decrease SHBG. Because of the different affinities of SHBG to testosterone and oestradiol, at low SHBG concentration, there is an increase in androgenic effects, while at high SHBG concentration, oestrogenic action predominates.

Disorders of Male Gonadal Function

Hypogonadism

Decreased function of the testes (spermatogenesis, testosterone secretion or both) is termed hypogonadism. If this occurs before puberty, it will lead to delayed puberty and may be associated with short stature. Decreased testicular function can be classified as primary or hypergonadotrophic hypogonadism where there is failure of the testes, and secondary or hypogonadotrophic hypogonadism where the dysfunction is in the hypothalamic/pituitary area (Table 20.2).

Table 20.2 Causes of hypogonadism

Primary or hypergonadotrophic hypogonadism
- Gonadal agenesis
- Gonadal dysgenesis
  — Male
    o Klinefelter's syndrome
  — Female
    o Turner's syndrome (45 XO)
    o Noonan's syndrome (46 XX)
    o 17α-hydroxylase deficiency in ovary
- Gonadal failure
  — Injury
  — Infection
  — Irradiation
  — Drugs — cytotoxic drugs, anabolic steroids
  — Varicocele in the male
- Chronic diseases
  — Liver disease
  — Renal failure

Secondary or hypogonadotrophic hypogonadism
- Hypothalamic disorders
  — Tumours
  — Kallmann syndrome
  — Strenuous exercise
  — Malnutrition
- Pituitary disorders
  — Tumours — prolactinomas
  — Panhypopituitarism
  — Metabolic disorders — haemochromatosis

The most common cause of developmental defect causing hypergonadotrophic hypogonadism is Klinefelter's syndrome, which is a chromosomal abnormality (47XXY) associated with dysgenesis of seminiferous tubules and high FSH. In testicular failure, spermatogenesis may only be impaired, in which case FSH will be elevated and LH will be normal. If this is acquired, it will lead to sterility or infertility without affecting secondary sexual characteristics or sexual function. This is seen in viral orchitis. If testosterone production is affected, it will lead to failure of spermatogenesis and decreased libido, impotence and reduced growth of facial, axillary and pubic hair. Gynaecomastia may develop due to an increase in the oestradiol/ testosterone ratio.

Hypogonadotrophic hypogonadism is due to a deficient secretion of pituitary hormones. A congenital cause is Kallmann syndrome, which is an X-linked disorder, characterised by isolated deficiency of GnRH secretion, associated with decrease or absence of smell and colour blindness. Tumours, such as craniopharyngioma can lead to hypogonadotrophic hypogonadism due to the destruction of the pituitary. Hyperprolactinaemia due to a prolactinoma may also cause hypogonadotrophic hypogonadism.

Investigation of hypogonadism involves measurement of basal hormones: testosterone, prolactin, FSH and LH. A high FSH/LH with low testosterone will indicate hypergonadotrophic hypogonadism, and a low or normal testosterone and low FSH/LH indicate hypogonadotrophic hypogonadism. Provocation or stimulation tests useful in these patients are hCG, GnRH or clomiphene stimulation tests. hCG binds to LH receptors and stimulates testosterone production. After injection of hCG, failure of testosterone to rise (after 3 days) is indicative of Leydig cell dysfunction. A GnRH test will help to distinguish between hypothalamic and pituitary dysfunction. In pituitary disease, there is little or no rise in FSH/LH, and in hypothalamic disorder, there is a delayed rise. Clomiphene has antioestrogenic effects and stimulates the secretion of GnRH. Following administration of clomiphene (100 mg), FSH and LH should rise by at least 20 IU/L. A blunted response is seen in hypothalamic disease, malnutrition, strenuous exercise or hyperprolactinaemia, conditions associated with abnormal GnRH pulses.

Management of hypogonadism involves treatment of the underlying disorder and hormone replacement therapy as indicated. Testosterone by injection or skin patches is necessary when there is low testosterone production.

Case 20.1

A 52-year-old man with diabetes was investigated for tiredness, loss of libido and weight gain. Results of investigations were as follows:

Serum		Reference Range
TSH (IU/L)	1.6	0.2–5.5
FT4 (pmol/L)	8.5	10.1–21.0
Testosterone (nmol/L)	5.2	10.2–25.0
FSH (IU/L)	0.8	2.0–10.0
LH (IU/L)	2.5	2.0–8.0
Prolactin (mU/L)	255	< 420

His renal functions tests were normal. A low testosterone with inappropriately low gonadotrophins indicates hypogonadotrophic hypogonadism. Low FT4 with inappropriately low TSH is indicative of secondary hypothyroidism. The possible diagnosis is hypopituitarism. With a history of diabetes and suspected hypopituitarism, further investigations for haemochromatosis were done and they showed increased iron saturation. Genetic analysis confirmed the diagnosis of haemochromatosis.

Gynaecomastia

Breast development in males is associated with an increase in oestrogen/androgen ratio. This may occur physiologically in the neonate, during puberty and in the elderly. Transient gynaecomastia is seen in about 70% of normal neonates and 50–70% of normal boys during puberty. Gynaecomastia in the elderly is caused by reduced testicular function and increased conversion of testosterone to oestrogens by the increasing adipose tissue. Causes of gynaecomastia are listed in Table 20.3. Use of anabolic steroids by body builders and other athletes causing gynaecomastia is now an increasing problem.

Investigation of gynaecomastia consists of biochemical tests to rule out liver, renal and thyroid diseases, (liver function tests, urea,

Table 20.3   Causes of gynaecomastia

Physiological
- Neonate
- Adolescent
- Old age

Pathological
- Testosterone deficiency
  — Congenital — e.g. Klinefelter's syndrome
  — Acquired — e.g. trauma
- Androgen insensitivity

Increased oestrogens
- Increased testicular production of oestrogens
  — Bronchogenic carcinoma secreting hCG
  — Testicular tumours
- Increased conversion of androgens to oestrogens
  — Liver diseases
  — Thyrotoxicosis

Drugs
- Oestrogens or oestrogenic drugs
- Anti-androgens — spironolactone
- Anabolic steroids
- Other mechanisms — methyldopa, cannabis

creatinine and thyroid function tests) and endocrine tests: FSH, LH, testosterone, hCG, oestradiol and prolactin.

Female Reproductive System

During the reproductive period, there is a highly coordinated feedback system between the hypothalamus, pituitary and ovary regulating the menstrual cycle (Figure 20.1). Menstrual cycle is between 23 and 39 days and is divided into the follicular (or proliferative) phase, ovulation and leutal (or secretory) phase. The start of the cycle is the first day of the menstrual bleed. In most women, the length of the leutal phase is constant between 13 and 15 days, but the follicular phase can vary. During the follicular phase, a small number of follicles start to develop.

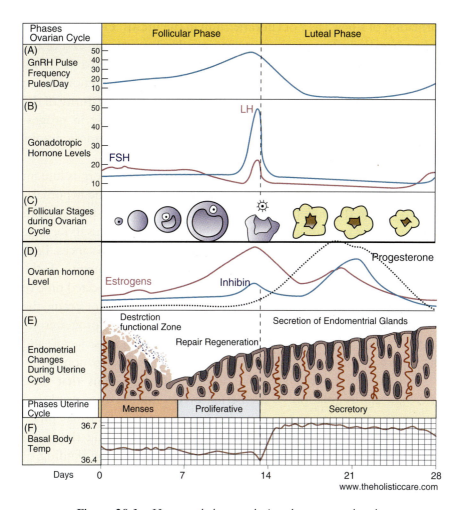

Figure 20.1 Hormonal changes during the menstrual cycle.

Of these, one becomes the dominant follicle and this grows rapidly during the follicular phase. At the end of the previous cycle, the secretion of inhibin A and ovarian steroids fall dramatically. This, together with changes in the pulsality of GnRH, causes an increase in FSH secretion. This increase in FSH stimulates the developing follicles, which secrete oestrogens and inhibin B. The rise in oestrogens and inhibin B cause a decrease in FSH. The rising concentration of oestrogens stimulates the formation of a new layer of endometrium in the uterus. Oestrogens also have a negative effect on the secretion of LH. Towards the end of the follicular phase, the rising level of osterogen causes a positive feedback on the pituitary, causing a massive increase in LH secretion and a smaller rise in FSH secretion. This LH surge triggers the maturity of the follicle and the oocyte is released, ovulation. A small rise in progesterone towards the end of the follicular phase seems to be important for the positive feedback effect on the pituitary.

Once the oocyte is released, the follicle becomes the corpus luteum and starts to secrete progestogens. Progesterone acts on the hypothalamus to reduce the pulsality of GnRH release and increase the set point of temperature regulation by 0.5°C. It also acts on the uterus to increase the secretion of mucus, specific proteins and vasoactive substances. Peak activity of the corpus luteum and the peak development of the endometrium are seen 8–9 days after the ovulation, the expected time of implantation of the fertilized oocyte. If pregnancy does not occur, the corpus luteum starts regressing and the secretion of progesterone declines. This causes spasm of the spiral arteries leading to ischaemia and the superficial layers of the endometrium are shed, menstruation.

Activin and inhibin are two closely related proteins that have effect on the regulation of menstrual cycle. Inhibin B is produced during the follicular phase; it peaks during the mid-follicular phase and inhibits FSH secretion. Inhibin A is secreted during the leutal phase and peaks at mid-leutal stage. Activin increases the binding of FSH to ovarian follicles. Inhibin and activin act in a paracrine fashion in regulating the menstrual cycle.

Ovaries secrete oestrogens, which are important for the development and maintenance of the female sex organs and secondary sex characteristics. In conjunction with progesterone secreted by the corpus luteum, they regulate uterine growth, breast growth, the menstrual cycle and help to maintain pregnancy.

Oestrogens are primarily secreted by the ovaries, with a small contribution from the adrenals. Oestrogen in circulation is primarily bound to SHBG with some non-specific binding to albumin. SHBG concentration is higher in females as oestrogen stimulates its production (Table 20.1).

Disorders of the Female Gonadal Function

Hypogonadism

Hypogonadism in the female, as in the male, can be due to primary (hypergonadotrophic) or secondary (hypogonadotrophic) causes (Table 20.2). Hypogonadism during the pre-pubertal stage will lead to delayed puberty.

Amenorrhoea/Oligomenorrhoea

Complete cessation of menses for 6 months is termed amenorrhoea and a reduction in the frequency or sparse menses is termed 'oligomenorrhoea'. Amenorrhoea is classified as primary when menses have never occurred by the time of expected menarchy and as secondary when the menses are absent in women who have previously had spontaneous periods. Physiological causes of amenorrhoea include pregnancy, menopause and prepuberty. The most common cause of amenorrhoea in women of childbearing age is pregnancy and this should always be excluded. A finding of abnormally high LH or abnormally low LH indicates pregnancy and this should be followed by hCG measurements. In some immunoassays using polyclonal antibodies, high LH is found due to cross-reaction of hCG. However, in modern immunometric assays, LH is undetectable in pregnancy due to the negative feedback of oestradiol and progesterone. The causes of amenorrhoea are given in Table 20.4.

The most common causes of amenorrhoea are polycystic ovarian syndrome (PCOS), hypothalamic amenorrhoea, hyperprolactinaemia and ovarian failure. PCOS and hyperprolactinaemia are described elsewhere. Hypothalamic amenorrhoea is often due to a functional abnormality without any anatomical or organic disease. Causes of functional hypothalamic amenorrhoea include excessive exercise, under nutrition and stress. These patients have normal or low gonadotrophins as a result of decreased secretion of GnRH. Low leptin (due to reduced adipose tissue and/or energy deficit), increased ghrelin and neuropeptide Y contribute to this. Of these,

Table 20.4   Causes of amenorrhoea

Primary Amenorrhoea
- Primary ovarian failure (hypergonadotrophic hypogonadism)
  — E.g. gonadal agenesis, Turner's syndrome

- Secondary ovarian failure (hypogonadotrophic hypogonadism)
  — E.g. tumours

Secondary Amenorrhoea
- Pregnancy
- Post 'pill' amenorrhoea
- Ovarian failure (hypergonadotrophic hypogonadism)
  — Physiological — menopause
  — Premature ovarian failure

- Hypogonadotrophic hypogonadism
  — Psychogenic
  — Weight loss, malnutrition
  — Hyperprolactinaemia
  — Infiltration/neoplasm or pituitary

- Excess sex steroid production
  — Androgens
    o Polycystic ovarian syndrome
    o Congenital adrenal hyperplasia
    o Virilising tumour

  — Oestrogen
    o Oestrogen-producing tumour

---

leptin is a critical factor. Chronic stress causes hypothalamic amenorrhoea due to increased stress hormones: cortisol, ACTH, prolactin, CRH, and catecholamines.

Ovarian failure (hypergonadotrophic hypogonadism) can be primary when there is gonadal dysgenesis, which may occur in association with chromosomal abnormalities such as Turner's syndrome (45X0). Features of Turner's syndrome include short stature, webbed neck and deformities of the chest and elbow. Phenotypically, these patients are female as the Müllerian duct differentiates normally. Premature ovarian failure is an autoimmune disorder, which may be associated with other autoimmune disorders such as Addison's disease. There is premature depletion of primordial oocytes and a secondary increase in gonadotrophins. Treatment with cytotoxic drugs or radiation may also lead to premature ovarian failure. Rarely, ovarian failure can be due to resistance to the action of gonadotrophins, resulting in hypergonadotrophic hypogonadism. This condition is probably due to a mutation in the FSH receptor.

The contraceptive pill inhibits gonadotrophins and this inhibition may persist for a varying period of time, even after stopping the pill.

- Investigation of amenorrhoea (Figure 20.2)

    A careful physical examination and some simple investigations are usually enough to make a diagnosis. Clinical examination should concentrate on the signs of secondary sexual characteristics and features of conditions such as Turner's syndrome. Evidence of hirsuitism should be looked for. Blood tests should include measurements of gonadotrophins, prolactin, oestradiol and thyroid function tests. Serum testosterone may be required in those patients with features of hirsuitism. Based on these investigations, patients should be classified into hypergonadotrophic, or hypogonadotrophic and those with normal hormonal values. In those with primary amenorrhoea and hypergonadotrophic hypogonadism, chromosomal analysis may be necessary. Pelvic ultrasound may be required to identify the presence of uterus and ovaries. These patients will require replacement therapy with oestrogens.

    In patients with hypogonadotrophic hypogonadism, a functional disorder of the hypothalamus should be excluded. In the majority of women with hypogonadotrophic hypogonadism, there will be no organic disease. In normal gonadotrophic anovulation, there may be abnormalities in the pattern of LH secretion such as that seen in polycystic ovarian syndrome (PCOS).

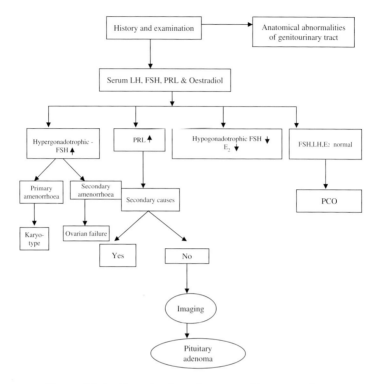

Figure 20.2   Investigation of women with amenorrhoea.

Case 20.2

A 30-year-old female consulted her GP because of irregular periods and galactorrhoea. Her periods were regular in the past and she has a 6-year-old son. Her GP requested some blood tests and the results were as follows:

Serum		Reference Range
TSH (IU/L)	1.6	0.2–5.5
FT4 (pmol/L)	15.2	10.1–21.0
Testosterone (nmol/L)	1.5	1.0–3.0
FSH (U/L)	5.8	2.0–10.0
LH (U/L)	4.5	2.0–8.0
Prolactin (mU/L)	3545	< 420

> *Her renal function tests were normal. She clearly has a high prolactin and the laboratory reported that there was no macroprolactin. Radiological investigation showed that she had a small adenoma in her pituitary. She was treated with cabergoline and her prolactin concentration decreased and her periods became regular.*

Hirsuitism and virilisation

Excessive hair growth in sites usually associated with males, i.e. face, lower abdomen, thigh, chest, etc., is classified as hirsuitism. Virilism refers to a manifestation of androgen excess such as clitoromegaly, receding hair in the temporal region, increased muscle mass, breast atrophy and deepening voice.

Hair follicles are of two types: vellus, which is androgen insensitive and terminal, which is androgen sensitive. In true hirsuitism, androgen-sensitive hair growth is increased. The amount of hair in an individual is determined by genetic factors, e.g. Orientals have less body hair than people of Mediterranean descent. The primary mechanism leading to hirsuitism and virilisation is increased androgenic activity either due to increased secretion or due to increased sensitivity of hair follicles to androgens.

The principal sources of androgens in females are shown in Figure 20.3. In normal women, circulating testosterone is derived from direct secretion by the ovary and by extraglandular conversion of androstenedione secreted by the ovary and adrenals. Adrenals are the major source of DHEA and DHEA sulphate. In women with hirsuitism, the circulating testosterone is usually increased with reduced SHBG concentrations, thereby increasing the free testosterone concentration. In women with mild hirsuitism, normal ovulation cycles and normal concentrations of androgens, the excess hair growth is explained on the basis of increased sensitivity of hair follicles to androgen.

The causes of hirsuitism and virilisation are listed in Table 20.5. The most common cause of hirsuitism is polycystic ovarian syndrome (PCO) (see below). Late-onset congenital adrenal hyperplasia (CAH)

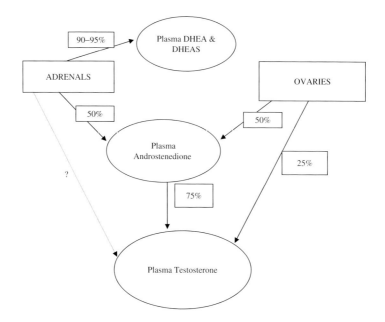

Figure 20.3 Sources of circulating androgens in normal women.

Table 20.5 Causes of hirsuitism and virilisation

Ovarian
  • Polycystic ovarian syndrome
  • Neoplasms
    — Germ-cell tumours
Adrenal
  • Congenital adrenal hyperplasia — adult-onset adrenal hyperplasia
    — 21- or 11β-hydroxylase deficiency
  • Cushing's syndrome
Drugs
  • Diazoxide
  • Anabolic steroids
  • Progestagens
  • Danazol
Miscellaneous
  • Hyperprolactinaemia
  • Acromegaly
  • Menopause
Idiopathic

---

can present with hirsuitism, and is due to mutations, which are less severe than the classical form CAH (see Chapter 21). Some centres estimate that up to 20% of hirsute women may have late onset CAH. Hirsuitism, together with menstrual irregularities, is seen in Cushing's syndrome. Mild hirsuitism is also observed in hyperprolactinaemia and acromegaly.

When serum testosterone is very high (> 5 nmol/L) and/or features of virilisation are present, androgen-producing tumours should be suspected. These, however, are rare.

Investigation of patients with hirsuitism depends on the clinical presentation. In the majority of patients, serum gonadotrophins, testosterone, SHBG (and calculation of free testosterone index) are all that is required. Measurement of 17-hydroxyprogesterone (17-OHP) and DHEA sulphate may be required if an adrenal cause is suspected. The diagnosis of late onset CAH depends on demonstrating high serum 17-OHP before and/or after 250 μg of ACTH. When testosterone concentration is high (greater than 5 nmol/L), it is very suggestive of virilising tumours and further investigations such as CT scan may be required.

Management of PCOS is discussed below. When other causes are found, specific treatment is indicated. In idiopathic hirsuitism without menstrual irregularities, removal of hair by physical methods may be the only viable option.

Polycystic ovary syndrome (PCOS)

This is one of the most common endocrine disorders. It is a common cause of hirsuitism and menstrual irregularity. It is defined as the association of hyperandrogenism and chronic anovulation without specific underlying diseases of the adrenal or pituitary glands. Hyperandrogenism manifests as hirsuitism, acne and increased serum concentration of androgens such as testosterone and androstenedione. The exact prevalence of PCOS is not known but it is an important cause of menstrual irregularity and hirsuitism. It may be present in up to 50–70% of patients presenting with either menstrual irregularity or hirsuitism. Typical clinical pictures include anovulation, which manifests as menstrual disturbance (amenor-rhoea, oligomenorrhoea, dysfunctional bleeding) and infertility, hyperandrogenism that presents as hirsuitism, acne or male pattern of alopecia. Obesity and impaired glucose intolerance are common. Up to 20% of patients with PCOS who are obese develop diabetes mellitus by the age of 40. These patients are also more prone to cardiovascular disease.

Biochemical abnormalities in PCOS include elevated serum LH and a normal or low FSH giving rise to elevated LH/FSH ratio. Women with PCOS generally have normal oestradiol concentrations for early follicular and mid-follicular phases of the cycle. Serum con-centration of testosterone, androstenedione and free testosterone index are increased. Some women with PCOS also have impaired secretion of growth hormone. Hyperprolactinaemia may be found in 30% of patients. One of the differential diagnoses of PCOS is late-onset congenital adrenal hyperplasia, which can be excluded by meas-urements of 17-OHP before and 60 minutes after an ACTH stimulation test.

Management of PCOS includes management of anovulation and hirsuitism. Hirsuitism can be managed non-pharmacologically by electrolysis, laser treatment, etc. or pharmacologically. Oral contra-ceptives, which stimulate the production of SHBG and cause a decrease in free testosterone, may improve hirsuitism. Anti-androgens such as spironolactone and cyproterone acetate are the most effective treatment. Recent studies suggest that treatment with metformin may improve the condition. For management of anovulation, see 'Infertility'.

Case 20.3

A 25-year-old secretary was referred for investigation of hirsutism and irregular periods. Her periods started when she was 14 years old and had been regular for nearly 7 years. For the past few years, she noticed increased hair growth in her thighs, lower abdomen as well as her face.

She was moderately obese (BMI of 29) and investigations showed the following results:

Serum		Reference Range
TSH (IU/L)	1.6	0.2–5.5
FT4 (pmol/L)	15.2	10.1–21.0
Testosterone (nmol/L)	3.2	1.0–3.0
FSH (IU/L)	5.5	1.0–15.0
LH (IU/L)	13	1.0–25.0
Prolactin (mU/L)	375	< 420
SHBG (nmol/L)	15	20–90

History of hirsutism with elevated testosterone, low SHBG and high LH/FSH ratio strongly indicate a diagnosis of PCOS. An ultrasound examination showed multiple cysts and confirmed the diagnosis.

Infertility

Infertility is a common clinical problem and it is defined as the failure of a couple to conceive after 1 year of regular unprotected intercourse. Infertility could be primary where conception has never occurred or secondary where there is a failure to conceive a second child or subsequent children. The causes of infertility can be either in the male or female and are listed in Table 20.6. Common causes of

Table 20.6   Causes of infertility

Sperm defects or dysfunction	30%
Failure of ovulation	25%
• Amenorrhoea or oligomenorrhoea	
Tubal infection/damage	20%
Unexplained infertility	25%
Endometriosis	5%
Coital failure or infrequency	5%
Cervical mucus defects or dysfunction	3%
(some couples have more than one cause)	

Table 20.7   Laboratory investigations of infertility

Male
  • Semen analysis — include sperm function tests
  • FSH/LH
  • Prolactin
Female
  • Mid-luteal progesterone (2–6 cycles)
  • Basal FSH/LH
  • Prolactin
  • Serology
    — Rubella
    — Hepatic HIV (both partners)
    — Chlamydia

primary infertility are endometriosis and sperm disorders. Tubal damage due to infection is a frequent cause of secondary infertility.

Investigation of infertility requires the examination of both partners (Table 20.7). Laboratory investigations should include semen analysis, that includes volume, sperm concentration, sperm motility and the percentage of normal sperms. In addition, other sperm function tests such as osmotic swelling tests, sperm cervical mucosa penetration test should be done. In men, other investigations, which may be occasionally necessary, include serum gonadotrophins, testosterone and prolactin. Women with normal menstrual cycle rarely have ovulatory disorders. If there is menstrual irregularity, amenorrhoea or oligomenorrhoea, this should be investigated. Evidence of ovulation can be obtained by the rise in basal body temperature in the latter half of the cycle or by an increase in serum progesterone to > 30 nmol/L on day 21 of the cycle. Serum progesterone < 10 nmol/L is very suggestive of anovulatory cycles. Other investigations include prolactin and gonadotrophins, especially when there is oligo- or amenorrhoea. If tubal damage is suspected, it should be investigated by hysterosalphingography.

Treatment of infertility depends on the cause. In anovulatory patients, ovulation can be induced by treatment with gonadotrophins or clomiphene citrate. Some cases may require assisted conception.

Assisted conception

Assisted conception is used to overcome infertility resulting from a variety of causes. This term includes donor insemination, *in vitro* fertilisation (IVF), intracytoplasmic sperm injection (ICSI) and gamete intrafallopian transfer (GIFT). Almost all these techniques require superovulation or controlled ovarian hyperstimulation to obtain multiple oocytes. This involves administration of clomiphene or gonadotrophins. Careful monitoring of serum oestradiol concentration and pelvic ultrasound should follow ovarian stimulation.

Pregnancy

Healthy pregnancy is characterised by major physiological and anatomical adaptations in the mother. The fertilised ovum enters the uterus 3 days after ovulation and becomes implanted into the uterine lining about 5 to 8 days later. During the early stages of development, a cavity develops. The cells on the outer lining of the cavity, trophoblasts, invade the uterine endometrium and develop as chorionic villi, which develop into the placenta. The developing embryo is bathed in amniotic fluid, which acts as a buffer for the foetus.

The placenta nourishes the foetus, eliminates waste products, and produces hormones, which are important to maintain the pregnancy. The placenta is composed of large villi of foetal blood vessels, separated from maternal blood by intervillous spaces. Movement of substances between the two circulations depends on concentration gradients and facilitated or active transport.

Placental hormones

1. *Human Chorionic Gonadotrophin (hCG)*

   Human chorionic gonadotrophin, a glycoprotein hormone produced by the trophoblasts, is important for the maintenance of the corpus luteum. It can be detected in maternal serum as early as 8 days after ovulation and in the urine a few days later. Serum hCG concentration rises rapidly during early pregnancy, and reaches a peak at about 8–10 weeks and falls after 10–12 weeks, but can be detected throughout pregnancy. Serum hCG is present in many forms with differing oligosaccharides; the hyperglycosylated form is the predominant form in early pregnancy and this form is important for implantation of the fertilized ovum. Serum and urine hCG is measured in the diagnosis of pregnancy, ectopic pregnancy and some tumours (Chapter 24).

2. *Human placental lactogen (hPL)*

Human placental lactogen, also known as human chorionic somatomammotropin, is a single polypeptide chain hormone of molecular weight 23,000. Its structure closely resembles human GH and prolactin. HPL increases with gestational age, and together with prolactin, prepares the mammary gland for lactation. Its metabolic actions are similar to that of GH (reduces glucose uptake and increases lipolysis). HPL is seldom measured clinically.

3. *Oestrogen*

The secretion of oestrogen increases during pregnancy, initially from the corpus luteum and then from the placenta. The major oestrogen during pregnancy is oestriol, which is synthesised by the placenta from androgens produced by the foetus. Foetal adrenals produce DHEA sulphate from cholesterol, and this is converted to 16α-OH DHEAS in the foetal liver. DHEAS from the adrenal and 16α-OH DHEAS are taken up by the placenta and converted to oestriol. Measurements of oestriol have been used in the assessment of foetal well being. Ultrasonography has now replaced this measurement.

4. *Progesterone*

Progesterone is produced initially by the corpus luteum and then by the placenta. They are important to support the endometrium to provide suitable environment for the foetus and suppress the contractility of the uterus.

Other hormones produced by the placenta include chorionic ACTH, PTHrP, calcitriol, CRH, relaxin, number of cytokine growth factors, activin and inhibin.

Metabolic changes in mother

During pregnancy, many metabolic and physiological changes occur in the mother. There is net retention of sodium and water, and total body fluid in the mother increases by 4–6 litres, of which 75% is in the ECF. Blood volume increases by about 1.5L; about 75% of which is plasma and 25% erythrocyte mass. The resulting haemodilution is reflected in the lowered haemoglobin concentration. Renal plasma flow and GFR increases, and the GFR in late pregnancy may be 50% higher than non-pregnant values. As a result of increased GFR and increased plasma volume, concentrations of urea and creatinine decrease. Due to the high oestrogens, synthesis of many proteins including coagulation factors, renin substrate, hormone binding proteins and acute phase proteins are increased. Increases in hormone binding proteins will increase total serum hormones. Serum uric acid tends to decrease in the early stages, but increase later due to increased tubular reabsorption. Glycosuria may be present due to changes in renal threshold. Plasma triglycerides, LDL and HDL are higher in pregnancy. Glucose tolerance may deteriorate during pregnancy.

Biochemical tests in pregnancy

Diagnosis of pregnancy is made by the detection of hCG in the urine using a convenient kit. Urine pregnancy tests become positive 10–12 days after ovulation, or at the time of the expected period. Highly sensitive serum assays that detect hyperglycosylated form can be used to detect hCG earlier than this, and a rising hCG or serum hCG greater than 25 IU/L indicates pregnancy.

Serum hCG is also useful in the diagnosis and management of ectopic pregnancy. A serum hCG concentration of less than

2000 IU/L in the absence of an intrauterine sac by ultrasonography, and/or a slower rate of rise in serum hCG (hCG doubles every 48 hours in normal pregnancy) is suggestive of ectopic pregnancy. Recent studies suggest that a low hyperglycosylated hCG may be a more reliable test. Some studies also suggest that serum progesterone below 72 nmol/L is highly suggestive of abnormal pregnancy. However, the sensitivity of this test is poor.

Pre-eclampsia is a multi-system disorder of pregnancy affecting the cardiovascular, renal, hepatic, coagulation and nervous systems. It is usually diagnosed by a combination of hypertension, oedema and proteinuria. Since hypertension is not specific for pre-eclampsia, other tests such as serum uric acid may be helpful. A high serum uric acid is associated with increased prenatal mortality.

Biochemical tests in maternal serum are also useful in detecting some foetal defects. Amniotic fluid and serum α-fetoprotein (AFP) are high in the presence of neural tube defects. However, ultrasonography is the method of choice for the detection of this disorder. Down's syndrome in the foetus can be detected by the measurement of maternal AFP, β-hCG and oestriol ('triple' test), AFP and oestriol are lower and β-hCG is high. However, this test has a false positive rate of 5%, and the detection rate is around 60%. Some centers measure inhibin A (Quad test) and this increases the detection rate to 85%, with a 5% false positive rate. Measurements of free hCG, pregnancy-associated plasma protein A (PAPA) together with nuchal translucency (by ultrasound) during the first trimester has an 85% detection rate. A definite diagnosis of Down's syndrome is by the analysis of foetal chromosomes in samples obtained by amniocentesis or chronic villus sampling.

Preterm labour is an important cause of premature birth. Early detection of those who may go into premature labour is helpful to plan management. Detection of foetal fibronectin in cervical and vaginal secretions is useful in detecting those at risk.

Biochemical and genetic analysis of foetal blood obtained antenatally can be used for the diagnosis of inherited disorders. Recently, techniques have been developed to detect foetal abnormalities using foetal DNA and foetal cells found in maternal circulation.

Hormonal Contraception

There are two types of hormonal contraception: combined preparation, which contain oestrogen and progesterone, and progesterone-only preparations. The combined preparation can be given orally, by skin patches, as a vaginal pessary or by injection. Oestrogens decrease the release of FSH by negative feedback, thereby inhibiting follicular development and ovulation. Progesterone in these preparations also decreases the pulse frequency of GnRH release, hence reducing the release of FSH and LH. A secondary mechanism of contraception of progesterone is by inhibiting sperm penetration by increasing the viscosity and decreasing the amount of cervical mucosa.

Low-dose progesterone-only preparations inhibit ovulation but not very effectively. Their main mechanism of action is by increasing the viscosity of the cervical mucosa. High-dose progesterone contraceptives inhibit ovulation more consistently.

Combined contraceptives increase the risk of deep vein thrombosis. They can cause a small increase in the risk of cancer and cardiovascular disease especially in people with pre-existing risk factors such as diabetes, metabolic syndrome, smoking, hypertension or family history of heart disease.

Oestrogen tends to increase HDL cholesterol and triglycerides and reduce LDL cholesterol whereas progesterone increase LDL and decrease HDL. In the combined preparations, these tend to cancel each other. Progesterone also tends to increase insulin resistance but this is considered not clinically significant. Oral contraceptives tend to increase blood pressure by increasing the synthesis of angiotensinogen. Progesterone-only preparations especially the injectable forms tend to decrease bone mineral density and increase the risk of osteoporosis probably by reducing oestradiol concentration.

Menopause/The Climacteric

Menopause refers to the permanent cessation of menstrual cycle and this usually happens around the age of 50 (range 40–60 years) ending the fertile phase of life. This transition takes several years when the cycles are longer and sparse. This transition period is referred to as perimenopause and during this period, FSH and LH increase and oestradiol starts to decrease. During the perimenopause and immediate postmenopausal period, some women may have symptoms such as flushing, night sweats and mood swings.

Metabolic changes associated with menopause are an increase in LDL cholesterol and triglycerides, a decrease in HDL cholesterol and a decrease in insulin sensitivity. There is an increased tendency for blood coagulation due to the increase in the concentration of proteins involved in clotting. Overall, the risk of cardiovascular disease is increased.

Menopausal women with symptoms are given hormone replacement therapy (HRT); combined oestrogen and progesterone can be given to women with a uterus while an oestrogen-only therapy can be offered to those who have had a hysterectomy. HRT is given orally, as a skin patch, gel or as a subcutaneous implant. In addition to alleviating symptoms, HRT reduces postmenopausal bone loss. HRT is associated with lower LDL cholesterol, increased clearance of remnant particles, and lower Lp(a), depending on the type of hormone used. HRT is associated with increased risk of deep vein thrombosis, cancer, cardiovascular disease and stroke.

Further Reading

1. ASRM Practice Committee. Current evaluation of amenorrhea. *Fertil Steril* 2008; 90:S219–S215.
2. Balen AH, Rutherford AJ. Management of infertility. *BMJ* 2007; 335:60.
3. Braunstein G.D. Gynecomastia. *N Engl J Med* 2007; 357:1229–1237.
4. De Vos M, Devroey P, Fauser CJM. Primary ovarian insufficiency. *The Lancet* 2010; 376:911–921.
5. Dickerson, EH, Raghunath AS, Atkin SL. Rational testing: Initial investigation of amenorrhoea. *BMJ* 2009; 339:10.113.
6. Gordon CM. Functional hypothalamic amenorrhoea. *N Engl J Med* 2010; 363:365–371.
7. Lain KY. Metabolic changes in pregnancy. *Clin Obste Gynaecol* 2007; 50:938–948.

8. Norman RJ, Dewailly D, Legro RS, Hickey TE. Polycystic ovary syndrome. *The Lancet* 2007; 370:685–697.
9. Sathyapalan T, Atkin SL. Rational testing: investigating hirsutism. *BMJ* 2009; 338:10.1136.

Summary/Key Points

1. Testosterone is the main androgen produced by the Leyding cells of the testis and is regulated by LH from the pituitary by a negative feedback mechanism. Spermatogenesis is regulated by FSH.
2. Sex hormones are bound to proteins, mainly SHBG and albumin.
3. Hypogonadism in the male can arise due to failure of the testes (hypergonadotrophic) or failure of the pituitary/hypothalamus (hypogonadotrophic). Important causes include gonadal failure due to infection, injury, irradiation and developmental defects like Klinefelter's syndrome or Turner's syndrome.
4. Gynaecomastia is physiological during puberty. Pathological causes include testosterone deficiency and rarely due to increase in oestrogen production from tumours. Use of anabolic steroids by body builders and athletes is now becoming an increasingly common cause.
5. Menstrual cycle consists of the follicular and leutal phases and ovulation. In the follicular phase, FSH stimulates the development of follicles, which secrete oestrogens. Rising oestrogens cause a surge in LH, which induces ovulation.
6. During the luteal phase, the follicle become the corpus luteum and secretes progesterone, which prepares the endometrium for implantation. If pregnancy does not occur, the fall in progesterone causes spasm of spiral arteries and shedding of endometrium, menses.
7. Amenorrhoea can be physiological (pregnancy, menopause and pre-pubertal) or pathological.
8. Common causes of pathological amenorrhoea are functional hypothalamic hypogonadism (secondary to nutritional disorders, heavy exercise or stress) hyperprolactinaemia, PCOS and ovarian failure.

9. Investigation of amenorrhoea requires measurement of FSH, LH, prolactin, oestradiol and, if necessary, thyroid function tests.

10. Androgens in the female are produced from the ovary (androstenedione and testosterone) and adrenal (androstenedione and DHEA) or by peripheral conversion of androstenedione and DHEA to testosterone.

11. Hirsutism defined as excessive male pattern of hair growth is commonly due to PCOS, congenital adrenal hyperplasia or, rarely, testosterone-producing tumours or drugs.

12. PCOS is a common endocrine disorder causing menstrual irregularities, infertility and hirsutism. It is associated with insulin resistance and obesity.

13. In pregnancy, the placenta secretes a large number of hormones including hCG, HPL, oestrogens, progesterone and CRH.

14. Measurement of hCG in early pregnancy is useful to detect pregnancy and to diagnose ectopic pregnancy or failure of pregnancy.

15. Screening for Down's syndrome in early pregnancy is done using serum markers (a combination of β-hCG, pregnancy associated plama protein A (PAPA), inhibin, AFP and oestradiol) and nuchal translucency (by ultrasound).

16. Menopause, cessation of menses, is associated with symptoms in some women and HRT can relieve these.

chapter 21

The Adrenal Gland

Introduction

The adrenal gland consists of two distinct areas, a yellow outer cortex and a grey inner medulla. The adrenal medulla synthesises catecholamines (see Chapter 12). Three discrete zones can be seen in the adrenal cortex; zona fasciculata, zona reticularis and zona glomerulosa. The adrenal cortex synthesises three types of steroids: glucocorticoids, mineralocorticoids and androgens. Glucocorticoids and adrenal androgens are synthesized in the two inner layers while aldosterone, the mineralocorticoid, is synthesized in the outer zona glomerulosa. The foetal adrenal has a fourth layer which is involved in the production of oestriol.

Glucocorticoids

Cortisol is the major glucocorticoid synthesised from cholesterol by the adrenal cortex. The major synthetic pathway of cortisol and other steroids are shown in Figure 21.1. Cortisol secretion is controlled by ACTH (Figure 21.2), which in turn is regulated by corticotropin-releasing hormone (CRH). ACTH and CRH in turn are regulated by negative feedback by cortisol. The action of CRH on the pituitary is potentiated by antidiuretic hormone/arginine vasopressin (AVP) produced by the hypothalamus. CRH acts via a cyclic AMP second messenger system, whereas antidiuretic hormone alters intracellular ion channels. ACTH increases cortisol synthesis within 3 minutes by stimulating cholesterol esterase. The long-term effects of ACTH

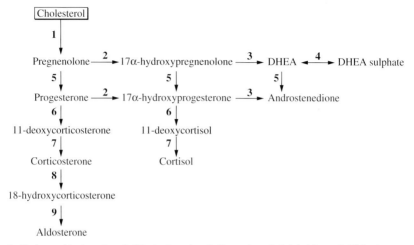

1. Cholesterol hydroxylase 2. 17α-hydroxylase 3. Desmolase 4. Sulphokinase 5. 3β-hydroxysteriod dehydrogenase 6. 21α-hydroxylase 7. 11β-hydroxylase 8 & 9. Aldosterone synthase

Figure 21.1   Synthesis of corticosteroids.

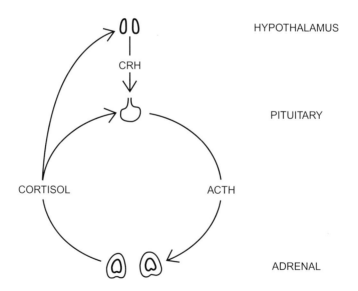

Figure 21.2   Feedback control of glucocorticoids — HPA axis.

include induction of transcription of the genes that encode steroido-genic enzymes. ACTH is secreted in a pulsatile fashion which is superimposed on a marked diurnal rhythm with a peak at 5 a.m. and a nadir at midnight. ACTH, which is transported unbound in plasma, has a half-life of about 10 minutes. Circulating cortisol has a slow and a fast negative feedback on the hypothalamic-pituitary secretion of ACTH and CRH. Fast feedback inhibits the release of CRH and the CRH mediated ACTH secretion. Slow feedback results from decreased synthesis of CRH. The secretion of CRH is stimulated by stress, which overrides any diurnal variation, and by the cytokines IL-1, IL-6 and tumour necrosis factor-α (TNF-α). This pathway is part of the link that exists between the neuroendocrine and immune systems. Cytokines can also stimulate cortisol secretion directly and this may explain the increase in cortisol during sepsis where there is no paral-lel increase in ACTH. The half-life of cortisol in circulation is about 100 minutes.

Actions of Glucocorticoids (Table 21.1)

Cortisol is essential for life and its effects are seen on cardiovascular, immunological and homeostatic systems and metabolism. Glucocorticoids bind to its receptor in the cytoplasm and this complex is translocated to the nucleus where it regulates gene

Table 21.1   Actions of glucocorticoids

Carbohydrate metabolism	— increases hepatic gluconeogenesis
	— increases hepatic glycogen synthesis
	— reduces glucose utilisation
Protein metabolism	— increases catabolism
Fat metabolism	— increases lipolysis
Water metabolism	— permissive effect — required to excrete a water load
Cardiovascular system	— sensitises arterioles to catecholamines and essential for maintenance of BP
Anti-inflammatory	

expression; upregulating some genes, such as those regulating anti-inflammatory proteins or gluconeogenesis and downregulate others such as pro-inflammatory genes. In addition, cortisol also has non-genomic effects via the stimulation of phosphatidyl inositol 3-kinase.

Glucocorticoids promote gluconeogenesis and liver glycogen synthesis and reduce glucose utilisation. Protein catabolism is increased which gives rise to more gluconeogeneic substrates and amino-acid uptake by peripheral tissues for protein synthesis is inhibited. The deposition of fat with a central distribution (face, neck and trunk) is increased. Glucocorticoids have important anti-inflammatory and immunosuppressive effects and these effects are used therapeutically to treat conditions such as rheumatoid arthritis. The immunosuppressive effect of glucocorticoids results from the gradual destruction of the lymphoid tissue followed by decrease in antibody production and the number of eosinophils, basophils and lymphocytes. The anti-inflammatory and immunosuppressive effects are through the modulation of production of cytokines. Glucocorticoids also have anti-allergic properties mediated by the inhibition of histamine synthesis by mast cells and basophils. The glucocorticoid activity of various natural steroids and synthetic steroids are compared in Table 21.2.

Table 21.2   Relative potencies of corticosteroids

	Glucocorticoid activity	Mineralocorticoid activity
Cortisol	1	1
Cortisone	0.7	0.7
Corticosterone	0.2	2
11-Deoxycorticosterone	0	20
Aldosterone	0.3	600
Fludrocortisone	10	200
Prednisone	4	0.7
Prednisolone	4	0.7
Dexamethasone	30	02
Trimacinolone	3	0
6-Methylprednisolone	5	0.5

Glucocorticoids have important effects on foetal development. Cortisol is essential for excretion of water load. In the absence of cortisol, there is a stimulation of ADH secretion, which prevents excretion of water load. Cortisol reduces bone formation and reduces calcium absorption. It increases sensitivity of heart and blood vessels to catecholamines.

Mineralocorticoids

Aldosterone, the major mineralocorticoid, is synthesised exclusively in the zona glomerulosa. The synthesis of aldosterone is outlined in Figure 21.1. The synthesis of aldosterone requires the enzyme aldosterone synthase that is present mainly in this zone. Deoxycorticosterone (DOC), 18-dehydroxy-DOC, corticosterone and cortisol have some mineralocorticoid activity (Table 21.2). The major control mechanism for the secretion of aldosterone is the renin–angiotensin system (Figure 21.3). Renin, a proteolytic enzyme secreted by the juxtaglomerular (JG) apparatus in the kidney, is

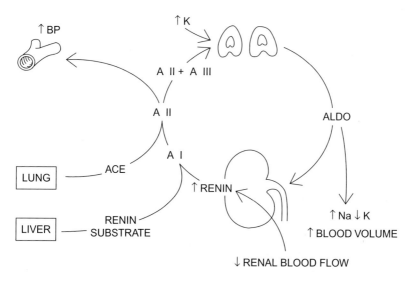

Figure 21.3   Regulation of aldosterone.

stimulated by the decrease in perfusion pressure to the JG apparatus, sympathetic stimulation, a decrease in Na^+ or Cl^- delivery to the macular densa and prostaglandins. Renin acts on circulating angiotensinogen, a protein produced by the liver and converts it to angiotensin I. This is converted to angiotensin II by the angiotensin-converting enzyme, which is found mainly in the lungs, but is also found in the vascular endothelium. Angiotensin II is a powerful vasoconstrictor and also acts on the zona glomerulosa to increase the synthesis and secretion of aldosterone. It also has a negative feedback effect on renin release. Other important factors that increase aldosterone secretion are ACTH and high plasma potassium concentration. ACTH is not important physiologically in the regulation of aldosterone secretion, but may have a more prominent role when the renin–angiotensin system is suppressed.

Actions of Aldosterone

Aldosterone increases sodium reabsorption in the kidney, colon and salivary gland. In the distal convoluted tubule, aldosterone binds to the mineralocorticoid receptor in the principal cells and increases the number of epithelial sodium channels, which is normally the rate-limiting step for the reabsorption of sodium. It increases potassium secretion via specific epithelial channels. It also stimulates the synthesis of the basolateral sodium–potassium–ATPase that generates the electrochemical gradient necessary for the movement of sodium and potassium. The mineralocorticoid receptors in the renal tubules are non-selective, mineralocorticoids and glucocorticoids have equal affinity. The specificity of this receptor for mineralocorticoids is achieved by the presence of the enzyme 11β-hydroxysteroid dehydrogenase type 2 that metabolises cortisol to the inactive cortisone which does not bind to the receptor. If the enzyme is inhibited, as for example by, glycyrrhizic acid present in liquorice, cortisol will bind to the mineralocorticoid receptor and cause increased sodium reabsorption. Aldosterone stimulates hydrogen secretion by the intercalated cells in the collecting ducts. It also stimulates sodium and water reabsorption by the colon, salivary glands and sweat glands.

Adrenal Androgens

Adrenal androgens are synthesised by the zona fasciculata and zona reticularis. The major adrenal androgens are androstenedione, and dehydroepiandrosterone (DHEA) and DHEA sulphate. Quantitatively, DHEA and DHEA sulphate are the most important adrenal androgens. ACTH stimulates the secretion of adrenal androgens. Adrenal androgens DHEA, DHEA sulphate and androstenedione are converted to testosterone and dihydrotestosterone by peripheral tissues mainly hair follicle, the sebaceous glands, the prostate and the external genitalia. In the adipose tissue, androgens are converted to oestrogens. Peripheral conversion accounts for most of the circulating testosterone in the female.

Transport of Steroid Hormones

Steroid hormones are transported in the blood mainly bound to proteins produced by the liver. About 95% of cortisol is bound to transcortin or cortisol binding globulin (CBG), albumin and sex hormone-binding globulin albumin (SHBG). Testosterone and dihydrotestosterone are carried by SHBG. DHEA and DHEA sulphate circulate primarily bound to albumin. Prednisolone is the only synthetic glucocorticoid that binds to CBG whereas other synthetic steroids like dexamethasone and methylprednisolone bind primarily to albumin. The functions of CBG and SHBG are to serve as circulating stores and to protect the hormones from inactivation. The concentration of CBG is increased by oestrogen and is decreased in chronic liver disease. Approximately 60% of aldosterone in circulation is bound to albumin. It is generally agreed that the free fraction and the albumin-bound fraction of steroid hormones are available for binding to receptors. However, this concept has been challenged recently as some plasma membrane receptors can bind to carrier protein–hormone complexes.

Metabolism of Steroid Hormones

Steroid hormones are primarily metabolised and conjugated in the liver and less than 2% of cortisol is excreted unchanged in the urine.

At normal cortisol concentration, CBG is almost fully saturated. If cortisol secretion increases, the binding capacity of CBG will be exceeded and the excretion of cortisol in the urine increases disproportionately.

Factors Affecting the Secretion and Metabolism of Steroids

Circadian rhythm is an important physiological factor affecting steroid hormones. Plasma cortisol concentration is highest in the morning and lowest at night (Figure 21.4). Acute stress increases cortisol as well as adrenal androgens, DHEA and DHEA sulphate. Chronic illness increases plasma cortisol and suppresses DHEA, DHEA sulphate and androstenedione concentration. At birth, more cortisone than cortisol is produced. The adult pattern of predominant cortisol production is seen within a few days of birth. In old age, the production and metabolic clearance of cortisol is 25% lower than that in young adults. However, stimulation of cortisol by ACTH remains normal. Age and familial factors affect plasma adrenal androgens, DHEA and DHEA sulphate, in men and women. Adrenal androgens start to rise at the age of 9, peak at the age of 25 and decline after 30 years. At the age of 70 and over, the adrenal androgen concentrations fall to prepubertal levels. Thus, it has been suggested that

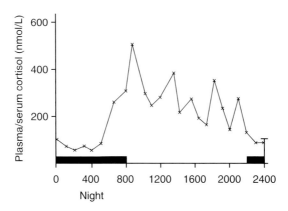

Figure 21.4   Variation in plasma cortisol concentration during the day.

DHEA and DHEA sulphate may have a role in ageing. Oestrogens increase plasma concentration of CBG and thus influence total cortisol concentration; free cortisol is unaffected. The clearance of cortisol and its production rate, as well as urinary metabolites are all reduced by oestrogens. Obesity increases the excretion of urinary metabolites of cortisol, adrenal androgens, DHEA and DHEA sulphate. Plasma cortisol and urinary free cortisol are not affected. In starvation, the production and clearance of cortisol are decreased. As cortisol is metabolised in the liver, cortisol concentrations tend to be elevated in chronic liver disease. Hypothyroidism reduces and hyperthyroidism increases the secretion and metabolism of cortisol without affecting the circulating concentration. Drugs, such as phenytoin, phenobarbitone, rifampicin, which induce hepatic mixed function oxidase enzymes, can lead to an increase in metabolism of cortisol and other steroids including synthetic steroids.

Tests for Adrenal Cortical Function

Serum Cortisol

Serum cortisol shows marked diurnal variation and therefore should be interpreted in relation to the time of day. In some immunoassays, synthetic steroids, such as prednisolone, can cause significant cross-reaction. As stress is a powerful stimulus for cortisol secretion, attempts should be made be reduce stress during blood collection for cortisol. Because of the diurnal variation, as well as the sporadic nature of cortisol secretion, single cortisol secretion measurements are of little value in diagnosis. However, the demonstration of diurnal variation is highly suggestive that there is no adrenal pathology. If stress can be excluded, loss of diurnal variation is suggestive of adrenal hyperfunction.

Salivary Cortisol

Salivary cortisol is a reflection of unbound/free cortisol in plasma, therefore it is a better test than plasma total cortisol. Furthermore, collection of saliva is less stressful. It is very useful in children, subjects who are reluctant to have venepuncture and in determining midnight cortisol values. Midnight salivary sample can be collected by the patient without admission to hospital. However, this test is not widely available.

Urinary Cortisol

Unbound cortisol is filtered at the glomerulus and excreted in the urine (urine 'free' cortisol). Urinary cortisol measurements eliminate the fluctuation in plasma cortisol and are therefore a better test of adrenal function. Cortisol binding globulin is near its saturation point at normal plasma cortisol concentrations. When there is hyperfunction of the adrenal cortex, the increase in free (unbound) cortisol is greater than the increase in total cortisol and this is reflected in the urine cortisol.

ACTH

Plasma ACTH concentration, like cortisol, fluctuates throughout the day. However, it is a useful test in the differential diagnosis of hypo- and hyperadrenal function, e.g. undetectable or low ACTH concentration in a patient with adrenal hyperfunction points to an adrenal adenoma (see later).

Dynamic Function Tests

• *Stimulation test*

   Stimulation tests are useful in demonstrating hypofunction of the adrenal gland.

(1) ACTH stimulation test

   This examines the reserve capacity of the adrenal gland. In the short ACTH stimulation test, serum cortisol concentration is measured before, 30 and 60 minutes after injection of 250 μg of synthetic ACTH, tetracosactrin (Synacthen). Serum cortisol should increase to > 550 nmol/L or the increment should be > 200 nmol/L. A normal response excludes adrenal hypofunction. In the prolonged or long ACTH (Synacthen) stimulation test, a slow release (depot) preparation of ACTH is given and the increase in cortisol is measured. This helps to differentiate between primary adrenal failure and secondary adrenal failure. In secondary adrenal failure, the adrenal glands atrophy as a result of the lack of ACTH. Prolonged stimulation is required before a response is seen.

(2) CRH stimulation test

Measurements of the ACTH and cortisol secretion following the injection of CRH is useful in the differential diagnosis of adreno-cortical hyperfunction.

(3) Insulin tolerance test

As hypoglycaemia is a potent stimulus for CRH and ACTH, this test is sometimes used to assess the hypothalamic–pituitary–adrenal axis reserve.

• *Suppression tests*

Suppression tests are useful in the diagnosis and differential diagnosis of adrenal hyperfunction. Dexamethasone is a potent synthetic steroid which suppresses CRH and ACTH and hence cortisol. Furthermore, it does not cross react in the cortisol assays. In the overnight dexamethasone suppression test, serum cortisol is measured after administration of 1 mg of dexamethasone at night. A decrease in cortisol concentration to 50 nmol/L or lower excludes adrenal hyperfunction. However, failure to suppress serum cortisol after dexamethasone may occur in obesity, subjects who are noncompliant, those who rapidly metabolise dexamethasone, alcohol abuse, and in acute illness or depression. In the 2-day version of the test, dexamethasone 0.5 mg is given 6 hourly for 2 days and serum cortisol is measured on the third morning. A decrease in cortisol concentration to 50 nmol/L or lower excludes adrenal hyperfunction. High-dose dexamethasone suppression test is used to differentiate the causes of Cushing's syndrome. In this test, 2-mg dexamethasone is given every 6 hours for 2 days followed by cortisol measurement. In patients with Cushing's syndrome due to pituitary adenoma, serum cortisol decreases by 50%.

Disorders of Adrenal Function

Adrenal Insufficiency

Adrenal insufficiency is usually classified as primary when the adrenal is destroyed or secondary when there is pituitary/hypothalamic disease causing suppression of the adrenals. The causes of adrenal insufficiency are outlined in Table 21.3. The most common cause of adrenal hypofunction is the therapeutic use of glucocorticoids leading to suppression of the pituitary–adrenal axis. These patients will develop symptoms of adrenal insufficiency when the steroids are withdrawn or if the steroids are not increased during an intercurrent illness. The second most common cause is adrenal atrophy due to autoimmune disease although infections such as tuberculosis and

Table 21.3   Causes of adrenal insufficiency

Primary
- Autoimmune adrenal atrophy
- Granulomatous disease — tuberculosis, sarcoidosis
- Adrenalectomy
- Neoplastic infiltration
- Metabolic disorders — haemochromatosis, amyloidosis, adrenoleukodystrophy
- Acquired immune deficiency syndrome (AIDS)
- Congenital adrenal hyperplasia
- Adrenal haemorrhage — Waterhouse-Frederickson syndrome

Secondary
- Chronic glucocorticoid administration
- Hypopituitarism — tumour
                       — post-surgical

AIDS are becoming increasingly common. About 50% of patients with autoimmune adrenal insufficiency have other autoimmune endocrine diseases such as hypoparathyroidism and hypothyroidism. These are called polyglandular autoimmune syndrome type 1 (associated with thyroid disease and diabetes mellitus) or type 2 (associated with hypoparathyroidism).

In primary adrenal insufficiency, the clinical features are due to the lack of mineralocorticoids and glucocorticoids (Table 21.4). Adrenal failure is usually of insidious onset but may present acutely when intercurrent illnesses or stress such as trauma is interposed in a patient with previously unrecognised adrenal insufficiency. Patients with acute adrenal insufficiency present with severe hypovolaemia, hypoglycaemia and shock. In the chronic state, they present with non-specific features like tiredness, weakness, anorexia, nausea and apathy. There will be hyperpigmentation due to increased ACTH secretion and the accompanying melanocyte-stimulating hormone (MSH). Hypoglycaemia, due to the unopposed action of insulin may be present. Postural hypotension is often found due to decreased ECF volume as a result of lack of mineralocorticoids. Lack of glucocorticoids reduces the sensitivity of the arterioles for catecholamines contributing to the hypotension. Hyponatraemia caused by the inability of the kidney to excrete a water load presumably due to an increase in ADH secretion, and hyperkalaemia due to lack of mineralocorticoids are features of adrenal failure. In secondary

Table 21.4 Clinical features of adrenal insufficiency

Tiredness, lethargy
Generalised weakness
Anorexia, nausea, vomiting
Weight loss
Dizziness
Postural hypotension
Pigmentation
Loss of body hair
Depression
Hypoglycaemia

hypoadrenalism, hyperpigmentation is absent and serum potassium is normal as mineralocorticoid secretion is well preserved. Hyponatremia is present due to inability to excrete a water load. They usually have other features of hypopituitarism.

Diagnosis

Laboratory findings in a patient with adrenal insufficiency include hyponatraemia, hyperkalaemia, mild metabolic acidosis, and increased plasma urea concentration. Basal plasma ACTH concentration will be high in primary adrenal failure. Basal 9 a.m. cortisol of < 50 nmol/L is strongly indicative of adrenal insufficiency and a value of > 550 nmol/L strongly excludes it. In most patients, adrenal insufficiency is diagnosed by ACTH stimulation test. Failure to adequately increase plasma cortisol after an injection of 250 μg of ACTH (to 550 nmol/L or >) is diagnostic of hypoadrenalism. A high plasma ACTH is indicative of primary adrenal failure. In secondary adrenal failure, plasma ACTH will be inappropriately low and a prolonged ACTH stimulation test will show an adequate response.

Management of hypoadrenalism

Adrenal insufficiency is a life-threatening condition and the treatment should be started as early as possible and treatment is life-long. Patients with adrenal insufficiency should be followed or monitored regularly and they should carry a medical alert bracelet or necklace to alert health care workers to administer steroids in an emergency. In patients with primary adrenal failure, mineralocorticoids and glucocorticoids should be replaced. During intercurrent illness, the glucocorticoid administration should be increased. Assessment of adequacy of treatment is difficult and consists of a combination of clinical and laboratory assessments. Measurement of renin activity is sometimes used as assessment of adequacy of mineralocorticoid replacement; a suppressed plasma renin activity indicates overreplacement and an elevated value suggests under replacement. Glucocorticoid replacement can be assessed clinically by general well-being, measurements of blood pressure, urinary free cortisol or cortisol measurement throughout the day (cortisol day curve).

Case 21.1

A 35-year-old lady was referred to the endocrinology clinic for investigation of tiredness, amenorrhoea, weight loss and low blood pressure. She complained that she had no energy to do even simple tasks.

Initial investigations showed the following results:

Serum		Reference Range
Sodium (mmol/L)	122	135–145
Potassium (mmol/L)	5.6	3.5–5.0
Bicarbonate (mmol/L)	19	23–32
Chloride (mmol/L)	98	90–108
Urea (mmol/L	16.5	3.5–7.2
Creatinine (μmol/L)	92	60–112
Glucose (mmol/L)	3.0	

History of tiredness and low blood pressure together with hyponatraemia and hyperkalaemia suggested adrenal failure and a short ACTH stimulation test was done. Her plasma cortisol before and 30 minutes after 250 ug of ACTH (Synacthen) were 210 and 275 nmol/L, respectively. This confirmed adrenal failure and she was started on hydrocortisone and fludrocortisone and her symptoms dramatically improved.

Adrenal Hyperfunction

Corticosteroid excess/Cushing's syndrome

Cushing's syndrome is the result of chronic exposure to high levels of glucocorticoids. Causes of Cushing's syndrome are listed in Table 21.5. Causes can be divided into ACTH dependent (pituitary adenoma or ectopic ACTH syndrome) or ACTH independent (adrenal tumours). The most common cause of Cushing's syndrome is exogenous administration of steroids. The clinical manifestations of Cushing's syndrome

Table 21.5   Causes of Cushing's syndrome

Iatrogenic
ACTH dependent
- Cushing's disease (pituitary adenoma)
- Ectopic ACTH syndrome
  — Bronchial carcinoid tumours
  — Small cell carcinoma of lung
  — Medullary thyroid carcinoma
  — Others

ACTH independent
- Adrenal adenoma
- Adrenal carcinoma

Table 21.6   Clinical features of Cushing's syndrome

Obesity — truncal
Hypertension
Hyperglycaemia and decreased glucose tolerance
Menstrual irregularity
Hirsuitism
Plethora
Rounded face
Thin skin
Easy bruising
Decreased libido
Striae
Psychiatric disturbances — depression, mania, euphoria
Osteoporosis or fracture
Weakness/proximal myopathy

include truncal obesity, hypertension, hypokalaemic alkalosis, carbohydrate intolerance, disturbances in reproductive function, osteopenia and neuropsychiatric symptoms (Table 21.6). Obesity is present in up to 93% of patients with Cushing's syndrome. Other clinical features which suggest Cushing's syndrome include mild proximal muscle weakness, easy bruising, thinness and fragility of the skin. The majority of the clinical features are due to excess glucocorticoids but some of the features may be due to the secretion of other hormones, e.g. androgens which give rise to hirsuitism. In Cushing's syndrome, cortisol precursors are increased, and these, together with the high cortisol concentration, can cause a significant mineralocorticoid effect causing sodium retention, hypertension, potassium wasting and hypokalaemic alkalosis. The classical features of Cushing's syndrome are fewer in patients who develop the condition rapidly. In such patients, muscle weakness, electrolyte disturbances (hypokalaemia) and pigmentation are more often seen.

Pseudo-Cushing's syndrome is a syndrome where patients have some clinical and biochemical features of Cushing's syndrome but they do not have Cushing's syndrome. This syndrome is very often seen in severe depression and in alcoholics.

- *Diagnosis of Cushing's syndrome*

  The cardinal biochemical features of Cushing's syndrome include excessive endogenous secretion of cortisol, loss of diurnal rhythm and loss of feedback of the hypothalamic–pituitary–adrenal axis. Diagnostic tests used in the first stage of diagnosis of Cushing's syndrome are based on these biochemical features. The initial biochemical test used in outpatients is the overnight dexamethasone suppression test. A morning serum cortisol of 50 nmol/L or lower after 1-mg dexamethasone given the previous night is highly effective in excluding Cushing's syndrome. However, there may be reasons for false positive results (see above). A raised urinary free cortisol excretion, which integrates the intermittent cortisol secretion, is consistent with Cushing's syndrome. However, one of the major problems associated with this test is the reliability of urine collection and it may be within the reference range in up to 10% of patients with Cushing's syndrome. Furthermore, urinary free cortisol is also raised in depressed patients, in patients with polycystic ovary syndrome and in various other pseudo-Cushing states. Urinary cortisol assays by immunoassays tend to overestimate cortisol due to cross reactivity of other steroids in the urine.

  The assessment of diurnal rhythm is a highly specific test. The patient needs to be admitted to the hospital to perform this test. However, the stress of hospitalisation and waking patients up in the middle of the night may increase cortisol secretion in normal subjects. To avoid false positive results due to anticipation, it is suggested that the patient should not be warned when the test is to be performed. A single undetectable midnight cortisol value is highly specific in excluding Cushing's syndrome.

Salivary cortisol is an attractive alternative to plasma cortisol. It reflects serum-free cortisol and collection of saliva does not induce a stress response. Late-night salivary cortisol is a useful screening test for Cushing's syndrome.

The 2-day low-dose dexamethasone test has high sensitivity and moderate specificity in diagnosing Cushing's syndrome. Variations in absorption and metabolism of dexamethasone, however, may affect the reliability of this test. In patients taking drugs that can induce liver enzymes, such as anticonvulsants like phenytoin, metabolism of dexamethasone may be increased causing false negative results.

Differentiation of pseudo-Cushing's syndrome from mild Cushing's syndrome can prove extremely difficult. In both these situations there is hypercortisolaemia, thus the 24-hour urinary free cortisol may not differentiate between the two conditions. It has been suggested that a combination of tests may be more successful, e.g. a CRH test after the 48-hour low-dose dexamethasone test. Cortisol response to insulin-induced hypoglycaemia test is useful in differentiating pseudo-Cushing's syndrome due to depressive illnesses. Other tests which have been shown to differentiate these two conditions include the use of loperimide, an opiate agonist, and desmopressin. Patients with alcohol-induced pseudo-Cushing's syndrome may show a detectable blood alcohol level and alcohol withdrawal usually returns the biochemical abnormality to normal in about 2 days.

- *Differential diagnosis of Cushing's syndrome*

A low or undetectable ACTH concentration is suggestive of adrenal adenoma and a very high ACTH concentration is suggestive of an ectopic ACTH syndrome. Presence of hypokalaemia is usually indicative of ectopic ACTH syndrome. However, up to 10% of patients with Cushing's disease may also show hypokalaemia. Cosecretion of other hormones such as calcitonin and α-fetoprotein (AFP) may indicate the presence of ectopic ACTH syndrome.

(1) High-dose dexamethasone test

The high-dose dexamethasone test is the classical test for the diagnosis of pituitary dependent Cushing's. The basis of this test is that the corticotroph tumour cells in Cushing's disease retain some responsiveness to the negative feedback effects of glucocorticoids, while in ectopic ACTH syndrome the feedback is lost. In Cushing's disease, 48 hours after high dose dexamethasone (2 mgs every 6 hours for 48 hours) plasma cortisol is suppressed by at least 50%. Diagnostic sensitivity and specificity have been reported to be 81% and 67% respectively.

(2) CRH test

Plasma ACTH or cortisol concentration after administration of CRH increases by 50% in patients with Cushing's disease whereas in ectopic ACTH syndrome, the response is less. The response seen in Cushing's disease is probably due to the greater expression of CRH receptor in the corticotroph cells. This test has sensitivity of 86% and a specificity of 95%.

(3) Invasive investigations — Inferior petrosal sinus sampling for ACTH

In this procedure, bilateral sampling from the inferior petrosal sinuses before and after administration of CRH is done. In Cushing's disease, there is a gradient between the central and peripheral ACTH concentrations whereas in non-pituitary ACTH-dependent Cushing's syndrome, this is not seen. The difficulty in the procedure is the availability of expertise to introduce the catheters in the right place.

(4) Imaging

Non-invasive investigations such as CT scan and magnetic resonance imaging (MRI) will localise pituitary adenomas. However, some adenomas are a few millimetres in diameter and may not be visualised. This is further complicated by the observation that up to 10% of patients may have pituitary incidentalomas, i.e. adenomas found on incidental scanning without any functional activity.

- *Management*

Management of Cushing's syndrome depends on the cause. In Cushing's disease, transphenoidal surgery and removal of the pituitary adenoma is the treatment of choice. However, there is a significant failure rate or relapse rate and bilateral adrenalectomy may be necessary to relieve symptoms. If bilateral adrenalectomy is done, the pituitary should be treated by radiation; otherwise, the tumour may increase in size due to lack of negative feedback leading to excessive ACTH secretion and subsequently hyperpigmentation (this is described as Nelson's syndrome). Patients after transphenoidal resection should be carefully followed up as they may require steroid replacement treatment until the suppressed adrenal glands recover. Sometimes, this may take as long as 12 months. In patients who cannot undergo surgery, metyrapone or other blocking agents such as mitotane, ketoconazole and etomidate could be used to suppress cortisol secretion.

Patients with adrenal adenoma require adrenalectomy and these patients require careful monitoring post-operatively. They may require steroid replacement until the remaining adrenal gland recovers from chronic suppression.

Case 21.2

A 32-year-old Nigerian student was referred by his GP for further investigation and management of hypertension. His blood pressure was 175/120 mmHg inspite of treatment with several anti-hypertensive drugs. At the hospital, the medical registrar thought he looked Cushinoid and did some initial investigations.

Serum		Reference Range
Sodium (mmol/L)	142	135–145
Potassium (mmol/L)	3.5	3.5–5.0
Urea (mmol/L)	4	2.5–7.2
Creatinine (μmol/L)	110	66–122
9 a.m. Cortisol (nmol/L)	625	180–720
Cortisol after overnight dexamethasone (nmol/L)	450	

Failure to suppress cortisol after overnight dexamethasone suggested the possibility of Cushing's syndrome. He was admitted for further investigation. The results of low dose and high dose dexamethasone tests were as follows:

	Cortisol (nmol/L)	
Before low dose	565	
After low dose	445	
After high dose	55	
ACTH (ng/L)	95	10.0–65.0

Failure to suppress cortisol after low dose dexamethasone confirmed Cushing's syndrome.

Administration of high dose dexamethasone reduced the plasma cortisol to 55 nmol/L. This, together with a high ACTH, suggested a pituitary adenoma. Pituitary imaging confirmed the presence of an adenoma. He underwent a trans-sphenoidal hypophysectomy. Six months after surgery a repeat overnight dexamethasone test was performed and his cortisol was < 55 nmol/L. This confirmed that the surgery was successful.

Congenital Adrenal Hyperplasia

Congenital adrenal hyperplasia (CAH), also called adrenogenital syndrome, is a group of inherited disorders of the biosynthetic pathway of the adrenal steroid hormones. As a result of the deficiency or a defect in the enzymes, cortisol biosynthesis is impaired leading to a decreased cortisol concentration and a compensatory increase in ACTH release, which causes hyperplasia of the adrenal cortex. Intermediate metabolites accumulate proximal to the block and substrates may be shunted towards the adrenal androgen pathway. The term 'adrenogenital syndrome' is used as this describes the effects of the increased adrenal androgens on the genitalia and secondary sex characteristics. In girls, this will lead to ambiguous gentilia at birth and in boys it may lead to precocious puberty or accelerated growth later on. Some defects may be associated with hypertension and others with salt-wasting. CAH may also be seen in

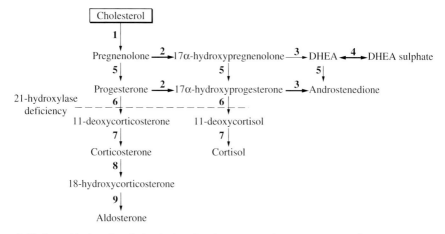

1. Cholesterol hydroxylase 2. 17α-hydroxylase 3. Desmolase 4. Sulphokinase 5. 3β-hydroxysteriod dehydrogenase 6. 21α-hydroxylase 7. 11β-hydroxylase 8 & 9. Aldosterone synthase

Figure 21.5 Enzyme defects in congenital adrenal hyperplasia.

adults with subtle abnormalities leading to hirsuitism, menstrual irregularities and sub-fertility in women.

Deficiency of the 21-hydroxylase enzyme (Figure 21.5) is the most common form of CAH with an incidence of 1:12,000 live births and accounts for over 95% of all cases. Serum 17α-hydroxy progesterone (17-OHP), 17-hydroxypregnenolone, progesterone, pregnenolone, DHEA sulphate and DHEA are all elevated. An increase in plasma androstenedione and testosterone is also seen as a result of the conversion of adrenal androgens in the peripheral tissues. Clinical presentation of CAH depends on the nature and severity of the enzyme deficiency and on the gender. About 50% of patients with 21-hydroxylase deficiency have inadequate aldosterone in addition to cortisol. These patients present with salt-wasting. Female infants with severe forms of CAH present with ambiguous genitalia and salt-wasting. If there is no salt wasting, female infants may be brought up as males until late in life. Milder forms of 21-hydroxylase deficiency present late in children with precocious puberty, clitoromegaly, accelerated growth and skeletal maturation. Females with very mild forms of CAH may present in adolescents or late life with menstrual irregularities, hirsutism and/or infertility. Male infants with severe forms of 21-hydroxylase deficiency may not be diagnosed soon after birth as the genitals are normal. They may present at 1–4 weeks with failure to thrive, vomiting, dehydration and hypotension due to salt wasting. Less severe forms may present late in childhood with pubic hair, phallic enlargement and accelerated skeletal growth.

Early onset cases with salt wasting will show hyponatraemia, hyperkalaemia and/or hypoglycaemia. Diagnosis can be established by measurement of serum 17-OHP. In late onset cases, an ACTH stimulation test with simultaneous measurement of 17-OHP may be necessary. In CAH, there will be an exaggerated rise in 17-OHP.

Treatment involves replacement with cortisol and a mineralocorticoid if there is salt-wasting. Treatment with glucocorticoids will suppress the excessive ACTH production and hence the production of androgens. Adequacy of treatment is monitored by measurement of serum androstenedione and 17-OHP.

Deficiency of 11β-hydroxylase is the second most common form of CAH and is associated with manifestations of virilisation, hypertension and elevated concentrations of plasma androstenedione, and DHEA sulphate. These patients develop hypertension due to high DOC. 11-Deoxycortisol concentrations are markedly elevated in this defect.

Deficiency of 3β-hydroxysteroid dehydrogenase is very rare and in severe forms of this disorder, female infants have pseudohermaphroditism and male infants present with incomplete masculinisation. Deficiency of C17, 20-lyase/17α-hydroxylase is also a rare syndrome, leading to pubertal failure in females and pseudohermaphroditism in males.

Hyperaldosteronism

Hyperaldosteronism may be either primary, when there is autonomous secretion of aldosterone, or secondary, when the hypersecretion is caused by a stimulus outside the adrenal gland activating the renin–angiotensin system. Secondary aldosteronism is a common finding and is present in patients with congestive heart failure, nephrotic syndrome, cirrhosis of the liver and other conditions associated with depletion of ECF volume. Diuretic use is a common cause of secondary hyperaldosteronism due to a decrease in ECF volume. All causes of increased aldosterone may cause hypokalaemic alkalosis. In secondary hyperaldosteronism, measurements of renin activity and aldosterone concentration are rarely required and if measured, both will be elevated. Renal artery stenosis as a cause of secondary hyperaldosteronism is discussed in Chapter 12.

Primary hyperaldosteronism

The causes of primary hyperaldosteronism are listed in Table 21.7. The commonest cause is an adenoma of the adrenal gland and this is called 'Conn's syndrome'. Primary hyperaldosteronism is characterised by hypertension and hypokalaemia and accounts for 1–4% of hypertensive patients. In any hypertensive patient with hypokalaemia, primary hyperaldosteronism should be excluded. However, many more hypertensive patients may have hypokalaemia due to diuretic treatment and before further investigations are undertaken secondary hyperaldosteronism due to diuretic treatment should be excluded.

Recent studies suggest that primary hyperaldosteronism may be present in up to 10% of patients with hypertension and many patients have normal serum potassium. Clinical manifestations of primary hyperaldosteronism are generally related to the effects of aldosterone over production. Increased retention of sodium by aldosterone leads to ECF volume expansion and hypertension. Increased tubular secretion of potassium and hydrogen ions leads to hypokalaemic alkalosis and features of hypokalaemia such as muscle weakness are also present.

Table 21.7   Causes of primary hyperaldosteronism

Adrenal adenoma (Conn's syndrome)
Bilateral adrenal hyperplasia
Glucocorticoid remediable hyperaldosteronism
Adrenal carcinoma producing aldosterone

- *Diagnosis*

Diagnosis of primary hyperaldosteronism depends on the demonstration of low renin and high aldosterone secretion. Initial investigations should include serum electrolytes, which will show hypokalaemic alkalosis and high or a high normal serum sodium concentration in primary hyperaldosteronism whereas in secondary hyperaldosteronism serum sodium concentration is usually at the lower end of the reference range. In the presence of hypokalaemia, urine potassium concentration > 30 mmol/L is suggestive of renal loss. Measurements of plasma renin activity (PRA) and aldosterone concentration at the clinic after the patient has rested comfortably for half an hour is adequate as an initial test. If the aldosterone:PRA ratio is greater than 2000, a diagnosis of primary hyperaldosteronism is almost certain and if the ratio is less than 800, this is highly unlikely.

Many antihypertensive drugs affect the renin:aldosterone ratio. Ideally, patients suspected of primary hyperaldosteronism should be taken off drugs listed in Table 21.8 and if necessary treated with bethanidine, doxazosin or prazosin and should be adequately hydrated before taking samples for the measurement of plasma renin activity and aldosterone. Any hypokalaemia should be corrected.

Table 21.8 Drugs which may affect the plasma renin activity (PRA) and aldosterone

Drugs	PRA	Aldosterone	Time to recover from the effect
ACE inhibitors	I	D	2 weeks
β-blockers	D	D	2 weeks
Ca channel blockers	I	D	2 weeks
Diuretics	I	I	2 weeks
Hypokalaemia	—	D	
NSAIDS	D	—	2 weeks
Oestrogens	I (substrate)[@]	—	6 weeks
Spironolactone	I	V	6 weeks

* I = Increase; D = Decrease; V = Variable
@Increases the substrate angiotensinogen.

However, this may not be practical in an outpatient setting and it is suggested that PRA:aldosterone ratio can be done as a screening test after stopping β blockers and ACE inhibitors. If the ratio is abnormal or equivocal, confirmatory tests are necessary. The basis of these confirmatory tests is the fact that in primary hyperaldosteronism, aldosterone secretion is not suppressed by volume expansion or pharmacological stimulus.

One such test is the saline infusion test in which subjects are given 1.25 L of 0.9% isotonic saline intravenously over 4 hours and plasma aldosterone is measured before and after the infusion. Failure to suppress aldosterone is indicative of primary aldosteronism. In the fludrocortisone suppression test after measuring basal aldosterone, fludrocortisone 0.1 mg is given 6 hourly together with 30 mmol of sodium (as slow sodium) three times a day for 4 days. Potassium supplements may be required to maintain plasma potassium. Plasma aldosterone is measured on day 4 and failure to suppress aldosterone is suggestive of primary hyperaldosteronism.

Once the diagnosis of primary hyperaldosteronism has been established, it is necessary to distinguish adenoma from bilateral adrenal hyperplasia. Localisation using imaging techniques such as a CT scan will usually demonstrate an adenoma. Adrenal scanning with iodocholesterol after suppressing glucocorticoid synthesis with dexamethasone is also helpful in localising the tumour. Sometimes, the tumours may be too small to be detected by imaging techniques. Biochemical tests used in the differential diagnosis of adenoma from bilateral hyperplasia include measurement of aldosterone and renin in the recumbent posture and after 2–4 hours of upright posture. Patients with adenoma usually show no change or a paradoxical fall in aldosterone in assuming the upright posture whereas patients with adrenal hyperplasia typically show a rise in plasma aldosterone. Most patients with adrenal adenoma also show elevated concentrations of aldosterone precursors, 18-hydroxycorticosteroids. In a small percentage of cases, it may be necessary to demonstrate high aldosterone concentration in one of the adrenal veins.

- *Management*

  Adrenalectomy is the treatment of choice for Conn's syndrome. If surgery is contraindicated or if there is bilateral hyperplasia, treatment with spironolactone or eplerenone may be used. A major side effect of spironolactone is gynaecomastia which is not seen with eplerenone.

Case 21.3

A 45-year-old lady was found to be hypertensive on routine examination. Before starting her on treatment her GP requested some basic investigations:

Full blood count — normal
Urine examination — was negative

Serum		Reference Range
Sodium (mmol/L)	144	135–145
Potassium (mmol/L)	3	3.5–5.0
Bicarbonate (mmol/L)	34	23–32
Chloride (mmol/L)	92	90–108
Urea (mmol/L)	4.2	3.5–7.2
Creatinine (μmol/L)	98	60–112

Hypertension and hypokalaemia raised the possibility of primary hyperaldosteronism. She was referred for further investigations to the hospital where a random plasma aldosterone and renin measurements were done and the results were:

Plasma aldosterone   1250 pmol/L (100–450)
Plasma renin activity   < 0.5 nmol/min/L (1.1–2.7)

A high aldosterone with low renin supports the diagnosis of primary hyperaldosteronism and the aldosterone renin ratio was > 2000 which is typical for primary hyperaldosteronism. Adrenal imaging showed an adenoma in her left adrenal and she underwent a left adrenalectomy. Three months after the operation, her blood pressure was normal.

Further Reading

1. Arlt W, Allolio B. Adrenal insufficiency. *The Lancet* 2003; 361: 1881–1893.
2. Bertagna X, Guignat L, Groussin L, Bertherat J. Cushing's disease. *Best Pract Res Clin Endocrinol Metab* 2009; 23:607–623.
3. Boscaro M, Arnaldi G. Approach to the patient with possible Cushing's syndrome. *J Clin Endocrinol Metab* 2009; 94:3121–3131.
4. Rossi GP. Prevalence and diagnosis of primary aldosteronism. *Curr Hypertens Rep* 2010; Jul 28.
5. Stowasser M, Taylor PJ, Pimenta E, Ahmed AH, Gordon RD. Laboratory investigation of primary aldosteronism. *Clin Biochem Rev* 2010; 31:39–56.
6. Young WF. Endocrine hypertension: Then and now. *Endocr Pract* 2010; 16:1–52.

Summary/Key Points

1. The adrenal cortex secretes cortisol (glucocorticoids), aldosterone (mineralocorticoid) and androgens (DHEA, DHEA sulphate and androstenedione).
2. Cortisol secretion is regulated by the pituitary/hypothalamus via ACTH and CRH, a negative feedback loop.
3. Aldosterone is produced by the outer layer of adrenal cortex, the zona glomerulosa, and is secreted in response to ECF volume status via the renin–angiotensin system. It is important for reabsorption of sodium to maintain ECF volume. Sodium reabsorption is accompanied by secretion of potassium and hydrogen ions.
4. Commonest cause of adrenal failure is autoimmune disease or infection. It causes hypotension, hyperkalaemia and hyponatraemia. This is a potential medical emergency diagnosed by the short ACTH stimulation test by demonstrating a failure for cortisol to increase after ACTH injection.
5. Glucocorticoid excess, Cushing's syndrome, is caused by exogenous steroids, pituitary adenoma or rarely adrenal adenoma or ectopic ACTH secretion. Diagnosis of Cushing's syndrome depends on

demonstration of loss of diurnal rhythm, high urine cortisol, failure to suppress after overnight dexamethasone or low dose dexamethasone administration. In patients diagnosed with Cushing's syndrome, plasma ACTH and a high dose dexamethasone test are required to establish the cause. CRH test with or without petrosal venous sampling may be required in some cases.

6. Congenital defects in the enzymes involved in the synthesis of cortisol/aldosterone cause congenital adrenal hyperplasia. The commonest is 21-hydroxylase deficiency. This can cause salt-wasting and virilisation in severe forms. Milder forms may present later in life with hirsuitism, menstrual irregularities and infertility. Diagnosis depends on demonstrating high 17-OHP either in the basal state or after ACTH stimulation.

7. Primary hyperaldosternisim is a common cause of secondary hypertension and is commonly due to an adenoma. It is diagnosed by demonstrating a high aldosterone:renin ratio. Patients with primary hyeraldosteronism may have hypokalaemia and an inappropriate loss of potassium in the urine (> 30 mmol/L) in addition to hypertension.

chapter 22

The Thyroid

Introduction

The thyroid gland, which develops from the floor of the buccal cavity and migrates down, consists of follicles lined by a layer of epithelial cells. The cells enclose an amorphous material called 'colloid' which is mainly composed of thyroglobulin and small quantities of hydrogenated thyroalbumin. The thyroid also contains parafollicular or 'C' cells, which secrete calcitonin. The thyroid gland secretes mainly thyroxine (T4) and a small amount of triiodothyronine (T3).

Actions of Thyroid Hormones

In most tissues (except the brain, red cells, retina, lungs, spleen and testes), thyroid hormones increase the metabolic rate by stimulation of mitochondrial metabolism and by increasing the membrane sodium–potassium–ATPase activity. As the activity of the sodium–potassium–ATPase in maintaining the sodium–potassium gradient across the cell membrane accounts for 15–40% of resting energy expenditure, increasing the activity of this enzyme increases the resting metabolic rate. They also have inotrophic and chronotrophic effects on the heart, thereby increasing the heart rate, cardiac output, and increasing the risk of atrial fibrillation in patients with hyperthyroidism. The sensitivity of β-adrenergic receptors to catecholamines is increased.

Thyroid hormones promote differentiation and growth, and are essential for foetal and neonatal development and sexual maturation. They also stimulate protein synthesis, carbohydrate metabolism and lipid metabolism. Stimulation of lipid metabolism leads to a fall in plasma cholesterol as degradation is increased to a greater extent than synthesis. Thyroid hormones increase gluconeogenesis and accelerate insulin degradation and thus in thyrotoxicosis, there may be deterioration in the control of diabetes mellitus. Bone turnover is stimulated, resorption more so than formation.

Thyroid hormones bind to nuclear receptors and influence gene expression. In addition, they also have a rapid non-genomic effect on cellular membrane.

Synthesis of Thyroid Hormones

Steps involved in the synthesis are shown in Figure 22.1. Follicular cells take up iodide from the blood stream against a concentration gradient, mediated by an iodide pump situated in the basal membrane of the follicular cells. Iodide transport into the follicular cells is the first rate

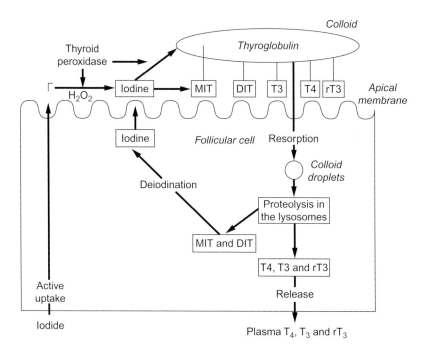

Figure 22.1   Synthesis of thyroid hormones.

limiting step in the synthesis of thyroid hormones. This step can be blocked by some anions such as pertechnetate (TcO_4^{-1}) and perchlorate (ClO_4^-). Iodide is then oxidised to iodine by thyroid peroxidise (TPO). Synthesis of thyroid hormones takes place in the thyroglobulin molecule, which contains many tyrosine residues. Thyroglobulin is synthesised within the follicular cells and stored in the colloid of the follicle. Tyrosine residues in the thyroglobulin are iodinated to monoiodotyrosine (MIT) and then to diiodotyrosine (DIT). Two DIT molecules condense and produce T4 while condensation of one MIT and one DIT forms T3. Oxidation of iodine, iodination of tyrosol groups and coupling of the DIT and MIT within the thyroglobulin molecule are all catalysed by the enzyme thyroid peroxidase. When iodide is sufficient, T4 is formed and when the supply of iodide is limited, T3 is synthesised predominantly. The thyroid hormones are synthesised at the follicular cell/colloid interface and stored in the colloid. When TSH stimulates the thyroid, the thyroglobulin molecules are taken into the lysosomes of the follicular cells by pinocytosis and T4 and T3 are released by proteolytic degradation of the colloid. TSH stimulates all the steps in the thyroid hormone synthesis — the uptake of iodide, thyroglobulin synthesis, colloidal uptake by follicular cells and the liberation of T4 and T3 from thyroglobulin. Antithyroid drugs such as propylthiouracil and carbimazole inhibit thyroid peroxidase. A large excess of iodine inhibits the adenyl cyclase response to TSH and the iodination of thyroglobulin.

Transport and Metabolism of Thyroid Hormones

The thyroid gland secretes approximately 100 nmol of T4 per day. Approximately 40% of this is converted by peripheral tissues to T3 and about 45% is deiodinated to yield the biologically inactive 3',5'-L-triiodothyronine (reverse T3, rT3). Almost 85% of the normal T3 production and almost all of the rT3 production is by peripheral deiodination of T4 rather than by direct secretion from the thyroid gland. Major organs involved in this peripheral conversion are liver and kidney. The enzymes responsible for peripheral conversion are the deiodinases, which contain selenium as selenocysteine. There are three forms of deiodinase. Deiodinase type 1 is present in the liver, kidney, thyroid and pituitary and converts T4 to T3 or rT3 and T3 to diiodthyronine (T2). Deiodinase type 2 is present in the thyroid, heart, brain, skeletal muscle and brown adipose tissue converts T4 to T3. Deiodinase type 3 is present in the foetal tissue and placenta and converts T4 to rT3 or T3 to T2.

T4 and T3 in the circulation are reversibly and almost completely bound to carrier proteins, thyroxine-binding globulin (TBG), thyroxine binding prealbumin (TBPA) and albumin. These proteins bind 99.97% of T4 and 99.7% of T3. The total T4 concentration is about 50 times that of T3 but the concentration of free T4 is only about three times that of free T3. The half-life of T4 is 5–7 days whereas that of T3 is 1–2 days. The free fraction of thyroid hormones is the biologically active form and the large protein bound fraction acts as a reservoir.

In the liver, T4 and T3 are conjugated to form sulphates and glucouronides which are excreted in the bile. T4 and T3 are also metabolised by deamination and deiodination and the products are subsequently metabolised.

Control of Thyroid Hormone Synthesis and Secretion

Plasma concentration of thyroid hormones is controlled by a tightly coordinated feedback loop involving the thyroid gland, hypothalamus and pituitary gland (Figure 22.2). The hypothalamus releases a tripeptide, thyrotrophin-releasing hormone (TRH) which acts on the pituitary thyrotroph to stimulate the synthesis and release of TSH. TSH, in turn, controls the synthesis and release of thyroid hormones and induces an increase in the size and number of follicular cells. TSH is a glycoprotein with two subunits — α-subunit which is identical to that of gonadotrophins and a specific β-subunit. Alteration in the glycosylation of TSH can affect its biological activity and half-life. Thyroid hormone alters the pituitary response to TRH and this depends on the intracellular concentration of T3, which is produced locally. An increase in thyroid hormones will

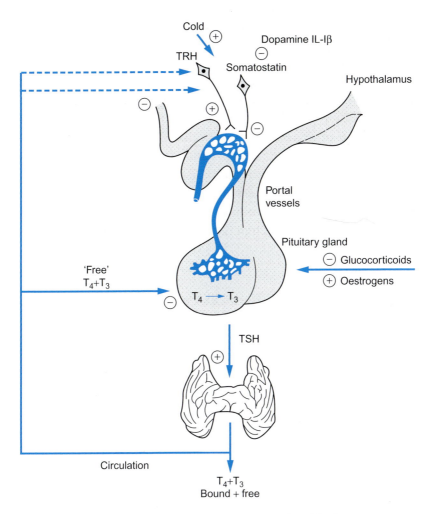

Figure 22.2 Hypothalamic pituitary thyroid axis.

reduce the number of TRH receptors on the pituitary cells and vice versa, thereby altering TSH secretion. A rise in thyroid hormones inhibits the pituitary response to TRH and a fall in thyroid hormones causes an increase in TRH and TSH secretions. TSH binds to a specific receptor on the surface of the follicular cell. This receptor is a single protein, belonging to the large family of

G protein-coupled receptors. Binding of TSH to the receptor stimulates cyclic AMP production via stimulation of guanine nucleotide-binding protein. The activation of the receptor leads to the release of thyroid hormones within an hour. Stimulation of the receptor also produces an IGF1-like factor, platelet-derived growth factor and other growth factors, which stimulate the growth and differentiation of thyroid epithelial cells. Thyroid hormone directly suppresses TSH secretion. This negative feedback loop is influenced by other factors. TSH is secreted in a pulsatile fashion with a diurnal variation, being highest around midnight and 4 a.m. and lowest midday. Glucocorticoid, cytokines, somatostatin, dopamine, oestrogens and environmental temperature are some of the other factors involved (Figure 22.2).

Factors Affecting Thyroid Hormone Concentration

Age

At birth, TSH and T4 concentrations rise rapidly, possibly due to a drop in the ambient temperature at birth. These remain elevated for three days. During childhood, TSH and free T4 are within the adult range and free T3 is higher than in adults. After puberty, no change is seen except in pregnant women.

Pregnancy

In pregnancy, there is a two- to three-fold increase in serum TBG concentration induced by oestrogens and a decrease in albumin concentration. TBG increases within a few weeks of conception and reaches a plateau by mid gestation. Total T4 and total T3 are therefore increased by 30–100%. In early pregnancy, free thyroid hormone concentration may show a slight increase and serum TSH show a decrease due to stimulation by the high concentration of human chorionic gonadotrophin (HCG) which has a weak TSH like activity. Thereafter, free T4 and free T3 tend to show a slight decrease.

Intraindividual variation

Thyroid hormone concentrations are tightly regulated and there is very little fluctuation within an individual.

Non-thyroidal illness (NTI) [sick euthyroid syndrome]

Patients with many diseases show changes in serum thyroid hormones without any evidence of thyroid disease. This is described as non-thyroidal illness or sick euthyroid syndrome. During systemic illness and fasting, the conversion of T4 to T3 is reduced, plasma concentration of total and free T3 falls, reverse T3 increases and TSH concentration falls. Although a lot has been written about the changes in thyroid function tests in non-thyroidal illnesses, the significance of these changes is not clear. Glucocorticoids and other drugs probably bring about the decreased conversion of T4 to T3 in peripheral tissues. A decrease in total T4 may also be seen in NTI. However, free T4 concentrations usually remain within the reference range. This discrepancy between a decrease in total T4 and free T4 is probably due to a change in the concentration and binding properties of serum thyroid hormone binding proteins induced by circulating inhibitors and drugs. The changes in free T4 in NTI are further complicated by method related problems. Abnormalities in serum TSH are thought to be due to the effect of endogenous and exogenous hormones such as glucocorticoids or dopamine, both of which suppress pituitary TSH secretion. Altered nutrition, and change in biological activity of TSH may contribute to the changes in circulating thyroid hormones in NTI. When patients recover from NTI, serum TSH shows a transient rise. Therefore, in ill patients, assessment of thyroid function is better left until they recover.

Some studies have suggested that the degree of abnormality in thyroid hormone concentration is a reflection of severity and therefore may predict outcome.

Drugs

Many drugs may alter thyroid function and thyroid function tests (see later).

Others

TSH shows a circadian variation, but the amplitude of this variation is small. Seasonal fluctuations in TSH, FT4 and FT3 are small, and exercise and environmental temperature have very little effect.

Laboratory Assessment of Thyroid Function

Serum TSH

Measurements of serum TSH are often the first line of investigation. TSH is measured by immunometric assays and depending on the lower detection limit, these assays are classified as first, second or third generation assays. Respective lower detection limits are 1.0 mU/L, 0.1 mU/L and 0.01 mU/L respectively. As the TSH concentration is regulated by feedback mechanisms, TSH concentration will be low in hyperthyroidism and high in primary hypothyroidism. Thus, apart from pituitary dysfunction, a normal TSH concentration signifies a normal thyroid function. A high TSH concentration is accompanied by a decrease in total or free T4 and is diagnostic of hypothyroidism. When a raised TSH is accompanied by a normal total or free T4, it is described as subclinical hypothyroidism.

A low or undetectable TSH concentration must be interpreted carefully as TSH concentrations are subnormal or undetectable in non-thyroidal illness especially in elderly subjects. Third-generation TSH assays can distinguish between the hyperthyroid state and NTI.

It is also important to note that serum TSH alone is not a reliable test to assess adequacy of treatment of hyper- or hypothyroid patients, as TSH is slow to respond. For example, when hyperthyroid patients are treated, it may take several months for the TSH to normalise even when the patient is euthyroid.

Total T4 and total T3

Serum concentrations of total T4 and T3 can be measured by immunoassays. However, these should be interpreted with caution as changes in binding proteins may cause abnormalities in total

Table 22.1 Causes of abnormal thyroid hormone-binding protein concentrations

Increase in TGB
- Genetic
- Pregnancy
- Oestrogens — oral contraceptives, oestrogens, tamoxifen
- Newborns

Decrease in TGB
- Genetic
- Androgens, anabolic steroids, high dose of glucocorticoids
- Protein-losing states — e.g. nephrotic syndrome
- Severe non-thyroidal illness
- Salicylate and phenytoin — affects TBG binding capacity
- Decrease in albumin — familial dysalbuminaemia

hormones (Table 22.1). Total hormone assays however are now replaced by free hormone assays.

Free T4 and free T3

Free thyroid hormones can now be measured fairly reliably. Free T4 and free T3 are raised in thyrotoxicosis, free T3 to a greater extent than free T4. In a small proportion of patients, only free T3 is elevated, and this is described as 'T3 toxicosis'. In hypothyroidism, free T4 is abnormal and free T3 may be within the reference range. Antibodies that bind to T4 and T3, antibodies to animal immunoglobulins, and other proteins such as rheumatoid factor and complement, may sometimes cause interference in these analyses.

TRH test

In this dynamic test, the TSH response to an intravenous injection of TRH is measured. Due to the feedback regulation, the TSH response to TRH is suppressed in the hyperthyroid state and is exaggerated in hypothyroid states. The TRH test is a valuable test to exclude thyroid disease. A normal TRH response virtually excludes

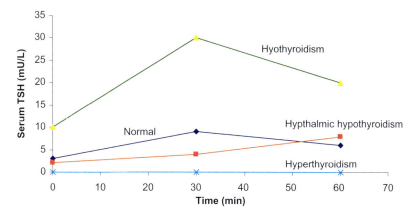

Figure 22.3   TSH response to 200 ug of TRH given intravenously (TRH test) in normal and different thyroid diseases.

any thyroid dysfunction (Figure 22.3). However, with the availability of sensitive TSH assays capable of detecting very low TSH concentration, the TRH test gives no more information than the basal TSH measurements. However, the TRH test is of value in the diagnosis of thyroid hormone resistance syndromes or TSH secreting adenomas. The TSH response to TRH is flat in TSH secreting pituitary adenoma whereas in thyroid hormone resistance, there is a brisk response. In hypothalamic hypothyroidism, there is delayed response.

Thyroxine binding globulin (TBG)

With the availability of free T4 and free T3 assays, measurement of TBG assay is not normally required in practice.

Thyroglobulin

Thyroglobulin is normally present in very small quantities in the circulation. This is a useful test in the follow up of patients treated for thyroid carcinoma. An increase in thyroglobulin concentration during follow up suggests recurrence of tumour. Thyroglobulin antibodies can cause interference in the assays.

Alpha subunit

The α-subunit of the TSH molecule is identical to that of FSH and LH. A raised concentration of alpha subunit is suggestive of TSH secreting pituitary tumours.

Antibodies to thyroidal antigens

- *Antibodies to thyroid peroxidase/thyroid microsomal antibodies*

  These antibodies, which probably play a pathogenic role in destructive autoimmune thyroid disease, are found in about 95% of patients with autoimmune hypothyroidism and 70–80% of patients with Grave's disease. Detection of these antibodies, especially in high titre, is helpful in Hashimoto's disease and ophthalmic Grave's disease.

  Raised levels of thyroid peroxidise (TPO) antibodies are found in 10–14% of apparently healthy individuals and a higher percentage in patients with other autoimmune disorders. The prevalence of the antibody is greater in females and it increases with age. Presence of these antibodies in 'healthy' subjects is a predictor of future development of thyroid disease. Furthermore, the presence of these antibodies is indicative of increased risk of miscarriage, infertility, *in vitro* fertilisation failure and maternal complications such as pre-eclampsia, and postpartum thyroiditis.

Antibodies to thyroglobulin

Thyroglobulin antibodies are found in patients with autoimmune thyroid disease. However, it is found less frequently than TPO antibodies.

Antibodies to TSH receptor/Thyroid-stimulating immunoglobulins

Hyperthyroidism of Grave's disease is mediated by an IgG class of autoantibody binding and stimulating the TSH receptor on the plasma membrane of the thyroid follicular cells. These antibodies could be detected in up to 60–90% of patients with Grave's disease. The value of measuring these in clinical practice is debatable. It has been argued that the presence of these antibodies helps in the diagnosis of Grave's disease. A decrease in these antibodies during treatment with antithyroid drugs suggests that remission may be maintained. These antibodies can cross the placenta and cause hyperthyroidism in the foetus. Measurement of these antibodies is therefore useful in pregnant women with Grave's disease.

Interpretation of Thyroid Function Tests

Table 22.2 summarises the patterns of serum thyroid hormones seen in practice and their interpretation. In hyperthyroidism, TSH is low and FT4 and FT3 are high, occasionally only FT3 may be high (T3 toxicosis). In hypothyroidism, TSH is high and FT4 is low, FT3 is often within the reference interval.

In interpreting thyroid function tests, it is important to remember that serum thyroid hormones show very little variation in an individual and therefore a change in thyroid hormone concentration signifies a change in thyroid status even though the concentration may be within the reference interval.

Table 22.2   Interpretation of serum thyroid hormone tests

TSH	FT4 low	FT4 normal	FT4 high
Low	NTI, treatment of hyperthyroidism	Subclinical hyperthyroidism, over-replacement with T4	Hyperthyroidism, over-replacement with T4
Normal	NTI, Hypopituitarism	Euthyroid state	Erratic T4 replacement
High	Hypothyroidism, Inadequate T4 replacement	Inadequate T4 replacement, subclinical hypothyroidism	Erratic T4 replacement, TSHoma, Thyroid hormone resistance

Table 22.3 Drugs that influence thyroid function and their mechanism

Decrease in TSH secretion
- Dopamine, Glucocorticoids, Octreotide

Decreased thyroid hormone secretion
- Lithium, Iodide, Amiodarone, Aminoglutethimide

Increased thyroid hormone secretion
- Iodide, Amiodarone

Altered T4 metabolism
- Phenobarbitone, Rifampacin, Phenytoin, Carbamazepine

Decreased conversion of T4 to T3
- Propothiouracil, Amiodarone, β-adrenergic antagonists, Glucocorticoids

Drugs that affect T4 absorption
- Ferrous sulphate, Cholestyramine, Aluminium hydroxide, Sucralfate, Calcium carbonate, Phosphate binders, Raloxifene, Proton pump inhibitors

Autoimmune disease or thyroiditis
- Interferon-α
- Sunitinib

Drugs and Thyroid

Drugs can cause changes in thyroid hormone concentrations by interfering with secretion, metabolism, transport and absorption. Table 22.3 shows how thyroid function can be altered by drugs and their possible mechanism.

Amiodarone, which is an anti-arrhythmic drug, contains large amounts of iodine (30–60 times the iodine intake) and can cause abnormalities in thyroid function tests as well as in the thyroid status Amiodarone inhibits the conversion of T4 to T3, causing high T4 and low T3. It can cause a transient increase in TSH in the early stages of treatment. It also inhibits iodine uptake by the thyroid cells and can cause hypo- or hyperthyroidism. Hyperthyroidism is due to thyroiditis or is iodine induced. Hypothyroidism is caused by the antithyroid effect of iodine. Amiodarone induced hyperthyroidism seem to be more frequent in iodine-deficient areas and hypothyroidism in iodine rich areas. It is recommended that patients on amiodarone should have their thyroid function tests done before starting treatment and thereafter every 6 months.

Lithium, an antidepressant drug, inhibits the synthesis and secretion of T4 and T3 and cause hypothyroidism in 20–30% of patients, goitre in 50% and subclinical hypothyroidism in 20% of patients. Patients with TPO antibodies are at increased risk of development of hypothyroidism. Lithium can also cause hyperthyroidism due to thyroiditis. Patients on lithium treatment should be monitored for thyroid function before and every 6 to 12 months.

Interferon-α, which is used in the treatment of hepatitis C and some malignancies, can cause thyroid dysfunction in up to 8% of patients. Abnormalities include autoimmune hypothyroidism, often subclinical, destructive thyroiditis, and Grave's hyperthyroidism.

Sunitinib, a tyrosine kinase inhibitor used in the treatment of some cancers can cause thyroid function abnormalities in 60–70% of patients and clinical hypothyroidism in up to 20% of cases.

Some drugs such as ferrous sulphate interfere with the absorption of thyroxine orally. In these patients, thyroxine should be taken at least 4 hours before or 4 hours after such medication. Hepatic cytochrome P450-inducing drugs such as phenytoin and carbamazepine increase the metabolism of ingested thyroxine; therefore, patients on these drugs require higher dosage.

Drugs such as heparin can cause interfere in the measurement of free T4. Heparin treatment can cause the release of free fatty acids and this can cause interference with the assay when the sample is left standing. Other drugs that interfere with the measurement of free T4 include furosemide, salicylate and carbamazepine.

In vivo Tests of Thyroid Function

The activity of the thyroid gland can be assessed by the administration of isotopes such as iodine[125] or technetium[99]. The uptake of these isotopes is increased in hyperthyroidism and decreased in hypothyroidism. However, the uptake may also be influenced by other factors such as the amount of iodine in the diet, ingestion of iodine containing drugs such as cough medicines, amiodarone, contrast media, etc. For these reasons, these tests are seldom used now.

Qualitative imaging of thyroid function using a scintillation camera after intravenous administration of an isotope such as technicium[99] pertechnetate is useful to identify nodules. This technique will allow differentiation between hot or active nodules from inactive or cold nodules. The latter are potentially malignant. Thyroid scanning is also useful in distinguishing Grave's disease, where the uptake is uniformly increased from multinodular goitre where the uptake is patchy. Ectopic thyroid tissue can also be detected by this method. Ultrasound is useful in assessing the size of the gland.

Disorders of the Thyroid Gland

Hyperthyroidism

Hyperthyroidism can be caused by a number of conditions (Table 22.4). The most common cause of hyperthyroidism is Grave's disease, which is an autoimmune disorder due to the presence of antibodies, which bind and stimulate the TSH receptor in the thyroid gland. The prevalence of hyperthyroidism is about 0.3–0.6% in the general population. Some studies have suggested that there may be as many cases of unsuspected hyperthyroidism in women. The lifetime risk of hyperthyroidism has been estimated to be 5% in women and 1% in men.

Table 22.4   Causes of hyperthyroidism

Graves disease	85%
Toxic multinodular goitre	6%
Toxic solitary adenoma	3%
Acute thyroiditis	3%
Hydatidiform mole/choriocarcinoma	
Hyperemesis gravidarum	
Drug induced thyrotoxicosis e.g. amiodarone	
Post partum thyroiditis	
Excessive ingestion of T4 and T3	
TSH-secreting adenoma	
Pituitary resistance to thyroid hormone	

Table 22.5   Clinical features of hyperthyroidism

Symptoms	Signs
Weight loss	Tremor
Fatigue	Goitre
Menstrual irregularity	Tachycardia/arterial fibrillation
Heat intolerance	Muscle weakness
Increased sweating	Proximal myopathy
Restlessness, agitation	Eye signs
Nervousness	
Palpitations	
Diarrhoea	

The signs and symptoms in hyperthyroidism are listed in Table 22.5. Biochemical diagnosis depends on demonstrating a suppressed TSH and a raised free T3 and free T4. In hyperthyroidism, due to TSH secreting pituitary adenoma, TSH will be inappropriately high with high FT4 and FT3.

Management

Treatment of hyperthyroidism is initially by antithyroid drugs such as carbimazole or propylthiouracil. These block the synthesis of thyroid hormones and will take 4–8 weeks to normalise serum FT4 and FT3 as these drugs do not affect the already synthesised hormones. During this period, β-adrenergic blockers which inhibit the conversion of T4 to T3 may be used to relieve symptoms. Treatment with antithyroid drugs for 18 months may result in remission in 30–50% of patients. Treatment is monitored by measuring TSH, FT4 and FT3. However, it is important to realise that TSH may take several weeks or months to normalise after FT4 and FT3 are normalised as it takes time for the pituitary to recover after prolonged suppression.

Treatment with radioiodine or partial thyrodiectomy may be necessary when drug treatment fails, if there is contraindication for the use of antithyroid drugs and in those who relapse.

Case 22.1

A 22-year-old woman from Vietnam consulted her GP feeling anxious. She had lost nearly 7 kg in weight in the preceding 3 months. On examination, she was found to be anxious with sweaty palms and a raised pulse rate of 100 per min. She had no goitre. Serum thyroid function results were:

Serum TSH < 0.01 mU/L (0.2–5.6)

FT4 38 pmol/L (12.0–25.0)

FT3 14.4 pmol/L (3.0–7.0)

The high FT4 and FT3 with suppressed TSH are characteristic of hyperthyroidism. A thyroid isotope scan showed uniformly increased uptake. Serum thyroid-stimulating immunoglobulins was elevated. She was diagnosed with Grave's disease and treated with carbimazole. Her symptoms improved and she became euthyroid in 3 months time.

Subclinical hyperthyroidism

This is defined as a low or undetectable serum TSH with FT4 and FT3 within the reference range. The incidence in the general population has been reported to vary from 0.7% to 16% depending on the TSH assay and the lower limit of cut-off value for TSH. The causes of subclinical hyperthyroidism are the same as for overt hyperthyroidism. These subjects are more likely to develop overt hyperthyroidism at the rate of 1–4% per annum. There is evidence that these patients have accelerated bone loss and are at increased risk of cardiovascular complications especially atrial fibrillation. Treatment of this condition remains controversial. Elderly subjects and patients at increased risk (cardiovascular disease or postmenopausal women) may be treated with antithyroid drugs.

Thyroid storm

Thyroid storm usually develops in those patients with untreated or incompletely treated pre-existing thyrotoxicosis. Crisis is precipitated by infection, surgery, trauma, diabetic ketoacidosis, toxaemia of pregnancy and parturition. The clinical picture is one of severe hypermetabolism, hyperpyrexia, sweating, severe tachycardia, arrhythmia, heart failure as well as changes in mental state including disorientation, agitation and even coma. Diagnosis is made clinically. Prompt treatment is necessary as the mortality can be up to 20%. Treatment includes high doses of β-blockers, intravenous sodium iodide, antithyroid drugs, glucocorticoids and general supportive measures.

Hypothyroidism

Hypothyroidism is a relatively common disorder occurring in 2–15% of the population, it is more frequent in women and the incidence increases with age. Severe hypothyroidism occurring in the newborn is termed 'cretinism'. The common causes of hypothyroidism are listed in Table 22.6 and the most common of these are Hashimoto's disease and atrophic destruction of thyroid gland, followed by iatrogenic hypothyroidism due to either surgery or radioactive iodine

Table 22.6   Causes of hypothyroidism

Primary hypothyroidism
- Chronic lymphocytic thyroiditis (Hashimoto's)
- Atrophic hypothyroidism
- Iatrogenic — post-surgery/radioiodine treatment
- Subacute thyroiditis
- Postpartum thyroiditis
- Congenital
- Dyshormonogenesis
- Iodine deficiency

Drug induced
- Lithium
- Iodine

Secondary hypothyroidism
- Pituitary disease
- Hypothalamic disease

Table 22.7   Clinical features of hypothyroidism

Symptoms
- Tiredness and lethargy, easy fatigue
- Cold intolerance
- Weight gain
- Hoarseness of voice
- Hair loss
- Constipation
- Depression

Signs
- Dry, coarse skin
- Bradycardia
- Slow relaxation of muscle
- High cholesterol
- Growth retardation
- Carpal tunnel syndrome

treatment. In iodine-deficient areas, goitre is predominant and hypothyroidism is relatively common in these places. Symptoms and signs of hypothyroidism are listed in Table 22.7. However, the features can vary and hypothyroidism should always be borne in mind in any medical patient. Hypothyroid patients also have hypercholesterolaemia, increased serum creatinine kinase activity and macrocytic anaemia.

Diagnosis of hypothyroidism is established by demonstrating high TSH concentration in the presence of low free T4. In secondary hypothyroidism, TSH will be inappropriately low in the presence of a low T4.

Case 22.2

A 59-year-old female consulted her GP feeling increasingly tired and lack of energy. Her GP requested thyroid function tests among other tests and the results were:

Serum TSH   58.0 mU/L (0.20–5.6)
FT4   7.0 pmol/L (12.0–25.0)

A raised TSH with low FT4 is diagnostic of hypothyroidism. Serum TPO antibodies were positive. She was treated with thyroxine and her symptoms improved.

Case 22.3

A 30-year-old man was referred to the endocrinology clinic with a history of increasing tiredness and fatigability. He had gained weight recently. Initial investigation by his GP showed:

$$\begin{array}{lll} \text{Serum TSH} & 0.1\ \text{mU/L} & (0.2\text{--}5.6) \\ \text{FT4} & 9.0\ \text{pmol/L} & (12.0\text{--}25.0) \end{array}$$

Examination at the clinic showed that he had a dry skin, pulse rate of 60/min. His reflexes were slow and he had loss of peripheral vision. A low TSH and a low FT4 are typical of secondary hypothyroidism. Loss of peripheral vision suggests a pituitary tumour which was confirmed by an MRI scan. He underwent surgery and post-operatively he was treated with thyroxine and hydrocortisone.

Subclinical hypothyroidism

This is defined as elevated TSH concentration in the presence of FT4 and FT3 in the reference range. The reported prevalence of this condition varies from 4–8% in the general population to 15–18% in those over the age of 60. These patients may have mild non-specific symptoms such as tiredness and depression. Patients with subclinical hypothyroidism have increased risk of cardiovascular disease than euthyroid subjects. A significant proportion of these patients go on to develop clinical hypothyroidism. Patients with TPO antibodies are especially prone to develop overt hypothyroidism. Although there is no firm evidence, most authorities (such as the UK thyroid association) recommend thyroxine replacement if the TSH is > 10 mU/L especially if TPO antibody is present. Patients with subclinical hypothyroidism who are not treated should have their thyroid function measured annually if they have TPO antibodies or otherwise every 3 years.

Case 22.4

A 58-year-old woman wanted her serum lipids checked. Her serum cholesterol was raised and her GP measured her thyroid function test to
exclude secondary hyperlipidaemia.

> Serum TSH   12 mU/L (0.2–5.6)
>   FT4   15 pmol/L (12.0–25.0)

*A raised TSH with FT4 with in the reference range indicated subclinical
hypothyroidism. Her serum TPO antibodies were positive. She was started on
a small dose of thyroxine and her serum TSH returned to normal and so did
her serum cholesterol.*

Management of patients with hypothyroidism

Patients with thyroid deficiency should be started on thyroxine treatment, initially in small doses and increased as necessary. The
effectiveness of treatment should be monitored by measuring TSH
concentration. The aim of treatment is to maintain serum TSH within
the reference range irrespective of the free T4 concentration.
However, it should be remembered that TSH might take several
weeks to normalise. Patients on long-term thyroxine treatment
should have an annual thyroid function test measured.

Screening for Thyroid Disease

Congenital hypothyroidism is a serious condition and if untreated can
lead to mental retardation (cretinism). Most common cause of congenital hypothyroidism is a defect in the development of the thyroid
gland. The prevalence of congenital hypothyroidism is approximately
1 in 3500 live births in the UK but higher in some countries. If
treated early with thyroxine replacement, these children develop normally. Many countries have a screening programme to detect this
condition. Screening is done by measuring TSH concentration as part
of the neonatal screening programme at 5 to 8 days of age (see
Chapter 26).

Thyroid dysfunction especially hypothyroidism is common in the elderly. The symptoms of hypothyroidism in the early stages can be non-specific and therefore it has been suggested that elderly people should be screened for thyroid disorders. However, the cost effectiveness of this has not been proven. Current practice is to screen if the patient comes into the general practitioners clinic, incidental screening rather than total population screening. However, the effect of non-thyroidal illness should be borne in mind in doing such screening.

Goitre

Goitre is an enlargement of the thyroid gland and can be multi nodular, diffuse uniform enlargement or may be a single nodule. Goitre is not uncommon during puberty. In areas of iodine deficiency, goitre is very common. Goitre may be associated with normal, hyper- or hypo-function of the thyroid gland. A solitary nodule is usually due to cysts, localised haemorrhage, adenoma or rarely carcinoma of thyroid. Fine needle aspirations is done to establish the diagnosis.

Tumours of the Thyroid Gland

Primary malignant tumours of the thyroid gland are follicular, papillary, medullary or anaplastic carcinoma. Medullary carcinoma originates from parafollicular cells and secretes calcitonin. Measurement of serum calcitonin is a useful tumour marker. There is a familial tendency for this malignancy and serum calcitonin is useful in detecting this type of cancer in family members. Serum thyroglobulin is often measured in patients with thyroid cancer to monitor recurrence after total thyroidectomy. However, it must be noted that presence of thyroglobulin antibodies may cause interference in the measurement of thyroglobulin concentration.

Further Reading

1. Adler SM, Wartofsky L. The nonthyroidal illness syndrome. *Endocrinol Metab Clin North Am* 2007; 36:657–672.

2. Dayan CM. Interpretation of thyroid function test. *Lancet* 2001; 367:619–624.
3. Devdhar M, Ousman YH, Burman KD. Hypothyroidism. *Endocrinol Metab Clin North Am* 2007; 36:595–615.
4. Dufour DR. Laboratory tests of thyroid function: Uses and limitations. *Endocrinol Metab Clin North Am* 2007; 36:579–594.
5. Nayak B, Hodak SP. Hyperthyroidism. *Endocrinol Metab Clin North Am* 2007; 36:617–656.
6. Ross DS. Serum thyroid stimulating hormone measurement for assessment of thyroid function and disease. *Endocrinol Metab Clin North Am* 2001; 30:245–264.

Summary/Key Points

1. Thyroid gland secretes predominantly T4.
2. Most of the T3 is produced from peripheral deiodination of T4 in the liver and kidney.
3. Thyroid hormone synthesis and secretion are regulated by TSH from the anterior pituitary which in turn is regulated by TRH from the hypothalamus. The secretion of TRH and TSH in turn are regulated by T4 and T3, a negative feedback loop.
4. T4 and T3 are carried bound to proteins. Only 0.03% of T4 and 0.3% of T3 are free in the circulation and these are the physiologically active fractions.
5. Thyroid function is assessed by measurement of serum TSH and FT4. FT3 may be necessary in cases of suspected hyperthyroidism especially in T3 toxicosis.
6. Serum thyroid function test should be interpreted with caution in pregnancy and when some drugs are administered.
7. In patients who are ill, abnormalities in thyroid function test are common and this is called non-thyroidal illness or sick euthyroid syndrome.
8. Hyperthyroidism is commonly due to auto immune disease (Grave's disease). Hyperthyroid patients show suppressed TSH and elevated FT4 and FT3.

9. When TSH is suppressed and FT4 and FT3 are within the reference range, it is called subclinical hyperthyroidism. Treatment of this condition is controversial.
10. Hypothyroidism is a common disorder especially in elderly women and it is most commonly due to Hashimoto's disease.
11. High TSH with a FT4 within the reference range is called subclinical hypothyroidism. Treatment is indicated if the TSH > 10 mIU/L.
12. Screening for thyroid disorder in the general population is not proven but incidental screening is common.

chapter 23

Clinical Biochemistry of Haematology

Introduction

Haemoglobin, a globular molecule of molecular weight 64,450, is made up of four subunits with each subunit consisting of a haem moiety conjugated to a polypeptide. The polypeptides form the globin portion of the haemoglobin. There are two pairs of identical polypeptides in each haemoglobin molecule: a pair of α or α-like chain and a pair of non–α-chain which may be β, χ, δ or ε. The haem group is the ferrous complex of protoporphyrin-IX (Figure 23.1), which can bind to oxygen reversibly. Other haem proteins in the body include myoglobin, which binds to oxygen in skeletal muscle, and cytochromes which are enzymes responsible for catalysing many oxidative processes in the body.

In a normal adult, 98% of the haemoglobin molecule is haemoglobin A that is composed of two α-chains and two β-chains. About 2.5% of the normal haemoglobin is haemoglobin A_2 where the β-chains are replaced by δ-chains. Predominant form of haemoglobin in the foetus is HbF, which consists of two α- and two γ-chains ($\alpha_2\gamma_2$).

Thalassaemias

Thalassaemias are a group of inherited disorders where there is reduced synthesis of one or the other type of globin chains. They are classified into α-, β-, $\delta\beta$- or $\chi\delta\beta$-thalassaemia according to which globin chain is produced in reduced amounts. In some thalassaemias, one or other of

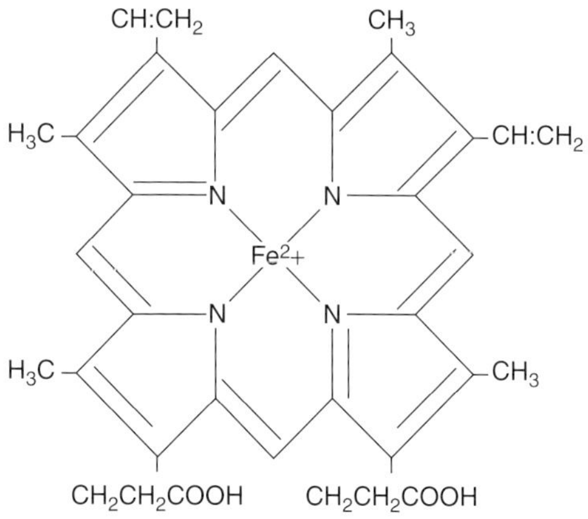

Figure 23.1   Structure of haem.

the globin chains is not synthesised at all and they are called $\alpha 0$- or $\beta 0$-thalassaemias. Thalassaemias are inherited in a simple Mendelian codominant fashion. Heterozygotes are usually symptomless although they can be recognised by simple haematological tests. More severely affected patients are either homozygous α- or β-thalassaemia. Clinically, thalassaemia is divided into major, minor or intermediate forms. Thalassaemia major is a severe transfusion dependent disorder. Thalassaemia minor is a symptomless carrier state and thalassaemia intermediate is characterised by splenomegaly and anaemia though not severe enough to require transfusion. β-thalassaemia is the most important type of thalassaemia and it is common in the Mediterranean, parts of North and South Africa, the Indian subcontinent and South East Asia. The molecular defect in β-thalassaemia results in absent or reduced β-chain production, and excess production of α-chain. In the absence of the partner chain, α-chain is unstable and precipitates in the red cell precursors. This causes interference in red cell maturation and variable degree of intramedullary destruction of red cell precursors. Those cells that do mature are destroyed prematurely and severe anaemia follows. Anaemia stimulates erythropoietin production causing massive expansion of bone marrow.

Although α-thalassaemia is more common, it poses less of a health problem and occurs widely throughout the Mediterranean region,

West Africa, Middle East, parts of Indian subcontinent, China and Thailand. In this disorder, the deficiency of α-chains leads to production of excess γ-chains or β-chains and the resulting molecules are haemoglobin Bart's and H, respectively. These do not precipitate in the bone marrow and hence erythropoiesis is not as severely affected as in β-thalassaemia.

The effects of thalassaemias include ineffective erythropoiesis, haemolysis and a variable degree of anaemia. The severity of the disease varies from clinically silent types to severe persistent anaemia. Furthermore, there is underutilisation of iron and increased absorption of dietary iron. With repeated transfusions for correction of anaemia these patients may develop haemosiderosis.

Haemoglobin Variants

Over 600 structural variants of haemoglobin have been described, most of which result from a single amino acid substitution. Many of them are harmless and have been discovered accidentally. However, several of these variants can cause clinical symptoms due to altered stability or altered functional properties of haemoglobin. The most common of these are the sickling disorders where HbA is replaced by HbS. The sickling disorders consist of a heterozygous state of haemoglobin S or the sickle cell trait (AS), the homozygous state or sickle cell disease (SS) and the compound heterozygote state for haemoglobin S together with other haemoglobin variants. Sickle cell disease is common in Africa, throughout the Mediterranean and Middle East. It is thought that the sickle cell gene gives resistance to infection with malarial parasite. It is inherited as an autosomal dominant trait.

Abnormal Derivatives of Haemoglobin

Methaemoglobin

When the iron in the haemoglobin is oxidised to ferric form, it results in a brown pigment, methaemoglobin. It is incapable of carrying oxygen and in healthy subjects, a small amount of haemoglobin, up to 1.5% may be present as methaemoglobin. This is due to continual oxidation of ferrous iron to ferric iron, which is then reduced by the enzyme system, NADH-methaemoglobin reductase. Congenital absence of this system can cause hereditary methaemoglobinaemia. The acquired form of methaemoglobinaemia may follow ingestion of nitrates, certain drugs or chemicals such as sulphonamides and amyl nitrite which is often used illegally to enhance sexual pleasure. When the methaemoglobin concentration exceeds 10%, there is cyanosis while concentration > 35% will lead to breathlessness. Methaemoglobinaemia can be treated with reducing agents such as methylene blue.

Sulphaemoglobin

Sulphaemoglobin is not a normal constituent of blood. It is sometimes found in concentrations of up to 1–10% after excessive use of sulphur-containing compounds such as sulphonamides, sulphosalazine and sumatriptan (drug used in migraine). Sulphaemoglobin is incapable of carrying oxygen, can cause a greenish or bluish discolouration of blood. Sulphaemoglobin cannot be converted back to normal haemoglobin. Drugs like nitrites can cause sulphaemoglobin if hydrogen sulphide is present.

Carboxyhaemoglobin

Carboxyhaemoglobin is a complex of carbon monoxide with haemoglobin. As carbon monoxide has 200 times more affinity for haemoglobin than oxygen, even the small quantities normally generated within the body can combine with haem to form carboxyhaemoglobin. In non-smokers, carboxyhaemoglobin accounts for 0.5–1% of haemoglobin and in heavy smokers, it can increase to 8–9%. Increased carboxyhaemoglobin is found in subjects who are exposed to car exhaust fumes and domestic appliances in which the fuel is incompletely burnt. Because of the tight affinity of carbon monoxide to haemoglobin, the oxygen-carrying capacity is reduced and patients with carbon monoxide poisoning have signs and symptoms of hypoxia and tissue anoxia such as shortness of breath, hyperventilation, headache, confusion, nausea, vomiting and diarrhoea. Symptoms are mild with carboxyhaemoglobin concentrations up to 20–30%. When concentrations exceed 70%, symptoms are severe and it becomes rapidly fatal. Tissues with high oxygen demand such as heart and brain are very sensitive to hypoxia. There is a general correlation between carboxyhaemoglobin concentration and symptoms but not invariably so because other factors such as length of exposure, metabolic activity, and other underlying diseases may affect the toxicity. The insidious onset of carbon monoxide poisoning causes neuropsychiatric manifestations, which may include personality changes and memory loss.

Treatment of carbon monoxide poisoning involves removal of the individual from the contaminated area and administration of oxygen. Carboxyhaemoglobin has a half-life of 5–6 hours at room air and decreases to 1.5 hours when breathing 100% oxygen. In severe cases, hyperbaric oxygen treatment is used. The half-life is reduced to 25 minutes under hyperbaric oxygen treatment.

Methaemalbuminaemia

When the iron in the haem is oxidised to ferric form haematin is formed. When haematin in circulation combines with albumin, methaemalbumin is produced. Methaemalbuminaemia is sometimes seen in acute necrotizing pancreatitis.

Porphyrias

Porphyrin Metabolism

Haem, which consists of protoporphyrin-IXα and ferrous ion, is synthesised in all cells but the synthetic rate is high in the bone marrow and liver. The steps involved in the synthesis of haem are outlined in Figure 23.2. The initial step, which takes place in the mitochondria, is the formation of α-aminolaevulinic acid (ALA) from succinyl CoA and glycine by ALA synthase. ALA is then converted to

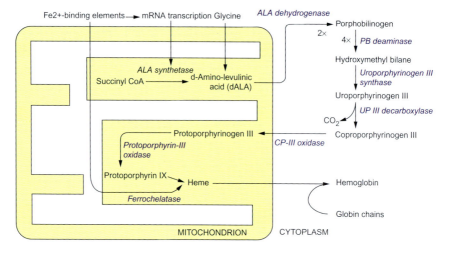

Figure 23.2   Synthesis of haem.

porphobilinogen (PBG) in the cytoplasm by ALA dehydratase (dehydrogenase) or PBG synthase. ALA is then converted by a series of steps to coproporphyrinogen III, which moves into the mitochondria, where it is converted to protoporphyrin-IX. Iron is incorporated into this to form haem. The formation of ALA is the rate-limiting step in the synthesis of haem. This enzyme (ALA synthase) is under negative feedback control by the end product haem. While PBG and porphyrin precursors are colourless, porphyrins are dark red in colour and intensely fluorescent.

Disorders of Porphyrin Metabolism

Porphyrias are a group of rare mainly inherited disorders involving the biosynthesis of haem, leading to accumulation of haem precursors. The clinical features depend on the type of intermediate that accumulates. Accumulation of ALA and PBG is associated with acute neurovisceral attacks while accumulation of porphyrins causes cutaneous lesions, as these are highly photosensitive agents. Thus, porphyrias can be divided into those presenting with acute attacks and those presenting with cutaneous manifestations (Table 23.1).

Table 23.1   Classification and characterisation of porphyrias

	Enzyme defect	Inheritance	Abnormal metabolite	Prevalence
Acute porphyrias				
AIP	Hydroxymethylbilane synthase	AD	Urine ALA × PBG	1 in 100,000
VP	Protoporphyrinogen oxidase	AD	Urine ALA × PBG	1 in 250,000
HC	Coproporphyrinogen oxidase	AD	Urine ALA × PBG	< 1 in 250,000
Cutaneous porphyrias				
Acute photosensitivity				
EPP	Ferrochelatase	AD	Red cell porphyrins	1 in 100,000
Skin Lesions				
PCT	Uroporphyrinogen decarboxylase	AD	Urine porphyrins	1 in 25,000
CEP	Uroporphyrinogen III synthase	AR	Urine porphyrins	< 1 in 10^6

*AD: autosomal dominant; AR: autosomal recessive; AIP: acute intermittent porphyria; VP: variegate porphyria; HC: hereditary coproporphyria; EPP: erythropoietic protoporphyria; PCT: porphyria cutanea tarda; CEP: congenital erythropoietic porphyria.

Acute porphyrias

Acute porphyrias are associated with accumulation of porphyrin precursors — ALA and PBG. The commonest form of acute porphyria is acute intermittent porphyria (AIP), which has a prevalence of 1 in 100,000. AIP, inherited as an autosomal dominant trait, is due to reduced activity of porphobilinogen deaminase (also known as hydroxymethylbilane synthase, HMBS), which converts PBG to hydroxymethylbilane. In the other two forms of acute porphyria, variegate porphyria (VP) and hereditary coproporphyria (HC), there is deficiency of protoporphyrinogen oxidase and coproporphyrinogen oxidase respectively (Figure 23.2). As a result, there is accumulation of protoporphyrinogen-IX and coproporphyrinogen-III respectively. These compounds inhibit HMBS, causing the accumulation of PBG and ALA.

- *Clinical manifestations*

  During an acute attack, patients present with abdominal pains accompanied by nausea, vomiting, constipation, tachycardia, hypertension, muscle weakness, sensory neuropathy and psychiatric symptoms. The exact mechanism of acute attack is poorly understood. Acute attacks are uncommon before puberty and are more common in females and less common after the menopause. They are usually precipitated by a wide variety of drugs (Table 23.2), starvation, alcohol, infection, stress and hormonal factors. Many of the drugs that precipitate attacks are inducers of hepatic cytochrome P450 dependent microsomal oxidases. Induction of these enzymes leads to depletion of free haem pool in the liver and this leads to increased production of PBG and ALA as a result of the negative feedback on ALA synthase. Only about 20% of patients with acute porphyria develop symptoms, while the majority remain asymptomatic throughout life.

- *Diagnosis*

  Diagnosis of acute porphyrias depends on the demonstration of increased excretion of PBG during an acute attack. As the excretion of PBG may be normal in between attacks, it is important to investigate these patients during the acute attacks. Normal

Table 23.2 Some common drugs that precipitate acute porphyrias

Barbiturate
Carbamazepine
Oestrogens
Progestogens
Sulphonamides
Tolbutamide
Valproate

excretion of PBG when symptoms are present excludes acute porphyria. In asymptomatic patients, diagnosis is based on quantitation of urinary PBG excretion for AIP, measurement of faecal porphyrins for HC and the detection of plasma fluorescence for VP. Differentiation between the types of acute porphyrias can be done by analysis of faecal porphyrins. In AIP, faecal porphyrins are normal, faecal coproporphyrin III is grossly elevated in HC and in VP faecal protoporphyrin-IX is raised. Assay of HMBS in erythrocytes is useful to detect latent AIP but haematological disorders may also affect the activity of this enzyme and there is overlap in enzyme activity between AIP and normal subjects. Genetic testing is essential in family studies but it is not useful in the diagnosis because there are a large number of molecular defects.

- *Treatment*

Treatment of the acute attack irrespective of the type of acute porphyria is the same and involves control of pain and supportive measures such as removal of the precipitating factors, administration of carbohydrate and/or haemarginine to suppress the haem pathway.

Case 23.1

A 20-year-old woman was admitted with severe acute abdominal pain. The pain was colicky in nature, started suddenly the previous day and accompanied by several bouts of vomiting. Her periods were regular and her last menstrual period was 3 weeks prior to admission. On examination her body temperature was normal, her blood pressure was 150/105 and apart from some tenderness in her abdomen there were no other signs. Initial investigations including pregnancy test, liver function tests, renal function tests and amylase were all normal. When her previous medical records were available, it was noticed that she had several similar episodes in the past few years and she has had an explorative laparotomy which did not reveal any abnormality. Acute intermittent porphyria was suspected and a urine sample was sent to the laboratory:

Urinary porphobilinogen	125 μmol/L (< 10)
Urinary ALA	very high
Urinary porphyrins	7500 nmol/L (20–330)

Presence of excess of PBG and ALA in her urine together with the acute presentation with abdominal pain and hypertension are all features typical of acute intermittent porphyria. The diagnosis was confirmed by analysis of the enzyme HMBS in red cells.

Porphyrias with cutaneous manifestations

- *Porphyrias with acute photosensitivity*

  In this type of porphyrias, patients develop symptoms such as burning and itching leading to erythema and bullae formation when exposed to sunlight. Between exposures, there is very little skin change. Erythropoietic protoporphyria (EPP) is the only porphyria that produces this picture. It is due to an inherited deficiency of ferrochelatase with an incidence of 1:200,000. The disease usually presents in early childhood and photosensitisation is due to accumulation of protoporphyrin-IX. As this protoporphyrin is insoluble in water, urinary porphyrin excretion is normal in EPP. Diagnosis of this condition depends on the demonstration of excess porphyrin in erythrocytes.

- *Porphyrias with skin lesions*

  Patients with these porphyrias have skin lesions such as bullae, scarring and pigmentation on sun-exposed areas. Avoidance of sunlight will reduce the incidence and severity of these lesions. They may also have fragile skin which heels poorly following minor injury. Four types of porphyrias can lead to skin lesions (Table 23.1); porphyria cutanae tarda (PCT), variegate porphyria (VP), hereditary coproporphyria (HC), and congenital erythro-poietic porphyria (CEP). The type of enzyme defect and the incidence is given in Table 23.1.

(1) Porphyria cutanae tarda (PCT)

PCT is the most common of the cutaneous porphyrias. It is due to a reduced activity of hepatic uroporphyrinogen decarboxylase and is precipitated by liver damage especially alcohol abuse, hep-atitis C infection, HIV infection, haemochromatosis, oestrogen treatment or exposure to chemicals such as hexachlorobenzene. Iron overload is common in this condition and removal of iron by venesection is a very effective form of treatment.

(2) Congenital erythropoietic porphyria (CEP)

This is a very rare autosomal recessive disorder, which presents in early childhood with severe mutilating skin lesions, haemolytic anaemia and enlarged spleen. This is due to uropor-phingogen III synthase deficiency. Treatment involves suppression of erythropoiesis by repeated blood transfusion together with chelation therapy.

The initial investigation of these patients is the measurement of urine and faeces for excess porphyrins. Normal porphyrins in urine and faeces exclude porphyria as the cause of skin lesions. All specimens with increased porphyrins should be further investi-gated to identify and quantitate the individual porphyrins.

In addition to porphyrias, there are other circumstances in which there may be increased excretion of porphyrin metabolites. In lead toxicity, there is raised urine ALA, coproporphyrin and red cell zinc protoporphyrin and in renal failure, there is increased plasma porphyrins (see Chapter 18).

Iron Metabolism

The capacity of iron to exist in both ferric and ferrous forms makes it a useful component of cytochromes, haemoglobin, myoglobin and respiratory chain enzymes. However, excess iron can damage tissues by producing free radical ions, which attack cellular membranes, proteins and DNA.

Iron content of healthy adults is approximately 0.9 mmol/kg body weight (50 mg of iron/kg) in men and 0.5–0.6 mmol/kg (30–35 mg/kg) in women. The iron content of premenopausal women is lower due to recurrent blood loss through menstruation. The majority of the iron (85%) is present as haemoglobin and myoglobin. The rest is stored in hepatocytes and reticuloendothelial macrophages as ferritin and haemosiderin (Figure 23.3). The average loss of iron from the body is 1–2 mg/day and it is higher in women during the reproductive age. The loss of iron occurs through the loss of mucosal cells and desquamation of skin. This loss is balanced by absorption of iron from the intestinal tract (Figure 23.3).

Iron Absorption

Iron is present as haem in red meat and as non-haem iron in unrefined cereals, vegetables and fruits. The average intake of iron is

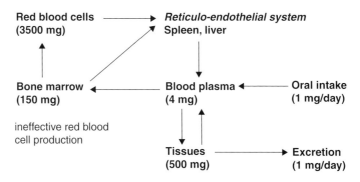

Figure 23.3   Metabolism of iron.

10–22 mg/day. In animal products, iron is present in the ferrous form and in vegetables, it is present as ferric form, mostly complexed to phytates, phosphates and other ions. Intestinal absorption of iron is very tightly regulated and the amount of iron absorbed is dependent on the iron requirements of the body. Iron is absorbed in the duodenum and the amount of iron absorbed is determined by the programming of the cells that line the duodenal crypts. The low pH of the stomach helps to dissolve the ingested iron and converts ferric iron to ferrous iron facilitated by the enzyme ferroreductase present in the brush border. Iron is then transported across the apical membrane by divalent metal transporter 1 (DMT1) through a proton coupled process. From the enterocyte, iron is transported across the basolateral membrane by ferroportin and oxidised to ferric iron by hephastin, a ferroxidase. The ferric iron binds to transferrin that enters the circulation. The mechanism of absorption of haem iron is less well understood; it is thought to be via a proton-coupled folate transporter (PCFT) or haem carrier protein (HCP1). Haem is broken down in the enterocyte and the iron is transported via ferroportin.

Intestinal absorption of iron is regulated in several ways. Absorption of iron is modulated by the amount of iron recently consumed in the diet. This is referred to as dietary regulator. Several days after high iron intake the enterocytes take up less iron. This is sometimes referred to as the mucosal block and this blocking action is probably due to the accumulation of intracellular iron. The mechanism by which iron absorption is regulated by iron stores is called the stores regulator in which the total body iron stores regulates the absorption via hepcidin, a protein produced by the liver. When iron stores are full, hepcidin is produced and it decreases iron absorption by binding to ferroportin and induces its internalisation and degradation. The absorptive capacity can increase by a factor of 2 or 3 when iron stores are depleted. The third mechanism of regulation is called the erythropoietic regulator where absorption responds to requirements for erythropoiesis probably via hepcidin. When the red cell production in the bone marrow is increased, intestinal absorption of iron is increased. It is important to note that absorption of iron is increased in anaemic states where the erythroid cells are destroyed within the bone marrow, i.e. ineffective erythropoiesis, but not when erythroid cells are destroyed in the periphery. Iron absorption is also increased in hypoxic states via an unidentified mechanism.

Iron Transport and Storage

Absorbed iron is transported in the circulation bound to transferrin. Transferrin synthesis by the liver varies inversely with body iron stores. Transferrin is normally 30–40% saturated.

Iron is stored in the liver and reticuloendothelial macrophages as ferritin and haemosiderin. Ferritin consists of an inner core of ferric iron and an outer protein shell. Although ferritin is an intracellular protein, small amounts of ferritin appear in the circulation and the plasma concentration of ferritin is related to the amount of iron stores. Haemosiderin is an insoluble complex that is formed from ferritin. When iron stores are increased, more haemosiderin is formed giving rise to siderosis.

Assessment of Iron Status

Serum iron and iron-binding capacity

Although serum iron concentration decreases in iron deficiency, it is of little diagnostic value by itself as the concentration of iron in the circulation fluctuates widely in normal individuals. These fluctuations can be up to 20% within a few minutes and up to 100% from day to day. Menstrual cycle, pregnancy, inflammatory diseases, trauma, liver disease, neoplasia and chronic inflammatory conditions affect serum iron concentration. The only indication for measuring serum iron is in suspected cases of iron poisoning.

Total iron-binding capacity is an indirect measure of transferrin concentration. Transferrin is normally 30–40% saturated and the saturation can be calculated from total iron-binding capacity and serum iron. It is decreased in iron deficiency anaemia due to increased synthesis of transferrin. Iron saturation is also low in pregnancy and chronic diseases. In haemochromatosis and iron overload states, iron saturation increases and is a very useful screening test for iron overload.

Serum ferritin

Serum ferritin concentration reflects tissue iron state, lower in iron deficiency and higher in iron overload states. However, it is an acute-phase protein and concentrations can be high in infections and liver disease even when there is iron deficiency. Serum ferritin is also high in some malignancies due to the presence of large amounts of acidic isoferritin in the malignant cells.

Soluble transferrin receptor

Transferrin receptor is a transmembrane glycoprotein found on cell surface and facilitates transport of iron into the cell by binding to the transferrin–iron complex. This complex is internalised and the iron is released. A cleavage product of this receptor, which lacks the cytoplasmic and transmembrane domain, is found in circulation. As the expression of transferrin receptor is increased in iron deficiency, serum soluble transferrin receptor is also increased. As serum soluble transferrin receptor is not affected by pregnancy, gender or anaemia of chronic disease this is a better test of iron deficiency than transferrin saturation. The ratio of serum soluble transferrin receptor to ferritin has been suggested by some as a superior marker of iron deficiency. However, serum soluble transferrin receptor assay is not widely available and it is expensive.

More recently, serum hepcidin has been examined as another marker of iron deficiency.

Disorders of Iron Metabolism

Iron deficiency

Iron deficiency is a common condition and it is an important public health problem. The most common cause of iron deficiency worldwide

Table 23.3   Causes of iron deficiency anaemia

Decreased intake
- Dietary deficiency
- Impaired absorption

Increased loss
- Gastrointestinal loss
  — Parasitic infestations — hook worm
  — Inflammatory bowel disease
  — Malignancy
- Menstrual loss — menorrhagia
- Genitourinary tract

Increased utilization
- Growth spurts
- Pregnancy — repeated pregnancies

is increased loss of blood from gastrointestinal loss due to parasitic infestations. In women, there is increased loss during menstruation. Other causes of iron deficiency are listed in the Table 23.3. In the developed world, important causes of gastrointestinal iron loss are inflammatory bowel disease and malignancy. When the iron stores get depleted, transferrin saturation decreases. The first biochemical evidence of iron deficiency is increased levels of free protoporphyrin and zinc protoporphyrin in red cells. The soluble transferrin receptor increases and frank anaemia with microcytosis is detected. Decreased reticulocyte haemoglobin is suggested to be a far superior test for iron deficient erythropoiesis. The symptoms and signs of iron deficiency include pallor, fatigue, poor exercise tolerance and decreased work performance. In addition, there may be direct effects of iron deficiency on the central nervous system, producing abnormalities in cognitive function and pica, which is characterised by consumption of non-nutritive substances. Severe long-standing iron deficiency may also lead to koilonychia, which refers to abnormally thin nails that have lost their convexity, becoming flat or even concave in shape, and Plummer Wilson syndrome, which is a triad of dysphagia, glossitis, and iron deficiency anaemia.

Treatment of iron deficiency anaemia consists of the removal of the cause and oral treatment with iron salts such as ferrous sulphate or gluconate. Some patients have difficulty in tolerating oral iron salts as they develop disturbing gastrointestinal symptoms such as constipation, nausea and abdominal pain. In these patients and those who fail to absorb oral iron, parenteral iron therapy may be given.

Anaemia of chronic disease

This describes anaemia seen in chronic diseases such as chronic infection and malignancy. This has some features common with iron deficiency anaemia. A defect in iron recycling leads to iron-deficient erythropoiesis. Recent studies suggest that the mechanism of this anaemia is hepcidin mediated. Liver produces increased amounts of hepcidin in response to inflammatory cytokines. Hepcidin decreases iron absorption from the gut by binding to ferroportin. In addition, it directly inhibits erythropoietin. Anaemia seen in chronic renal failure is due to decreased production of erythropoietin. Anaemia of chronic disease can be normocytic or microcytic. Tests which are likely to suggest anaemia of chronic disease are (i) normal or high ferritin (ii) normal or low iron-binding capacity (iii) normal serum soluble transferrin receptor. In the future, serum hepcidin measurement may be more appropriate to diagnose this condition.

Diseases of iron overload

Iron overload can be of two types. In the first type, erythropoiesis is normal but iron is deposited in the parenchymal cells of liver, heart, etc. An example of this condition is hereditary haemochromatosis. In the second type, there is increased catabolism of erythrocytes, and iron accumulates in reticuloendothelial macrophages at first and only later in parenchymal cells.

1. *Hereditary haemochromatosis*

Hereditary haemochromatosis (HCC) is the most prevalent mono-allelic genetic disease among the Caucasian population with a frequency of 0.4–1.0%. It is inherited as an autosomal recessive fashion and leads to increased absorption of iron. It is now known that this is due to a mutation in the haemochromatosis gene (HFE). Protein produced by the gene regulates hepcidin via a poorly understood mechanism. In HCC, the defective protein leads to inadequate or low hepcidin and hence increased iron absorption.

There is a wide spectrum of clinical presentation. A small proportion of patients have no clinical or biochemical evidence of iron overload suggesting that other genetic or environmental factors may affect the phenotypic expression. Patients with HCC regularly absorb 2–3 times more iron than normal and symptoms do not develop until adult life. Excess iron is deposited in parenchymal cells of liver, heart, pancreas, joints, pituitary gland and parathyroid gland. Initially symptoms are non-specific including fatigue, arthralgia, erectile dysfunction and increased skin pigmentation. As the disease progresses, it may lead to enlargement of the liver and cirrhosis. There is increased incidence of hepatocellular carcinoma. Iron deposition in the heart can cause cardiomyopathy. Endocrinopathies such as diabetes mellitus, hypopituitarism, hypogonadism and hypoparathyroidism may develop due to the deposition of iron in these tissues. Patients with HCC are also susceptible to infection.

Initial investigation in suspected cases is transferrin saturation. A transferrin sturation > 55% in men and > 50% in women suggests iron accumulation. A high serum ferritin is also very suggestive of iron overload. Iron overload is confirmed by increased iron content of liver or by quantitative phlebotomy. Family members can be identified by HLA typing and by testing for mutations in the HFE gene. There is controversy as to whether there should be a screening programme for HCC in the Caucasian population.

Treatment of HCC consists of therapeutic phlebotomy, reduction in iron, alcohol and vitamin C intake.

Case 23.2

A 52-year-old man with diabetes was investigated for tiredness, loss of libido and weight gain. Results of investigations were as follows:

Serum		Reference range
TSH (IU/L)	1.6	0.2–5.5
FT4 (pmol/L)	8.5	10.1–21.0
Testosterone (nmol/L)	5.2	10.2–25.0
FSH (IU/L)	0.8	2.0–10.0
LH (IU/L)	2.5	2.0–8.0
Prolactin (mU/L)	255	< 420

His renal function tests were normal. A low testosterone with inappropriately low gonadotrophins shows that he has hypogonadotrophic hypogonadism. Low FT4 with inappropriately low TSH is indicative of secondary hypothyroidism. The possible diagnosis is hypopituitarism. With a history of diabetes and suspected hypopituitarism, further investigations for haemochromatosis were done and they showed the following:

Serum iron saturation	65%
Serum ferritin	3252 µg/L (15-300)

These results support the diagnosis of haemochromatosis. Genetic analysis confirmed the diagnosis.

---

2. *Haemochromatosis not due to HFE gene mutation*

- African iron overload

  This condition is present in up to 10% of some rural African population and results from a predisposition to iron overload that is exacerbated by excessive intake of dietary iron. This condition is particularly seen among Africans who drink a traditional beer brewed in non-galvanised steel drums. Iron overload is not due to a mutation of the HFE gene but it appears to be due to an unidentified iron loading gene. The pattern of iron deposition is different from that of haemochromatosis. Iron is deposited in Kupffer cells as well as in hepatocytes and the pattern resembles that seen in transfusional siderosis. Cirrhosis and occasionally hepatocellular carcinoma are major manifestations, while cardiomyopathy and diabetes are less common. Transferrin saturation does not always reflect the true extent of iron overload in these patients.

3. *Other forms of hereditary haemochromatosis*

Several other mutations causing HCC have been described. One such mutation is in the hepcidin gene. Collectively, these conditions are rarer than HCC.

4. *Transfusional siderosis*

Repeated transfusions to treat chronic anaemias such as thalassaemia and sickle cell disease can lead to iron overload. Each unit of blood contains 200–250 mg of iron and there is no mechanism to excrete this excess iron resulting in transfusional siderosis. Iron is deposited in the reticuloendothelial cells and then in the parenchymal cells. It will ultimately lead to the same symptoms as in other iron overload disorders. Body iron content can be determined by quantitative liver biopsy, as serum ferritin concentrations are less accurate. The usual treatment for this disorder is administration of desferroxamine, a chelating agent.

Further Reading

1. Andrews NC. Disorders of iron metabolism. *N Engl J Med* 1999; 341:1981–1995.
2. Anderson GJ, Frazer DM, McLaren GD. Iron absorption and metabolism. *Curr Opin Gastroenterol* 2009; 25:129–135.
3. Sherwood RA, Rippard MJ, Peters TJ. Iron homeostasis and the assessment of iron status. *Ann Clin Biochem* 1998; 35:693–708.

4. Worwood M. What is the role of genetic testing in diagnosis of haemochromatosis? *Ann Clin Biochem* 2001; 38:3–19.

Summary/Key points

1. Haemoglobin consists of two α- and two β-chains. Reduced production of the globin chain due to a genetic defect leads to thalassaemia. Inherited disorders producing abnormal haemoglobin leads to haemoglobinopathies, of which sickle cell disease is the commonest.

2. Porphyrias are a group of inherited disorders of porphyrin synthesis and are usually divided into acute and cutaneous porphyrias. In acute porphyrias, PBG and ALA are elevated and they present with acute symptoms precipitated by factors such as drugs, starvation or infection. Diagnosis is established by demonstrating increased ALA and PBG during the acute attack. Cutaneous porphyrias may be present with photosensitivity and diagnosis depends on analysis of urine and faeces for porphyrins.

3. Iron is an important element in the body and is an integral part of haemoglobin. Iron absorption is tightly regulated by hepcidin, a protein produced by the liver. Iron is transported bound to transferrin and stored as ferritin or haemosiderin.

4. Iron status can be assessed by serum iron and iron binding capacity, ferritin or serum soluble transferrin receptor. Ferritin is also an acute phase protein and is not useful if there is inflammation.

5. Iron deficiency anaemia is a common problem and it can lead to a variety of symptoms. It is caused by loss of blood from the body, either gastrointestinal loss or menstrual loss. It leads to microcytic, hypochromic anaemia. A high iron binding capacity and low ferritin are indicative of iron deficiency.

6. Anaemia of chronic disease is due to stimulation of hepcidin production by cytokines leading to reduced iron absorption and erythropoiesis.

7. Hereditary haemochromatosis is an inherited disorder of iron absorption leading to deposition of iron in pancreas, heart,

endocrine organs and joints. It is due to a mutation in the HFE gene, which affects hepcidin production leading to unregulated iron absorption. Serum iron binding capacity and serum ferritin are the initial tests for the diagnosis of suspected haemochromatosis. Liver iron content and genetic studies may be necessary to confirm the diagnosis. Treatment is by regular phlebotomy.

chapter 24

Malignancy and Clinical Biochemistry

Metabolic Aspects of Cancer

Malignancy may cause metabolic effects as a result of one of the following: direct local effects of the tumour and its metastases, indirect effects, secretion of various products by the tumour, metabolic activity of the tumour or non-specific systemic response.

Direct Metabolic Effects of Tumour

Tumour or its metastases may cause obstruction or tissue damage. Obstruction to the bile duct by tumour, e.g. carcinoma of the head of pancreas or lymph nodes will lead to obstructive jaundice with elevations in ALP and GGT. Intestinal obstruction as a result of tumour will cause fluid and electrolyte disturbances. A tumour in the liver may cause obstruction to bile flow in one part of the liver giving rise to increase in ALP and GGT but without jaundice, as the rest of the biliary tree is adequate to allow excretion of bilirubin. Obstruction of the urethra by carcinoma of the prostate will lead to renal impairment. Malnutrition may be caused by obstruction of the oesophagus and respiratory failure may result from obstruction of the bronchus.

The destruction of normal tissue may lead to biochemical changes. Destruction of the pituitary will lead to hypopituitarism and diabetes insipidus. Addison's disease may result from destruction of both adrenals by secondaries from lungs or breast cancer. Destruction of the bone by secondaries may cause hypercalcaemia and increased ALP.

Tumours in CNS may cause an increase in CSF protein. Fluid collection such as pleural effusion and ascites may arise as result of malignant invasion.

Malignant tumours may also cause biochemical change as a result of cellular activity and rapid growth of the tumour. Such biochemical changes include: (i) high serum urate due to high turnover of nucleic acids; (ii) high plasma LD, especially seen in lymphomas or leukaemia where elevation of LD2/LD3 is seen (In ovarian and testicular tumours, an increase in LD1 and LD2 is seen); and (iii) lactic acidosis, as a result of increased glycolysis.

Metabolic abnormalities may arise in malignancy as a result of response of tissues to damage caused by tumours. Production of cytokines such as IL-6, TNF-α and IL-1 will lead to acute phase response which is manifested by increase in acute phase proteins such as CRP and a decrease in serum zinc and iron (see 'Acute Phase Response' Chapter 14).

Cancer Cachexia

Cancer cachexia, a syndrome characterised by generalised wasting and weakness, is seen in about 20% of cancer patients. It is a hypercatabolic state, not correctable by feeding. Patients with cancer cachexia have anorexia, hypoalbuminaemia, oedema, anaemia and poor response to infections. Features of cancer cachexia are similar to those seen in chronic illness such as tuberculosis. There is loss of muscle and fat. Breakdown of fats, which is the main source of fuel, is increased, but there is no ketosis as in starvation. Patients with cancer cachexia have glucose intolerance and insulin resistance. Gluconeogenesis may be increased due to increased lactate produced by anaerobic glycolysis by the tumour. This lactate cycle is an inefficient process and may contribute to the negative energy balance.

The exact pathogenesis of cachexia is not fully understood. It is not related to the size of the tumour. It is thought to be due to production of cytokines such as TNF-α, interferon γ, IL-1 and IL-6 by the tumour.

Treatment of cancer cachexia is difficult. Parenteral or enteral feeding is required but not always helpful. Megestrol acetate, a progestational drug, may help to improve appetite. Factors, which may contribute to cachexia, are listed in Table 24.1.

Paraneoplastic Syndromes

These are syndromes, which develop in association with tumours but not due to a direct effect of the tumour or its metastases. It may be seen in 7–15% of patients with malignancy. These syndromes are caused either by secretion of chemicals (hormones, cytokines etc) (Table 24.2) or due to reaction by the body in response to the tumour such as production of antibodies which may affect normal tissues. Removal of the

Table 24.1 Factors contributing to cancer cachexia

Decreased food intake
  • Anorexia
  • Obstruction to GI tract
Malabsorption
Increased Loss
  • Loss of protein from tumour
Increased consumption
  • Anaerobic glycolysis by tumour
  • Tumour product such as TNF-α causing increased energy expenditure

Table 24.2 Some paraneoplastic endocrine syndromes

Tumour	Product	Syndrome
Small cell carcinoma of bronchus	ACTH, Vasopressin, hCG	Cushing's syndrome SIADH Gynaecomastia
Squamous cell carcinoma	PTHrP	Hypercalcaemia
Breast carcinoma	Calcitonin	—
Carcinoid tumours	ACTH, Vasopressin	Cushing's syndrome SIADH
Renal adenocarcinoma	PTHrP	Hypercalcaemia
Mesenchymal tumour	IGF-2	Hypoglycaemia

tumour relieves the syndrome in the first case but not in the latter. It is not unusual for patients to present with features of paraneoplastic syndrome before they show any features of the malignancy.

Hormones produced by tumours are the best examples of paraneoplastic syndromes. Production of hormones by tumours arising from cells, which normally do not secrete hormones, is usually described as 'ectopic' hormone secretion. However, this term is not preferred as it is now known that many tissues are capable of hormone production and these hormones may have local paracrine or autocrine function. Not all hormones produced by tumours cause syndromes because the concentration may be low, the hormone may be in a precursor form, inactive form or only a portion of the hormone is produced (e.g. subunits of FSH and LH are produced by many tumours).

Hypercalcaemia

Hypercalcaemia is a frequent complication of malignancy. Hypercalcaemia in malignancy is either due to osteolytic lesions of the bone or due to secretion of chemicals from the tumour, humeral hypercalcaemia of malignancy (HHM). In hypercalcaemia due to osteolytic lesions, substances such as transforming growth factors, TNF-β, TNF-α, IL-1 and IL-6 are produced and these act locally to increase bone resorption. Hypercalcaemia of myeloma is another example of osteolytic lesion caused by local production of TNF-β, IL-1 and IL-6.

In HHM, the tumour produces PTHrP, which increases bone resorption and tubular reabsorption of calcium. Carcinomas, especially those of the respiratory tract, breast, renal, ovarian and bladder are most often associated with HHM. In HHM, there may be secondary deposits in bone as well.

Increased production of 1,25-dihydroxyvitamin D by lymphoma cells can also cause hypercalcaemia (see Chapter 6).

Syndrome of inappropriate ADH secretion (SIADH)

SIADH is frequently seen in small-cell carcinoma of the lung (SCCL). These tumours produce ADH (AVP) as well as associated peptides such as neurophysin. ADH is increased in about 50% of patients with

SCCL and a third of these develop SIADH and hyponatraemia (see Chapter 2). Carcinoid tumour, pancreatic adenocarcinoma and thymic tumours are also associated with SIADH.

Cushing's syndrome

Secretion of ACTH or rarely CRH by tumours can lead to Cushing's syndrome. Tumours associated with Cushing's syndrome are small-cell carcinoma of the lung (SCCL), bronchial carcinoid tumours and phaeochromocytoma. SCCL very often express genes for peptide hormones such as ACTH, antidiuretic hormone (ADH), calcitonin, oxytocin and GIP. About 50% of patients with SCCL have increased plasma ACTH but in only a minority, about 5%, develop Cushing's syndrome. The secretion of ACTH precursors may explain the lack of development of Cushing's syndrome in many patients with SCCL.

Clinical features of Cushing's syndrome associated with malignancy are usually different. As the disease develops rapidly, classical features of Cushing's syndrome are not seen. Hypokalaemic alkalosis and glucose intolerance may be prominent. Pigmentation may be seen due to the effects of ACTH precursors, which have MSH-like activity (Chapter 21).

Case 24.1

A 65-year-old man was admitted with muscle weakness, weight loss, and increased pigmentation. He was a smoker, who has smoked 20–30 cigarettes a day for nearly 40 years. Initial investigations showed the following results:

Serum		Reference Range
Sodium (mmol/L)	144	135–145
Potassium (mmol/L)	2.1	3.5–5.0
Bicarbonate (mmol/L)	39	23–32
Chloride (mmol/L)	92	90–108
Urea (mmol/L	7.6	3.5–7.2
Creatinine (μmol/L)	120	60–112
Glucose (mmol/L)	15.2	

These results show a severe hypokalaemic alkalosis. Together with hyperglycaemia and history of weight loss and heavy smoking, a provisional diagnosis of ectopic ACTH syndrome was made and further investigations were done:

Serum		Reference Range
Random serum cortisol (nmol/L)	2560	140–690 (9 a.m. range)
Urine 24 h free cortisol excretion (nmol)	1650	< 265
Plasma ACTH (ng/L)	650	20–40

These results show increased cortisol production with a very high ACTH and the diagnosis of ectopic ACTH syndrome is likely. A high-dose dexamethasone was done next and there was no suppression of serum cortisol. Imaging studies showed a bronchial tumour.

Tumour-associated hypoglycaemia

Large mesodermal tumours such as fibrosarcomas and mesotheliomas as well as carcinoma of the liver and adrenal cortex may be associated with hypoglycaemia. The cause of the hypoglycaemia is the unregulated production of IGF-II, which suppresses GH secretion. This in turn leads to reduced IGF-I and IGFBP-3, the main binding protein for IGF-I. Reduced binding protein allows IGF-II to cross capillary matrix and act on tissue receptors (Chapter 10).

Occasionally in Hodgkin's lymphoma, hypoglycaemia may develop due to insulin antibodies that bind and activate insulin receptors.

Oncogenic osteomalacia

This syndrome is associated with mesenchymal tumours such as haemoangiopericytomas, which are seen in bone and soft tissue of extremities. The tumours are small and often benign. Patients present with muscle weakness and bone pain and there is profound hypophosphataemia, renal phosphate loss and evidence of osteomalacia. Patients may present with these features without evidence of tumour which may only be found sometimes years later. Removal of the tumour causes complete reversal of biochemical and clinical features.

These tumours secrete fibroblast growth factor-23, which causes renal phosphate loss and inhibition of 1,25-dihydroxyvitamin D production. PTH concentration unlike in classical vitamin D deficiency is often normal.

Other syndromes (Table 24.3)

Other rare endocrine related syndromes associated with tumours include acromegaly due to secretion of GH or more commonly GHRH, and polycythaemia due to erythropoietin production. Other paraneoplastic syndromes are listed in Table 24.3.

Metabolic Complications

Tumour lysis syndrome

A metabolic complication arising as a result of treatment is tumour lysis syndrome. This is due to the massive necrosis of tumour cells as a result of treatment with cytotoxic drugs leading to hyperkalaemia, hyperphosphataemia, hyperuricaemia, hypocalcaemia and often acute renal failure. This syndrome is most often seen with tumours such as lymphomas and leukaemias that are sensitive to chemotherapy.

Table 24.3 Other paraneoplastic syndromes

Fever
Immune-mediated renal disease
Dermatological disorders
   • Aacanthosis nigricans
Neuromuscular disorders
   • Dermatomyositis
   • Polymyositis
   • Myasthaenia gravis
   • Peripheral neuropathy
   • Subacute cerebellar degeneration
Erythrocytosis
Leukocytosis — leukaemoid reaction
Acromegaly

Tumour lysis syndrome can be prevented or its effects minimised by adequate hydration during chemotherapy, treatment with hypouricaemic drugs (allopurinol) and by regular monitoring of electrolytes and renal function.

Other metabolic complications

Treatment with cisplatin or carboplatin can lead to hypomagnesamia due to renal magnesium wasting. Renal potassium wasting and severe hypokalaemia may result from treatment with amphotericin that is used to treat fungal infection in patients with malignancy.

Tumour Markers

Tumour markers are substances that are measured in tissue or body fluids in order to detect malignancy and monitor the progress of treatment. Tumour markers may be structural molecules, secretory products and enzymes, or non-specific markers of cell turnover. Most, not all, of these markers are found in non-malignant cells and therefore, interpretation of results should be done with caution. Structural molecules, which are used as tumour markers, include carcinoembryonic antigen (CEA), glycoproteins (mucins) found on the cell surface of tumours (e.g. CA19-9, CA15-3, CA-125), β_2 microglobulin and cytokeratins. Cytokeratins are filaments of the cytoskeleton of the cell and examples include tissue polypeptide antigen and tissue polypeptide-specific antigen. Secretory products used as tumour markers include α-fetoprotein (AFP), hCG and prostate specific antigen (PSA). Non-specific markers of cell turnover include neopterin, a breakdown product of pteridine metabolism, thymidine kinase and tumour-associated trypsin inhibitor. Some of the tumour markers and their potential applications are listed in Table 24.4.

An ideal tumour marker should be detectable only when malignancy is present, be specific for the type and site of malignancy, correlate with the amount of malignant tissue present and respond rapidly to changes in tumour size. Such an ideal tumour marker should therefore be useful in screening, diagnosis, assessment of

Table 24.4 Tumour markers and their potential applications

Marker	Tumour
α-Fetoprotein	Hepatic and germ cell tumours
Human chorionic gonadotrophin (hCG)	Germ cell tumours, choriocarcinoma
Carcinoembryonic antigen (CEA)	Colorectal carcinoma
CA125	Ovarian carcinoma
CA15-3	Breast malignancy
CA19-9	Gastrointestinal and pancreatic tumours
Calcitonin	Medullary carcinoma of thyroid
Prostatic-specific antigen	Prostate carcinoma
Thyroglobulin	Follicular or papillary carcinoma of thyroid
Paraprotein	Myeloma
Chromogranin	Neuroendocrine tumours
Neurone-specific enolase	Small cell carcinoma of lungs and APUD tumours
Placental alkaline phosphatase	Germ cell tumours
Des-gamma-carboxyprothrombin	Hepatocellular carcinoma
β_2 Microglobulin	Multiple myeloma

prognosis, detection of residual disease, monitoring of response to treatment and for follow up. At present, there is no marker that fulfils all these criteria. Only hCG comes close to fulfilling these criteria.

Serum concentration of tumour markers may be affected by the production rate, which depends on the rate of growth of the tumour, and vascularity of the tumour, and clearance, excretion and metabolism. Most tumour markers are unsuitable for screening as they have poor specificity. The low prevalence of malignancy in the population gives rise to a large number of false positives and small tumours may not produce an elevation of the tumour marker. Tumour markers are often used to make a diagnosis in patients with symptoms. However, even in this group, the usefulness of tumour marker is limited as the tumour marker may be elevated due to non-malignant conditions (e.g. AFP may be elevated in non-malignant liver diseases) and some tumours may not produce tumour markers. Thus, a high concentration of a tumour marker makes the diagnosis likely but does not prove it. Tumour markers are most often used for determining prognosis and in monitoring treatment. In general, the serum concentration of a marker at presentation is related to the tumour mass and the fall of tumour marker after treatment gives some indication of the success or failure of the treatment procedure. The time for the tumour marker to reach normal concentration depends on the initial concentration of the tumour marker, half-life of the tumour marker, and the nature of treatment. However, following treatment with cytotoxic drugs there may be transient elevations of tumour markers as a result of release from cells. Tumour markers are useful in detecting recurrence. A significant rise in serum tumour marker on more than one occasion indicates a high likelihood of a recurrence. However, serum tumour marker within the reference range does not exclude recurrence as cellular composition of the tumour may change after treatment, causing the tumour marker to remain low despite recurrence.

Alpha fetoprotein (AFP)

Alpha fetoprotein (AFP), a heterogeneous glycoprotein of molecular weight 70 kD, is synthesised by the yolk sac, foetal gastrointestinal tract and the liver. In the foetus, it is the major serum protein and may function like albumin. AFP is found in monomeric, dimeric and trimeric forms. The very high serum concentration in the foetus causes diffusion of AFP into the amniotic fluid and maternal serum. Maternal serum AFP can be used for screening of neural tube defects and Down's syndrome (see Chapter 20). Serum AFP concentration is highest at birth and then gradually decline to reach adult values in 8 to 12 months. AFP is useful in the diagnosis and monitoring of primary hepatocellular carcinoma, hepatoblastoma and germ cell tumours.

Serum AFP concentration in liver tumours varies widely. It is elevated in 80% of cases with hepatocellular carcinoma and in 100% of patients with hepatoblastoma. In some countries where hepatoma is common, measurement of AFP followed by ultrasound examination has been used as a screening procedure for early detection of hepatomas. In countries where hepatoma is less common, AFP and ultrasound can be used to detect hepatomas in high risk patients such as those with cirrhosis, persistent hepatitis B virus infection and haemochromatosis. Recent studies, however, have shown that the positive predictive value of AFP in the diagnosis of hepatocellular carcinoma is lower in patients with viral aetiology compared to those without viral aetiology (70% vs. 90%). Serum concentration of AFP is related to tumour size in hepatocellular carcinoma. Serum AFP concentrations may be elevated in non-malignant conditions (usually conditions associated with hepatic necrosis and regeneration such as viral and other forms of hepatitis, portal cirrhosis and primary biliary cirrhosis). Extrahepatic cholestasis and hereditary tyrosinaemia may also cause increased concentrations. However, in most of these conditions the serum concentration remains < 200 μg/L. When the value is > 500 μg/L, the diagnosis of hepatocellular carcinoma is almost certain. Specificity and sensitivity varies with the concentration of serum AFP. At high concentrations specificity is high (99%) but sensitivity is low (20%). Serum concentration AFP is also of some prognostic value. Patients with values > 1000 μg/L usually have poor prognosis. Serial monitoring of AFP is of value in detecting response to treatment. Serum AFP is also elevated in 5–24% of malignancies of pancreas, stomach and colon.

Based on the reaction to lectin, AFP can be separated into three fractions, L1, L2 and L3. Of these, AFP-L3 is useful in hepatocellular carcinoma. L3 isoform is associated with malignancy and L1 isoform with non-malignant liver diseases. The ratio of L3 to total AFP is useful to diagnose hepatocellular carcinoma in patients who are at high risk of developing hepatocellular carcinoma (e.g. hepatitis C carriers).

AFP is a useful marker for non-seminomatous germ cell tumours. About 60% of teratoma of the testes and yolk sack tumours of the ovaries produce AFP. Serum AFP is important in staging and monitoring treatment of these patients. Failure of AFP to return to normal after treatment, persistent elevations or a long half life (> 5 days) is indicative of residual tumours or metastatic disease. An increase in serum concentrations following treatment indicates recurrence of the tumour provided other causes such as chemotherapy-induced liver damage has been excluded. The adequacy of therapy is assessed by serial estimations of marker concentrations.

Carcinoembryonic Antigen (CEA)

CEA is an acid glycoprotein present in the cell surface involved in intracellular adhesion and is normally produced by the foetus. Serum CEA is higher in smokers than non-smokers. CEA is raised in colorectal cancer, other gastrointestinal tract malignancies as well as in breast, bronchial, bladder and ovarian tumours. In patients with colorectal cancer, CEA is elevated in 10% of patients with localised lesions and in about 70% of patients with disseminated disease. Elevated CEA may also be found in non-malignant conditions such as normal pregnancy, inflammatory bowel disorders including ulcerative colitis and Crohn's disease, pancreatitis, alcoholic cirrhosis and in any acute insult. Because of its poor sensitivity and specificity, it cannot be used for screening for colorectal cancer but it is of value in prognosis and monitoring. Several studies have shown that patients with high CEA concentrations before surgery have a poor prognosis probably reflecting large tumour mass or the presence of unsuspected liver metastases. The half-life of CEA is about 3 days. After removal of the tumour elevated concentrations should return to normal in 4–6 weeks. Failure of serum CEA to fall to normal values suggests residual tumour. During follow up of patients, serum CEA measurements are useful to detect early recurrence. Serum CEA concentration may be elevated 3–6 months before clinical signs develop. A falling serum CEA concentration during monitoring suggests response to treatment whereas a rising concentration indicates the need for a change of treatment.

CEA has also been measured in fluids such as pleural and peritoneal effusions where the concentrations are usually greater than in serum. In general, high concentrations in these fluids suggest malignancy but it is not diagnostic. Occasionally, CEA is measured in CSF and an increase in CSF relative to serum suggests meningeal involvement. However, CEA is elevated in only 50% of cases of cerebral metastases. Due to the poor specificity and sensitivity, a normal CEA concentration does not exclude malignancy.

Human Chorionic Gonadotrophin

Human chorionic gonadotrophin (hCG) is a glycoprotein secreted by the syncytiotrophoblasts of the normal placenta. hCG consists of two subunits, α and β. α-subunit is identical to the α-subunit of the other pituitary glycoprotein hormones (FSH, LH and TSH). The β-subunit has some similarities to that of LH. hCG is also secreted by germ cell tumours and small amounts are secreted by the normal anterior pituitary gland. hCG may be present in several forms in the serum and urine. These forms include the intact molecule, the free α- and β-subunits, hyperglycosylated hCG, hyperglycosylated β-hCG-subunit and the inactive breakdown products, nicked hCG and β core fragment. Nicked hCG is produced by leucocyte elastase breaking down peptide bonds in the β-subunit of hCG. This in turn is degraded to the β-subunit core fragment in the kidney. Hyperglycosylated hCG is the predominant form in early pregnancy and in invasive mole, choriocarcinoma and other invasive malignancies.

The production of the subunits of hCG is under separate genetic control. In early pregnancy, the free β-subunit is produced together with intact hCG and in later pregnancy, free α-subunit predominates. β hCG peaks at about 8–10 weeks of pregnancy, but production of the α-subunit continues to increase and appears to be a function of the mass of the placenta. Most cancer patients produce both intact molecule and the free β-subunits.

hCG is most useful as a tumour marker in trophoblastic tumours, in non-seminomatous testicular tumours and in some seminomas. Choriocarcinoma is a malignant tumour of the chorionic villi developing from a hydatidiform mole, which is a benign proliferation of the chorionic villi. Prevalence of hydatidiform mole in Europe is 1 in 2000 pregnancies whereas in South East Asia, it is much more common accounting for 1 in 200 deliveries. Hydatidiform mole can develop into an invasive mole or choriocarcinoma in about 3% of patients. Very high serum hCG concentration in early pregnancy is very suggestive of a molar pregnancy. Molar pregnancies are treated by evacuation. These patients are followed up regularly by serial hCG measurement. Failure of serum hCG to fall to undetectable levels in 1–2 weeks is indicative of invasive mole or choriocarcinoma. Invasive moles and choriocarcinoma produce hyperglycosylated form of hCG where as hydatidiform moles secrete normal hCG. It is important to note that the ability of commercial assays to detect different forms of hCG especially the hyperglycosylated form varies; this has implications for the monitoring of these patients.

Serum hCG is also a useful marker in germ cell tumours. Germ cell tumours are usually divided into seminoma and non-seminomatous germ cell tumours. The presence of AFP classifies a tumour as non-seminomatous tumour, usually teratoma. Increased hCG concentrations are seen in 75% of teratomas and combined tumours and 30% of seminomas. Serum hCG concentration is an indicator of prognosis; very high values indicate a poor prognosis. Serum hCG and AFP are useful markers in monitoring treatment and follow up.

Many non-trophoblastic malignancies such as ovary, breast, endometrial, colon, lung and bladder produce hyperglycosylated β unit and the β core fragment. The presence of these substances is a poor prognostic sign.

The simultaneous measurement of serum and CSF hCG is of help in the detection of cerebral metastases of germ cell tumours. If the serum:CSF ratio of hCG is < 10:1 (normal ratio > 60:1) it is diagnostic of brain metastases. Ratios in between 10:1 and 60:1 are very suggestive of cerebral metastases.

β_2 *Microglobulin*

β_2 Microglobulin is a small molecular weight protein of 11,000 molecular weight found on the surface of all cells. It is the light chain of the class 1 major histocompatibility antigen. It is freely filtered and almost completed reabsorbed and catabolised in renal tubules. Serum concentration therefore increases in renal failure. In B cell malignancies, particularly those that do not produce paraproteins, it is a useful indicator of tumour mass and growth rate. It is also an important prognostic marker in multiple myeloma. Serum concentration <4 mg/L is associated with good prognosis while concentrations >20 mg/L with poor prognosis. Serum β_2 microglobulin is increased in many other malignancies, but this increase is non-specific and does not correlate with tumour type or severity of disease. In acute leukaemia, CSF concentration of β_2 microglobulin, if elevated in relation to serum, indicates CSF involvement. β_2 microglobulin is also increased in collagen vascular disorders such as Sjogren's syndrome and rheumatoid arthritis. Treatment with corticosteroids and immunosuppresants tend to decrease the concentration.

Paraproteins

Quantitation and typing of paraproteins help to establish whether the condition is malignant or benign. Serum concentration of paraprotein indicates the tumour burden. Serial measurements are useful in monitoring therapy and to follow the course of the myeloma. Quantitation of Bence-Jones protein is also of value in monitoring response to treatment and to follow the progression of multiple myeloma (see Chapter 14).

Calcitonin

Calcitonin, a peptide hormone produced by parafollicular C cells of the thyroid, is increased in medullary carcinoma of the thyroid, which has a familial origin in 20% of cases. It is a useful tumour marker in the diagnosis and monitoring of this tumour and as a screening test of family members of patients. The familial form is transmitted as an autosomal dominant form or as a component of MEN II. Other tumours such as small cell carcinoma of the lung, breast carcinoma and carcinoid tumours can also produce calcitonin. In medullary carcinoma of the thyroid, serial measurements are of value in monitoring the progress of the disease and the detection of recurrence. In the early detection of the tumour in family members, basal calcitonin concentration may not be sufficiently high. Stimulation tests such as the pentagastrin stimulation test is recommended. In this test, serum calcitonin is measured before and 0, 2, 5, 15 and 30 minutes after pentagastrin (0.5 μg/kg body weight) injection. Calcitonin concentration rises markedly in medullary carcinoma.

Neurone Specific Enolase (NSE)

Enolase, a glycolytic enzyme, is present in two forms: $\chi\chi$ which are specific for neurones and $\alpha\alpha$ which are found in glial cells. The $\chi\chi$ form (NSE) is found in tumours of neuroendocrine or neuroectodermal origin such as small cell carcinoma of lung and neuroblastoma. Serum NSE is of little value in the diagnosis of these tumours but may have a role in monitoring. It is also increased in medullary carcinoma of thyroid, neuroblastomas, carcinoid tumours, endocrine tumours of pancreas and melanoma.

Placental Alkaline Phosphatase (PLAP)

Placental and placental-like isoenzyme of ALP are oncofoetal antigens and together with hCG or AFP are of value in the diagnosis and monitoring of germ cell tumours. It is increased in 50% of seminomas and 60% of dysgerminomas but not in teratomas. It is especially of value in the 10% of patients where it is the only marker. PLAP is not increased in non malignant conditions of testis, but is increased in smokers. The half-life of PLAP is less than 3 days and is of value in monitoring the response to treatment and the progression of the disease.

Thyroglobulin

Thyroglobulin, a large molecular weight protein synthesised by the follicular cells of the thyroid (see Chapter 22), is a useful marker for follicular and papillary carcinoma of the thyroid. These tumours are treated by total thyrodiectomy and thyroglobulin measurements are helpful in detecting recurrence. Anti-thyroglobulin antibodies present in serum may interfere with the assay.

Carbohydrate Antigen CA125

This is a high molecular weight protein present in tissues of coelomic epithelial origin. CA125 is present in the female reproductive tract, gastrointestinal tract, pleura and peritoneum. It is increased in a proportion of epithelial tumours of the ovary. High values are seen in 50–60% of serous adenocarcinoma and in only 15% of mucinous adenocarcinoma. If CA125 is elevated before treatment, it is a reliable marker in monitoring treatment. Raised CA125 is also found in malignancies of gastrointestinal tract, lung and breast and in non-malignant conditions such as endometriosis, chronic liver disease, pelvic inflammatory disease and in inflammation of the pleura and peritoneum. In familial ovarian cancer, yearly measurements of CA125 in unaffected family members is useful in early detection of malignancy. In postmenopausal women, CA125 as a screening test is undergoing a randomised trial. Preliminary results from this trial suggests that this test is useful in screening when combined with ultrasound. High CA125 values are indicative of poor prognosis. In premenopausal women, the concentration is variable depending on the stage of menstrual cycle; therefore, it is less reliable as a diagnostic test.

Carbohydrate Antigen CA15-3

This is a large molecular weight (400 kDa) protein of epithelial mucin. Serum CA15-3 is raised in a large number of malignant and non-malignant diseases of the lung, gastrointestinal tract, reproductive system and liver. However, CA15-3 is most often increased in carcinoma of breast and useful in detecting metastases, to monitor response to treatment and to detect recurrence. Proportion of patients with elevated CA15-3 increases with increasing stages of breast cancer, from 10% in Stage I disease to 85–100% in Stage IV. It is also increased in benign breast disease and in 10% of patients with liver disease.

Carbohydrate Antigen 19-9

This is a high molecular-weight mucin. It represents a sialyated form of the Lewis blood group antigen and as such it is not found in about

10% of the population who lack this blood group antigen. It is increased in pancreatic, gastrointestinal and hepatobiliary tumours. It is increased in up to 80% of patients with Stage II pancreatic carcinoma. In colorectal carcinoma, the proportion of cases with elevated CA19-9 varies from < 10% in Stage A to 60% in Stage D. Benign conditions such as cirrhosis, alcoholic hepatitis, chronic pancreatitis, acute hepatic necrosis, cholecystitis and obstructive jaundice may cause an increase in serum 19-9. It is useful mainly in monitoring the treatment of pancreatic tumours.

Prostate Specific Antigen (PSA)

Prostate specific antigen is a serum protease found normally in seminal fluid where it dissolves seminal coagulum. It is predominantly produced by the glandular epithelium of the prostate, but is also produced by paraurethral and perianal glands as well as mammary glands. PSA concentration in semen is a million fold greater than in serum and its secretion is dependent on androgens. In the circulation, PSA is predominantly found as a complex with α_1-antichymotrypsin, the free fraction accounting for a variable proportion but usually < 10%. It may also be complexed with α_2 macroglobulin and this form is not usually measured by most assays.

Serum PSA is elevated in carcinoma of the prostate, up to 15% of patients with benign prostatic hyperplasia (BPH), in prostatitis, prostate ischaemia, urinary retention and in acute renal failure. Digital rectal examination as well as cytoscopy may increase PSA and it is recommended that blood sample should be taken before such examinations.

Serum PSA is elevated in up to 50% of patients with prostatic tumour confined within the prostatic capsule and in 90% when the tumour has metastasised. Serum PSA correlates with tumour size and extent. It is useful in monitoring treatment and in the detection of recurrence. The half-life of PSA is 2–3 days and serum PSA becomes undetectable within a month of radical prostatectomy. After radiotherapy it takes longer to fall, taking about a year and with anti-androgen treatment it will fall within 6 months. In patients on anti-androgen treatment a low or falling concentration does not necessarily reflect tumour elimination, as the anti-androgens will suppress expression of PSA.

The use of serum PSA for screening is controversial as it has inadequate sensitivity and specificity. Up to 30% of patients with carcinomas may be missed and a large number of patients without disease will be subjected to unnecessary anxiety and investigation. In order to improve the detection of prostatic cancer, several suggestions have been made.

- Reference range: PSA rises with age and is also different in different ethnic groups. Age and ethnic group-specific reference ranges may improve the sensitivity and specificity of detection of prostate cancer.
- PSA density: As the serum PSA is related to size of the prostate gland, calculation of PSA density (serum PSA in relation to the size of the prostate as determined by transrectal ultrasound) is higher in malignancy than in BPH.
- PSA velocity: The rate of rise in PSA with time is greater in malignancy. A rise in serum PSA by 20% a year is suggestive of malignancy. However, this should be interpreted in relation to analytical as well as biological variation, which can be up to 20% and recent studies have not confirmed the usefulness of PSA velocity as a specific test.
- Free and bound PSA: The proportion of bound PSA is higher in malignancy. PSA index which is free/total PSA as a percentage is < 17% in malignancy and > 17% in BPH. This index is of value when the PSA concentration is between 4 and 10 μg/L.

Recent large-scale trials have not shown benefit in screening for prostate cancer with serum PSA. Prostate cancer in many instances is a slow growing tumour and may not manifest clinically. Other tumours progress rapidly. At present, there is no method to distinguish these from slow-growing tumours. Use of PSA as a screening test will result in large numbers of 'unnecessary' intervention with potential complications arising as a result of treatment.

Miscellaneous

Several enzymes and other tumour products may be useful as tumour markers.

Lactate dehydrogenase (LD), a marker of cell proliferation, has been used in ovarian and testicular tumours, lymphoma and leukaemia. Tumours of the neuroendocrine system such as small cell carcinoma of the lungs and prostate carcinoma may produce creatine kinase isoenzyme BB. Presence of BB isoenzyme is usually suspected when the CKMB value in an immunoinhibition assay is greater than that of the total CK. Lysozyme (muramidase) is an enzyme secreted by the myelomonocytic cells and in leukaemia of this cell type serum concentrations are high. This enzyme may damage the renal tubules and cause hypokalaemia and hypomagnesaemia.

Des-γ-carboxyprothrombin (DCP), also called PIVK-II (protein induced by vitamin K deficiency or antagonists) is a form of prothrombin which is undercarboxylated. DCP is increased in patients with hepatocellular carcinoma and has a sensitivity of 50% and specificity of 85%, similar to that of AFP. A combination of AFP and DCP has a better diagnostic value. DCP is increased in those taking warfarin and with vitamin K deficiency.

Chromorgranin A is a protein found in the secretory granules of the cells of neuroendocrine origin. It is released along with the secretory products of the cell. Serum chromogranin A has been suggested as a tumour marker for carcinoid tumour and phaeochromocytoma. However, it is also increased in pancreatic cancer, kidney and liver failure and inflammatory bowel disease.

Carcinoid Syndrome (Tumour)

Carcinoid tumours are tumours of enterochromaffin cells that produce serotonin or 5-hydroxytryptamine (5HT). The most common sites for this tumour are the appendix and rectum. Very often these tumours are found incidentally on histological examination. Usually these tumours are benign. Carcinoid syndrome occurs in about

10% of patients with carcinoid tumours and only when the venous return from the tumour enters the systematic circulation. When the venous return passes through the liver, 5HT will be metabolised. In midgut tumours, symptoms will appear only after metastases appear in the liver. Carcinoid tumours may also occur in the bronchus and rarely in the ovary and testes; these tumours may be associated with the syndrome in the absence of metastases.

Classical clinical features of carcinoid syndrome include flushing, bronchial constriction, diarrhoea and cardiac valvular lesions. The symptoms are due to serotonin excess as well as due to other amines: histamines, bradykinin, prostaglandins, and other vasoactive peptides. Some patients may also have pellagra-like features due to deficiency of tryptophan, which is predominately metabolised to serotonin.

Serotonin (5HT) is derived from the amino acid tryptophan (Figure 24.1) and it is a powerful smooth muscle stimulant and vaso-constrictor. About 1–3% of dietary tryptophan is normally converted to 5-hydroxytryptophan (5HTP) by tryptophan hydroxylase. 5HTP is then decarboxylated to 5HT. 5HT is metabolised by monoamine oxidase to 5 hydroxyindoleacetic acid (5HIAA). In carcinoid syndrome, up to 50% of dietary tryptophan may be metabolised by this pathway. Production and metabolism of serotonin varies in relation to the site of origin of the tumour. Tumours from the midgut such as ileal tumours produce 5HT. Tumours of the foregut, for example bronchial tumour, secrete primarily 5HTP as they lack the decarboxylase enzyme.

Diagnosis

Diagnosis of carcinoid syndrome depends on the demonstration of increased excretion of 5HIAA in the urine. Patients should avoid foods such as pineapples, avocados, bananas, walnuts, chocolate, etc. which may produce spurious elevations of 5HIAA. Measurements of 5HT or 5HTP in the urine or platelets has also been used. Serum chromogranin A is another test, but it lacks specificity. The best method to localise the tumour is by somatostatin receptor scintigraphy where radiolabelled octreotide is given before scintigraphy.

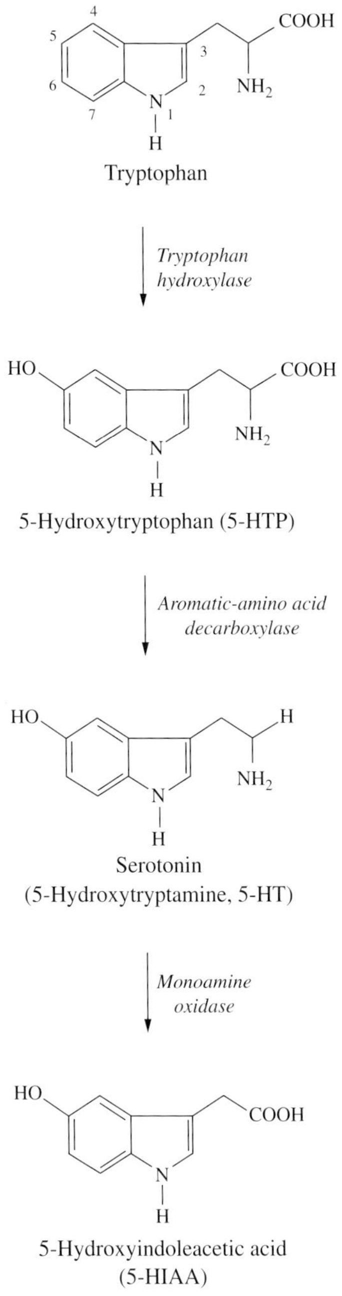

Figure 24.1   Metabolism of serotonin (5-hydoxytryptamine).

Management

Surgical removal is the treatment of choice. If this is not possible, palliative treatment with codeine, diphenoxylate or loperamide to control diarrhoea is recommneded. Octreotide, a somatostatin analogue, is effective in giving symptomatic relief.

Neuroblastomas

Neuroblastoma is one of the most common malignant tumours in children. More than 80% of the neoplasms are found in children younger than 5 years in age. These tumours are characterised by rapid growth and widespread metastases. Most of the tumours (> 90%) produce excessive catecholamines and its metabolites. Measurements of catecholamines or its metabolites (VMA, HVA or dopamine) are necessary to establish the diagnosis. Increased dopamine excretion is particularly characteristic of neuroblastoma.

Phaeochromocytoma

See Chapter 12.

Further Reading

1. Diamandis EP, Hoffman BR, Sturgeon CM. Practice Guidelines for the use of Tumour Markers. *Clin Chem* 2008; 54:1935–1939.
2. Diamandis EP, Fritsche HA, Lilja H, Chan DW, Schwartz MK. *Tumour Markers: Physiology, Pathobiology, Technology and Clinical Applications* 2002. AACC Publications.
3. Gullett N, Rossi P, Kucuk O, Johnstone PA. Cancer-induced cachexia: A guide for the oncologist. *J Soc Integr Oncol* 2009; 7:155–169.
4. Pannall P, Kotasck D. *Cancer and Clinical Biochemistry* 1997. ACB Venture Publications.
5. Srirajaskanthan R, Shanmugabavan D, Ramage JK. Carcinoid syndrome. *Br Med J* 2010; 341:c3941.

Summary/Key points

1. Tumours can cause metabolic effects due to direct effect: by obstruction or destruction of tissues (e.g. hypopituitarism by pituitary tumours).

2. Many patients with malignancy have metabolic effects due to substances produced by the tumour entering the circulation. These paraneoplastic syndromes may be due to ectopic hormone production (e.g. Cushing's syndrome, SIADH, humoral hypercalcaemia of malignancy (HHM) and oncogenic osteomalacia).

3. Patients with malignancy are often wasted (cancer cachexia) and this is due to a combination of reduced food intake and an increase metabolic requirement.

4. Other metabolic complications such as tumour lysis syndrome, hypomagnesaemia may arise during treatment.

5. Tumour markers are substances produced by tumour cells and secreted into blood and other body fluids. These are in general useful in monitoring treatment and detecting recurrence. As most tumour markers are non specific, they are not useful as a diagnostic test.

6. AFP, which is normally produced by the foetus, is increased in hepatic diseases especially hepatocellular carcinoma and in germ cell rumours of the ovary and testes. AFP is also increased in many non-malignant diseases of the liver.

7. hCG produced by the placenta is a good tumour marker for invasive mole and choriocarcinoma. When hydatidiform mole is detected, follow up is necessary to detect those who develop choriocarcinoma. Measurement of hyperglycosylated form of hCG is a better marker than total hCG.

8. AFP and hCG are useful in monitoring germ cell tumours.

9. Carcinoid tumours are tumours of the enterochromaffin cells usually of the gut and occasionally lung. These secrete 5-hydroxytryptamine, which is metabolised to 5-hydroxyindoleacetic acid (HIAA). Urinary 5HIAA is useful in diagnosing and monitoring these patients.

chapter 25

Therapeutic Drug Monitoring and Toxicology

Therapeutic Drug Monitoring

Therapeutic drug monitoring (TDM) has now become an essential part of clinical medicine. TDM is the use of drug measurements in body fluids to aid in the management of patients receiving drug treatment. When drugs are administered to patients, the dosage of the drug should be adjusted so that it gives the optimum desired therapeutic effect without causing side effects or toxic effects. As there is a large variation in the therapeutic response between individuals for a given dose of drug, it is necessary to monitor and adjust the dosage according to the individual's needs. The dosage can be adjusted according to clinical effects (e.g. changes in blood pressure in patients treated with antihypertensive drugs) or by measuring an effect of the drug (e.g. blood glucose concentration or lipid concentration in patients on antidiabetic drugs and lipid-lowering drugs respectively). Laboratory tests can also help to detect toxic effects of drugs, e.g. electrolyte abnormalities during cisplatin treatment and hypoglycaemia during antidiabetic treatment. In some situations, adjustment of dosage is much more difficult as the action of the drug cannot be readily assessed clinically or by simple laboratory tests, e.g. in the prophylaxis of epilepsy or mania. Dosage adjustment is also difficult when the toxic effects cannot be detected until they become irreversible, e.g. aminoglycoside antibiotics.

The large interindividual variation in the response to a given dose of drug depends on pharmacokinetic and pharmacodynamic factors. Pharmacokinetic factors include absorption, distribution, metabolism and excretion of the drug. Pharmacodynamics describes the interaction of the drug or the active metabolite with the target sites/receptors and the consequences of the interaction. The factors affecting the response of drugs are shown in Figure 25.1.

After administration of a drug, the amount of drug reaching the circulation vary due to variation in absorption. This is described as bioavailability, and it depends on the formulation of the drug (how easily it dissolves), concurrent administration of other drugs or food, and factors in the gastrointestinal tract such as motility and pH. Even after intramuscular injection, bioavailability may vary depending on factors such as blood supply at the injection site. Once absorbed, the drug will first reach the liver where a significant amount may be removed. This is called 'first pass metabolism effect'.

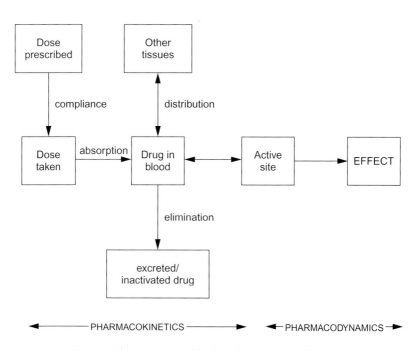

Figure 25.1   Factors affecting the response of a drug.

Drugs with high hepatic clearance include tricyclic antidepressants and lignocaine. Once the drug reaches the systemic circulation, the distribution of the drug within the body is described by the apparent volume of distribution, which depends on the solubility of the drug and the binding of the drug to proteins and tissues. The apparent volume of distribution may vary between individuals according to the size of the individual, the amount of adipose tissue and presence of other diseases (e.g. ascites). Once the drug is distributed, it will be metabolised and excreted. These, together are called the clearance of the drug. Factors that affect clearance include body weight, surface area, plasma protein binding, renal function, liver function and cardiac output. Many drugs are protein bound and the protein binding varies from virtually nil for lithium to 90% or greater for phenytoin. It is generally thought that it is the free fraction that is responsible for the clinical effects of the drug. The effects of the drug at the tissue level may vary according to other factors such as blood supply, interindividual variation, receptor binding of the drug, rate of diffusion across membranes and concomitant factors, e.g. presence of hypokalaemia may affect the binding of digoxin to its receptor. These pharmacodynamic factors limit the value of TDM.

Criteria for Therapeutic Drug Monitoring

The criteria that must be fulfilled before TDM can be considered useful for a given drug are listed in Table 25.1.

Table 25.1   Criteria required for valid therapeutic drug monitoring

Poor correlation between dose and effect
Narrow therapeutic index
  • i.e. narrow range of concentration between therapeutic and toxic effects

Absence of clinical effects or simple laboratory tests for the effects of drugs
Good correlation between plasma concentration and effect
  • Low pharmacodynamic variation
  • No active metabolites
  • Reversible action at receptor site

Correlation between dose and effect

If the pharmacological effect can be predicted from the dose, TDM is unnecessary.

Narrow therapeutic index

If the concentration range over which toxic effects and/or therapeutic effects are seen is narrow, TDM is helpful. For drugs with a high therapeutic index such as penicillin, large doses can be given without serious side effects. For drugs with a narrow therapeutic index, such as anticoagulants or antiepileptic drugs, the margin between pharmacological and toxic concentration is very small and monitoring is important to avoid toxic effects.

Absence of clinical or other easily measurable effects

TDM is of little value if clinical or simple laboratory measurements can be used to monitor the therapeutic effects, e.g. blood glucose concentration for monitoring treatment with antidiabetic drugs.

Correlation between plasma concentration and efficacy

There must be a good correlation between the plasma concentration and the therapeutic effect. If the correlation is poor, it is very difficult to use the plasma concentration of a drug to guide the dosage. In cases where an active metabolite has pharmacological effects, it may be necessary to measure the parent compound as well as the metabolites, e.g. in the case of tricyclic antidepressants, the parent drug and its methyl derivatives are active.

Compliance

One of the most basic problems in therapeutic drug monitoring is compliance. This is especially true for drugs that have to be taken for a very long time or life long, for example antiepileptic drugs and when the drugs are used for prophylaxis.

Measurement of Drug Concentration

Drug concentrations are usually measured in plasma or serum. Many drugs are protein bound and changes in protein concentration therefore will affect total plasma concentration without any effect or a smaller effect on the 'free' drug, which is the active fraction. In some instances it may be necessary to measure free drug concentration in plasma. However, this is technically difficult and an alternative is to use saliva. The concentration of drug in saliva is a reflection of free drug in plasma. Furthermore, saliva is easier to collect especially in children.

Samples for TDM should be taken at the appropriate time. In most cases, the sample should be taken at the time the plasma concentration has reached a steady state. However, if toxicity is suspected, waiting to attain steady state may be inapproprite. The time taken for each drug to reach steady state depends on the half-life of the drug. The steady state of the drug is usually reached after five half-lives. After three half-lives, 90% of the steady state value is reached and this is the minimum time to take samples after starting treatment or adjusting the dose. Some drugs like digoxin and phenobarbitone have long half-lives and it may take 2 weeks or more before a new steady state is reached when the dosage of these drugs are altered. The sample for drug measurement should be taken at the appropriate time in relation to the last dose. It is usual to take the sample pre-dose, i.e. the trough concentration when trying to assess the efficacy of the drug. In diagnosing toxicity, measurement of peak concentrations may be necessary. Samples for peak concentrations are taken usually 1 hour after intramuscular injection or 1–2 hours after an oral dose. The timing of the sample is discussed further in relation to each drug.

The drug concentration should be interpreted carefully as there may be other factors which determine the final therapeutic effects. For example, in the presence of hypokalaemia, a digoxin concentration within the 'therapeutic range' may be toxic. Drugs for which TDM has been shown to be useful are listed in Table 25.2 together with their pharmacokinetic data.

Table 25.2   Summary of common drugs and their pharmacokinetic data

Drug	Half-life (hours)	Protein-binding (%)	Route of elimination	Active metabolite	Sample timing	Indication for TDM	Target range
Phenytoin	20–40	90	Liver		Pre-dose	Yes	10–20 mg/L
Carbamazepine	8–24	70–80	Liver		Pre-dose	Yes	4–10 mg/L
Phenobarbitone	100	50	Liver, Kidney (30%)		Anytime	Occasionally	10–140 mg/L
Ethosuximide	40–60	< 5	Liver			Not useful	40–100 mg/L
Lamotrigine	20–30	55	Liver			Possibly	1–15 mg/L
Vigabatrin	5–7	< 5	Kidney			Not useful	
Sodium valproate	7–16	84–95	Liver		Pre-dose	No	
Digoxin	36–48	20–30	Kidney		8–12 hours post-dose	Yes?	0.8–2.0 μg/L
Lithium	10–35	0	Kidney			Yes	0.5–0.8 mmol/L
Theophylline	5–13	50–65	Liver	Yes		Yes	5–20 mg/L
Caffeine	3–7	35	Liver	Yes		Possibly	
Cyclosporine	6–24	98	Liver	Yes	Pre-dose	Yes	
Tacrolimus	4–41	75–99	Liver		Pre-dose	Yes	
Aminoglycoside (gentamycin)	2–3	<10	Kidney		Peak and trough	Yes	Peak 5–10 mg/L Trough < 2 mg/L
Amiodarone	10–100 days	> 95	Liver	Yes		Occasionally	0.5–2.5 mg/L
Methotrexate	5–9	50	Kidney			Yes	

Individual Drugs

Phenytoin

Phenytoin is widely used in the treatment of generalised tonic-clonic and partial seizures. Approximately 90% of phenytoin is bound to plasma proteins in normal individuals. Protein binding is lower in patients with renal or hepatic disease, neonates and during pregnancy. Co-administration of other drugs may displace phenytoin from the protein binding sites, thereby reducing the total concentration while increasing the free fraction. Phenytoin is metabolised by the hepatic mixed function oxidase system. However, this system has limited capacity and becomes saturated at therapeutic concentrations and the relationship between dose and plasma concentration becomes non-linear. When the plasma concentration is close to saturation, a small increase in dose will cause a large increase in plasma concentration leading to toxicity. There is large intraindividual variation in phenytoin concentration at which saturation of this enzyme occurs (Figure 25.2). A change in the metabolic capacity, for example during an intercurrent illness, may also cause a large increase in plasma concentration for a given dosage. Phenytoin has a low therapeutic ratio. One of the toxic effects of phenytoin is fits, thus mimicking the disease for which it is given. Other toxic effects include neurotoxicity, which includes nystagmus, nausea, vomiting, tremor, ataxia as well as hepatotoxicity, and blood abnormalities. Long-term phenytoin treatment may cause osteomalacia and megaloblastic anaemia.

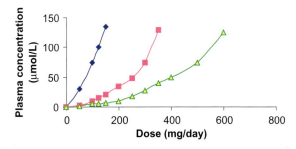

Figure 25.2   Relationship between plasma phenytoin concentration and dose in different patients.

Table 25.3 Indications for measuring phenytoin concentration

Initiation of therapy
Status epilepticus
Unexpected deterioration of seizure control
Suspected phenytoin toxicity
Simultaneous administration of drugs which
    induce or inhibit metabolising enzymes
During pregnancy

There are many indications for monitoring phenytoin concentrations during therapy (Table 25.3). During the initiation of therapy, monitoring of plasma concentration is required to adjust the dosage and to provide good control of fits without toxic effects. In status epilepticus, intravenous treatment may be required and careful monitoring of concentration may be necessary. Drug measurement is also indicated when there is unexpected deterioration in seizure control, which may have been due to poor compliance or changes in pharmacokinetics. In patients suspected of phenytoin toxicity, phenytoin concentration should be measured. Plasma phenytoin measurement is required if drugs, which may induce or inhibit enzyme systems responsible for metabolising phenytoin or drugs that can displace phenytoin from binding protein, are administered concurrently. During pregnancy, there is increased clearance of phenytoin and regular monitoring is advisable.

Toxic symptoms usually appear when plasma concentration exceeds 20 mg/L (80 μmol/L). However, some patients require much higher concentrations to achieve effective control of fits whereas others are adequately controlled at much lower concentrations.

Carbamazepine

Carbamazepine is used in the treatment of epilepsy, manic–depressive illness and trigeminal neuralgia. Carbamazepine is metabolised by the liver and induces its own metabolism and therefore, during chronic administration, its half-life is reduced. The metabolism of carbamazepine is also influenced by other anticonvulsants such as phenytoin. Toxic effects of carbamazepine include neurological as well as metabolic complications, e.g hyponatraemia. Monitoring of plasma concentration is complicated by the fact that one of the metabolite may be pharmacologically active. Monitoring plasma concentration is necessary to guide treatment, especially when the control of fits is difficult.

Phenobarbitone and primidone

These are useful second-line drugs to treat epilepsy in difficult cases. Primidone is converted largely to phenobarbitone, which accounts for its pharmacokinetic effects. Phenobarbitone is metabolised by the liver as well as excreted by the kidneys (30%). In plasma, 40–60% of phenobarbitone is protein bound and its half-life is very long, typically 100 hours. Primidone has a much shorter half-life. As most of the anticonvulsant activity of primidone is via its metabolites, it is questionable whether primidone should be measured. Treatment with these drugs causes tolerance such that higher concentrations may be required for therapeutic effects during long-term therapy. As there is poor correlation between plasma concentration and effects, monitoring of phenobarbitone or primidone concentration is of limited value. Measurement of plasma concentration may be useful in monitoring compliance, detecting drug interaction and in children with febrile convulsions to ensure that adequate concentration have been achieved. The timing of the sample is unimportant because of the long half-life.

Ethosuxamide

Ethosuxamide, which is used in the treatment of petit mal, shows considerable interindividual variation in clearance and plasma concentration cannot be predicted from the dose. It is not protein bound, is metabolised primarily by the liver and has a long half-life of 40–60 hours. As effects of this drug can be monitored by other means (e.g. electroencephalographic telemetry), the need for concentration monitoring is not entirely proven. Measurements may be useful to detect compliance or when there is poor response to a given dose.

Lamotrigine

Lamotrigine, used in conjunction with other anti-epileptic drugs is primarily metabolised by the liver and has a half-life of 20–30 hours. About 50% of the drug in plasma is protein bound. Clearance is significantly affected by other anticonvulsant drugs such as phenytoin and carbamazepine, which cause enzyme induction, thereby reducing the half-life of lamotrigine. Thus, plasma concentration measurement may be useful, especially in patients taking other drugs.

Vigabatrin

Vigabatrin, one of the newer anticonvulsants used as second line treatment, has very little protein binding, is excreted mainly by the kidney and has a short half-life. There is no relationship between plasma concentration and clinical effect, therefore monitoring plasma concentrations is unlikely to be of value, other than to assess compliance.

Sodium valproate

Valproate, often used in the treatment of myoclonic seizures, is highly protein bound. It is primarily cleared by the liver, and has a relatively short half-life. However, the major metabolite has a longer half-life. It can cause serious hepatotoxicity in a small proportion of patients. This is probably an idiosyncratic reaction and not related to plasma concentrations. There is no clear correlation between plasma concentration and its effects and thus routine monitoring of valproate is not necessary. However, it may be useful in the event of failure of therapy in spite of adequate dosage.

Digoxin

Digoxin is a cardiac glycoside used for its positive ionotrophic effect and its ability to alter conduction through the atrioventricular node. Digoxin is most useful in the control and prevention of tachycardias especially atrial fibrillation. Although digoxin has been used extensively in the past for the treatment of congestive cardiac failure, currently it is used relatively infrequently as other much better drugs are available. In plasma, about 20–25% of digoxin is bound to proteins. Due to its high binding affinity to its receptor, the sodium–potassium pump, the distribution of digoxin is extensive and it has a very large volume of distribution. Digoxin is primarily cleared by the kidney and its half-life is 26–48 hours. In patients with poor renal function, half-life may be over 100 hours. The time taken to reach steady state plasma concentration is about a week (five half-lives) in patients with normal renal function and 3 weeks in patients with poor renal function. In most people, < 20% is metabolised, but in a small proportion of patients, a significant fraction may be converted to metabolites some of which may be active. These metabolites, active and inactive, may be measured by digoxin immunoassays to a variable extent and this may give rise to discrepancy between plasma concentration and the pharmacological effects. Samples for digoxin measurement should be taken at least 6 hours after the last dose as it takes several hours for distribution of digoxin within the body. Toxic effects of digoxin include cardiac arrhythmias and conduction defects.

A number of factors such as age, hypokalaemia, hypomagnesaemia, hypercalcaemia and thyroid function affect the response of digoxin (Table 25.4). Binding of digoxin to the sodium–potassium pump is increased in hypokalaemia as potassium also binds to the same binding site. The tissue sensitivity of digoxin is increased in hypothyroidism and decreased in hyperthyroidism due to alteration in the number of sodium–potassium pumps induced by thyroid hormones. Thus, plasma concentration of digoxin does not always

Table 25.4 Factors affecting response to digoxin

Increased sensitivity	Decreased sensitivity
Hypokaleamia	Hyperkalaemia
Hypercalacaemia	Hypocalcaemia
Hypothyroidism	Hyperthyroidism
Hypoxia/acidosis	Neonates
Hypomagnesaemia	

correlate with clinical effects. In addition to renal function, one of the reasons for the variation in response to digoxin is due to polymorphism of the p-glycoprotein gene. This protein is involved in the transport of digoxin in the gastro intestinal tract and renal tubules. Drugs that influence the concentration of digoxin probably act via competitive inhibition of p-glycoprotein. Table 25.5 gives a list of drugs that affect digoxin concentration. In general, therapeutic effects of digoxin are not seen when the concentration is < 1 nmol/L (0.8 μg/L) and toxic effects are common when the concentration is > 3.8 nmol/L (3.0 μg/L). Another complicating factor in the interpretation of plasma digoxin concentration is the interference in immunoassays from endogenous digoxin-like substances, found in neonates, pregnant women, and in patients with renal failure and severe liver failure.

Table 25.5   Drugs that interact with digoxin

Drugs that increase digoxin concentration
- Diuretics
  — Spironolactone, amiloride, triamterene
- Antiarrhythimcs
  — Quninidine, amiodarone
- Calcium antagonists
  — Verapamil
- HMGCoA redutase inhibitors
  — Atorvastatin
- Antibiotics
  — Erythromycin, clarithromycin, roxithromycin
- Benzodiazepines
  — Alprazolam

Drugs that decrease digoxin concentration
- Rifampacin
  — Induces p-glycoprotein mediated tubular secretion
- Liquid antacids
  — Reduces digoxin absorption

Drugs that increase the effect of digoxin
- Diuretics
  — Via hypokalaemia

Monitoring of plasma digoxin concentration may be of value (a) patients with poor renal function, (b) when the initial response is poor, (c) to test for compliance and (d) in cases of suspected digoxin toxicity.

Lithium

Lithium is used in the management of mood disorders, especially in bipolar (manic–depressive) disorder. When taken orally, peak concentrations are seen within 2–4 hours and it is rapidly distributed within 6–10 hours. Lithium is not bound to plasma proteins and is excreted unchanged in the urine, kidneys accounting for > 95% of the elimination of lithium. Renal handling of lithium is similar to that of sodium. Most of the filtered lithium is reabsorbed in the proximal tubules. Sodium depletion, reduced salt intake and diuretics increase the reabsorption of lithium in the proximal tubules. Diuretics, by reducing the reabsorption in the distal segments of the renal tubules, cause an increase in sodium reabsorption in the proximal tubules and hence, increase lithium reabsorption. The half-life varies from about 10 hours in young patients to about 40 hours in the elderly and in patients with poor renal function. The time taken to reach steady state varies from 2 to 5 days.

Side effects of lithium such as polyuria and thirst are not uncommon. Lithium reduces the sensitivity of the distal tubular cells to ADH, causing an acquired nephrogenic diabetes insipidus leading to polyuria. Some patients also develop thyroid abnormalities with an increase in TSH and a fall in free T4. Thus, regular monitoring of thyroid function is recommended. Lithium toxicity is seen when concentration exceeds 1.4 mmol/L and includes neurological features (restlessness, ataxia, dysarthria and convulsions), acute renal failure and coma. As lithium is excreted entirely by the kidney, renal failure will further exaggerate any toxic effects. Plasma lithium concentration > 2.5 mmol/L is associated with significant mortality.

Because of the short half-life and variable absorption, plasma concentration cannot be predicted from the dose. There is good correlation between serum concentration and the therapeutic/toxic effects and the therapeutic index of lithium is very narrow. For these reasons, monitoring of plasma concentrations is important when starting patients on lithium treatment. Plasma concentration between 0.5 and 0.8 mmol/L taken 12 hours after the last dose seems to be effective. Measurement of serum lithium is also indicated when patients are started on diuretics, when toxicity and non-compliance are suspected or after changing dosage. Regular monitoring of renal function and thyroid function are also necessary. Blood samples for lithium measurement should be taken at steady state and this should be 10–12 hours after the last dose.

Theophylline

Theophylline, a bronchodilator used in the treatment of asthma, chronic obstructive pulmonary disease and neonatal apnoea, is about 50–60% protein bound in plasma. It is metabolised primarily by the liver and the metabolism may be affected by other diseases, other drugs, dietary factors and smoking. In infants, it is metabolised to caffeine, which is also active. The half-life of theophylline is reduced in smokers due to induction of hepatic metabolism. In infants, neonates and patients with cirrhosis, the half-life is usually longer. Because of the marked interindividual variation in pharmacokinetics for a given dose of theophylline, therapeutic drug monitoring is useful.

Serious toxic effects of theophylline include cardiac arrhythmias, tremor, seizures, rhabdomyolysis and acute renal failure. The therapeutic range for theophylline is well defined and there is a linear relationship between plasma concentration and its effect on respiratory function. Monitoring of theophylline is useful when optimising dosage, in suspected cases of theophylline toxicity, and to assess compliance.

Immunosuppressants

The most commonly used immunosuppressants in transplantation are cyclosporin and tacrolimus. Cyclosporin, which acts by inhibiting the activation of T lymphocytes by binding to calcineurin, shows marked variation between individuals in relation to pharmacokinetics. About 55% of the drug in blood is in the red cells, 10% in leucocytes and the rest in plasma. In plasma, 98% is bound to protein, mainly to lipoproteins. As the drug is lipophilic, it is widely distributed in the body and its apparent volume of distribution is high. Cyclosporin is usually measured in whole blood taking advantage of the higher concentration. Cyclosporin is metabolised in the liver by cytochrome P450 3A4 isoform and excreted in the bile. The half-life varies from 6 to 24 hours. Some of the metabolites have immunosuppressant effect. The precise contribution of these metabolites to the overall activity of the drug is not known. However, the blood concentration of cyclosporin and its pharmacological effects are correlated. Abnormalities of liver function or co-administration of other drugs, which affect cytochrome P450 activity, may alter the blood concentration. The toxic effects of cyclosporin include renal damage, mild liver damage and hypertension.

Tacrolimus (FK506) is pharmacokinetically similar to cyclosporin, but far more potent. Tacrolimus is metabolised by the liver. It is present in higher concentrations in red cells therefore, whole blood is used for analysis. Toxic effects of tacrolimus include hepatotoxicity, cardiotoxicity and some neurotoxicity.

Monitoring of cyclosporin and tacrolimus are useful during the early stages of immunosuppression. Frequent measurements, every day or every 2 days may be useful in determining the effective dose with minimal toxicity. After the initial stages, it is recommended that blood concentrations should be measured two or three times a week for the first 3–6 months and every few months thereafter. Further monitoring may be required if there is a change in patients' clinical condition.

Sirolimus (rapamycin) is a relatively new immunosppressant. It is not nephrotoxic but toxic to bone marrow and also causes arthropathy. Monitoring is required to check compliance and to check pharmacokinetic variation.

Aminoglycosides

Aminoglycosides are powerful antimicrobial agents used in the treatment of severe infections. They are relatively poorly bound to proteins, are freely filtered and excreted almost entirely by the kidneys. Because of the diurnal variation in GFR, higher concentrations are observed in the morning compared to the evening. The half-life of aminoglycosides is 2–3 hours, but may be prolonged up to 100 hours in patients with poor renal function. There is large intra- and interindividual variation, in the rate of absorption, volume of distribution and half-life. Thus, plasma concentration may vary widely for the same dose. Because of this high pharmacokinetic variability and potential nephrotoxicity, therapeutic drug monitoring is recommended, particularly in patients with severe infections, impaired renal function, patients on prolonged therapy, and in patients who are receiving other drugs which may potentiate toxicity.

Serum creatinine should be measured before starting therapy and repeated every few days during treatment. Assessment of GFR by calculated or measured creatinine clearance may be useful in adjusting the dosage. Initially, peak and trough concentrations may need to be monitored and during long term treatment, trough concentration every 3 or 4 days is adequate. More frequent monitoring may be required if there are features of toxicity, or if other drugs that may potentiate toxicity are given.

Amiodarone

Amiodarone, a frequently used anti-arrhythmic drug contains a large amount of iodine. It is highly protein bound, has a large volume of distribution and is metabolised by liver to desethylamiodarone. Its half-life varies from 10–40 days initially to 100 days during chronic treatment. In addition to neuropathy, liver toxicity and photosensitive reactions, amiodarone can cause thyroid dysfunction, hypo- and hyperthyroidism, due to its large iodine content. Although the pharmacokinetic variability, narrow therapeutic index and poorly defined therapeutic end point suggest that it will be useful to monitor amiodarone, the consensus opinion at the moment is that monitoring may not be useful except to detect compliance or confirm toxicity.

Regular monitoring of liver and thyroid functions every 6 months is recommended to reduce serious side effects.

Methotrexate

Methotrexate inhibits dihydrofolate reductase, thereby preventing the formation of tetrahydrofolate necessary for nucleic acid synthesis. It is used in the treatment of malignant tumours such as choriocarcinoma, in the conservative treatment of ectopic pregnancy and as an immuno-suppressant in rheumatoid arthritis, psoriasis and psoriatic arthropathy. In the treatment of malignant tumours, it is given in high doses by intravenous infusion over 12 hours. Monitoring plasma concentration may be necessary in these patients. Plasma methotrexate should fall rapidly after the infusion. If it does not, treatment with folinic acid may be required. Two blood samples should be obtained over a period of hours to calculate the half-life. A prolonged half-life and high plasma concentrations are indications for folinic acid rescue. Hepatotoxicity is a complication of methotrexate treatment and liver function tests as well as N-terminal propeptide of collagen type III are commonly monitored to detect liver damage and liver fibrosis respectively.

Tricyclic antidepressants

Tricyclic antidepressants are widely used in the treatment of various forms of depression. Genetic variation in the expression of liver enzymes involved in the metabolism of tricyclics causes significant effect on the relationship between dose and plasma concentration. In future, methods may be available to identify the different phenotypes of the drug metabolising enzyme (CYP2D6) so that effective treatment could be instituted to avoid toxicity. Measurement of plasma concentrations is useful for the antidepressants imipramine, desipramine and nortriptyline. TDM of these drugs may be required to evaluate compliance, to assess potential toxicity and in those taking enzyme-inducing drugs such as carbamazepine or phenytoin.

Pharmacogenetics

Response to a given drug varies between people and one of the factors contributing to this variation is the genetic variation in the absorption, metabolism and action of drugs. Pharmacogenetics is the study of such genetic variation. Inherited variation in drug-metabolising enzymes may have profound effects on the efficacy of a drug or the development of toxicity. Such variation have been described for several of the enzymes in the cytochrome P450 enzyme super family, CYP2D6, CYP2C19, are examples. The enzyme CYP2D6 is responsible for metabolism of many drugs including antiarrthymics, antihypertensives and antidepressants. More than 50 different alleles have been described for this enzyme and these mutations may result in decreased, normal or increased activity of the enzyme. This enzyme is absent in 5–10% of the Caucasian population due to mutation. Another mutation that leads to reduced activity of the enzyme is common among Asians. Mutation in these genes can cause reduced or absent enzyme activity, leading to the development of toxicity when a standard dose of drug is given. Individuals who are homozygous for such mutations are sometimes described as slow metabolisers and they require smaller doses than others. Sometimes, duplication of the gene results in rapid metabolism; these individuals are called ultrarapid metabolisers and require higher dosage. Another clinically important polymorphism is seen in the enzyme thiopurine methyltransferase (TMPT), which metabolises azathioprine. Subjects with reduced activity of TPMT are susceptible to the toxic effects of this drug. It is now standard practice to measure TPMT in anyone who is prescribed azathioprine to prevent toxicity.

Table 25.6 gives some drugs for which pharmacogenetic analysis may be helpful. Pharmacogenetics is a rapidly expanding field. In the near future, the dosage of drugs will be tailored to the individual patient depending on the geno- and/or phenotype of the drug metabolising enzyme.

Clinical Toxicology

Acute and chronic poisoning with drugs and other chemicals is a common cause of hospital admissions; up to 50–60% of acute admissions may be related to drug toxicity. Toxicity may arise as a result of accidental overdose, which is especially common in children, suicide

Table 25.6 List of genes and corresponding drugs suitable for pharmacogentic analysis

Gene	Drugs
CYP2C9	Warfarin, Celecoxib
CYP2C19	Omeprazole, Diazepam, Clopidogrel
CYP2D6	Tamoxifen, Risperidone
NAT	Rifampacin, Isoniazid
TPMT	Azathioprine
UGT1A1	Irinotecan
HLA-B*5701	Abacavir
VKORC1	Warfarin

or homicide. Toxicity may also be precipitated by interaction with other factors.

Management of Poisoned Patients

Management of poisoned patients includes non-specific general measures and specific measures. General measures include maintenance of ventilation, cardiac output and attempts to prevent further absorption of the toxin. Gastric lavage or administration of activated charcoal will reduce intestinal absorption while washing of the skin will reduce cutaneous absorption. In addition, attempts may be made to remove the drug from the body; this is especially useful if the route of elimination of the drug is known, e.g. if the drug is known to be eliminated by the kidney, haemodialysis or haemoperfusion may be indicated. For some of the poisons and drugs, specific antidotes are available and in these instances, identification and quantitation of these drugs are important to decide on the treatment and to monitor the effectiveness of the treatment.

Paracetamol

Paracetamol is a commonly used analgesic drug that may cause severe hepatotoxicity and renal toxicity if consumed in large doses. It is the most common cause of acute liver failure in many countries. In therapeutic doses, paracetamol is conjugated with sulphate or glucouronic acid in the liver to non-toxic metabolites (glucouronide 50–60% and sulphate 30%), which are eliminated by the kidneys (Figure 25.3). Less than 10% is metabolised by a cytochrome P450 mixed function oxidase (CYP2E1 and CYP1A2) to a highly reactive intermediate (N-acetyl-p-benzoquinone imine (NAPQ)). This undergoes conjugation with glutathione and is then converted to non-toxic products such as cysteine and mercapturic acid conjugate (Figure 25.3). In paracetamol overdose, the sulphation pathway becomes saturated and a greater proportion is metabolised through the P450 mixed function oxidase pathway. When the tissue stores of glutathione are depleted, the reactive metabolite causes hepatotoxicity and nephrotoxicity. The specific treatment of paracetamol poisoning therefore involves the administration of glutathione or glutathione substitutes. The most commonly used one is N-acetylcysteine (NAC),

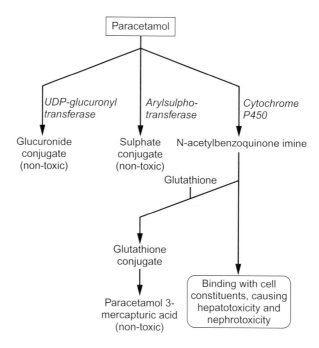

Figure 25.3   Metabolism of paracetamol.

which in addition to replenishing hepatic glutathione stores, enhances conjugation with sulphate. The timing of the administration of NAC is critical (see below).

Features of paracetamol toxicity can be described in three stages. In the first stage of paracetamol toxicity, symptoms are mild and non-specific such as nausea and vomiting. Hepatic necrosis is not evident until at least 48 hours later. However, with very severe overdose, coma and metabolic acidosis may develop. In the second stage, which develops 24 to 72 hours after ingestion, abdominal pain and tenderness over the liver may develop due to hepatic necrosis; at this stage, liver function tests may be abnormal: prolonged prothrombin time, increased transaminase activity and bilirubin concentration. Increased prothrombin time is an early indication of toxicity and is also the best marker of severity of toxicity. Acute renal failure may develop at this stage. In the third stage, which develops 3 to 5 days after ingestion, hepatic failure with coagulation defect, hypoglycaemia and acute renal failure are seen. The risk of paracetamol toxicity is increased by a number of factors. Chronic alcohol intake induces CYP2E1 and increases the production of toxic metabolites. Fasting, starvation and malnutrition reduce hepatic glutathione and drugs such as carbemazapine and phenytoin, which induce CYP2E1, increase the risk of toxicity. The likelihood of development of hepatic toxicity is related to the plasma concentration and a nomogram relating the serum paracetamol concentration and the time after ingestion to the probability of hepatic necrosis is available (Figure 25.4). Toxicity from chronic ingestion of paracetamol may occur at low plasma concentration. Blood samples for measurement of paracetamol should be taken at least 4 hours after the ingestion. When the time of ingestion is not known, it may be useful to calculate the half-life from plasma concentrations measured in two or more samples taken 2–3 hours apart. Hepatotoxicity is more likely if the half-life is greater than 4 hours and fulminant hepatic failure and coma is likely if the half-life is greater than 12 hours. Serial measurements may be required if the preparation containing paracetamol is a slow release preparation which may take more than 4 hours for absorption to be complete. Delayed absorption may also occur if there is concomitant administration of drugs that reduce gastric motility or

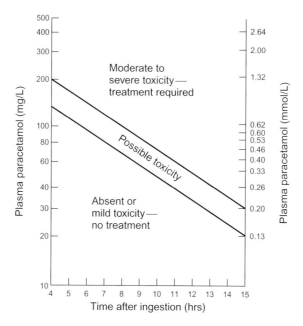

Figure 25.4   Nomogram showing the risk of toxicity and serum concentration of paracetamol in relation to the time of ingestion.

intestinal motility. The nomogram may not be reliable in patients who are likely to have reduced hepatic glutathione concentration, e.g. alcoholics, fasting or malnourished patients and patients treated with hepatic microsomal enzyme-inducing drugs.

In patients who use paracetamol over a prolonged period of time, a high anion gap metabolic acidosis is occasionally seen. Patients who are malnourished or chronic alcoholics are at particular risk of this. The acidosis is due to accumulation of oxoproline, which is a metabolite in the γ-glutamyl cycle, which produces glutathione (Figure 25.5). Depletion of glutathione stores by paracetamol leads to accumulation of oxoproline.

Management

Management of paracetamol overdose involves general measures: gastric lavage or activated charcoal to prevent absorption, which is

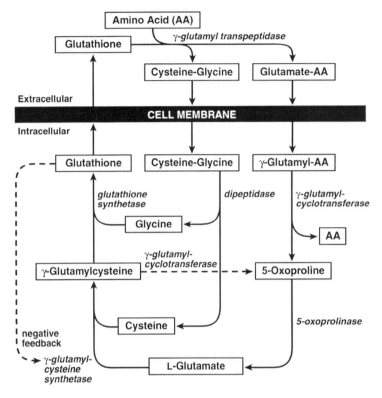

Figure 25.5 The γ-glutamyl cycle.

effective if started in the first 2–4 hours after ingestion and adequate hydration to reduce nephrotoxicity. Specific antidote, NAC, which reduces toxicity by replenishing the hepatic glutathione stores, should be given after measurement of plasma paracetamol concentration. The maximum efficacy of NAC is seen when administered within 8 hours of ingestion. However, even up to 48 hours after ingestion, it is useful. Most authorities recommend the use of NAC in suspected cases of paracetamol poisoning even if the plasma drug concentration is not available or if the time of ingestion is not known. Regular monitoring of renal function, blood glucose and liver function tests may be necessary to detect hepatic and renal damage. Other general measures include administration of glucose to avoid hypoglycaemia, and vitamin K administration in patients with suspected liver involvement.

In patients who develop acute fulminant hepatic failure, liver transplantation is the treatment of choice. Indications for transplantation are pH < 7.3, severe encephalopathy, prothombin time (INR) > 100 seconds and serum creatinine > 300 μmol/L.

Prognosis: Patients who survive the acute effects make a full recovery. Prognosis of patients who develop complications is poor. If the blood pH < 7.3 or if they develop acute renal failure or severe encephalopathy, prognosis is poor.

Salicylate

Salicylate is a widely available analgesic, antipyretic and anti-inflammatory drug. Salicylate poisoning, due to accidental or intentional overdosage, is less common now due to more public awareness. Aspirin (acetylsalicylic acid) is rapidly absorbed from the gastrointestinal tract and hydrolysed rapidly to salicylate. The half-life of aspirin is about 15 minutes. Salicylate is eliminated by conjugation with glycine and to some extent with glucuronic acid. When the salicylate concentration is very high, these pathways may become saturated and the serum salicylate concentration may increase disproportionately to the dose. At high concentrations, the half-life of salicylate is prolonged to 15–30 hours as compared to 2–3 hours. At high doses, salicylate is excreted unchanged in the urine. Salicylate causes a wide variety of metabolic disturbances. Most of the actions except platelet inhibition, which is due to aspirin, are caused by salicylate. Salicylate directly stimulates the respiratory centre, causing hyperventilation and respiratory alkalosis. It uncouples oxidative phosphorylation, leading to increased heat production, oxygen consumption and high metabolic rate. Salicylate causes a metabolic acidosis by several mechanisms. In toxic concentrations, salicylate (i) uncouples oxidative phosphorylation and interferes with energy production, (ii) causes renal failure, (iii) increases lipolysis and production of ketones. The acid–base disturbance in salicylate overdose depends on age and severity of toxicity. In adults and in older children, respiratory alkalosis is the predominant feature. In very severe cases, there may be a mixed respiratory alkalosis and metabolic acidosis with metabolic acidosis predominating in about 15% of cases. In patients with mixed disturbances, anion gap may be normal. In patients with metabolic acidosis, the mortality is significantly higher. In children under the age of 4, respiratory alkalosis is not common and metabolic acidosis predominates. Respiratory acidosis as a result of central nervous system depression may occasionally be seen. The symptoms of salicylate intoxication include tinnitus, hyperthermia, hyperventilation, nausea, vomiting and CNS disturbance such as lethargy and disorientation leading to coma and fits.

Measurement of serum salicylate concentration is important to assess the severity of toxicity. A nomogram (Figure 25.6) is available to predict the likelihood of toxicity depending on the concentration of salicylate and the time of the ingestion. The nomogram does not apply to slow-release preparations of aspirin.

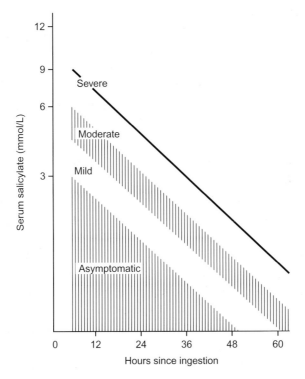

Figure 25.6 Nomogram showing the risk of toxicity and serum concentration of salicylate in relation to the time of ingestion.

Management

If the plasma salicylate concentration is > 3.6 mmol/L (> 500 mg/L), treatment is indicated. There is no specific antidote for salicylate but the elimination of salicylate via the kidneys could be enhanced. In acidic urine, salicylic acid is not ionised and is reabsorbed by the tubules. If the urine is alkaline, salicylic acid is ionised and tubular reabsorption is decreased, enhancing urinary excretion. Thus, the rationale for treatment of salicylate poisoning involves adequate hydration and alkalinisation of urine by sodium bicarbonate infusion to maintain an alkaline diuresis (urine pH of 7.5 or greater). Blood pH and electrolytes should be monitored regularly to avoid severe alkalosis and hypokalaemia. General measures to reduce absorption by gastric lavage are also important. In very severe toxicity, when the plasma concentration exceeds 6.5 mmol/L (900 mg/L), renal impairment may result and haemoperfusion, haemofiltration or dialysis may be required to remove the salicylate.

Case 25.1

A 16-year-old girl was brought to the A&E department following an overdose of aspirin after a domestic argument. She had vomited several times before admission. On admission, she was found to be hyperventilating, sweating with a temperature of 39.0°C. Investigations showed the following results:

Serum		Reference Range
Sodium (mmol/L)	140	135–145
Potassium (mmol/L)	3.2	3.5–5.0
Bicarbonate (mmol/L)	12	23–32
Chloride (mmol/L)	92	90–108
Urea (mmol/L	8.5	3.5–7.2
Creatinine (μmol/L)	98	60–112
Glucose (mmol/L)	3.2	
pH	7.4	7.35–7.45
pCO_2 (kPa)	3.2	4.5–6.0

> *A normal pH together with a low bicarbonate and a high anion gap (39 mmol/L) suggests that this is mixed picture of respiratory alkalosis and metabolic acidosis. Plasma salicylate was found to be 5.2 mmol/L confirming salicylate overdose. She was treated with alkaline diuresis and made a full recovery.*

Alcohols

Ethanol

Drug overdose is often complicated by the simultaneous ingestion of alcohol, which potentiates the toxic effects of many drugs. Measurement of blood alcohol concentration therefore may be necessary to explain the delay in recovery of patients who have taken a drug overdose. It has also been used in the management of patients with head injuries where the effects of alcohol may make it difficult to assess the severity of brain damage due to head injury. Central nervous system effects of alcohol vary depending on the blood concentration. It varies from euphoria and decreased inhibition to disorientation, poor coordination, coma and death. The degree of central nervous system dysfunction varies depending on the previous exposure of the individual to alcohol. In heavy alcohol users, tolerance develops. Furthermore, central nervous system actions are more pronounced when the alcohol concentration is increasing than when it is declining.

Ethanol is metabolised principally by the liver. Its effects on liver are described elsewhere. There is no specific treatment for alcoholic toxicity. The aim of treatment is supportive until the alcohol is metabolised by the body. Haemodialysis may be necessary if the blood alcohol concentration is very high (> 400 mg/dL, 86 mmol/L) or if there is metabolic acidosis. In addition to maintaining airways patency, glucose should be given to prevent or treat hypoglycaemia.

Methanol

Alcoholics may accidentally consume methanol, which is used as a solvent for a number of commercial products. Methanol is oxidised by alcohol dehydrogenase in the liver to formaldehyde, which is converted to formic acid, by aldehyde dehydrogenase. Accumulation of formic acid causes a severe metabolic acidosis (see Chapter 4). Optic neuropathy leading to blindness is a feature of methanol toxicity, the mechanism of which is not clear. The central nervous system effects of methanol are less severe than those of alcohol.

Treatment of methanol intoxication includes the administration of ethanol to inhibit the metabolism of methanol, sodium bicarbonate to correct the acidosis, folate administration to enhance folate mediated metabolism of formate and the use of haemodialysis to enhance clearance of methanol and formate. Recently, a drug fomepizole has been successfully used on the treatment of methanol toxicity.

Case 25.2

A 43-year-old man was brought to A&E by the police who found him unconscious at the road side. He was responding to painful stimuli and was found to be hyperventilating. On admission, blood gases and other emergency investigations were carried out and they showed the following:

Serum		Reference Range
Sodium (mmol/L)	141	135–145
Potassium (mmol/L)	3.9	3.5–5.0
Bicarbonate (mmol/L)	8	23–32
Chloride (mmol/L)	92	90–108
Urea (mmol/L	10.5	3.5–7.2
Creatinine (μmol/L)	115	60–112
Glucose (mmol/L)	5.2	
pH	6.96	7.35–7.45
pCO_2 (kPa)	1.8	4.5–6.0

He had severe metabolic acidosis as shown by a low pH with a low pCO_2. A normal blood glucose excluded diabetic ketoacidosis and his renal function was not severe enough to account for the acidosis. His measured plasma *osmolality was 375 mosmol/kg and his calculated osmolarity was 316 mmol/L. The osmolar gap was therefore 59 mmol/L. This suggested the presence of some other osmotically active substance. Due to the severity of metabolic acidosis, methanol or ethylene glycol poisoning was suspected. His serum calcium was not low, which is against ethylene glycol poisoning. Analysis of his blood for alcohols showed high methanol concentration confirming methanol toxicity.*

Carbon Monoxide

See Chapter 23.

Further Reading

1. Brent, J. Fomepizole for Ethylene Glycol and Methanol Poisoning. *N Eng J Med* 2009; 360: 2216–2223.
2. Hallworth M, Watson I. *Therapeutic Drug Monitoring and Laboratory Medicine* 2008. ACB Venture Publications: London.
3. Henry JP. In Marshall W, Bangert SK (eds.) *Clinical Biochemistry — Metabolic and Clinical Aspects* 2008. pp. 836–856. Churchill Livingstone: London.
4. Kao LW, Nañagas KA. Carbon monoxide poisoning. *Med Clin North Am* 2005; 89:1161–1194.
5. Rowden AK, Norvell J, Eldridge DL, Kirk MA. Acetaminophen poisoning. *Clin Lab Med* 2006; 26:49–65.
6. Watson I, Proudfoot A. *Poisoning and Laboratory Medicine* 2002 ACB Venture Publications: London.

Summary/Key Points

1. Variation in response to a given dose of drug between individuals is due to variation in pharmacokinetic (absorption, distribution, metabolism and excretion) and pharmacodynamic (interaction of drug with target site and consequent effects) factors.

2. Measurement of drug concentrations in blood or occasionally in saliva is useful in the management of patients receiving drug treatment.

3. Measurement of drug level are not required if the therapeutic effect or toxic effect can be easily monitored either by measuring a clinical effect (e.g. blood pressure) or a biochemical effect (e.g. blood glucose).

4. Therapeutic drug monitoring (TDM) is indicated when the therapeutic index is narrow (i.e. concentration range over which therapeutic or toxic effect are seen is narrow), when there are no easily measurable clinical/biochemical effects, if there is a good correlation between blood concentration and therapeutic effect and to monitor compliance.

5. In order to interpret drug concentration, the sample must be taken at the appropriate time. For many drugs, the trough level (i.e. pre-dose) is recommended except when toxicity is suspected.

6. Some common drugs for which TDM has been recommended include phenytoin digoxin, lithium, immunosuppressants, theophylline and aminoglycosides.

7. For some drugs, the effects of the drug (therapeutic or toxic) is influenced by other factors, e.g. hypokalaemia and hypothyroidism will enhance tissue sensitivity to digoxin, whereas hyperkalaemia, hypocalcaemia and hyperthyroidism will reduce sensitivity.

8. Genetic variation in the enzymes involved in the absorption, metabolism of drugs account for the variation in response between individuals. Pharmacogenetics is the study of such variation and may help to optimise the effects of the drug without risking toxicity. Analysis of thiopurine methyltransferase (TMPT), which metabolises azathioprine, is such an example.

9. Drug analysis may be required in cases of over dosage if specific treatment is available.

10. Management of patients admitted with drug overdose is mainly supportive.

11. Paracetamol is toxic when taken in large doses due to the formation of a toxic metabolite, which causes liver damage. Treatment

with N-acetylcysteine (NAC), which replenishes hepatic glutathione stores, will prevent liver damage. Measurement of paracetamol concentration is important to decide on treatment.

12. Salicylate overdose can cause a metabolic acidosis, renal damage or respiratory alkalosis. Excretion of salicylate can be enhanced by alkaline diuresis. Measurement drug concentration is important to decide on treatment.

chapter 26

Inherited Metabolic Diseases

Introduction

Inherited metabolic diseases usually present in childhood but are being increasingly recognised in later life. Individually, each of the diseases is rare but collectively, they form an important group. These disorders collectively account for nearly 50% of deaths in the first year of life and the clinical biochemistry laboratory plays an important role in the early recognition of these disorders. Inherited metabolic disorders are single gene disorders where there is a mutation of a gene controlling a particular protein. The mutation results in either deletion or substitution of an amino acid or deletion of a complete protein. Many of the inherited metabolic disorders are due to such mutations affecting enzymes that control metabolic processes. As illustrated in Figure 26.1, the formation of a metabolite, hormone or other similar substances depend on a series of steps, which are catalysed by different enzymes. Each of these enzymes is regulated by a gene. A mutation in one of the genes will result in reduced activity of this enzyme or complete absence of this enzyme (enzyme Z in Figure 26.1). This will lead to a reduction in concentration of the product of this enzyme activity (product D) and in the accumulation of metabolites proximal to this step (metabolites C, B & A in Figure 26.1). Furthermore, in many metabolic reactions, there is a negative feedback regulation and the absence of the product would increase the initial reactions, causing the precursor metabolites to increase further. In some instances, this will have an effect on other minor metabolic pathways leading to the production of other metabolites not normally produced or produced in small quantities (product E, Figure 26.1). The clinical effects of such

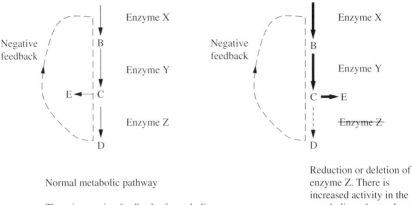

Figure 26.1   Effect of mutation in gene controlling an enzyme in a metabolic pathway.

disorders are due to reduction in the products, increase in precursors, or formation of unusual metabolites or a combination of them.

The diagnosis can be made either by measuring the enzyme activity concerned, the concentration of the final metabolite in question, the accumulation of the precursor metabolites or the abnormal metabolites.

There are numerous inherited metabolic disorders and some of them are very rare. In this chapter, a few of these will be described in detail.

Screening for Inherited Metabolic Disorders

In many countries, programmes have been established to screen neonates for a variety of inherited metabolic disorders. In the UK,

Table 26.1   Criteria for the selection of inherited metabolic disorders for screening

1. The condition should be an important public health problem.
2. There should an acceptable treatment available.
3. The condition should not be apparent clinically at the time of screening.
4. A suitable test should be available.
5. Natural history of the condition should be known.
6. The test should be acceptable to the population.
7. The test should be cost effective.

neonatal screening for phenylketonuria (PKU), and hypothyroidism were established several decades ago. Criteria for the selection of disorders for screening are listed in Table 26.1.

In the UK, the current policy is to screen for phenylketonuria (PKU), congenital hypothyroidism, cystic fibrosis, medium-chain acyl CoA dehydrogenase deficiency (MCAD) and haemoglobinopathies between five to eight days after birth. The scheme is voluntary in the UK, but mandatory in some states of USA. Blood is taken by heel prick from babies onto a filter paper and sent to the laboratory for analysis. A positive result in any of the tests is followed by a repeat test on the same sample, followed by analysis of a second sample to confirm the diagnosis. In some cases, e.g. cystic fibrosis initial screening is followed by a more definitive test. Once the diagnosis is established, local paediatricians institute treatment and follow up.

Prenatal screening/diagnosis

When an inherited metabolic disorder is suspected in the foetus, prenatal testing can be done to confirm the diagnosis. Previously affected babies or family history of inherited metabolic disorders are some indications for prenatal testing. Testing usually involves collecting either a chorionic villus sample or amniotic fluid sample followed by culture of foetal cells before doing an enzyme or genetic analysis. More recently, the use of fetal cells in maternal circulation and fetal DNA in maternal blood, have been used as alternatives for the diagnosis of inherited metabolic disorders in the foetus. These non-invasive methods will increase and become routine in the future.

Phenylketonuria

Phenylketonuria (PKU) is an autosomal recessive inherited metabolic disorder, which has an incidence of 1:10,000 births in the UK. The incidence in other countries varies from 1:5000 to 1:20,000. It is reported to be rare in Asians and the Mongoloid race. In PKU, there is an absence or deficiency of one of the enzymes involved in the metabolism of phenylalanine. Metabolism of phenylalanine is outlined in Figure 26.2. Phenylalanine is of dietary origin and is mainly metabolised to tyrosine by phenylalanine hydroxylase in the liver. This enzyme requires tetrahydrobiopterin as a cofactor. Tyrosine then undergoes further metabolism to thyroxine, catecholamines and melanin, is incorporated into tissue

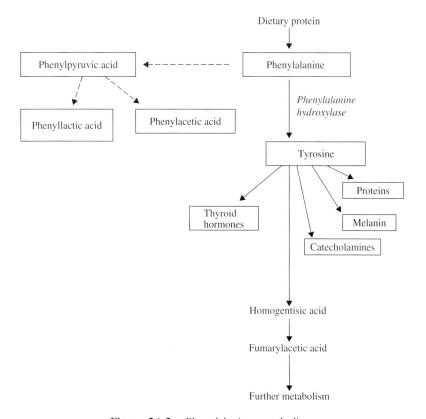

Figure 26.2   Phenylalanine metabolism.

proteins or undergoes further metabolism and enters the tricarboxylic acid (TCA) cycle. In PKU, the enzyme phenylalanine hydroxylase is absent or reduced in activity. This leads to an accumulation of phenylalanine, which is metabolised by an alternative pathway leading to the production of phenylpyruvic acid, which then undergoes metabolism to phenyllactic, and phenylacetic acids, which are normally present in negligible quantities. In PKU, these acids are excreted in excess quantities. Patients with PKU have severe learning difficulties and irreversible brain damage thought to be caused by the excess phenylalanine interfering with brain amino acid metabolism and neurotransmitter synthesis. The excess phenylalanine may also interfere with the metabolism of tyrosine to melanin as the enzyme tyrosinase is competitively inhibited by phenylalanine.

An infant with PKU is normal at birth as phenylalanine *in utero* is transferred across the placenta to the mother. After birth, the phenylalanine starts to accumulate and delayed development is noticed at 6 months and progressive mental retardation occurs. In addition, fits, aggressive behaviour and microcephalopathy may also occur. Due to the effect of excess phenylalanine on tyrosinase, there is a tendency for hypopigmentation and children with classic PKU have blue eyes and fair hair.

The diagnosis depends on the demonstration of high phenylalanine concentration in blood. There is a neonatal screening programme to detect this condition early.

If detected early, further mental retardation may be prevented. Management of these patients involves restricting the phenylalanine intake in the diet. Special diets are available in which the protein is replaced by an amino acid mixture low in phenylalanine. These patients should be continuously monitored to maintain phenylalanine concentration below 300 μmol/L in the first year of life and thereafter below 500 μmol/L. The low cutoff value at the first year of life is recommended, as the development of brain is rapid during this period. Dietary restrictions should be continued throughout life especially during pregnancy as maternal hyperphenylalaninaemia has a teratogenic effect on the foetus even if the foetus does not have PKU. The major problem in management is compliance as the diet is unpalatable. Since phenylalanine is an essential amino acid, a certain amount must be provided in the diet. Under normal circumstances, tyrosine is not an essential amino acid but when the intake of phenylalanine is restricted, tyrosine becomes a conditionally essential amino acid and adequate quantities of tyrosine should also be provided in the diet. Treated children grow up normally although there is some evidence that they may not attain full potential.

About 50% of all cases are due to classical PKU where the enzyme phenylalanine hydroxylase is almost absent. In less severe forms, there is partial deficiency of this enzyme and these are referred to as PKU variants. In others, there is a defect in the enzyme dihydropteridine reductase (DHPR) leading to a reduction in its cofactor tetrahydro-biopterin. In DHPR deficiency, there is severe progressive neurological disease due to failure of synthesis of neurotransmitters. Dietary restriction alone has no beneficial effect in DHPR deficiency. In PKU variants where enzyme activity may be up to 6% of normal, blood phenylalanine concentrations may not be very high and these children may not have phenylketones in urine.

Screening

In many countries, there is a screening programme for the detection of PKU. In this programme, blood is taken from all newborn babies 5–8 days after birth. This delay is important to allow introduction of feed containing phenylalanine to detect the hyperphenylalaninaemia. Blood is usually taken by heel prick onto a filter paper and sent to a central laboratory for testing. In the central laboratory, the filter spot is punched out, the sample eluted and the phenylalanine concentration measured by tandem mass spectrometry. The cutoff value for phenylalanine concentration has been selected to give 100% sensitivity. However, this makes it less specific and many babies will have false positive results. If the result is greater than the cutoff value, a repeat analysis is done on the same blood spot and if the second result is also high, a repeat blood sample is analysed.

Diagnosis of the Heterozygous Carrier State

Detection of heterozygotes is essential for genetic counselling. In heterozygotes, the activity of phenylalanine hydroxylase is lower than normal and can be detected by a phenylalanine loading test. However, it is now common practice to use molecular biology techniques to detect heterozygotes.

Maternal Phenylketonuria

If women with phenylketonuria become pregnant, the foetus, even if it is only a carrier of the disease, will be affected by high maternal phenylalanine levels which can cross the placenta and cause congenital heart disease, growth retardation, microcephaly and mental retardation. It is therefore important for a mother with PKU to lower her phenylalanine levels at the time of conception and throughout pregnancy.

Cystic Fibrosis

Cystic fibrosis is the commonest inherited metabolic disease in Caucasians affecting 1 in 2500–3000 live births. The incidence in Asians is 1 in 90,000 and in African Americans it is 1 in 14,000. About 1 in 25 Caucasians, 1 in 90 Asians and 1 in 65 Africans are carriers. Cystic fibrosis is due to a defect in the cystic fibrosis transmembrane conductance regulator (CFTR), which is a chloride transporter. Cystic fibrosis is a generalised disorder affecting all exocrine secretions, as the chloride channel is present in all cell membranes. The biochemical consequences of this defect are blockage of salt and water transport into and out of cells, resulting in viscous secretions and blockage in the lungs, pancreas and other organs. Although more than 1000 mutations have been identified in CFTR gene, the most common (> 70%) abnormality is the deletion of a single amino acid, phenylalanine (ΔF 508), which results in a functional change in the protein and impairs the transport of chloride.

Most affected individuals are diagnosed in childhood but some may be detected in adult life. The affected organs are the respiratory tract, pancreas, intestinal tract and genitourinary tract. Due to the increased viscosity of the mucous secretions in the respiratory tract, children often present with repeated, recurrent respiratory infections leading to irreversible lung damage and bronchiectasis. Pulmonary disease is the major cause of death. Deficiency of the exocrine pancreatic function leads to malabsorption. Failure of the exocrine pancreas in the neonatal life can lead to meconium ileus and small bowel obstruction may occur in older children. Over 90% of male patients are infertile due to obstructive azoospermia. Fertility is diminished in female patients.

Diagnosis of Cystic Fibrosis

Diagnosis is established by demonstrating high chloride and/or sodium concentrations in sweat collected following stimulation by pilocarpine iontophoresis.

Management

The aim of treatment is to control infection, to promote the clearance of mucus from the respiratory tract and to improve nutrition. Mucus is cleared by postural drainage. Chest infection, when it occurs, is treated aggressively with antibiotics. Regular and prophylactic antibiotic treatment may be necessary. If pancreatic enzyme deficiency is present, enzyme supplements are given. Adequate nutrition should be maintained. With these measures, the survival of children with cystic fibrosis has increased over recent years. Recent studies show that early diagnosis and treatment can lead to improvements in the health of these patients.

Screening

Due to the blockage of the pancreatic duct, trypsinogen enters the circulation and can be detected in the circulation as immunoreactive trypsin (IRT). Elevated levels of serum IRT are indicative of cystic fibrosis. Screening by measuring immunoreactive trypsin in plasma has been instituted in the UK as part of the neonatal screening programme. Screening is important as recent studies show that early diagnosis can lead to improvement in the health and prognosis. If the initial test shows elevated IRT, the diagnosis is confirmed by the detection of the mutation. Prenatal diagnosis is possible if both parents are known to be heterozygotes for cystic fibrosis.

Glycogen Storage Diseases

Glucose-6-phosphatase Deficiency — Glycogen Storage Type I

Glucose-6-phosphatase is an enzyme, which converts glucose-6-phosphate to glucose during gluconeogenesis and glycogenolysis. (Figure 26.3) In normal subjects, blood glucose is maintained by glycogenolysis and gluconeogenesis during fasting. In this disorder, due to the absence of glucose-6-phosphatase, hypoglycaemia develops in the fasting state. Glucose-6-phosphate is converted to glycogen, which accumulates in the liver causing enlargement of the liver. Further-more, the block in gluconeogenesis also leads to the accumulation of lactate leading to lactic acidosis as the accumulation of glucose-6-phosphate inhibits the conversion of lactate to pyruvate. Hyperlipidaemia with raised VLDL and LDL concentrations and hyperuricaemia are also frequently found.

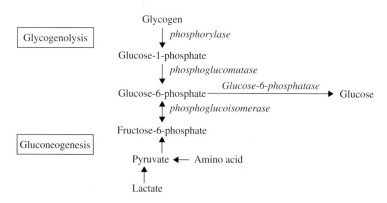

Figure 26.3   Metabolic abnormalities in glycogen storage disease type 1.

There is increased triglyceride synthesis due to reduced gluconeogenesis and low insulin levels. Hyperuricaemia is due to a combination of increased production as a result of increase in the pentose phosphate pathway and decreased excretion due to the lactic acid inhibiting the secretion of uric acid. Platelet dysfunction and bleeding tendency are also seen due to accumulation of glycogen in the platelets. The diagnosis of this condition is suspected in children presenting with hypoglycaemia during fasting and is established by the glucagon stimulation test. Failure of the blood glucose to rise after an intramuscular injection of glucagon indicate the diagnosis. The diagnosis is confirmed by measuring enzyme activity in a liver biopsy specimen.

Other Glycogen Storage Diseases

Glycogen storage disease type II or Pompe disease is due to maltase deficiency (α1,4-glucosidae). This enzyme present in lysosomes breaks down glycogen. In this disorder, glycogen accumulates and causes cardiomyopathy and muscle hypotonia.

In type III glycogen storage disease (Cori's disease), there is a deficiency of the glycogen-debranching enzyme that is involved in the breakdown of glycogen branches. Abnormal glycogen is deposited in the liver, muscle and heart. These children have enlarged liver, fasting hypoglycaemia and muscle weakness.

Type IV glycogen storage disease (Andersen disease) is due to deficiency of the glycogen-branching enzyme and abnormal glycogen accumulates in the liver, muscle and heart.

Deficiency of phosphorylase enzyme in muscle gives rise to type V glycogen storage disease or McArdle's disease. These patients develop fatigue after exercise due to the inability to mobilise muscle glycogen and may develop myoglobinuria as a result of muscle breakdown.

Type VI glycogen storage disease (Hers' disease) is due to a deficiency of liver phosphorylase. These patients develop fasting hypoglycaemia.

Type VII glycogen storage disease (Tarui's disease) is due to a deficiency of phosphofructokinase in muscle and leads to fatigue and muscle pain during exercise.

Galactosaemia

Galactosaemia is a condition where there is a defect in the metabolism of galactose, which is derived from lactose. Galactose is normally metabolised (as shown in Figure 26.4) to glucose. Galactose is essential for the formation of cerebroside and glycoproteins. Galactosaemia can be caused by three of the principle enzymes involved in galactose metabolism, namely galactokinase, galactose-1-phosphate uridyltransferase and UDP-galactose-4-epimerase. Of these, galactose-1-phosphate uridyltransferase deficiency is the commonest. Absence or reduced activity of the enzyme leads to accumulation of galactose-1-phosphate, which is toxic to cells of liver, kidney and brain. Galactose concentration is increased in blood and is excreted in the urine.

Infants with this disorder present with feeding difficulties, vomiting, failure to thrive and hypoglycaemia, usually in the first few weeks after birth. Other features include enlarged liver, jaundice, ascites, cataracts, mental retardation and renal tubular defects (Fanconi syndrome). The conversion of galactose to galacticol by the enzyme aldolase reductase in the lens is thought to cause cataracts. Accumulation of galacticol results in swelling of the lens.

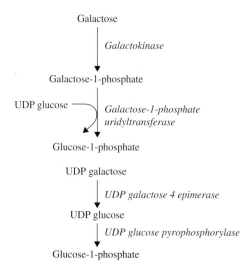

Figure 26.4   Metabolism of galactose.

In children with this disorder, galactose can be detected in the urine. The diagnosis is established by measurement of galactose-1-phosphate uridyltransferase in erythrocytes. In some countries, galactosaemia is included in the neonatal screening programme.

Deficiency of the enzyme UDP-galactose 4-epimerase produces a picture similar to that seen in galactose-1-phosphate uridyltransferase deficiency. In deficiencies of galactokinase, galactose-1-phosphate is not formed and clinical features of classical galactosaemia are not present except cataract, which is caused by metabolism of galactose to galacticol. Genetic tests are available to detect the mutation and this test is helpful in detecting carriers.

Treatment is by the elimination of lactose and galactose from the diet. Even with restricted diets, some patients have long-term neurological problems.

Case 26.1

A 9-day-old baby boy was admitted to hospital for investigation of jaundice and not feeding well. The baby was born at full term was normal at discharge from the postnatal ward. He was noticed to be slow in feeding at 7 days after birth and jaundice was noticed a day later. On admission, he was found to be lethargic and jaundiced. There were no abnormalities detected in the examination of the cardiovascular system, nervous system and the lungs.

A urine sample was sent to the laboratory and it showed the presence of reducing substances. Galactosaemia was suspected and a sample of blood was sent for enzyme analysis, which confirmed the diagnosis. Once he was started on a lactose-free diet, his symptoms improved and he started gaining weight.

Maple Syrup Urine Disease

This is a disease caused by a defect in the enzyme complex branched-chain α-ketoacid dehydrogenase, which is involved in the metabolism of branched-chain amino acids, leucine, isoleucine and valine. An inability to metabolise the oxoacids from these branched-chain amino acids leads to accumulation of these oxoacids, which are excreted in the urine giving its characteristic odour. Clinical features include failure to thrive, feeding difficulties, lethargy and neurological dysfunction. The child may present with hypoglycaemia and metabolic acidosis. The diagnosis can be established by measurement of branch chain amino acids, which are high, and by the presence of allo-isoleucine which is not normally present. Diagnosis is confirmed by measuring the enzyme activity in leucocytes or in cultured fibroblasts. Management involves restriction of branch chain amino acids in the diet.

Fatty Acid β-oxidation Defects

Oxidation of fatty acids is a complex metabolic process (Figure 26.5). Fatty acids are first transported across the mitochondrial membrane

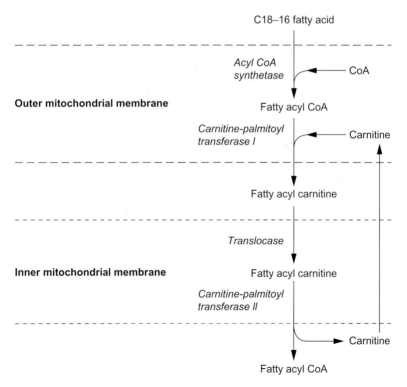

Figure 26.5 Transport of fatty acids into the mitochondria.

by carnitine and further oxidation takes place in the mitochondria. The enzyme carnitine palmitoyltransferase I (CPTI) is involved in this transport and the fatty acid is released into the mitochondria as acyl-CoA, which enters the spiral of β-oxidation. Short (C6) and medium-(C6–C12) chain fatty acids enter the mitochondria directly without the carnitine carrier. Important enzymes involved in the metabolism of fatty acids are long chain acyl-CoA dehydrogenase (LCAD) which reduces the length of long-chain fatty acids to medium-chain fatty acids (C12–C14). Medium-chain acyl-CoA dehydrogenase (MCAD) takes over and reduces the fatty acid length to short-chain fatty acids (C4–C6) which are further degraded by short-chain acyl-CoA dehydrogenase (SCAD).

A large number of defects in fatty acid oxidation have been described. The most common of these is deficiency of MCAD, which presents after the age of 2 months and usually before 4 years but may present later (even in early adult life) with hypoglycaemia during fasting. Young children depend on fat metabolism during fasting and failure of this in MCAD leads to hypoglycaemia, hypoketonuria and hyperammonaemia. Diagnosis can be established by analysis of urine for organic acids, and analysis of plasma for carnitines, which is also useful in asymptomatic patients. Many countries including the UK have started screening for this condition by measuring medium chain acyl carnitine (octanoylcarnitine) in blood spot as part of the neonatal screening programme.

Organic Acidaemias

Organic acids are continually produced in the body during metabolism of lipids, carbohydrates and amino acids. Normally, they do not accumulate in the body as they are rapidly metabolised or cleared in the urine. If they are produced in excess or if they are not metabolised due to inherited enzyme defects, organic acids will accumulate and cause acidosis and other life threatening features. A large number of inherited disorders can lead to organic acidaemia and collectively they are more common than PKU. Inborn errors leading to organic acidaemia include defects in amino acid metabolism (e.g. MSUD, methylmalonic aciduria), defects in fatty acid oxidation and congenital lactic acidosis where raised lactic acid is the major biochemical abnormality (due to mitochondrial disease). Patients with organic acidaemia usually have metabolic acidosis and hyperammonaemia.

Hyperammonaemia

Ammonia is produced in all organs during metabolism, large amounts are produced in muscle particularly during exercise. Most of the ammonia produced is converted to glutamine that is taken up by kidneys and intestinal calls. Intestinal epithelial cells oxidise the glutamine and the nitrogen is released as alanine or ammonia. These are taken up by the liver and detoxified in the urea cycle. Ammonia in high concentrations is toxic and can cause brain damage and if severe may be fatal. Hyperammonaemia can be produced by many inherited disorders, particularly urea cycle defects as well as acquired disorders such as liver disease (Table 26.2).

Ornithine Transcarbamylase (OTC) Deficiency

Of the urea cycle defects, OTC deficiency is the most common with an incidence of about 1 in 60,000 live births. It is an X-linked disorder and complete deficiency is usually fatal in the first few days

Table 26.2 Causes of hyperammonaemia

Inherited disorders
- Urea cycle disorders
  — Ornithine transcarbamylase (OTC) deficiency
- Organic acid disorders
  — E.g. fatty acid–β-oxidation defects

Acquired causes
- Liver failure
- Drugs
  — Valproate
- Toxins
  — Aflatoxin

---

of life. Partial OTC deficiency can present at any age during times of
metabolic stress, e.g. starvation, anorexia and pregnancy. Urea cycle
defects should be thought of when there is hyperammonaemia
without acidosis.

Other inherited disorders of metabolism are described in other
chapters in this book.

Further Reading

1. Baulny HG, Abasie V, Feillet F, Parscau L. Management of phenylke-
    tonuria and hyperphenylaaninemia. *J Nutr* 2007; 137:1561S–1563S.
2. Clague A, Thomas A. Neonatal biochemical screening for disease. *Clin
    Chim Acta* 2002; 315:99–110.
3. Green A, Morgan I. *Neonatology and Clinical Biochemistry* 2003. ACB
    Venture Publications London.
4. Kompare M, Rizzo WB. Mitochondrial fatty-acid oxidation disorders.
    Semin Pediatr Neurol 2002; 15:140–149.
5. O'Sullivan BP, Freedman SD. Cystic fibrosis. The *Lancet* 2009;
    373:1891–1904.
6. Rinaldo P, Lim JS, Tortorelli S, Gavrilov D, Matern D. Newborn screen-
    ing of metabolic disorders: Recent progress and future developments.
    Nestle Nutr Workshop Ser Pediatr Program 2008; 62:81–93.
7. Wilcken B, Wiley V. Newborn screening. *Pathology* 2008; 42:104–115.

Summary/Key Points

1. Individual inherited metabolic disorders (IMD) are rare, but col-
    lectively, these are an important cause of mortality and morbidity.
    IMD is usually due to a mutation which causes the production of
    no enzyme or a functionally inactive enzyme. This causes a reduc-
    tion in the enzyme product, increased concentration of precursors
    and/or increase in alternative metabolites. The clinical effects of
    IMD are due to these effects.
2. Phenylketonuria (PKU) is due to a defect in the metabolism of
    phenylalanine, most commonly due to deficiency of phenylalanine
    hydroxylase. Accumulation of phenylalanine causes the clinical

features of PKU. Screening for PKU in the neonatal period is now common in many countries. Reduction in the intake of phenylalanine, if started early, will prevent the development of mental retardation.

3. During pregnancy in patients with PKU, strict control of plasma phenylalanine is important to prevent foetal complications even if the foetus is only a carrier of the disease.

4. Cystic fibrosis is a common IMD in Caucasians and is due to mutation causing a defect in the cystic fibrosis transmembrane conductance regulator (CFTR), a chloride transporter. The biochemical consequence is blockage of salt and water secretion in the exocrine glands leading to viscous secretion in the lungs, pancreas and other tissues.

5. Patients with cystic fibrosis develop lung disease, pancreatic insufficiency and infertility. Control of infection, replacement of pancreatic enzymes and adequate nutrition are important in the management of these patients.

6. Screening of neonates for cystic fibrosis using immunoreactive trypsin is now common and the diagnosis is confirmed by the detection of the mutation.

7. Glycogen storage diseases are a group of disorders where there is a deficiency of one of the enzymes involved in the synthesis or degradation of glycogen. The most common is type I where there is deficiency of glucose-6-phosphatase. In these disorders, hypoglycaemia is common.

8. Galactosaemia is due to a defect in one of the enzymes involved in galactose metabolism, the most common being galactose-1-phosphateuridyltransferase. These patients present with hypoglycaemia and failure to thrive.

9. Medium-chain acyl-CoA dehydrogenase deficiency (MCAD) is the commonest of the fatty acid oxidation disorders. This disorder is now screened in the neonates by measuring octanoylcarnitine.

chapter 27

Disorders of Purine Metabolism

Introduction

Ribonucleosides and deoxyribonucleosides are the building blocks for nucleic acids, RNA and DNA. Each nucleotide is derived from a purine or pyrimidine base with a pentose sugar and phosphate group. Purine and pyrimidine ribonucleoside triphosphates (e.g. ATP, GTP) are vital constituents of all cells. They act as key substrates, cofactors and regulatory molecules in every aspect of metabolism. Purine compounds have an essential role in membrane signal transduction as neurotransmitters, vasodilators, and mediators of platelet aggregation and hormone action. Thus, disorders of purine and pyrmidine metabolism can result in a wide range of clinical syndromes.

Purine Nucleotide Synthesis and Metabolism

The first step in purine synthesis is condensation of pyrophosphate with ribose-5-phosphate to form phosphoribosyl (PRPP) pyrophosphate, catalysed by the enzyme phosphoribosyl pyrophosphate synthase (Figure 27.1). The next step involves the incorporation of an amino group from glutamine into PRPP, catalysed by the enzyme amidophosphoribosyl transferase. This is a rate-limiting step and this enzyme is inhibited by feedback regulation of purine nucleotides. Inosine monophosphate (IMP) is formed after several further steps. Adenosine and guanosine monophosphates are derived from IMP through interconversions. Adenine and guanosine nucleosides are

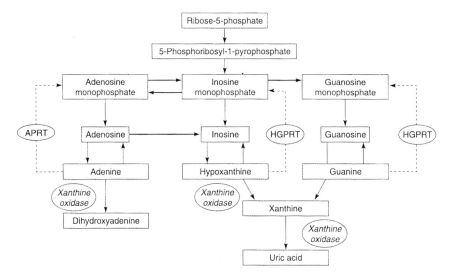

Figure 27.1   Outline of purine synthesis and metabolism.

then used as precursors for the corresponding nucleotides, which form the building blocks for DNA and RNA. Mononucleotides may be further phosphorylated to become ATP, GTP, etc.

Catabolism of nucleotides begins with the removal of ribose-linked phosphate catalysed by purine 5′-nucleotidase. Inosine and guanosine are converted to hypoxanthine and guanine by purine nucleoside phosphorylase. These are then converted to xanthine by xanthine oxidase and guanase respectively. Adenosine is converted to inosine by adenosine deaminase. Xanthine oxidase is responsible for conversion of hypoxanthine to uric acid. As the formation of nucleotides is energetically expensive, there is a salvage pathway to recycle purine bases. This salvage pathway converts guanine and hypoxanthine to their respective nucleotides catalysed by hypoxanthine–guanine phosphoribosyl transferase (HGPRT). Adenine is salvaged by adenine phosphoribosyl transferase (APRT) (Figure 27.1).

Intracellular purines are derived exclusively from endogenous metabolism as dietary purines are rapidly degraded to uric acid in the intestinal mucosa. Pyrimidines on the other hand may be absorbed from the diet and utilised.

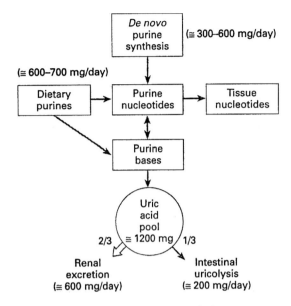

Figure 27.2   Purine metabolism.

Metabolism of Uric Acid

Uric acid is the end product of purine metabolism in humans and some higher apes. In other mammals, uric acid is converted into allantoin by uricase, which is absent in humans. Uric acid derived from the catabolism of purine nucleotides (about 300–600 mg daily) and from dietary purines (600–700 mg/day) forms the uric acid pool (Figure 27.2). Approximately 75% of the uric acid is excreted in the urine and the remainder is secreted into the gastrointestinal tract, where it is degraded to allantoin and other compounds by bacterial enzymes.

Renal handling of uric acid (Figure 27.3)

Renal handling of uric acid is complex and it involves four components: glomerular filtration, proximal tubular reabsorption, tubular secretion and post-secretory reabsorption. Glomerular

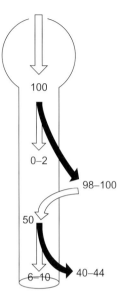

Figure 27.3   Renal handling of urate.

filtration is almost complete as there is no significant protein binding of serum uric acid *in vivo*. In the proximal tubule, there is
almost complete reabsorption of urate, by an active mechanism,
which is linked to sodium reabsorption. Very little uric acid reaches
the distal parts of the nephron, where about 50% of filtered load is
secreted. 80% of this secreted uric acid is reabsorbed. The amount
of uric acid finally excreted in the urine is equivalent to approximately 10% of the filtered load. Uric acid excretion depends on age,
gender and diet. Low-dose aspirin blocks urate secretion and causes
hyperuricaemia, while high-dose salicylate and probenecid blocks
the reabsorption and causes a net increase in uric acid excretion.
Urate excretion is influenced by circulatory volume; an increase in
circulatory volume increases urate excretion causing hypouricaemia.
Hypouricaemia in pregnancy is, at least partly, due to increase in
ECF volume. Many substances, endogenous as well as drugs, can
influence uric acid excretion. Factors affecting and handling of uric
acid are listed in Table 27.1.

Table 27.1   Factors affecting renal tubular handling of urate

Factors that decrease excretion
- Endogenous
  — Lactate
  — β-hydroxybutyrate
  — Acetoacetate
- Exogenous
  — Salicylate (low dose)
  — Diuretics
  — Cyclosporin
  — Ethambutol

Factors that increase excretion
- Endogenous
  — Pregnancy
- Exogenous
  — Saline infusion
  — Salicylate (high dose)
  — Probenecid
  — Phenylbutazone
  — Radiocontrast agents
  — Mega doses of vitamin C
  — Lead

Inherited Disorders of Purine Metabolism

Lesch-Nyhan syndrome (Hypoxanthine–Guanine Phosphoribosyl Transferase Deficiency)

Deficiency of the purine salvage enzyme hypoxanthine-guanine phosphoribosyl transferase (HPRT) is associated with increased production of purines, hyperuricaemia, gout and a neurological syndrome. Patients affected with this condition have mental retardation and a striking behavioural disturbance characterised by self-mutilation. Lesch- Nyhan syndrome is an X-linked disorder, which is only fully expressed in males. The spectrum of the disorder varies from complete deficiency of the enzyme to partial defect, which is associated with only overproduction of uric acid and therefore may present in adolescence or early adulthood. The spectrum of neurological features varies considerably, depending on the expression of the enzyme protein. In some, the activity is only moderately reduced with different kinetic properties. A number of mutations in the gene for HPRT have been described.

Diagnosis of HPRT deficiency involves the demonstration of increased excretion of uric acid. Plasma uric acid may not be always increased as the clearance of uric acid in children is high. The measurement of enzyme activity in red cells will establish the diagnosis. Genetic analysis may be required for carrier detection and prenatal diagnosis.

Management of this condition is to control the high uric acid concentration by allopurinol and adequate fluid intake. These measures will prevent gout and urate stone formation, but not the neurological features. The long-term prognosis in patients with complete absence of enzyme is poor. Other inherited disorders of purine metabolism and their manifestations are listed in Table 27.2.

Plasma Uric Acid

Plasma/serum uric acid shows wide variation in healthy subjects. It is higher in men than women (Figure 27.4) and tends to rise with age. Serum uric acid is higher in obese subjects and in more affluent social classes. Diet rice in protein and alcohol increases uric acid.

Hypouricaemia

Hypouricaemia can arise either due to decreased production or increased excretion. Inherited disorders causing decreased production include deficiency of xanthine oxidase, phosphoribose synthase deficiency and purine nucleoside phosphorylase deficiency (PNP). PNP deficiency is associated with immunodeficiency (Table 27.3). Increased excretion of uric acid may be due to tubular transport disorders such as Fanconi syndrome or uricosuric drugs.

Table 27.2   Inherited disorders of purine and pyrimidine metabolism

Enzyme		Features	Metabolism
HPRT	X-linked	Lesch-Nyhan syndrome Gout Neurological features	$U_{UA}\uparrow\uparrow$ $P_{UA}\uparrow$
APRT	Autosomal recessive	Renal calculi	Uadenine $\uparrow$ DHA $\uparrow$
PRPS (overactivity)	X-linked	Gout Neurological features	$P_{UA}\uparrow\uparrow$ $U_{UA}\uparrow\uparrow$
XDH	Recessive	Xanthine stones acute renal failure	$P_{UA}\downarrow$ $U_{UA}\downarrow$ or absent oxypurines $\uparrow\uparrow$
PNP		Immune deficiency	Deoxyguanasine or deoxyadenosine $\uparrow\uparrow$
ADA	Autosomal recessive	Immune deficiency	Deoxyadenosine$\uparrow\uparrow$
Familial Juvenile Hyperuricaemia (FJUN)	Autosomal dominant	Failure to thrive Renal stones	$\uparrow U_{UA}\uparrow$

*HPRT: Hypoxanthine phosphoribosyltransferase; APRT: Adenine phosphoribosyltransferase; PRPS: Phosphoribosyl pyrophosphate synthetase (overactivity); XDH: Xanthine oxidase; PNP: Purine nucleoside phosphorylase; ADA: Adenosine deaminase; DHA 2,8: Dihydroxyadenine; U_{UA}: Uric uric acid; P_{UA}: Plasma uric acid.

Hyperuricaemia

Hyperuricaemia can be due to increased production or decreased excretion and it can be primary or secondary (Table 27.4). Inherited metabolic disorders such as Lesch-Nyhan syndrome are examples of primary cause of increased production. In a significant number of hyperuricaemic subjects, uric acid production is increased without a clear mechanism, idiopathic hyperuricaemia. Secondary causes of hyperuricaemia are those where there is increased breakdown of nucleic acids (e.g. myeloproliferative disorders).

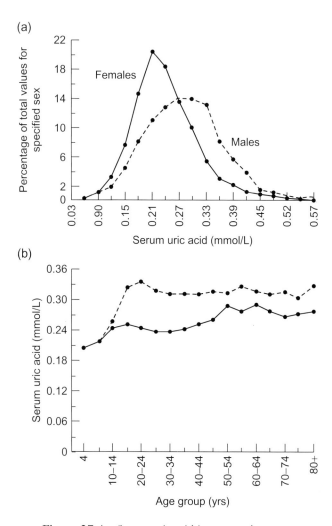

Figure 27.4   Serum uric acid in men and women.

Decreased excretion is a more common cause of hyperuricaemia. In majority of patients with gout (80–90%), uric acid clearance is reduced due to an unknown mechanism (idiopathic). An inherited disorder with a defect in excretion is familial juvenile hyperuricaemia, which is inherited as an autosomal dominant trait. It can present in childhood or early adult life with renal stones, gout or renal failure.

Table 27.3   Causes of hypouricaemia

Decrease production
- Enzymic defects
  - Xanthine oxidase
  - PP-ribose-P synthetase
  - Purine nucleoside phosphorylase
- Severe hepatic disease
- Acute intermittent porphyria
- Drugs
  - Allopurinol

Increased excretion
- Defect in tubular transport
  - Fanconi syndrome
  - Wilson's disease

Drugs — NSAID (non-steroidal anti-inflammatory drugs) — phenylbutazone
- Uricosuric drugs

Renal disease is the most common secondary cause of reduced excretion. Hyperuricaemia has been suggested to be an independent risk factor for cardiovascular disease.

Gout

Gout is a metabolic disease in which symptoms and signs result from tissue deposition of crystals of monosodium urate monohydrate from hyperuricaemic fluids. Features of gout include acute inflammatory arthritis, tenosynovitis, chronic arthritis, renal stones, chronic renal disease and hypertension. It is characterised by recurrent attacks of acute arthritis in which crystals of monosodium urate are found in leucocytes in the synovial fluid of affected joints. Urate crystals may also occur in tissues surrounding the joint. Gout is associated with hyperuricaemia. The risk of developing gout increases with increasing urate concentrations (Table 27.5). However, hyperuricaemia does not always cause gout and only a minority of patients with hyperuricaemia develop gout. In patients with gout, serum uric acid is or has been elevated.

Table 27.4   Causes of hyperuricaemia

Increased urate production
- Primary
  — Idiopathic
  — Inherited metabolic disorders
    o Lesch-Nyhan syndrome (hypoxanthine-guanine phosphoribosyl transferase deficiency)
    o Increased of activity of phosphoribosyl pyrophosphate synthetase (PRPS)
    o Glucose-6-phosphatase deficiency (glycogen storage disease type I)
    o Increased activity of xanthine oxidase

- Secondary
  — Increased intake of purines
  — Increased nucleic acid breakdown
    o Malignant disease
    o Cytotoxic drugs
    o Psoriasis
    o Myeloproliferative disorders
    o Lymphoproliferative disorders (leukaemia, lymphoma)
    o Carcinomatosis
    o ATP degradation — alcohol, hypoxia

Decreased renal excretion
- Primary
  — Idiopathic
  — Familial juvenile hyperuricaemia
- Secondary
  — Renal disease
  — Increased reabsorption/decreased secretion
    o Thiazide diuretics drugs
    o Salicylate (low dose)
    o Lead
    o Organic acids — lactic acid, keto acids

Table 27.5   Incidence of gout in relation to serum uric acid concentration in males

Serum uric acid (nmol/L)	5 year cumulative incidence (%)
< 0.41	0.6
0.42–0.47	2.0
0.53–0.58	9.8
> 0.59	30.5

Gout is more common in men (the male:female ratio is 8:1) and is rarely seen in premenopausal women in whom serum uric acid is lower than in men. The peak incidence is seen between the fourth and sixth decades. It is more common in certain races, e.g. New Zealand Mauri population. Risk factors for gout include obesity, alcohol consumption, lead exposure, renal insufficiency, the use of diuretics and a family history of gout. It often presents as an acute arthritis of a single joint, but commonly progresses to affect many joints. Metatarsophalangeal joint of the big toe is involved in majority of cases and is the site of first attack in 50% of cases. Other joints affected include ankle, knee, wrists, elbows and small joints of the hands and feet. Affected joints are hot and inflamed and this is accompanied by fever and leucocytosis. Attacks last a few hours to a few weeks and usually resolve completely. If untreated, attacks recur with increasing frequency and chronic gout may develop. Chronic gout is characterised by deposits of sodium urate around joints, leading to joint swelling, bone erosion and joint destruction. Topi (deposits of sodium urate) occur in many places, including the external ears and around joints. Renal disease may complicate gout and may take the form of (a) acute uric acid nephropathy with deposition of urate in renal collecting ducts and ureters, (b) chronic renal disease and (c) urate nephrolithiasis.

Gout is sometimes classified as primary (or idiopathic) and secondary when hyperuricaemia is due to an identifiable cause (Table 27.4).

Gout, however, is uncommon in secondary hyperuricaemia. In 90% of patients with gout, there is relative underexcretion of urate by the kidneys in relation to plasma urate and in the other 10% there is overproduction. Dietary intake of high purine containing foods (high meat diet) and alcohol contribute to hyperuricaemia and gout. The contribution of dietary factors is illustrated by the low incidence of gout during the World Wars and by the frequent occurrence of gout in higher social classes. Genetic polymorphism in the gene coding for human urate transporter 1 (URAT1) and glucose and fructose transporters are associated with hyperuricaemia and gout.

Pathogenesis

There may be long periods of asymptomatic hyperuricaemia. Acute attacks may be precipitated by direct trauma, local metabolic changes, sudden lowering of serum urate, high alcohol intake, dietary excess, severe dieting or diuretic therapy. Monosodium urate microcrystals are formed in the joints and are engulfed by polymorphonuclear leucocytes, where they cause damage to lysosomal membranes resulting in the release of lysosomal enzymes and generation of free radicals such as superoxide. This, in turn, results in the release of cytokines such as interleukin 1, which cause the inflammatory response.

Diagnosis

Diagnosis of gout is primarily clinical and should be differentiated from several other arthritic conditions. Findings of topi or monosodium urate crystals in leucocytes from synovial fluid confirm the diagnosis. These crystals are typically 2–10 mm long and show negative birefringence under polarised light. A high serum uric acid concentration is very suggestive of gout in patients with characteristic clinical features. A normal serum uric acid does not exclude the diagnosis of gout as hyperuricaemia may have occurred in the past. Consistently low serum uric acid concentration, however, excludes the diagnosis.

Management

Treatment of the acute attack includes non-steroidal inflammatory drugs such as indomethacin and colchicine. Prevention of further attacks involves reduction of serum urate by reducing the production or increasing the excretion. Production can be reduced by dietary measures such as a reduction in high purine diet (meat, seafood, yeast, lentils, etc.), reduction in alcohol intake and by drugs such as allopurinol that inhibits xanthine oxidase. Urocosuric drugs such as probenecid can increase excretion. Long-term (life long) treatment should be considered if there are recurrent and troublesome acute attacks, presence of topi, bone or cartilage damage, renal disease or uric acid stones in young patients with family history.

Further Reading

1. Cameron JS, Simmonds HA. Hereditary hyperuricaemia and renal disease. *Semin Nephrol* 2005; 25:9–18.
2. Doherty, M. New insights into the epidemiology of gout. *Rheumatology* 2009; 48:ii2–ii8.
3. Lee SJ, Terkeltaub RA, Kavanaugh A. Recent developments in diet and gout. *Curr Opin Rheumatol* 2006; 18:193–198.
4. Löffler M, Fairbanks LD, Zameitat E, Marinaki AM, Simmonds HA. Pyrimidine pathways in health and disease. *Trends Mol Med* 2005; 11:430–437.
5. Simmonds HA, Duley JA, Fairbanks LD, McBride MB. When to investigate for purine and pyrimidine disorders. Introduction and review of clinical and laboratory indications. *J Inherit Metab Dis* 1997; 20:214–226.

Summary/Key Points

1. Purines and pyrimidines are the building blocks of nucleic acids. Purines (adenosine, inosine and guanine) are synthesised *de novo*. When these are metabolised, uric acid is produced.

2. Inherited disorders of purine pathway can lead to abnormalities in uric acid. One such disorder is Lesch-Nyhan syndrome, which is due to a deficiency of an enzyme involved in the salvage pathway (hypoxanthine–guanine phosphoribosyltransferase (HPRT)). This syndrome is associated with hyperuricaemia, mental retardation and self-mutilation.

3. Uric acid is excreted by the kidney by a complicated process of filtration, complete reabsorption, secretion followed by reabsorption.

4. Hyperuricaemia can arise either due to excessive production (e.g. treatment of malignancies) or due to decreased excretion of uric acid. Gout is an arthropathy due to deposition of uric acid crystals in synovial joints.

5. Hyperuricaemia can be treated with urocosuric drugs such as allopurinol.

chapter 28

Clinical Biochemistry in Paediatrics and Geriatrics

Paediatric Clinical Biochemistry

Investigation in the paediatric age group gives rise to several special problems. These include (i) interpretation of results, which depends on the use of appropriate reference ranges, (ii) difficulty in getting adequate samples for analysis and, (iii) some special investigations for conditions, which are predominantly seen in the paediatric age group. Over the years, with improved medical and nursing care, an increasing number of premature infants with very low birth weight are surviving. Thus, there is a need for the laboratory to be able to do tests on very small volumes of blood samples.

Many investigations, especially in the newborn period, are carried out on capillary blood samples obtained by heel prick. Urine collection is also a problem in children and special techniques have to be adapted especially if timed collections are required. In older children, where this is more difficult, it may be more appropriate to use random urine specimens.

Reference Ranges

For correct interpretation of results, it is important to use appropriate reference ranges. For many analytes, the reference ranges vary with age. Examples of investigations affected by age include plasma calcium, alkaline phosphatase and phosphate. For other measurements such as creatinine clearance, which depend on the development of the child, use of chronological age may not be appropriate. It is recommended that such measurements should be corrected for either body surface area or height and weight.

Physiological Changes

Development of nephrons starts from six weeks of gestation and is complete by 36 weeks. Glomerular function at birth is low and rapidly increases during the first six months of life. Thereafter, there is a steady increase to reach adult values by the age of 14 years. Total body water content of newborn infants is 75% of body weight compared to 60% in adults, and in premature infants, it can be as high as 85%. In contrast to adults, there is more fluid in the extracellular compartment than in the intracellular compartment. During the first few weeks of life, extracellular space decreases; this is associated with an increased urine sodium excretion. During the first 12 months, extracellular fluid decreases further and reaches adult values as does total body water. At the same time, intracellular water increases. Newborn infants have limited tubular function and the counter current mechanism is not fully developed and the consequent ability to alter urine osmolality is impaired. The maximum urine osmolality achievable is < 700 mOsmol/kg.

The surface area of neonates is large in relation to the size and consequent insensible water loss is proportionately higher; this can increase by a further 50% during phototherapy. Fluid and sodium requirements of infant are higher (about 150 mL/kg body wt compared to 30–40 mL/kg in adult; 3–6 mmol/kg compared to < 1 mmol/kg in adult).

The renin–angiotensin–aldosterone (RAA) system is active during the first week of life.

Abnormalities in Fluid and Electrolyte Balance

Hyponatraemia

The causes and mechanism of hyponatraemia are similar to that in adults. In the neonate, fluid overload may result due to excessive fluid administration to mothers during labour or during the postnatal period. There may be inappropriate ADH secretion due to respiratory or cerebrovascular diseases. Hyponatraemia may also develop due to increased loss of sodium, e.g. gastroenteritis or increased loss of sweat sodium due to cystic fibrosis. Hyponatraemia may be found in 40% of premature babies due to inability of the immature kidney to conserve sodium. Congenital adrenal hyperplasia should be kept in mind as a possible cause.

Hypernatraemia

Hypernatraemia may be caused by water depletion, which is common, or by excess sodium administration which is less common. Insensible water loss in infants especially in premature infants is greater because of the greater surface area, increased skin blood flow and increased metabolic rate. Tendency to hypernatraemia is further aggravated by the inability of the immature kidneys to conserve water. Hypernatraemia may also occur in infants fed with formula milk with high sodium content.

Neonatal Jaundice

Serum bilirubin concentration increases during the first week of life in almost all babies. In some infants, the bilirubin may reach clinically significant values and neonatal jaundice is one of the most common causes of admission to hospital. Jaundice during the newborn period can be classified into unconjugated or conjugated hyperbilirubinaemia. The importance of unconjugated hyperbilirubinaemia is the potential for the development of neurotoxicity (kernicterus) especially in premature and low birth weight infants. Kernicterus is a neurological syndrome that develops from bilirubin neurotoxicity. Unconjugated bilirubin is carried in plasma bound to albumin. If the binding capacity of albumin is exceeded, free bilirubin appears in the circulation. Sulphonamides, furosemide, radiocontrast media, salicylate, hormones and free fatty acids can disturb the binding and will increase free bilirubin, which is lipid soluble. Free bilirubin is able to cross the blood brain barrier and cause irreversible brain damage probably by uncoupling oxidative phosphorylation. Risk of kernicterus is higher in premature infants, infants with hypoalbuminaemia or if the binding capacity of albumin is reduced.

Physiological jaundice of the newborn

Serum bilirubin concentration reaches a peak within 1–7 days of birth and remains elevated for approximately 2 weeks. In premature infants, this may last up to 4 weeks. The incidence of jaundice is higher in premature infants and is also thought to be higher in certain ethnic groups, e.g. the orientals. The bilirubin is predominantly unconjugated and the following factors contribute to the increase in bilirubin concentration:

1. In the newborn, there is a relative polycythaemia and this rapidly decreases during the first few days of life resulting in increased production of bilirubin.
2. Contribution from ineffective erythropoiesis and non-red blood cell sources is higher.
3. Decreased ability to conjugate the bilirubin due to low activity of UDP-glucuronyltransferase (UGT1A1) enzyme in the liver. The production of bilirubin during early neonatal period is approximately three times that in adults and the neonatal liver cannot cope with this.
4. Increased absorption of bilirubin from the gut due to a glucuronidase enzyme in the meconium. This enzyme breaks down conjugated bilirubin, and the resulting bilirubin is absorbed passively.
5. Breast-fed infants are exposed to pregnanediol in the breast milk. Pregnanediol inhibits bilirubin conjugation.

Pathological causes of unconjugated hyperbilirubinaemia

Increased bilirubin concentration appearing early in life, rapidly rising bilirubin, and persistent jaundice are all usually pathological (Table 28.1).

Table 28.1 Causes of neonatal jaundice

Unconjugated hyperbilirubinaemia
- Physiological jaundice
- Haemolytic disease
  — Rh incompatibility
  — ABO haemolytic disease
- Inherited defects in erythrocytes
  — Glucose-6-phosphate dehydrogenase deficiency
  — Pyruvate kinase deficiency
  — Spherocytosis
  — Sickle cell diseases
  — Thalassemia
- Prenatal infection
- Hypothyroidism
- Inborn errors of bilirubin metabolism
  — Crigler-Najjar syndrome

Conjugated hyperbilirubinaemia
- Infection
  — Cytomegalovirus
  — Hepatitis
- Inherited metabolic disorders
  — Galactosaemia
  — α_1-Antitrypsin deficiency
  — Tyrosinaemia
- Biliary atresia

Haemolytic disease of the newborn results from maternal–foetal incompatibility of rhesus or ABO blood groups. Rhesus incompatibility is now less common following the use of rhesus immunoglobulins. Inherited defects in erythrocytes such as glucose-6-phosphate dehydrogenase deficiency can lead to haemolytic disease especially if they are exposed to chemicals, drugs, or infection. Haemolysis is more severe in the mediterranean and oriental type. Congenital infections, such as syphilis and toxoplasmosis may result in increased erythrocyte turnover and haemolysis. Congenital hypothyroidism may present with prolonged jaundice. Other rare causes of hyperbilirubinaemia include inborn errors of bilirubin metabolism such as Crigler-Najjar syndrome (see Chapter 15).

Breast milk jaundice

Infants who are breast-fed are at increased risk of neonatal jaundice. Factors that contribute to this are:

1. Presence of an unusual metabolite of progesterone-pregnane-3-α20 β-diol in breast milk. This compound inhibits UGT1A1.
2. Increased enterohepatic circulation of bilirubin due to (a) higher β-glucuronidase activity in breast milk and (b) delayed colonisation of the gut. Bacterial colonisation is important for the conversion of bilirubin to stercobilinogen.

Lucey-Driscoll syndrome: This is a rare benign disorder characterised by transient unconjugated hyperbilirubinaemia due to an unknown inhibitor of *UGT1A1* in maternal serum. Phototherapy and treatment with phenobarbitone are effective forms of treatment.

Management of unconjugated hyperbilirubinaemia

The concentration of bilirubin at which treatment is indicated depends on the maturity of the infant, time after birth and the presence of risk factors for kernicterus, e.g. if there is sepsis, hypoalbuminaemia, acidosis or haemolytic disease treatment is initiated at a lower bilirubin concentration. For instance 72 hours after birth, treatment is initiated at 230 μmol/L in a high risk infant compared to 340 μmol/l in a low risk infant. Initial treatment is by phototherapy with UV light, which causes isomerisation of bilirubin to soluble isomers which are excreted in bile and urine.

If the response to phototherapy is inadequate, insufficient or if there is rapid rise in serum bilirubin (> 15–20 μmol/L/hr), exchange transfusion may be required. Exchange transfusion has potential complications such as hyperkalaemia, hypocalcaemia (due to anticoagulants in the transfused blood), hypoglycaemia and metabolic acidosis. Infants given phototherapy should be adequately hydrated as there is increased insensible loss of water.

Conjugated hyperbilirubinaemia

In this condition, more than 30% of the total bilirubin is conjugated. Important causes include biliary atresia and idiopathic neonatal hepatitis (Table 28.1). Biliary atresia is a heterogeneous group of disorders involving the extrahepatic or intrahepatic bile ducts and is caused by intrauterine destruction of bile ducts possibly by infections due to agents such as cytomegalovirus or rubella virus. Neonatal hepatitis can cause hyperbilirubinaemia and it will be associated with abnormalities in other liver function tests. Prolonged parenteral nutrition may lead to cholestatic jaundice. The aetiology of this condition is not known. Children with inborn errors of metabolism such as galactosaemia, tyrosinaemia and α_1-antitrypsin deficiency may present with jaundice.

Investigation of hyperbilirubinaemia

Jaundice in the newborn should be investigated if it occurs early, prolonged or severe. Investigations should include the type of bilirubin, (conjugated or unconjugated) and detection of possible haemolytic disorders.

Case 28.1

A 4-day-old baby boy born at 36-week gestation was admitted with worsening jaundice. When the baby was discharged from hospital the day after birth, he was noted to have a cephalohematoma. His stools were reported to be yellow. On examination, he was found to be markedly jaundiced and a resolving cephalohematoma was noted. Neurological examination was normal.

The total bilirubin concentration was reported to be is 370 μmol/L and the conjugated bilirubin concentration was 30 μmol/L.

These results showed that there was severe unconjugated hyperbilirubinaemia and further tests were done to investigate the cause. His blood group was A+ while his mother's blood group was O+. Direct Coomb's test was positive showing there were haemolysis of red cells due to presence of antibodies. An exchange transfusion was done and his bilirubin decreased.

Disorders of Calcium and Phosphate Metabolism

In utero calcium and phosphorus are actively transported across the placenta. The calcium concentration in the foetus is higher than that in the maternal circulation and the active transfer against this concentration gradient is probably mediated by parathyroid hormone related peptide (PTHrP). The foetus acquires large amounts of calcium during the last trimester of pregnancy. Thus, premature infants are very prone to develop calcium and phosphate disorders. Plasma calcium concentration, both total and ionised concentrations, fall after birth and reach lowest values 1–2 days after birth and rises again and reaches a plateau by day five. The degree of decrease in plasma calcium after birth is related inversely to gestational age and is more marked in infants born to diabetic mothers and those with perinatal asphyxia.

Hypocalcaemia

Hypocalcaemia is relatively common in the neonatal period and may present as jittery movements, convulsions and occasionally apnoea. Hypocalcaemia is due to the immaturity or inability of the parathyroid glands to mobilise bone calcium and immaturity of the 1α-hydroxylase system. This is compounded by the absence of calcium from food especially in premature infants. Administration of large volumes of blood containing citrate and phosphate further exaggerate the hypocalcaemia. Causes of neonatal hypocalcaemia are listed in Table 28.2. Hypocalcaemia occurring at 5–10 days of age is due to hypoparathyroidism. Hypoparathyroidism can be transient due to hypercalcaemia in the mother or due to congenital absence of parathyroid glands.

Hypocalcaemia occurring in the first 4 days of life is seen in infants born to mothers with diabetes, hyperparathyroidism or pre eclampsia, and in infants who are premature, infants with perinatal stress or perinatal trauma. The fall in serum calcium, which occurs in early, postnatal period is exaggerated and may cause symptoms. This is thought to be due to inadequate PTH response. Usually this condition is self-limiting.

Table 28.2   Causes of hypocalcaemia in the paediatric age group

Early neonatal hypocalcaemia
- Low birth weight infants
- Perinatal asphyxia
- Infants of diabetic mothers

Late neonatal hypocalcaemia
- Phosphorous overload
- Renal failure
- Hypoparathyroidism

Hypocalcaemia in older infants
- Critical illness
- Hypoparathyroidism
- Vitamin D deficiency and/or mineral deficiency
- Hypomagnesaemia

Hypocalcaemia occurring after 4–5 days, late-onset hypocalcaemia, may be caused by hypoparathyroidism or as a result of high phosphate intake. Neonatal vitamin D deficiency can present as hypocalcaemia. When the mother is severely D deficient, infants are born with vitamin D deficiency.

Phosphate overload used to be a common cause of hypocalcaemia in the neonate. Cow's milk contains six times more phosphorus than human milk and the capacity of the neonatal kidney to excrete phosphate is limited, leading to hyperphosphataemia and consequent hypocalcaemia. Introduction of humanised cow's milk with reduced phosphate has led to a decrease in this type of hypocalcaemia.

Hypocalcaemia later in the newborn period can be caused by critical illness or hypoparathyroidism. Hypocalcaemia is a feature of sick neonates and the exact mechanism is not clear. Hypomagnesaemia is a treatable cause (see Chapter 7). DiGeorge syndrome is a rare inherited cause of hypoparathyroidism due to agenesis of parathyroid gland associated with susceptibility to infection due to thymic aplasia and immune deficiency.

Metabolic bone disease/rickets of prematurity

Osteopenia may be present in up to 50% of premature infants and the incidence of clinical rickets has been reported to be as high as 13% in low-birth weight babies. The aetiology of rickets of prematurity is multifactorial. A major factor is deficiency of calcium and phosphorus, especially the latter as the amount of calcium and phosphate present in human milk is inadequate for the extra needs of the premature infant. Other factors contributing to the rickets are immaturity of the 1α-hydroxylase enzyme in the kidney together with limited absorptive capacity of the intestinal tract. In certain communities, such as in South Asians in UK, vitamin D deficiency may contribute. Rickets of prematurity is treated by increased calcium and phosphate intake and vitamin D supplementation. Treatment with the active metabolite, calcitriol, may be necessary.

Rickets during childhood

The major cause of rickets in this period is nutritional vitamin D deficiency (see Chapter 9). Children may present with tetany and convulsions due to hypocalcaemia and older children may present with characteristic waddling gait.

Inherited forms of rickets are less common and these are associated with hypophosphataemia where there is a renal tubular defect and/or defect in vitamin D metabolism. The X-linked hypophosphataemic rickets is a rare disorder occurring in 1:20,000 live births associated with hyperphosphaturia. Other forms include vitamin D dependent rickets Type I, which is due to a decrease or absence of renal 1α-hydroxylase enzyme and vitamin D-dependent rickets Type II, which is due to vitamin D receptor abnormalities.

Neonatal hyercalcaemia

Hypercalcaemia is uncommon in infants and neonates. In the neonate, hypercalcaemia is most often associated with hypophosphatemia due to inadequate dietary intake of phosphate especially in premature infant. Hypophosphataemia stimulates the synthesis of calcitriol, which increases calcium absorption. This should be treated with phosphate and calcium supplements as phosphate alone may cause rapid uptake into the bone and hypocalcaemia may develop. Other causes of hypercalcaemia in infant are similar to those in adults. Neonatal hyperparathyroidism is a rare disorder due to a mutation in the calcium-sensing receptor.

Acid-Based Disorders in the Neonate

Acidosis may occur either due to respiratory or metabolic causes. Respiratory causes include hypoxia, which may occur during birth due to CNS depression and respiratory distress syndrome (see Table 28.3). Metabolic acidosis may be due to accumulation of lactic acid as a result of hypoxia. Unexplained metabolic acidosis should lead to investigation of inherited metabolic disorders such as amino acid disorders and organic acidaemias.

Table 28.3   Causes of acidosis in the neonate

Respiratory acidosis
- Asphyxia
- Respiratory depression
  — CNS abnormalities
  — CNS depressant drugs
- Mechanical abnormality
  — e.g. respiratory distress syndrome
  — Diaphragmatic hernia

Metabolic acidosis
- Renal failure
- Lactic acidosis
  — Tissue hypoxia
  — Inborn errors of metabolism — e.g. glucose-6-phosphatase deficiency
  — Sepsis
- Inherited metabolic disorders
  — Organic acidaemias

Respiratory Distress Syndrome

Respiratory distress may be caused by a variety of conditions. The most common cause in premature infants is hyaline membrane disease due to reduced surfactant synthesis. Surfactant, a mixture of lecithin, phosphatidyl glycerol and phosphatidyl inositol helps to maintain patency of the alveoli by reducing the surface tension of the alveolar wall. The synthesis of surfactant begins by the 20th week and increases slowly up to the 34th week and then rapidly with maturation of the alveolar cells. The rate of synthesis of surfactants is sensitive to cold, hypoxia and acidosis. The incidence of hyaline membrane disease is inversely related to the gestational age of the child; up to 50% of babies born at 26–28 weeks have this disease. Infants with hyaline membrane disease develop hypoxia and respiratory acidosis within 4 hours of birth and this is later complicated by lactic acidosis. This disease has become less common due to the use of synthetic and natural surfactants immediately after birth and the use of antenatal steroids. Other causes of respiratory distress during the neonatal period include pneumonia, meconium aspiration and non-pulmonary causes such as patent ductus arteriosis.

Disorders of Growth

Failure to thrive

This is defined as a significant interruption in the expected growth during early childhood. Causes of failure to thrive are usually classified as organic and non-organic. However, often there is contribution from both organic and non-organic factors in a given child. Non-organic causes, such as psychosocial deprivation, accounts for a large proportion of cases of failure to thrive. Some causes of failure to thrive are listed in Table 28.4.

Short stature

Short stature is usually defined as height, which is greater than 2 SD below the mean or 2.5 percentile for age, and sex-matched normal children. Causes of short stature and retardation in growth are listed

Table 28.4 Causes of failure to thrive

Non-organic causes
- Psychosocial factors
- Prenatal factors — maternal malnutrition
- Postnatal — maternal neglect, poverty

Organic causes
- Prematurity
- Inadequate energy intake
  - Mechanical problem
  - Chronic infections — poor appetite
- Inadequate use of ingested energy
  - Vomiting
  - Malabsorption syndromes
- Increased metabolic demand
  - Chronic infections, malignancy

Malnutrition
- Inadequate food — protein energy malnutrition
- Malabsorption

Inherited metabolic disorders

Chronic systemic disease
- Renal failure
- Cardiac disease
- Hepatic disease
- Pulmonary disease

Endocrine
- Hypothyroidism
- Hypopituitarism

in Table 28.5. Of the many causes of short stature, growth hormone deficiency is only a rare cause of short stature. Growth hormone deficiency rarely presents before the third month of life. In older infants and children, it is recognised by reduction in growth velocity. Growth hormone deficiency can be due to pituitary disorders or hypothalamic GHRH deficiency.

Growth hormone deficiency can be diagnosed by measuring growth hormone concentrations during early sleep, following exercise or during provocation tests such as glucagon stimulation, arginine infusion or insulin tolerance test.

Table 28.5   Causes of short stature and growth retardation

Constitutional delay
Familial short stature
Intrauterine growth retardation
Emotional deprivation
Nutritional
- Malnutrition
- Malabsorption — e.g. coeliac disease

Systemic disease
- Renal failure
- Cardiac disease

Endocrine
- GH deficiency
- Hypothyroidism
- Cushing's syndrome
- Congenital adrenal hyperplasia

Chromosomal abnormalities
- E.g. Turner's syndrome

Skeletal disorders
- E.g. achondroplasia

Genetic causes
- Down's syndrome
- Turner's syndrome

Neonatal Hypoglycaemia

The foetus is entirely dependent on the mother for glucose. At birth, when the glucose supply is abruptly withdrawn, blood glucose decreases to approximately 2.8 mmol/L at 2 hours and then slowly rises and stabilises by 72 hours. This transient decrease is exaggerated in prematurity, asphyxia and sepsis. During this period, normal infants are able to mobilise ketones, which can be utilised by the neonatal brain. If the decrease in blood glucose is persistent or accompanied by symptoms, it should be further investigated. Neuroglycopenic symptoms of hypoglycaemia in a full-term baby include weakness, lethargy, confusion, in coordination, sweating, jitteriness, apnoea, cyanosis, hypotonia and convulsions. Premature babies are deficient in glycogen stores and therefore are more likely to become hypoglycaemic. This is more likely to be harmful as they are less able to mobilise ketones. However, premature babies are less likely to show symptoms because of their immature nervous system. Furthermore, hypoglycaemia induced blood flow increase may cause intraventricular haemorrhage. Therefore, it is important to diagnose hypoglycaemia early in at-risk babies. Repeated hypoglycaemic episodes increase the risk of permanent neurological damage. There is no agreement as to the definition of hypoglycaemia in neonates. It is generally accepted that a value less than 2.2 mmol/L between 3 and 24 hours and a value < 2.5 mmol/L after 24 hours should be considered as hypoglycaemia. However, it must be stressed that the value at which treatment is necessary will depend on the individual circumstances and individual neonate.

Causes of neonatal hypoglycaemia are listed in Table 28.6. Babies born to diabetic mothers are at risk of hypoglycaemia. However, these

Table 28.6   Causes of neonatal hypoglycaemia

Factors predisposing to hypoglycaemia
- Prematurity
- Sepsis
- Maternal diabetes
- Perinatal asphyxia/hypoxia
- Inadequate nutrition

Endocrine
- Hyperinsulinaemia
- Cortisol deficiency
- Growth hormone deficiency
- Hypopituitarism

Inborn errors of metabolism
- Carbohydrate metabolism
  — Glycogen storage disease
  — Galactosaemia
  — Hereditary fructose intolerance
- Fat metabolism
  — Fatty acid oxidation defects
- Amino acid metabolism
  — Tyrosinaemia
  — Maple syrup urine disease

babies are usually identified early and hypoglycaemia is averted by appropriate glucose infusion. Birth asphyxia and postnatal hypoxia are commonly associated with hypoglycaemia especially in very low birth weight babies. Glucose utilisation is increased because of increased metabolic rate and increased anaerobic metabolism. Hypothermia, sepsis and inadequate nutrition in premature babies are other predisposing causes of hypoglycaemia.

Hypoglycaemia associated with inappropriately high circulating insulin concentration, which used to be called nesidioblastosis, should be suspected if the glucose requirement is > 10 mg/kg/min. The relationship between glucose and insulin in infants, especially low-birth weight infants, is not well defined. As their insulin receptors are immature, they require higher insulin concentration to activate the receptor. Hyperinsulinaemic hypoglycaemia is due to mutations in the genes for β cell ATP-dependent K channel, sulphonylurea receptor 1 or hepatocyte nuclear factor-4α (HNF4A). Hyperinsulinaemic hypoglycaemia could be associated with Beckwith-Weidman syndrome, a condition of unknown aetiology associated with hyperplasia of many organs and microcephaly. Exogenous administration of insulin or sulphonylurea (due to Munchausen syndrome by proxy) is a rare cause of hypoglycaemia associated with hyperinsulinaemia. Short-chain 3-hydroxy acyl-CoA dehydrogenase deficiency can also lead to hyerinsulinism. Cortisol deficiency leading to hypoglycaemia may occur in abnormalities of the development of adrenal gland, congenital adrenal hyperplasia, hypopituitarism and isolated ACTH deficiency. One of the diagnostic pointers to cortisol deficiency is the presence of conjugated hyperbilirubinaemia. Inborn errors of metabolism leading to hypoglycaemia are discussed in Chapter 26.

Diagnosis

Urgent investigation is needed in infants who have repeated hypoglycaemia or who require excessive amounts of intravenous glucose to maintain blood glucose concentration. Blood samples should be taken at the time of hypoglycaemia for plasma free fatty acids, β-hydroxybutyrate, insulin, C-peptide, cortisol, growth hormone, amino acids and urine should be collected for organic acids and amino acids. Low serum free fatty acids and ketones suggest hyper-insulinaemia, whereas high free fatty acids with low ketones suggest a fatty acid metabolism defect. If fatty acids and ketones are high, further investigations for a defect in carbohydrate and amino acid metabolism are required. If the liver is enlarged, investigations for galactosaemia or other glycogen storage diseases should be under-taken. A flow chart for the investigation of neonatal hypoglycaemia is shown in Figure 28.1.

Hypoglycaemia in Infancy and Childhood

Causes of hypoglycaemia during infancy and childhood are endocrine diseases and inborn errors of metabolism (Table 28.6). Idiopathic ketotic hypoglycaemia is a common cause of hypoglycaemia beyond infancy occurring between 18 months and 7 years. This is associated with intercurrent illness presenting with recurrent episodes of hypo-glycaemia. The exact aetiology is uncertain. Hypoglycaemia responds to treatment and the condition improves with age and is rarely seen after puberty. This is a diagnosis of exclusion.

Case 28.2

A 3-year-old boy was admitted with convulsions. His mother reported that prior to his fits he was off food and was suffering from an upper res-piratory tract infection for 3 or 4 days. Initial investigations showed:

Serum		Reference Range
Sodium (mmol/L)	138	135–145
Potassium (mmol/L)	4	3.5–5.0
Bicarbonate (mmol/L)	18	23–32
Chloride (mmol/L)	102	90–108
Urea (mmol/L)	8.7	3.5–7.2
Creatinine (μmol/L)	70	40–60
Glucose (mmol/L)	< 2.0	

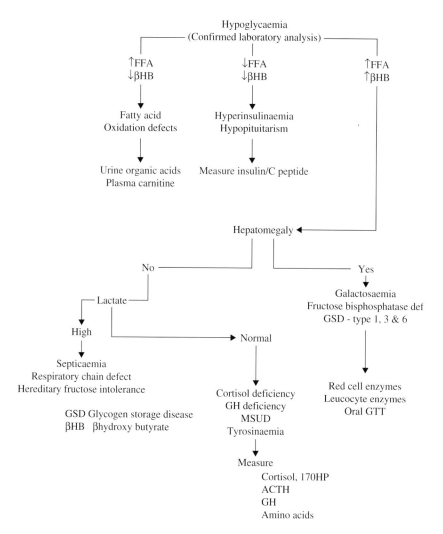

Figure 28.1   Flow chart for investigation of neonatal hypoglycaemia.

Blood was taken for further investigation and he was treated with intravenous glucose. He rapidly responded and his blood glucose returned to normal. Further investigations of the sample at the time of hypoglycaemia showed:

β-Hydroxybutrate   5.8 mmol/L (< 0.3)
Plasma insulin   < 5.0 mIU/L (fasting < 10 mIU/L)

> *His plasma insulin was appropriate for his blood glucose concentration and the ketonaemia was due to a normal response for fasting. His organic acids were normal. He was diagnosed as having idiopathic ketotic hypoglycaemia.*

Clinical Biochemistry of Old Age

Ageing is characterised by the loss of adaptation of an individual. Homeostatic mechanisms become less sensitive, slower, less accurate and less well sustained with age-associated loss of adaptability. Mortality rates increase with age and this is the biological marker of senescence. Mortality rates fall from infancy throughout childhood to around 12 or 13 years of age and then start to rise again. If deaths from accidents and violence are excluded, the rise in age-specific mortality rate is continuous throughout adult life. Thus, there is no biological justification in separating old age or the elderly from the rest of the population. Ageing is a consequence of interaction between environmental and lifestyle factors and intrinsic or genetic factors.

Diseases in old age are characterised by multiple pathology, non-specific or cryptic presentation, rapid deterioration if untreated, and higher incidence of secondary complications. Reduced adaptability of old age causes high incidence of secondary complications both of the disease and the treatment. Elderly people also require rehabilitation as they have less functional reserve to convalesce spontaneously. Environmental factors play an important role in the genesis and management of illness in the elderly due to loss of adaptability.

Age-Related Physiological Changes

Lean body mass declines with age and there is an associated fall in oxygen consumption. However, body weight remains constant owing to an increase in body fat. The age-related changes in body composition have implications for pharmacokinetics, nutrition, and function of various organs. Blood volume decreases with age and there is a slight decrease in haemoglobin concentration. Some of the

Table 28.7   Biochemical tests for which reference ranges change with age

Albumin	Decreases
Globulin	Increases
Potassium	Increases
Urea	Increases
Creatinine	Increases
Uric acid	Increases
ALP	Increases

common biochemical measurements are age-related, whereas others are independent of age (Table 28.7). Glomerular filtration rate falls with age by approximately 1% per year over the age of 40. This is greater than the change in lean body mass and therefore serum urea and creatinine concentrations tend to rise. Tubular reabsorptive and secretory capacities are also reduced. Response to acid load is impaired. Response to antidiuretic hormone is reduced and water conservation is thus less efficient. This, together with impaired thirst, leads to increase risk of water depletion. There is age-related decreases in hepatic mass, which may lead to reduction in hepatic metabolism of some drugs. There is reduced intestinal calcium absorption due to decreased sensitivity to vitamin D. In the skeleton, major changes take place in old age and this is discussed in Chapter 9.

Total body potassium is low in old age, partly due to the reduction in lean body mass and partly due to reduced potassium intake. Reduced body potassium becomes important when there are additional losses such as during diuretic therapy. Thus, elderly people are more likely to become hypokalaemic during diuretic therapy and usually require potassium supplements.

Because of an inability to conserve sodium and water with increasing age, electrolyte disturbances are common in the elderly. In the elderly population, the causes of hyponatraemia are similar to that in younger age group except that it is more common. Diuretics are a common cause of hyponatraemia in the elderly. Hypernatraemia is also a common finding in the elderly due to reduced thirst and the inability to conserve water.

Disorders of the thyroid gland are a very common problem in the elderly. The prevalence of hypothyroidism in individuals who are over 60 years old has been estimated to be 4.4% or higher. Nutritional disorders are common in the elderly due to a combination of factors such as reduced income, inability to go out and increased incidence of diseases. Incidence of type 2 diabetes mellitus increases with age and so does cardiovascular disease. An increasing problem in the elderly is dementia which can be due to many causes, one of the most common being Alzheimer's disease.

Further Reading

1. Dennery PA, Seidman DS, Stevenson DK. Neonatal hyperbilirubinaemia. *N Engl J Med* 2001; 344:581–590.
2. Green A, Morgan I. *Neonatology and Clinical Biochemistry* 2003. ACB Venture Publications: London.
3. Lteif AN, Schwenk WF. Hypoglycaemia in infants and children. *Endocrinol Metab Clin North Am* 1999; 28:619–646.
4. Straussman S, Levitsky LL. Neonatal hypoglycaemia. *Curr Opin Endocrinol Diabetes Obes* 2010; 17:20–24.
5. Wang X, Chowdhury JJ, Chowdhury NR. Bilirubin metabolism: Applied physiology. Curr Pediatr 2006; 16:70–74.

Summary/Key Points

1. Body composition and organ function is different at the extremes of ages. Reference ranges derived from healthy adults may not be applicable in the very young and the elderly. Laboratory results should be interpreted in relation to age-specific reference ranges.
2. Serum unconjugated bilirubin rises after birth in all infants due to increased production of bilirubin and due to the immaturity of the liver enzymes to handle this. This physiological jaundice is exaggerated in premature infants and or if there is sepsis, dehydration, hypoxia and poor nutrition. Pathological causes of unconjugated

hyperbilirubinaemia include haemolytic diseases, hypothyroid or rarely inherited disorders of bilirubin metabolism.

3. Conjugated hyperbilirubinaemia is caused by infection, inherited metabolic disorders like galactosaemia, and bilirary tract atresia.

4. Unconjugated hyperbilirubinaemia can lead to kernicturus, a neurological damage due to 'free unconjugated bilirubin' reaching the brain. Jaundice is initially treated by phototherapy under UV light. If this is inadequate or fails, exchange transfusion may be necessary to prevent kernicterus.

5. Blood glucose concentration falls at birth and this can be exaggerated in prematurity, in babies born to diabetic mothers, infection, hypoxia or those with inadequate nutrition. There is no agreed cut off value of blood glucose for the diagnosis of hypoglycaemia in neonates.

6. Other causes of hypoglycaemia include hyperinsulinaemia, endocrine disorders (e.g. adrenal insufficiency) and inherited metabolic disorders of fatty acids, and amino acid metabolism.

chapter 29

Cerebrospinal Fluid and the Nervous System

Introduction

Cerebrospinal fluid (CSF) is a clear fluid that occupies the ventricles, the central canal of the spinal cord and the subarachnoid space (the space between the arachnoid mater and the pia mater, the middle and the third layer of meninges (brain covering)). It is produced by the choroid plexus (about 70%), which is found in cerebral ventricles, and by the blood vessels in the brain and ventricular walls (30%). CSF, which is formed in the lateral ventricles, flows through the third and fourth ventricles and then into the subarachnoid space. CSF is reabsorbed by the arachnoid villi into the circulation. There is some evidence that CSF may be reabsorbed into the lymphatic channels especially in the neonate. Circulation of CSF takes approximately 1 hour. CSF has several functions including cushioning the brain from shock, providing nutrients, removing waste metabolites, defence against pathogens and an important role in the homeostasis and metabolism of the CNS. In adults, the total volume of CSF is about 150 mL and is produced at the rate of 500–600 mL per day. The turnover is therefore rapid, about four times a day. CSF is formed from the plasma by filtration, diffusion, facilitated diffusion and active transport.

CSF is usually sampled from the subarachnoid sac surrounding the cauda equine, below the termination of the spinal cord by lumbar puncture. Occasionally, it is sampled by cisternal and ventricular puncture.

CSF is normally clear and colourless. The electrolyte composition is similar to that of plasma except that protein concentration is only about 0.1–0.4 g/L (Table 29.1). Proteins enter the CSF from plasma by pinocytosis or by specific carriers. The small difference in electrolyte composition between CSF and plasma is probably due to the difference in protein concentration between the two fluids. The composition of the CSF varies slightly, depending on the site from which it is sampled. There is a gradual increase in concentration of components from the ventricles down to the lumbar region.

Turbidity in CSF is due to the presence of blood, proteins, bacteria or leucocytes. Yellow colouration (xanthochromia) is an indication of previous haemorrhage, hyperbilirubinaemia or high protein content (see later).

Glucose

Glucose enters the CSF by a specific membrane carrier transport system. The concentration of glucose in CSF is normally about 60–80%

Table 29.1   Composition of CSF compared to serum

	CSF	Serum
Sodium (mmol/L)	138	140
Potassium (mmol/L)	2.8	4.0
Calcium (mmol/L)	1.1	2.4
Chloride (mmol/L)	119	103
Bicarbonate (mmol/L)	22	26
Glucose (mmol/L)	3.3	5.0
Lactate (mmol/L)	1.6	1.0
Total protein (g/L)	0.350	70
Creatinine (μmol/L)	105	88
β_2 Microglobulin (mg/L)	1.1	1.9
Neuron-specific enolase (μg/L)	0.5	—
S-100 protein (pmol/L)	406	—
Myelin basic protein (μg/L)	< 0.5	—
Lactate dehydrogenase (U/L)	10	150

Adapted from Watson MA and Scott MG, *Clin Chem* 1995; 41:343–360.

of that in blood. During the first 6 months of life, CSF glucose approaches that of serum. CSF glucose does not rise above 17 mmol/L probably due to saturation of the transport system. In severe hyperglycaemic states, the ratio of CSF to serum glucose is lower than normal. As it takes about 4 hours for CSF glucose to reach equilibrium with blood glucose, CSF glucose will lag behind blood glucose when the latter changes rapidly.

As the CSF glucose is a proportion of blood glucose, interpretation of CSF glucose requires the simultaneous measurement of blood glucose. A low CSF glucose relative to blood glucose (a ratio less than 0.6) is seen in bacterial meningitis, tuberculous meningitis and hypoxia whereas it is normal in viral meningitis. The exact mechanism of low CSF glucose in these situations is not entirely clear, but it is thought to be due to increased consumption by white cells, bacteria and other cells and impaired glucose transport system into the CSF. However, the diagnostic value is rather low. CSF glucose may be low in some cases of non-bacterial meningitis and a normal CSF glucose does not exclude bacterial meningitis.

Lactate

CSF lactate represents metabolism within the CNS and it is not related to blood lactate. Increased CSF lactate may be seen in hypoxia, in some meningeal diseases especially bacterial meningitis but not in viral meningitis. In inherited disorders of pyruvate metabolism, CSF lactate and pyruvate are elevated and are of some diagnostics value.

Proteins

CSF protein concentration increases from about 100 (50–150) mg/L in the ventricular fluid to about 300 (150–450) mg/L in the lumbar fluid. In neonates, the concentration may be as high as 1300 mg/L due to the increased vascular permeability. About 80% of the CSF protein originates from plasma by passive diffusion along the blood–brain barrier and 20% is synthesised within the CNS. An exception is transferrin, which may be transferred by a receptor-mediated process as well as synthesised by choroid plexus. Factors influencing the diffusion of proteins into CSF are the permeability of the blood–brain barrier and the concentration in plasma. Most proteins in plasma could be found in CSF. In healthy subjects, albumin forms 50–60% of proteins in CSF. Permeability of the CSF barrier tends to increase with age and hence CSF protein concentration is higher in the elderly.

About 25% of transferrin in the CSF is in the asialylated form. In the circulation, desialylated transferrin is rapidly removed by asialoreceptors in the reticuloendothelial system, and therefore not normally detected in plasma. As no such mechanism exists in the CNS, desialylated transferrin accumulates in CSF. Presence of asialylated transferrin is therefore a marker of CSF and is a useful test in detecting CSF rhinorrhoea. Prealbumin is another protein present in concentrations 5–10 times greater than expected from its plasma concentration due to synthesis by the choroid plexus.

CSF protein concentration can be increased due to one of the following mechanisms:

1. Increased permeability: This may be due to high intracranial pressure, brain tumour, intracerebral haemorrhage, traumatic injury, infection such as meningitis or encephalitis or autoimmune disorders such as Guillain-Barré syndrome. CSF protein concentration is often measured as an aid to the diagnosis of meningitis when it can be very high (up to 5 g/L). However, a normal CSF protein does not exclude meningitis.

2. Reduced flow of spinal CSF: Very high protein concentration in lumbar CSF is seen when there is obstruction to the CSF circulation above the site of sampling. Increased equilibration with plasma is the likely explanation.

3. Increased intrathecal synthesis of immunoglobulins: Conditions associated with increased immunoglobulin synthesis includes multiple sclerosis, viral infections (subacute sclerosing panencephalitis, meningitis, and encephalitis), neurosyphilis, polyneuritis, bacterial meningitis, SLE and sarcoidosis. In these disorders, lymphocytes that infiltrate the demyelinating site synthesise IgG. As there are only a small number of B-cell clones in the CSF a small numbers

of discrete oligoclonal bands is seen on electrophoresis of CSF. Increased synthesis of immunoglobulins can be demonstrated by measuring the CSF/serum ratio of albumin and IgG. Increased permeability will result in increase in the ratio of both proteins while local synthesis will increase the CSF/serum, ratio of IgG, but not that of albumin. However, this test is no longer in use and instead demonstration of oligocolonal bands in CSF is used.

Oligoclonal banding is defined as the demonstration of two or more bands in CSF without corresponding bands in serum. The presence of identical bands in CSF and serum indicates a systemic process such as Guillain-Barré Syndrome and HIV infection. Oligoclonal bands are seen in 80–90% of cases of multiple sclerosis. Oligoclonal bands can be detected in many diseases (Table 29.2).

Other proteins

Myelin basic protein (MBP) is an 18.5-kDa protein present in myelin of the central nervous system. This is released into CSF during inflammation or cell destruction. In active demyelinating conditions, CSF

Table 29.2 Some conditions associated with CSF oligoclonal bands

Multiple sclerosis
Acute disseminated encephalomyelitis

HIV related disorders
- Neurosyphilis
- HTLV-1–associated myelopathy

Lyme disease
Subacute sclerosing panencephalitis
Neurosarcoidosis
Sjogren's syndrome
Systemic lupus erythematosis (SLE)
Chronic focal encephalitis
Stiff-person syndrome
Carcinomatous meningitis
Creutzfeldt-Jakob disease (CJD)
Vasricelle zoster encephalitis

MBP is increased and is useful in differentiating active from inactive disease.

CRP concentrations in CSF parallel those in plasma and high CSF concentration of CRP is found in bacterial meningitis due to increased plasma concentration and permeability.

A phosphoprotein associated with microtubules is the tau protein. CSF concentration of this protein is increased in a variety of degenerative disorders of the nervous system. In Alzheimer's disease, the tau protein undergoes anomalous phosphorylation and forms paired helical filament (PHF) structures. Increased PHF-tau proteins are consistently found in Alzheimer's disease. This protein is also increased in Creutzfeldt-Jakob disease.

$A\beta_{42}$ is a fragment of the amyloid protein β produced by proteolytic cleavage. $A\beta_{42}$ is present in all cells of the CNS. CSF concentration of $A\beta_{42}$ is increased in Alzheimer's disease and together with tau protein has diagnostic value in the diagnosis of this disease.

S100 is a calcium binding protein synthesised by astroglial cells of the CNS. Measurement of S100 in CSF is useful in brain hypoxia, Creutzfeldt–Jakob disease and dementia. Other brain-specific proteins include 14-3-3γ.

Biochemical Analysis of CSF in CNS Disorders

Acute Infections

Acute bacterial meningitis is a life-threatening disorder requiring urgent diagnosis and treatment. A common clinical problem is to differentiate it from viral meningitis. Important investigations in this regard are the analysis of CSF for cells and bacteria by microscopy and culture. Biochemical tests aid in the diagnosis but are not a substitute for microbiological examination. The CSF protein is higher in bacterial meningitis than in viral meningitis and the glucose is less than 40% of that in plasma. However, these tests are insensitive and non-specific. CSF lactate, which is independent of blood lactate, is increased in bacterial meningitis and is directly related to the number of leucocytes in CSF. A CSF lactate concentration greater than 3.5 mmol/L appears to be a sensitive test for bacterial meningitis but may remain elevated after starting successful treatment. CSF concentration of CRP of 100 mg/L or greater is highly suggestive of bacterial meningitis.

Case 29.1

An 18-year-old student presented to the A&E department with severe headache, vomiting and decreased energy. She was pyrexial and examination showed signs of meningismus (triad of neck stiffness, photophobia (intolerance of bright light) and headache). A lumbar puncture was done and sent for analysis:

CSF glucose	1.8 mmol/L
Plasma glucose	5.2 mmol/L
CSF protein	1.1 g/L (< 0.3g/L)
Microscopy:	Neutrophils +++

Gram staining showed gram negative bacteria and later culture confirmed infection with meningococcus. She was treated with antibiotics and made a full recovery.

In this patient, a low CSF glucose and a high CSF protein were highly suggestive of bacterial meningitis. However, diagnosis depends on the demonstration of bacteria.

Subarachnoid Haemorrhage

When bleeding from a vessel on the surface of the brain occurs (e.g. from an aneurysm), blood is found between the surface of the brain and the pia-arachnoid membrane and is called subarachnoid haemorrhage (SAH). Blood being an 'irritant' causes chemical meningitis. SAH causes severe headaches. Diagnosis of SAH is important to prevent re-bleeding and the associated morbidity and mortality. The diagnosis is mainly based on demonstration of blood in the subarachnoid space by computerised tomography (CT), which has high sensitivity, especially within the first 24 hours. In patients presenting late demonstration of blood, haemoglobin or its breakdown products in CSF is helpful. Red blood cells can be demonstrated in CSF up to 7–10 days after the bleed. Oxyhaemoglobin released from red cells can be demonstrated by spectrophotometry in CSF from 4–10 hours. Bilirubin will form from haemoglobin after a bleed and can be demonstrated from about 24 hours after the bleed. Yellowish condition of CSF is described as xanthochromia. When the bilirubin concentration is low, xanthochromia may not be visible to the naked eye but can be demonstrated by spectrophotometry as a peak at 450–460 nm. The presence of red cells, oxyhaemoglobin and/or bilirubin in CSF is indicative of SAH. Sometimes methaemoglobin can be found after SAH. The presence of protein in the CSF at concentration > 1.5 g/L will also give xanthochromia. Presence of red blood cells due to trauma during the lumbar puncture can be excluded by collecting three successive tubes of CSF: in traumatic punctures, the blood-stain will be lower in the third tube. Demonstration of bilirubin or methaemoglobin is strongly suggestive of SAH, but the absence of these does not exclude it.

Damage to the Central Nervous System

CSF analysis is sometimes helpful in damage to CNS, e.g. perinatal hypoxia, after cardiac resuscitation and cerebral ischaemia. It is particularly useful to predict recovery and to assess the extent of damage. Biochemical tests used in these situations include CSF lactate, LD, CK isoenzyme BB, neurone-specific enolase, S-100 protein and myelin basic protein. CSF lactate and LD are non-specific and not very reliable. CKBB in CSF is specific but its activity rapidly decreases. Other markers are specific for neuronal damage but are non-specific for the type of pathology.

Multiple sclerosis

Multiple sclerosis (MS) is a debilitating neurological disease affecting young adults, especially females. In this disease, the myelin sheaths surrounding the axons are damaged leading to demyelination and scarring. MS is thought to be an organ-specific autoimmune disorder. Most consistent laboratory feature of MS is the demonstration of local synthesis of IgG within the CNS. This can be demonstrated as an oligoclonal band. The gamma globulin pattern in normal CSF is the same as that of serum. Examination of CSF for oligoclonal band should always be done in parallel with serum analysis. The presence of oligoclonal band in CSF that is absent in serum is an indication of the intrathecal synthesis of IgG. Oligoclonal bands are present in about 80–90% of cases of MS.

CSF rhinorrhoea

Following trauma, surgery or infection, CSF may leak into the nose and present as a nasal discharge. This is a potentially life-threatening condition as infection may rapidly develop. Presence of desialylated transferrin in such watery discharge indicates the presence of CSF.

Other neurological disorders

CSF examination is also of some value in the diagnosis of Alzheimer's disease, the most common form of dementia in the elderly. The presence of soluble $A\beta_{42}$ protein and tau protein are indicative of Alzheimer's disease.

Further Reading

1. Watson MA, Scott MG. Clinical utility of biochemical analysis of cerebrospinal fluid. *Clin Chem* 1995; 41:343–360.

Summary/Key points

1. The choroid plexus produces CSF. The electrolyte composition is similar to that of plasma except the protein concentration is lot less. CSF glucose is about 60% of that of plasma.

2. In bacterial meningitis, CSF:plasma glucose ratio is 0.4 or lower and the protein concentration is higher than normal. However, these are not diagnostic. Diagnosis depends largely on microbiological examination.

3. Immunoglobulins may be synthesized within the brain and can be detected as oligoclonal bands. The presence of oligoclonal bands is suggestive of multiple sclerosis.

4. Measurement of CSF tau protein and $A\beta_{42}$ may be helpful in the diagnosis of Alzheimer's disease.

5. Detection of desialylated transferrin is useful in the diagnosis of CSF rhinorrhoea.

Self-Assessment Questions

True or False Questions

Please answer **true** or **false** for each of the options A–E.

1. Haemolysis of the blood sample can account for elevated plasma activity of:
    A. aspartate transaminase.
    B. lactate dehydrogenase.
    C. alanine transaminase.
    D. alkaline phosphatase.
    E. creatine kinase.

2. If plasma is left in contact with cells overnight after proper specimen collection, the plasma:
    A. glucose concentration will rise.
    B. potassium concentration will rise.
    C. bicarbonate concentration will fall.
    D. phosphate concentration will fall.
    E. lactate dehydrogenase activity will rise.

3. In a population of 1000, which of the following statements about a diagnostic test for diabetes mellitus, which has a sensitivity of 90% and a specificity of 80% is/are true?
    A. If the prevalence of diabetes mellitus is 10%, 90 subjects will show a false positive result.
    B. If the prevalence of diabetes mellitus is 60%, the predictive value of a positive result is 75%.
    C. If the prevalence of diabetes mellitus is 10%, the predictive value of a positive result is 16%.

D.   The false positive rate in non-diabetic subjects will be 20%.
E.   A negative result will be seen in 20% of diabetics.

4.   Indicate which of the following statements about factors affecting test results are true or false:

A.   Plasma total protein concentration will be higher if the sample is taken in the recumbent posture compared to that taken while standing.

B.   Serum creatine kinase values are higher in healthy black subjects compared to Caucasians.

C.   Serum creatinine values fall with age.

D.   Plasma sodium concentration in a healthy subject varies by 5 mmol/L between days.

E.   Serum sodium results from a patient taken on 2 days were 130 and 135 mmol/L. If the analytical SD was 1 mmol/L, this difference is likely to be due to analytical variation.

5.   If in a healthy population of 1000 subjects the plasma sodium concentration has a mean value of 140 mmol/L and a standard deviation of 3 mmol/L, then the plasma sodium will:

A.   be less than 134 mmol/L in approximately 25 of the subjects.

B.   be less than 131 mmol/L in approximately 10 of the subjects.

C.   be greater than 143 mmol/L in approximately 50 of the subjects.

D.   be between 137 and 143 mmol/L in approximately 320 of the subjects.

E.   have a conventional reference interval of 134–146 mmol/L.

6.   Indicate which of the following statements are true or false:

A.   In a healthy adult, 50% of total body water is extracellular.

B.   In infants, 60% of total body water is extracellular.

C.   In a healthy, adult minimum water requirement is < 250 ml/d.

D.   Plasma proteins contribute significantly to plasma osmolality.

E.   Water loss via the skin is an important mechanism of regulating water balance.

7.   A disproportionate increase in the serum urea concentration relative to that of creatinine would be expected in patients:

A.   with renal failure immediately after haemodialysis.

B.   who have sustained gastrointestinal haemorrhage.

C.   with renal disease who are on a low protein diet.

    D.   who are volume depleted.

    E.   with fulminate hepatic failure complicated by renal failure.

8.   Recognised causes of hyponatraemia include:

    A.   treatment with diuretics.

    B.   Addison's disease.

    C.   treatment with carbamazepine.

    D.   psychogenic polydipsia.

    E.   primary hyperaldosteronism.

9.   Which of the following causes should be considered in a patient presenting with polyuria:

    A.   Diabetes mellitus.

    B.   Hypokalaemia.

    C.   Psychogenic polydipsia.

    D.   Hypocalcaemia.

    E.   Diabetes insipidus.

10. Concentration of urea in plasma is often increased:

    A.   after a high protein meal.

    B.   after losing 300 ml of blood by postpartum haemorrhage.

    C.   after severe trauma.

    D.   in patients with fluid depletion.

    E.   in mild congestive cardiac failure.

11. A raised plasma urea concentration is seen in:

    A.   patients with gastrointestinal bleeding.

    B.   subjects on high protein diet.

    C.   patients with ECF volume depletion.

    D.   pregnancy.

    E.   steatorrhoea.

12. Water depletion:

    A.   is common in the elderly.

    B.   leads to increased haematocrit.

    C.   is seen in head injury patients.

    D.   causes an increase in urine sodium excretion.

    E.   when severe should be corrected slowly.

13. In untreated diabetes insipidus:

    A.   failure to respond to exogenous arginine–vasopressin suggests that the condition is nephrogenic.

     B.   hyponatraemia may occur.
     C.   proteinuria is usual.
     D.   the plasma osmolality is low.
     E.   the urine osmolality is low.

14. The loss of large amounts of fluid from a small intestinal fistulae may cause:
     A.   raised plasma urea concentration.
     B.   raised plasma sodium concentration.
     C.   reduced skin turgor.
     D.   tachycardia.
     E.   postural hypotension.

15. In a patient presenting with persistent vomiting due to pyloric stenosis:
     A.   plasma total calcium will be low.
     B.   plasma bicarbonate will be low.
     C.   plasma potassium is likely to be high.
     D.   urine sodium concentration is likely to be < 20 mmol/L.
     E.   secondary hyperaldosteronism is likely.

16. Which of the following statements are true:
     A.   The amount of total body water as percentage of body weight is higher in males than in females.
     B.   A 70-kg man has approximately 14 L of ECF.
     C.   The volume of interstitial fluid in a 70-kg man is approximately 12 L.
     D.   More than 90% of albumin in the body is in the vascular compartment.
     E.   In the neonate, 60% of total body water is intracellular.

17. Causes of water depletion include:
     A.   a high salt intake in healthy adults.
     B.   reduced intake of water in the elderly.
     C.   Addison's disease.
     D.   osmotic diuretics.
     E.   thiazide diuretics.

18. The syndrome of inappropriate secretion of antidiuretic hormone (SIADH):
     A.   may be associated with encephalitis.
     B.   is usually associated with elevated plasma urea.
     C.   is usually associated with peripheral oedema.

D.   is usually associated with abnormal renal and adrenal functions.

E.   can be caused by non-malignant disease of the lung.

19.   Which of the following statements is/are true about intravenous fluids:

A.   5% Dextrose solution has an osmolality of 200 mOsmol/kg.

B.   Hartman's solution contains sodium bicarbonate.

C.   100 ml of 20% dextrose contains 80 kilocalories.

D.   0.9% saline solution has 90 mmol/L of sodium.

E.   0.9% saline solution is an irritant to the veins because it is acidic.

20.   Loss of sodium and water from the ECF is likely to result in:

A.   increased plasma sodium concentration.

B.   increased pulse rate.

C.   postural hypotension.

D.   a high plasma urea/creatinine ratio.

E.   an increased haematocrit.

21.   The osmolality of a plasma sample was 325 mOsmol/kg. The concentrations of sodium, potassium, glucose and urea in this plasma were 136, 4.0, 3.0 and 5.0 mmol/L respectively. These results indicate the:

A.   patient may have taken alcohol.

B.   concentration of albumin is very low (< 20 g/L).

C.   patient is suffering from diabetic ketoacidosis.

D.   plasma lipids may be high.

E.   patient may have paraproteinaemia.

22.   Plasma sodium concentration:

A.   may be low in hyperlipdaemia.

B.   is high in patients with sodium retention.

C.   may be low in patients with congestive cardiac failure.

D.   if within the reference range excludes volume depletion.

E.   may be low in patients treated with a diuretic.

23.   In a patient with pancreatic fistula:

A.   anion gap will be normal.

B.   metabolic acidosis is likely.

C.   intravenous saline will correct the acid–base disturbance.

D.   plasma potassium is likely to be high.

E.   plasma urea is likely to be high.

24.   Hypokalaemia is a feature of:

A.   vomiting

B.   diuretic therapy

    C. hyperaldosteronism
    D. metabolic acidosis
    E. cholera

25. Consequences of hypokalaemia include:
    A. cardiac arrhythmia
    B. metabolic acidosis
    C. polyuria
    D. muscle weakness
    E. paralytic ileus

26. A low plasma potassium concentration is a recognised feature of:
    A. acute oliguric renal failure.
    B. hypopituitarism.
    C. treatment with spironolactone.
    D. treatment with thiazide diuretics.
    E. villous adenoma of the rectum.

27. Causes of hyperkalaemia include:
    A. Addison's disease.
    B. Bartter's syndrome.
    C. obstructive uropathy.
    D. magnesium depletion.
    E. syndrome of hyporeninaemic hypoaldosteronism.

28. Hypokalaemia:
    A. may be caused by carbenoxolone.
    B. may develop during the treatment of diabetic ketoacidosis.
    C. produces peaked T-waves on the ECG.
    D. is a common complication of diuretic therapy.
    E. may be treated with intravenous infusion of sodium bicarbonate.

29. Hypokalaemic alkalosis is associated with:
    A. treatment with diuretics.
    B. carcinoma of lung.
    C. essential hypertension.
    D. a low pCO_2.
    E. Fanconi syndrome.

30. Increased loss of potassium in the urine is frequently seen:
    A. in chronic renal failure
    B. following spironolactone administration

    C.   in metabolic alkalosis

    D.   in liquorice addiction

    E.   during the early stages of diabetic ketosis.

31.   The following may cause influx of potassium into cells:

    A.   intravenous infusion of insulin.

    B.   metabolic acidosis.

    C.   elevated plasma aldosterone levels.

    D.   intravenous infusion of sodium bicarbonate.

    E.   digoxin overdose.

32.   A metabolic acidosis with a normal anion gap is seen in:

    A.   salicylate poisoning.

    B.   methanol poisoning.

    C.   ureterosigmoid anastomosis.

    D.   renal tubular acidosis.

    E.   renal failure.

33.   Recognised causes of a raised plasma bicarbonate include:

    A.   chronic respiratory disease with carbon dioxide retention.

    B.   potassium depletion.

    C.   primary hyperaldosteronism.

    D.   prolonged nasogastric aspiration.

    E.   salicylate poisoning.

34.   Causes of metabolic acidosis include:

    A.   diabetes mellitus.

    B.   ethylene glycol poisoning.

    C.   prolonged nasogastric aspiration.

    D.   tissue hypoxia.

    E.   pancreatic fistula.

35.   In a patient with acidosis:

    A.   the urine may be alkaline.

    B.   a low arterial pCO_2 suggests a metabolic cause.

    C.   plasma chloride can be a useful investigation.

    D.   there is a shift of potassium into cells.

    E.   high anion gap excludes salicylate poisoning.

36.   In metabolic acidosis:

    A.   hypokalaemia is a usual finding.

    B.   urinary potassium excretion is usually increased.

    C.   the oxygen dissociation curve is shifted to the right.

    D.   erythrocyte 2,3-DPG concentration is increased.

    E.   the ionisation of plasma calcium is decreased.

37.   A low plasma bicarbonate concentration would be expected following:

    A.   ammonium chloride ingestion.

    B.   starvation for 24 hours.

    C.   hysterical hyperventilation.

    D.   salicylate overdose.

    E.   the onset of anuria.

38.   Proximal renal tubular acidosis:

    A.   is accompanied by hypokalaemia.

    B.   can be diagnosed by doing an acid load test.

    C.   may be associated with aminoaciduria.

    D.   may result from heavy metal poisoning.

    E.   is one of the causes of hyperchloraemic acidosis.

39.   Plasma chloride measurement is useful in patients:

    A.   with ureteric transplantation into the intestine.

    B.   with renal tubular acidosis.

    C.   given acetazolamide.

    D.   with pyloric stenosis.

    E.   with acute respiratory alkalosis.

40.   In respiratory failure due to chronic obstructive airways disease:

    A.   administration of (40%) oxygen may cause elevation of $PaCO_2$.

    B.   the $PaCO_2$ is always raised.

    C.   complete correction of hypoxia cannot be achieved with mechanical ventilation.

    D.   an elevated plasma bicarbonate concentration indicates a bad prognosis.

    E.   the cause of hypoxia is largely due to ventilation–perfusion (V/Q) imbalance.

41.   The Fanconi syndrome is characterized by:

    A.   aminoaciduria.

    B.   hypophosphataemia.

    C.   hyperkalaemia.

    D.   glycosuria.

    E.   hypercalcaemia.

42. Nocturia may be associated with:
    A. chronic renal failure.
    B. the recovery phase of acute tubular necrosis.
    C. hypokalaemia.
    D. hypercalcaemia.
    E. diabetes mellitus.

43. Complications of acute tubular necrosis include:
    A. sepsis.
    B. hyponatraemia.
    C. hypokalaemia.
    D. hypophosphataemia.
    E. pulmonary oedema.

44. The following are characteristic findings in untreated chronic renal failure:
    A. hypocalcaemia.
    B. hyperglycaemia.
    C. hyperuricaemia.
    D. urine of constant osmolality.
    E. raised serum alkaline phosphatase activity.

45. Creatinine clearance:
    A. in a healthy subject is higher than inulin clearance.
    B. is affected by diet.
    C. can be assessed from plasma creatinine alone.
    D. is useful in distinguishing prerenal from renal causes of renal failure.
    E. is proportional to body size.

46. Prerenal acute renal failure is characterised by:
    A. oliguria.
    B. presence of urine casts.
    C. a high plasma urea concentration disproportionate to plasma creatinine.
    D. high urinary sodium concentration.
    E. urine osmolality > 500 mOsm/kg.

47. The following change(s) may occur in patients with chronic renal failure:
    A. metabolic acidosis.
    B. hypophosphataemia.
    C. hypoparathyroidism.

    D.   urine fractional excretion of sodium of 0.5%.

    E.   a progressive increase of serum creatinine concentration.

48. Acute tubular necrosis is associated with:
    A.   a maximally concentrated urine.
    B.   immediate reversibility on restoring renal perfusion.
    C.   a spot urine sodium of greater than 40 mmol/L.
    D.   gentamicin administration.
    E.   crush injuries.

49. The renal excretion of creatinine:
    A.   is decreased during the early stages of diabetic nephropathy.
    B.   is reproducible to within 1% in successive 24-hour urine collections from a healthy adult.
    C.   is proportional to muscle mass.
    D.   is higher after administration of cimtidine.
    E.   is lower in vegetarians.

50. Factors contributing to the development of renal osteodystrophy include:
    A.   reduced synthesis of calcitriol.
    B.   aluminium retention.
    C.   high plasma parathyroid hormone concentration.
    D.   metabolic acidosis.
    E.   hypermagnesaemia.

51. In the oliguric phase of acute renal failure not attributable to inadequate fluid intake:
    A.   fluid intake should be restricted.
    B.   the urine sodium concentration is very low (< 20 mmol/L).
    C.   the urine usually has a high osmolality relative to plasma (ratio > 1.5).
    D.   plasma potassium concentration tends to rise.
    E.   a high protein diet is therapeutic.

52. Patients with chronic renal failure:
    A.   can rapidly excrete a large acid load.
    B.   will become more uraemic when sodium depleted.
    C.   cannot raise their urine osmolality to 900 mmol/kg when water-depleted.
    D.   should not be given vitamin D_3 analogues when hyperphosphataemic.
    E.   typically develop a normochromic normocytic anaemia.

53. Investigations, which may be helpful in management of a patient with renal stones, include:
    A. serum calcium concentration.
    B. urine oxalate excretion.
    C. urine calcium excretion.
    D. test for malabsorption.
    E. urine pH.

54. Abnormal renal tubular function may be seen in:
    A. Wilson's disease.
    B. Fanconi syndrome.
    C. lithium toxicity.
    D. heavy metal poisoning.
    E. multiple myeloma.

55. Hypercalcaemia may commonly occur in:
    A. osteoporosis.
    B. sarcoidosis.
    C. magnesium deficiency.
    D. primary hyperparathyroidism.
    E. steatorrhoea.

56. The following are characteristic findings in primary hyperparathyroidism:
    A. metabolic alkalosis.
    B. hyperchloraemia.
    C. low renal phosphate threshold.
    D. low serum calcitriol concentration.
    E. high serum alkaline phosphatase.

57. Indicate which of the following statements are true:
    A. Patients with Paget's disease are usually normocalcaemic.
    B. Patients with osteoporosis usually have a high serum alkaline phosphatase activity.
    C. In type 1 vitamin D-dependent rickets, there is an abnormal target tissue response to vitamin D.
    D. Malignancy is the most common cause of hypercalcaemia in a hospital population.
    E. Hypomagnesaemia is associated with hypokalaemia.

58. The level of plasma phosphate is usually reduced below normal in:
    A. hypoparathyroidism.
    B. osteomalacia.

    C.   refeeding syndrome.

    D.   chronic renal failure.

    E.   thyrotoxicosis.

59.   Which of the following factors may affect plasma total calcium concentration?

    A.   Chronic metabolic acidosis.

    B.   Dietary intake of calcium.

    C.   Gender.

    D.   Age.

    E.   Plasma albumin concentration.

60.   Hypophosphataemia:

    A.   is common during treatment of diabetic ketoacidosis

    B.   will shift the oxygen dissociation curve to the right.

    C.   may be caused by the shift of phosphate into the intracellular compartment.

    D.   needs treatment if very severe and prolonged.

    E.   always indicates a calcium disorder.

61.   A plasma calcium of 2.7 mmol/L (2.25–2.50), and a plasma inorganic phosphate of 0.7 mmol/L (0.8–1.3) is compatible with:

    A.   primary hyperparathyroidism.

    B.   osteoporosis.

    C.   tertiary hyperparathyroidism.

    D.   chronic renal failure.

    E.   osteomalacia.

62.   Actions of parathyroid hormone (PTH) include:

    A.   reduction in TmP/GFR (tubular maximum for phosphate reabsorption).

    B.   increase in renal calcium reabsorption.

    C.   activation of 1-hydroxylase enzyme in the kidney.

    D.   stimulation of osteoclast-mediated bone resorption.

    E.   activation of 24-hydroxylase enzyme in the kidney.

63.   Factors stimulating the production of calcitriol include:

    A.   calcitonin.

    B.   low serum calcium concentration.

    C.   low serum phosphate concentration.

    D.   parathyroid hormone.

    E.   high serum 25-hydroxyvitamin D (calcidiol).

64. Hypoglycaemia is recognised in association with:
    A. Cushing's syndrome.
    B. acromegaly.
    C. post-gastric surgery.
    D. pancreatic islet cell tumour of the β cells.
    E. metastatic carcinoma of bone.

65. In a patient admitted with severe diabetic ketoacidosis, there will be:
    A. net loss of body sodium.
    B. high plasma anion gap.
    C. low blood $PaCO_2$.
    D. low plasma potassium concentration.
    E. high plasma free fatty acid concentration.

66. Fasting hypoglycaemia is associated with:
    A. alcoholism.
    B. acute pancreatitis.
    C. gliclazide therapy.
    D. renal failure.
    E. glucagonoma.

67. Complications of the treatment of diabetic ketoacidosis include:
    A. hypoglycaemia.
    B. respiratory arrest.
    C. hyperkalaemia.
    D. hyperphosphataemia.
    E. cerebral oedema.

68. In the investigation of a patient with hypoglycaemia, which of the following tests are useful?
    A. Prolonged oral glucose tolerance test.
    B. Plasma insulin.
    C. ACTH stimulation test.
    D. Test for malabsorption.
    E. Liver function tests.

69. Which of the following conditions may cause hypertension and hyperglycaemia?
    A. Cushing's syndrome.
    B. Conn's syndrome.
    C. Pheochromocytoma.

    D.   Congenital adrenal hyperplasia.

    E.   Polycystic ovarian syndrome.

70. Features of hyperosmolar non-ketotic coma include:
    A.   a more common occurrence among type I (insulin dependent) diabetics.
    B.   urine hypo-osmolality.
    C.   an increased risk of thomboembolism.
    D.   a large plasma anion gap attributable to lactate and pyruvate.
    E.   severe cerebral dehydration.

71. Increased risk of atherosclerosis is associated with increased serum concentrations of:
    A.   high-density lipoprotein.
    B.   low-density lipoprotein.
    C.   lipoprotein (a).
    D.   apolipoprotein A-I.
    E.   apolipoprotein B.

72. Plasma cholesterol:
    A.   is transported from the peripheral tissue to the liver as high-density lipoprotein.
    B.   concentration is increased in hypothyroidism.
    C.   in the fasting state is mainly derived from liver.
    D.   concentration increases with age.
    E.   will decrease in patients treated with ezetimibe.

73. High-density lipoprotein in plasma:
    A.   consists of 80–90% triglyceride.
    B.   is negatively correlated with the risk of cardiovascular disease.
    C.   is higher in males compared to females.
    D.   is the carrier of cholesterol from peripheral tissues to the liver.
    E.   is increased after a diet rich in carbohydrate.

74. Atherogenic factors include:
    A.   hyperchylomicronaemia.
    B.   increased plasma LDL concentration.
    C.   increased plasma VLDL concentration.
    D.   increased plasma HDL–cholesterol concentration.
    E.   increased plasma total cholesterol concentration.

75. Which of the following statements is/are true about lipids:
    A. Chylomicrons are the lightest and largest particle of the lipoproteins.
    B. Chylomicrons are the main carriers of cholesterol.
    C. VLDL carries triglycerides from the liver to the peripheral tissues.
    D. LDL is removed mainly (> 50%) by non-receptor–mediated pathways.
    E. HDL is secreted by the liver.

76. Hyperlipidaemia may be seen in:
    A. hypothyroidism.
    B. excess alcohol intake.
    C. treatment with thiazide diuretics.
    D. nephrotic syndrome.
    E. glucocorticoid treatment.

77. Plasma creatine phosphokinase activity is usually increased in:
    A. cholecystitis.
    B. Duchenne's muscular dystrophy.
    C. cystic fibrosis.
    D. galactosaemia.
    E. myocardial infarction.

78. Assay of serum amylase activity is useful for the diagnosis of:
    A. Wilson's disease.
    B. acute pancreatitis.
    C. muscular dystrophy.
    D. myocardial infarction.
    E. megaloblastic anaemia.

79. Causes of increased plasma amylase activity include:
    A. acute pancreatitis.
    B. perforated peptic ulcer.
    C. acute peritonitis.
    D. parotitis.
    E. treatment with opiates.

80. Causes of increased plasma gamma-glutamyl transpeptidase (GGT) activity include:
    A. cholestasis.
    B. pregnancy.

    C.   renal disease.

    D.   Paget's disease of bone.

    E.   alcoholism.

81. Low plasma cholinesterase activity:
   A.   is an indicator of synthetic capacity of the liver.
   B.   is seen after organophosphorus insecticide poisoning.
   C.   may lead to prolonged apnoea after injection of suxamethonium.
   D.   is a feature of Alzheimer's disease.
   E.   is seen in nephrotic syndrome.

82. Causes of increased serum alkaline phosphatase activity include:
   A.   Paget's disease of bone.
   B.   pregnancy.
   C.   cholestasis.
   D.   muscle trauma.
   E.   carcinoma of the bronchus.

83. Simultaneous increase in serum alkaline phosphatase (ALP) and gamma-glutamyl transferase (GGT) may be seen in:
   A.   third trimester of pregnancy.
   B.   carcinoma of the head of the pancreas.
   C.   primary biliary cirrhosis.
   D.   renal osteodystrophy.
   E.   malignant deposits in bone.

84. Causes of increased serum gamma-glutamyl transferase (GGT) include:
   A.   pubertal growth spurt.
   B.   chronic alcoholism.
   C.   chronic renal failure
   D.   a space-occupying lesion in the liver.
   E.   treatment with carbamezapine.

85. Increased serum alkaline phosphatase (ALP) may be seen in
   A.   neonates.
   B.   osteoporosis.
   C.   vitamin D deficiency.
   D.   viral hepatitis.
   E.   carcinoma of the bronchus.

86. Which of the following statements about enzymes are true or false:
   A.   Serum CK is lower in people of African origin compared to Caucasians.

B. CK-MB isoenzyme is the predominant form in skeletal muscle.

C. Serum AST is found in skeletal muscle, cardiac muscle, liver and red cells.

D. Concentration of GGT is higher in the kidney than in the liver.

E. Serum ALT is cleared by glomerular filtration.

87. Increase in plasma total protein concentration is commonly found in:
    A. burns.
    B. acute hepatitis.
    C. paraproteinaemia.
    D. nephrotic syndrome.
    E. primary immunoglobin deficiency.

88. During the early period (up to 12–18 hours) after severe trauma:
    A. the resting metabolic rate is elevated.
    B. the blood glucose concentration is elevated.
    C. there is increased sensitivity to insulin.
    D. plasma albumin increases.
    E. there is water retention.

89. Which of the following criteria should be fulfilled for the diagnosis of myelomatosis:
    A. High serum alkaline phosphatase activity.
    B. Hyperproteinaemia.
    C. Presence of malignant plasma cells.
    D. Presence of pathological immunoglobulins in serum and/or urine.
    E. Presence of pathological bone fracture without other causes.

90. α_1-Antitrypsin:
    A. is an acute phase protein.
    B. deficiency can lead to liver cirrhosis.
    C. has a low molecular weight enabling it to pass into body fluids.
    D. possesses protease activity.
    E. deficiency can cause pulmonary emphysema

91. Oestrogen-containing contractive pills will tend to increase serum:
    A. albumin.
    B. total thyroxine.
    C. cortisol.
    D. copper.
    E. fasting triglycerides.

92. Concentrations of the following proteins increase after severe injury:
    A. transferrin.
    B. albumin.
    C. fibrinogen
    D. thyroxine-binding globulin.
    E. α_1-antichymotrypsin.

93. In multiple myeloma:
    A. plasma sodium may be low.
    B. plasma calcium is likely to be low.
    C. serum β_2 microglobulin is of prognostic value.
    D. serum ALP will be raised.
    E. there is proximal renal tubular damage.

94. In uncomplicated obstructive jaundice, there is likely to be:
    A. excess urobilinogen in the urine.
    B. decreased faecal urobilinogen.
    C. bilirubinuria.
    D. markedly increased plasma alkaline phosphatase (ALP) activity.
    E. markedly increased plasma aspartate transaminase (AST) activity.

95. Results of tests which may indicate liver disease in the absence of jaundice include increased:
    A. serum creatinine kinase (CK) activity.
    B. serum gamma-glutamyl transpeptidase (GGT) activity.
    C. alanine transaminase (ALT) activity.
    D. serum alkaline phosphatase (ALP) activity.
    E. urinary urobilinogen excretion.

96. Which of the following abnormal results will be expected in a patient, with extrahepatic cholestatic jaundice?
    A. High serum cholesterol concentration.
    B. High serum concentration of bile acids.
    C. High serum alkaline phosphatase (ALP) activity.
    D. Impaired oral glucose tolerance.
    E. Increased urobilinogen excretion.

97. Unconjugated bilirubin:
    A. is water soluble.
    B. in plasma is elevated in Gilbert's syndrome.

C. is the end product of haemoglobin breakdown only.
D. the plasma concentration is elevated in patients with haemolysis.
E. is carried in circulation bound to albumin.

98. In primary biliary cirrhosis:
A. plasma cholesterol will be elevated.
B. plasma IgM will be increased.
C. the plasma alkaline phosphatase will be normal.
D. there will be bilirubin in the urine.
E. mitochondrial antibody test will be positive.

99. Results consistent with a diagnosis of drug-induced haemolytic jaundice include:
A. presence of bilirubin in the urine.
B. absence of urobilinogen in the urine.
C. increased plasma conjugated bilirubin concentration.
D. increased plasma haptoglobin concentration.
E. presence of haemoglobin in the urine.

100. Steatorrhoea is a feature of:
A. chronic pancreatitis.
B. gluten-induced enteropathy.
C. post-hepatic biliary obstruction.
D. protein-energy malnutrition.
E. Crohn's disease.

101. In a patient presenting with severe abdominal pain due to acute pancreatitis:
A. an increased plasma amylase is essential for the diagnosis.
B. plasma calcium may be low.
C. a low PaO_2 (< 7.5 kPa) indicates a poor prognosis.
D. acute renal failure may develop.
E. hypoglycaemia may be present.

102. Which of the following measurements are useful in the assessment of nutritional status of an individual?
A. Plasma total protein concentration.
B. Urinary excretion of creatinine.
C. Serum insulin-like growth factor 1 (IGF-1).
D. Plasma cortisol.
E. Plasma prealbumin.

104. In patients who are on prolonged intravenous feeding, which of the following tests should be regularly monitored to avoid metabolic complications?
      A.   Plasma cortisol.
      B.   Plasma potassium.
      C.   Plasma glucose.
      D.   Liver function tests.
      E.   Plasma urea.

105. Two days after major surgery:
      A.   ADH secretion will be suppressed.
      B.   plasma glucose will be high.
      C.   there will be a tendency to lose sodium in the urine.
      D.   metabolic rate will be increased.
      E.   secretion of GH will be suppressed.

106. Which of the following will stimulate growth hormone secretion in normal subjects?
      A.   Stress.
      B.   Glucose.
      C.   Exercise.
      D.   Arginine.
      E.   Insulin-like growth factor.

107. In a patient suspected of panhypopituitarism, which of the following investigations are required:
      A.   Measurement of GH and cortisol during an insulin-induced hypoglycaemia test.
      B.   TRH test.
      C.   Glucose tolerance test with GH measurement.
      D.   Prolonged (3-day) tetracosactrin (Synacthen) test.
      E.   Metyrapone test.

108. In healthy pregnancy,
      A.   plasma sodium tends to be lower.
      B.   renal threshold for glucose increases.
      C.   hemoglobin concentration tends to fall.
      D.   serum creatinine decreases.
      E.   plasma triglycerides are higher than in the non-pregnant state.

109. In a 28-year-old man investigated for infertility, serum FSH was 22 IU/L (2.0–10.0), LH was 18 IU/L (2.0–8.0), serum testosterone was 3.2 nmol/L (10.2–25.0) and serum SHBG was 15 nmol/L (20–90). Which of the following statements are true or false?
   A. Fertility can be restored with testosterone replacement.
   B. Results indicate primary testicular failure.
   C. He may be taking anabolic steroids.
   D. Hyperprolactinaemia may explain these findings.
   E. Clomiphene administration will increase serum testosterone.

110. A raised FSH and LH, together with a low oestradiol, may be caused by:
   A. polycystic ovarian syndrome.
   B. congenital adrenal hyperplasia.
   C. premature ovarian failure.
   D. hypothyroidism.
   E. pituitary microadenoma secreting prolactin.

111. Which of the following condition(s) is/are associated with secondary hyperaldosteronism?
   A. Cirrhosis of liver.
   B. Phaeochromocytoma.
   C. Congestive heart failure.
   D. Salt-losing nephropathy.
   E. Syndrome of inappropriate antidiuretic hormone (ADH) secretion (SIADH).

112. Congenital adrenal hyperplasia (CAH) due to 21-hydroxylase deficiency:
   A. accounts for about 5% of the cases of CAH.
   B. can lead to salt loss in 10% of the cases.
   C. leads to increased secretion of 17α-hydroxyprogesterone.
   D. may manifest in adult life as menstrual irregularities.
   E. is a non-virilizing form due to deficiency of androgens.

113. Which of the following statements are true about glucocorticoids?
   A. Glucocorticoids inhibit bone formation.
   B. In its absence a water load can't be excreted efficiently.
   C. Endogenous production can be inhibited by ketoconazole.
   D. Hypoglycaemia may result if there is deficiency of glucocorticoids.
   E. Glucocorticoids increase uptake of amino acids by skeletal muscle.

114. Which of the following features suggest a diagnosis of Cushing's syndrome?
    A.   A serum cortisol of 25 nmol/L at 9 a.m. after administration of 1-mg dexamethasone the previous night.
    B.   A undetectable salivary cortisol at midnight.
    C.   Hirsutism.
    D.   Failure of ACTH to increase after administration of CRH.
    E.   Hypokalaemia.

115. A plasma cortisol of 340 nmol/L after an overnight dexamethasone may be seen in:
    A.   Cushing's disease.
    B.   carcinoma of bronchus secreting ACTH.
    C.   stress.
    D.   depression.
    E.   alcoholism.

116. Causes of simultaneously elevated serum TSH (thyroid-stimulating hormone) and FT4 (free thyroxine) concentrations include:
    A.   Graves' disease.
    B.   primary hypothyroidism.
    C.   TSH secreting pituitary tumour.
    D.   pituitary resistance to thyroid hormones.
    E.   erratic thyroxine replacement in a hypothyroid patient.

117. Elevated plasma total thyroxine concentration may be seen in:
    A.   normal pregnancy.
    B.   patients treated with androgens.
    C.   nephrotic syndrome.
    D.   chronic renal failure.
    E.   patients treated with phenytoin.

118. Thyroid-stimulating hormone (TSH)
    A.   is a glycoprotein with α and β chains.
    B.   assay in plasma is the best test to diagnose early primary hypothyroidism.
    C.   is useful in monitoring treatment of patients with primary hypothyroidism.
    D.   secretion is constant thoughout the day.
    E.   is often increased in elderly ill patients in the absence of thyroid disease.

119. In iron deficiency anaemia during pregnancy,
   A. intestinal absorption of iron decreases.
   B. iron store is however adequate because of ineffective erythropoiesis.
   C. there is hypochromic microcytic anaemia.
   D. plasma TIBC is low.
   E. plasma ferritin concentration is subnormal.

120. Manifestations of acute hepatic porphyrias include:
   A. dermatitis induced by exposure to infrared light.
   B. gradual darkening of urine on standing.
   C. protoporphyrin accumulation during the latent phase.
   D. neurological disturbances during the acute phase.
   E. hirsutism, hyperpigmentation and erythrodontia.

121. In a patient with anaemia due to chronic long standing infection, serum:
   A. hepcidin will be low.
   B. iron concentration will be high.
   C. TIBC will be low.
   D. transferrin concentration will be low.
   E. ferritin concentration will be low.

122. Which of the following statements are true of false?
   A. Pheochromocytoma is usually associated with elevation of urinary metanephrines.
   B. Carcinoid syndrome is associated with secretion of 5-hydroxy-tryptamine metabolites.
   C. Both AFP (α-fetoprotein) and HCG (human chorionic gonadotrophin) can be detected in the plasma of a pregnant woman.
   D. Serum carcinoembryonic antigen (CEA) is elevated in pregnancy.
   E. β-Human chorionic gonadotrophin (HCG) is detectable in the serum of most teratoma patients.

123. Human chorionic gonadotrophin (HCG):
   A. is highest at 8–10 weeks of pregnancy.
   B. if hyperglycosylated in a patient with trophoblastic tumor indicates an invasive tumour.
   C. is useful in the management of testicular teratoma.
   D. is not detectable in non-pregnant subjects.
   E. is elevated in all germ cell tumours.

124. Increases in the following serum enzyme activity may be seen 4 days after paracetamol overdose:
    A.   aspartate aminotransferase (AST).
    B.   alanine aminotransferase (ALT).
    C.   alkaline phosphatase (ALP).
    D.   amylase.
    E.   cholinesterase.

125. Measurement of which of the following drugs is required for management of patients suspected of overdose with that drug?
    A.   Lithium.
    B.   Salicylate.
    C.   Paracetamol.
    D.   Nitrazepam.
    E.   Propanolol.

126. Measurement of which of the following drugs is frequently required for monitoring treatment?
    A.   Lithium.
    B.   Phenytoin.
    C.   Nitrazepam.
    D.   Penicillin.
    E.   Gentamicin.

127. Which of the following abnormalities may be seen in a patient who has taken an overdose of salicylate?
    A.   Metabolic acidosis.
    B.   Respiratory alkalosis.
    C.   Increased serum ALT.
    D.   Hypercapnea.
    E.   A normal anion gap.

128. Which of the following factors may affect the interpretation of serum digoxin concentration?
    A.   Serum potassium concentration.
    B.   Thyroid function.
    C.   Serum calcium concentration.
    D.   Serum magnesium concentration.
    E.   Serum sodium concentration.

129. In a 26-year-old female who took an overdose of paracetamol 48 hours prior to admission:
    A.   serum ALT will be elevated.
    B.   prothrombin time will be prolonged.
    C.   plasma glucose will be elevated.
    D.   history of chronic alcohol intake indicates a poor prognosis.
    E.   an arterial pH of < 7.3 is a sign of poor prognosis.

130. Early diagnosis is important to prevent irreversible clinical consequences in:
    A.   maple syrup urine disease.
    B.   neonatal hypothyroidism.
    C.   phenylketonuria.
    D.   muscular dystrophy.
    E.   galactosaemia.

131. Galactosaemia:
    A.   results in decrease in the production of galactitol.
    B.   is usually associated with jaundice, hepatomegaly and aminoaciduria.
    C.   is untreatable.
    D.   can be confirmed by a measurement of the activity of hexose-1-phosphate uridyl transferase in erythrocytes.
    E.   is associated with a defect in epithelial transport of neutral amino acids.

132. Which of the following statements are true?
    A.   Maple syrup urine disease leads to early death.
    B.   Amniotic fluid cells can be cultured for detection of enzyme deficiencies.
    C.   Cystinuria causes permanent brain damage.
    D.   Galactosaemia is associated with jaundice and hepatomegaly.
    E.   Prenatal diagnosis is important for medium-chain acyl CoA dehydrogenase (MCAD) deficiency.

133. In Wilson's disease, there will be:
    A.   increased urinary copper.
    B.   abnormal liver function tests.
    C.   renal failure.
    D.   aminoaciduria.
    E.   renal tubular acidosis.

134. Elevation of serum urate:
    A. may be due to deficiency in activity of hypoxanthine guanine phosphoribosyl transferase (HGPRT).
    B. is more common in females than in males.
    C. can promote formation of renal stones.
    D. is related to high protein intake.
    E. can be due to thiazide diuretics.

135. Which of the following plasma constituents has different reference values between adults and children (2–14 years)?
    A. Alkaline phosphatase.
    B. Creatinine.
    C. Glucose.
    D. Calcium.
    E. Phosphate.

136. Which of the following diseases are more common in the elderly?
    A. Diabetes mellitus.
    B. Grave's disease.
    C. Chronic renal failure.
    D. Osteoporosis.
    E. Wilson's disease.

137. Which of the following plasma reference values change with age (in adults)?
    A. Alanine aminotransferase (ALT).
    B. Cholesterol.
    C. Alkaline phosphatase.
    D. Calcium.
    E. Glucose.

138. Causes of unconjugated hyperbilirubinaemia in infants include:
    A. glucose-6-phosphatase deficiency.
    B. galactosaemia.
    C. biliary atresia.
    D. hypothyroidism.
    E. Lucey-Driscoll syndrome.

139. Hypocalcaemia during the neonatal period is seen in:
    A. infants born to diabetic mothers.
    B. infants given high phosphate feeds.
    C. low-birth weight infants.

D. vitamin D deficiency.

E. DiGeorge syndrome.

140. Hypoglycaemia during the neonatal period:
    A. is seen in infants with severe hypoxia.
    B. if associated with hyponatraemia indicates congenital adrenal hyperplasia.
    C. if associated with ketones excludes hyperinsulinaemia.
    D. if associated with lactic acidosis indicates galactosaemia.
    E. is seen in infants born to diabetic mothers.

141. In a child presenting with short stature, which of the following investigations will be of help?
    A. Serum TSH.
    B. Renal function tests.
    C. Urine reducing substances.
    D. Sweat test.
    E. Serum anti-endomysial antibodies.

142. The concentration of glucose in CSF is:
    A. normally about 60% of that in plasma.
    B. increased in subarachnoid haemorrhage.
    C. decreased in bacterial meningitis.
    D. of little diagnostic value without simultaneous measurement of blood glucose.
    E. very low in viral meningitis.

143. Protein concentration of CSF taken from a lumbar puncture:
    A. is high in neonates compared to adults.
    B. is increased if there is a block in the spinal canal.
    C. is increased in multiple sclerosis due to an increase in IgM.
    D. is high in Guillain–Barré syndrome.
    E. if normal excludes bacterial meningitis.

144. A 30-year-old man complains of episodes of double vision. He is admitted to hospital following a fit, but remained confused for several hours. Papilloedema was absent and a lumbar puncture was performed? The CSF findings were:

    Pressure   : normal
    Protein   : 200 mg/L   (150–450 mg/L)
    Glucose   : 0.2 mmol/L   (> 3.0 mmol/L)
    Cell count : not increased
    Culture   : sterile

Indicate whether the following statements are true or false:

A. Bacterial meningitis is a likely diagnosis.
B. The plasma glucose concentration is probably elevated in this patient.
C. Plasma glucose, insulin and C peptide would be helpful investigations.
D. Insulinoma is a possible diagnosis.
E. Multiple sclerosis is a likely diagnosis.

145. A 60-year-old man was complaining of cough, weakness, weight loss and skin pigmentation. Investigations showed the following:

Plasma results concentration   Reference range

Sodium	138 mmol/L	(135–145)
Potassium	2.5 mmol/L	(3.5–5.0)
Chloride	90 mmol/L	(103–110)
Bicarbonate	36 mmol/L	(23–32)
Urea	6.9 mmol/L	(2.5–7.5)
Glucose (fasting)	9.0 mmol/L	(3.0–5.5)
Albumin	36 g/L	(37–47)
Cortisol 24.00 hr	860 nmol/L	(70–345)
Cortisol 09.00 hr	900 nmol/L	(140–690)

Indicate whether the following statements are true or false:

A. The results are compatible with Cushing's syndrome.
B. He has diabetes mellitus.
C. An appropriate further investigation would be a short ACTH stimulation (Synacthen) test.
D. Measurement of plasma ACTH concentration would be valuable.
E. Measurement of plasma C peptide concentration would be valuable.

146. A 65-year-old man develops jaundice with weight loss but no pain. Physical examination is normal apart from deep icterus. The patient has also noticed his urine was dark in colour. Plasma test results showed the following:

Bilirubin	233 μmol/L (< 19)
Alanine aminotransferase (ALT)	55 U/L (< 45)
Alkaline phosphatase	570 U/L (38–126)

Indicate whether the following statements are true or false:

    A.   The change in urine colour is due to the presence of urobilinogen.

    B.   The serum γ-glutamyltransferase (GGT) will be raised.

    C.   These findings are compatible with drug-induced jaundice.

    D.   These findings are compatible with haemolytic jaundice.

    E.   These findings are compatible with a tumour in the head of the pancreas.

147.   A 53-year-old woman is admitted via casualty with confusion. For the past 2 weeks, she has become progressively more lethargic and complains of headaches. The day before admission she had vomited. She had a grand mal fit in casualty and treated with IV valium. Physical examination and X-ray suggested the presence of a lung tumour.

    Laboratory tests showed the following:

Plasma	Sodium	105 mmol/L	(135–145)
	Potassium	3.7 mmol/L	(3.5–5.0)
	Urea	3.5 mmol/L	(3.2–7.1)
	Osmolality	215 mmol/kg	(275–285)
Urine	Osmolality	425 mmol/kg	
Blood	Glucose	4.6 mmol/L	(3.0–6.0)

Indicate whether the following statements are true or false:

    A.   These findings are compatible with dehydration due to reduced fluid intake.

    B.   The findings suggest an Addisonian crisis.

    C.   The results are consistent with syndrome of inappropriate anti-diuretic hormone production (SIADH).

    D.   The urine sodium would be very low.

    E.   One litre of 5% dextrose intravenously over 2 hours would be appropriate.

148–150.   A 30-year-old gardener was admitted to hospital with a 3-month history of loss of appetite and abdominal pain. He had severe diarrhoea for 3 weeks. The following results were obtained from the Emergency Laboratory:

Serum

Sodium	138 mmol/L	(135–145)
Potassium	2.9 mmol/L	(3.5–5.0)
Chloride	115 mmol/L	(103–110)

HCO_3	11 mmol/L	(22–30)
Urea	1.8 mmol/L	(3.4–7.2)
Creatinine	76 μmol/L	(60–120)
pH	7.29	(7.35–7.45)
pCO_2	3.5 kPa	(4.6–5.9)
pO_2	11.0 kPa	(10.5–13.0)

148. These results show that:
  A. the glomerular filtration rate is reduced.
  B. there is a partially compensated metabolic acidosis.
  C. ventilation is impaired.
  D. the cation–anion gap is normal.
  E. renal bicarbonate regeneration is defective.

149. The high blood H^+ ion concentration in this patient is a consequence of:
  A. chloride retention.
  B. loss of bicarbonate in diarrhoea.
  C. failure of urinary H^+ excretion.
  D. potassium depletion.
  E. respiratory disease.

150. The following factors may have contributed to the patient's hypokalaemia:
  A. potassium shift into the ICF compartment due to acidosis.
  B. secondary hyperaldosteronism.
  C. hypoxia.
  D. gastrointestinal loss and low dietary intake of potassium.
  E. impending renal failure.

151–153. A 14-year-old boy was admitted to hospital in a comatose state. A history was obtained from parents, of painful micturition, loss of appetite and vomiting. He was unconscious but responded to painful stimuli and had rapid respiration rate, and there was a smell of acetone on his breath.

Sodium	135 mmol/L	(135–145)
Potassium	5.2 mmol/L	(3.58–5.0)
Chloride	99 mmol/L	(103–110)
HCO_3	7 mmol/L	(22–30)
Urea	13.5 mmol/L	(3.4–7.2)

Glucose	42 mmol/L	(< 5.0)
pH	7.10	(7.35–7.45)
pCO_2	3.9 kPa	(4.6–5.9)
pO_2	11.5 kPa	(10.5–13.0)

151. In this patient:
   A. sodium deficiency would be expected.
   B. the ICF volume will be depleted.
   C. plasma osmolality will be normal.
   D. the plasma sodium concentration is low-normal because the hyperglycaemia induces a water shift from the ICF to ECF.
   E. cation–anion will be increased.

152. The patient may be potassium depleted because of:
   A. vomiting.
   B. increased protein breakdown.
   C. aldosterone deficiency.
   D. urinary excretion of potassium.
   E. osmotic diuresis.

153. The treatment of this patient would involve:
   A. intravenous insulin at a rate of 50 U/hour.
   B. restoring ECF volume by giving 1.5 L of 5% dextrose in the first hour.
   C. intravenous bicarbonate infusion.
   D. infusion of glucose when plasma glucose reaches 12–15 mmol/L.
   E. intravenous potassium supplements when plasma potassium is below 5 mmol/L.

154. The following results were obtained from a 16-year-old girl who was admitted to hospital with acute abdominal pain and a recent history of weight loss and polyuria.
   Plasma test results showed the following:

Sodium	128 mmol/L	(137–144)
Potassium	5.8 mmol/L	(3.3–4.4)
Chloride	94 mmol/L	(96–106)
Bicarbonate	11 mmol/L	(23–30)
Urea	9.0 mmol/L	(3.2–7.1)
Creatinine	210 μmol/L	(44–88)
Blood Glucose	37.0 mmol/L	

Indicate whether the following statements are true or false:
- A. This patient has a metabolic alkalosis.
- B. The anion gap is increased.
- C. The raised blood glucose could be explained by the intraveous administration of dextrose.
- D. A test for ketones in the urine would be useful.
- E. These findings are compatible with diabetic ketoacidosis.

Answers to True or False Questions

1.	TTFFF		35.	TTTFF
2.	FTTFT		36.	FFTFF
3.	FFFTF		37.	TTFTT
4.	FTFFF		38.	TFTTT
5.	TTFFT		39.	TTTTF
6.	FTFFF		40.	TFFFT
7.	FTFTF		41.	TTFTF
8.	TTTTF		42.	TTTTT
9.	TTTFT		43.	TTFFT
10.	TFTTF		44.	TFTTT
11.	TTTFF		45.	TTTFT
12.	TFTFT		46.	TFTFT
13.	TFFFT		47.	TFFTT
14.	TFTTT		48.	FFTTT
15.	FFFTT		49.	FTTTT
16.	TTTFF		50.	TTTTF
17.	FTFTF		51.	TFFTF
18.	TFFFT		52.	FTTFT
19.	FFTFF		53.	TTTTT
20.	FTTTT		54.	TTTTT
21.	TFFTT		55.	FTFTF
22.	TFTFT		56.	FTTFT
23.	TTFFT		57.	TFFTT
24.	TTTFT		58.	FTTFF
25.	TFTTT		59.	TFFFT
26.	FFFTT		60.	TFTTF
27.	TFTFT		61.	TFTFF
28.	TTFTF		62.	TTTTF
29.	TTFFF		63.	FTTTF
30.	FFTTT		64.	FFTTF
31.	TFTTF		65.	TTTFT
32.	FFTTF		66.	TTTTF
33.	TTTTF		67.	TTFFT
34.	TTFTT		68.	TTTTT

69.	TFTFT		107.	TTFFF
70.	FFTFT		108.	FFTTT
71.	FTTFT		109.	FTFFF
72.	TTTTT		110.	FFTFF
73.	FTFTF		111.	TFTTF
74.	FTTFT		112.	FFTTF
75.	TFTFF		113.	TTTTF
76.	FTTTT		114.	TFTFT
77.	FTFFT		115.	TTTTT
78.	FTFFF		116.	FFTTT
79.	TTTTT		117.	TFFFF
80.	TFFFT		118.	TTTFF
81.	TTTFF		119.	FFTFT
82.	TFTFT		120.	FTFTF
83.	FTTFF		121.	FFTTF
84.	FTFTT		122.	TTTTF
85.	TFTTT		123.	TTTFT
86.	FFTTF		124.	TTFFF
87.	TFTFF		125.	TTTFF
88.	FTFFT		126.	TTFFT
89.	FFTTF		127.	TTFTF
90.	TTTFT		128.	TTTTF
91.	FTTTT		129.	TTFTT
92.	FFTFT		130.	TTTFT
93.	TFTFT		131.	FTFTF
94.	FTTTF		132.	TTFTF
95.	FTTTT		133.	FTFTT
96.	TTTFF		134.	TFTTT
97.	FTFTT		135.	TTFFT
98.	TTFFT		136.	TFTTF
99.	FFTFT		137.	FTTFF
100.	TTTFT		138.	TTFTT
101.	FTTTF		139.	TTTTT
102.	FTTFT		140.	TFTFT
104.	FTTTT		141.	TTFTT
105.	FTFFF		142.	TFTTF
106.	TFTTF		143.	TTFTF

144.	FFTTF		150.	FTFTF
145.	TTFTF		151.	TTFTT
146.	FTTFT		152.	TFFTT
147.	FFTFF		153.	FFFTT
148.	TTFTF		154.	FTFTT
149.	FTFFF			

Extended Matching Questions

1. **Match each set of results with a diagnosis given below:**

Patient	Serum Calcium (2.21–2.60) mmol/L	Serum Phosphate (0.80–1.50) mmol/L	Alkaline Phosphatase (38–126) u/L
1	2.83	0.60	75
2	2.10	0.50	220
3	2.35	1.10	575
4	2.40	0.90	60
5	2.05	1.80	79

A. Primary hyperparathyroidism.
B. Osteomalacia.
C. Paget's disease of the bone.
D. Pseudohypoparathyroidism.
E. Osteoporosis.

2. **Match the composition of the fluids to the type of fluids A–E below.**

	Sodium	Potassium	Chloride	Bicarbonate
1	5	50	4	0
2	70	10	120	0
3	138	4.5	100	24
4	140	5	100	25
5	120	8	55	35

A. Gastric fluid.
B. Interstitial fluid.
C. Urine.
D. Pancreatic juice.
E. CSF.

3. **Match the results to the diagnosis:**

1	pH	7.10	(7.35–7.45)
	pCO_2	2.2 kPa	(4.6–5.9)
	anion gap	40 mmol/L	(15–20)

2	Sodium	151 mmol/L	(135–145)
	Potassium	4.0 mmol/L	(3.5–5.0)
	Urea	15.0 mmol/L	(3.4–7.2)
	Glucose	50.0 mmol/L	(< 5.0)
	Osmolality	375 mOsm/kg	(285–295)
3	pH	7.35	(7.35–7.45)
	pCO_2	9.0 kPa	(4.6–5.9)
4	Sodium	155 mmol/L	(135–145)
	Potassium	4.2 mmol/L	(3.5–5.0)
	Urea	10.0 mmol/L	(3.4–7.2)
	Creatinine	90 mmol/L	(60–110)
5	Plasma bicarbonate	16 mmol/L	(22–30)
	Urine pH	7.00	

A. Diabetic hyerosmolar coma.
B. Chronic respiratory acidosis.
C. Acute hypoventilation.
D. Diabetic ketoacidosis.
E. Distal renal tubular acidosis.
F. Water depletion.

4. **For each set of results below, choose the *SINGLE* most appropriate diagnosis from the list A–H. Each option may be used once, more than once or not at all.**

	pH (7.35–7.45)	pCO_2 (4.3–6.0) kPa	pO_2 (11.7–15.3) kPa	Bicarbonate (22–28) mmolL
1	7.19	3.9	14.1	11
2	7.00	8.0	27.7	14
3	7.34	8.0	8.3	33
4	7.5	6.5	11.5	37
5	7.49	3.6	15.1	20

A. Acute hypoventilation.
B. Cardiac arrest.
C. Chronic obstructive airways disease.
D. Diabetic ketoacidosis.
E. Gilbert's syndrome.
F. Hypokalaemia.

    G. Hysterical hyperventilation.

    H. Normal.

5.   **For the results below, select the *SINGLE* most appropriate value for the following parameters. Each option may be used once, more than once or not at all.**

    1. Plasma osmolarity (mmol/L)

    2. Anion gap (mmol/L)

    3. Creatinine clearance (ml/min)

    4. Blood hydrogen ion concentration (nmol/L)

    5. Total cholesterol:HDL cholesterol ratio

Sodium	147 mmol/L
Potassium	3.0 mmol/L
Chloride	100 mmol/L
Bicarbonate	10 mmol/L
Urea	20.0 mmol/L
Creatinine	200 μmol/L
Glucose	30.0 mmol/L
Total cholesterol	7.2 mmol/L
HDL cholesterol	0.90 mmol/L
LDL cholesterol	5.4 mmol/L
pH	7.10
Urine creatinine	10 mmol/L
Urine volume	1440 ml/24-hr

    A. 2   I.   80

    B. 5   J.   120

    C. 8   K. 150

    D. 15   L. 200

    E. 20   M. 250

    F. 30   N. 330

    G. 40   O. 350

    H. 50   P. 400

6.   **For each description below, choose the *SINGLE* most likely answer from the list of options A–O. Each option may be used once, more than once or not at all.**

    1. Deficiency may lead to neonatal jaundice.

    2. Deficiency leads to increased uric acid concentration.

    3. Increased in Paget's disease of bone.

4. Different isoenzymes present in brain and muscle.
5. Raised in mumps.

A.	Alanine aminotransferase (ALT)	I.	Glucose-6-phosphate dehydrogenase (G6PD)
B.	Alkaline phosphatase (ALP)	J.	Hydroxymethylglutaryl CoA reductase (HMG-CoA reductase)
C.	Amylase	K.	Hypoxanthine-guanine phosphoribosyl transferase (HGPRT)
D.	Aspartate aminotransferase (AST)	L.	Lactate dehydrogenase (LDH)
E.	Cholinesterase	M.	Lipoprotein lipase
F.	Creatine kinase (CK)	N.	Medium-chain acyl-CoA dehydrogenase (MCAD)
G.	Galactose-1-phosphate uridyltransferase	O.	Phenylalanine hydroxylase
H.	Gamma-glutamyl transferase (GGT)		

7. **Reference ranges:**

Sodium	135–145 mmol/L
Potassium	3.5–5.0 mmol/L
Urea	2.5–7.0 mmol/L
Creatinine	60–125 mmol/L
Osmolality	275–295 mmol/kg

For each set of results below, choose the *SINGLE* most likely condition from the list of options A–O. Each option may be used once, more than once or not at all.

1. 45-year-old man with polyuria and polydipsia:

Serum sodium	128 mmol/L
Serum potassium	3.9 mmol/L
Serum urea	3.0 mmol/L
Serum creatinine	90 mmol/L
Serum osmolality	270 mmol/kg
Urine osmolality	125 mmol/kg

2. 78-year-old woman three days after a total hip replacement:
    Serum sodium   128 mmol/L
    Serum potassium   3.2 mmol/L
    Serum urea   3.0 mmol/L
    Serum creatinine   80 μmol/L
    Serum osmolality   270 mmol/kg
    Urine osmolality   625 mmol/kg

3. 50-year-old woman with thirst and polyuria
    Serum sodium   126 mmol/L
    Serum potassium   6.9 mmol/L
    Serum urea   9.3 mmol/L
    Serum creatinine   148 μmol/L
    Serum osmolality   315 mmol/kg

4. 35-year-old man with dizziness
    Serum sodium   126 mmol/L
    Serum potassium   6.9 mmol/L
    Serum urea   9.3 mmol/L
    Serum creatinine   148 μmol/L
    Serum osmolality   280 mmol/kg

5. 35-year-old man with hypertension
    Serum sodium   142 mmol/L
    Serum potassium   2.7 mmol/L
    Serum urea   5.5 mmol/L
    Serum creatinine   98 μmol/L
    Serum osmolality   298 mmol/kg

A.	Addison's disease	I.	Drip arm sample
B.	Chronic renal failure	J.	Hypopituitarism
C.	Conn's syndrome	K.	Syndrome of inappropriate ADH secretion (SIADH)
D.	Cushing's syndrome	L.	Pyloric stenosis
E.	Diabetes insipidus	M.	Steroid 21-hydroxylase deficiency causing congenital adrenal hyperplasia
F.	Diabetes mellitus	N.	Water depletion
G.	Diarrhoea	O.	Water intoxication
H.	Diuretic therapy		

8. **For each set of situations below, choose the *SINGLE* most appropriate investigation from the list of options A–K. Each option may be used once, more than once or not at all.**
    1. 35-year-old woman who complains of increased weight, abdominal striae, difficulty climbing stairs, a fasting glucose of 9.8 mM and a round-looking face.
    2. 33-year-old woman complaining of prognathism, increased shoe and ring size, and a fasting glucose of 9.8 mmol/L.
    3. 44-year-old man complains of polyuria, polydipsia and nocturia.

Serum sodium	138 mmol/L
Serum potassium	3.9 mmol/L
Serum urea	8.0 mmol/L
Fasting plasma glucose	7.0 mmol/L

    4. Routine visit to GP by a 55-year-old man with diabetes mellitus.
    5. 25-year-old man with postural hypotension

Serum sodium	117 mmol/L
Serum potassium	5.9 mmol/L
Serum urea	8.0 mmol/L
Serum glucose	3.1 mmol/L

    A. Blood gases.
    B. Combined pituitary function test.
    C. Dexamethasone suppression test.
    D. Glucose.
    E. Glucose tolerance test.
    F. HbA1c.
    H. Liver function tests (LFTs).
    I. Short Synacthen test.
    J. Urea and electrolytes.
    K. Water deprivation test.
    G. Insulin tolerance test.

9. **For each of the blood results below, choose the *SINGLE* most likely condition from the list of options A–F. Each option may be use once, more than once or not at all.**

	Calcium (mmol/L)	Phosphate (mmol/L	PTH pmol/L	ALPU/L
i)	1.8	0.7	125	250
ii)	1.8	2.6	1.2	90
iii)	2.45	1.0	3.0	90
iv)	3.0	0.7	8.0	100
v)	3.1	1.2	0.9	270
Ref Range	2.15–2.55	0.8–1.4	1.1–6.8	30–130

A. Carcinoma of breast
B. Hypoparathyroidism
C. Osteomalacia
D. Osteoporosis
E. Paget's disease
F. Primary hyperparathyroidism

---

10. **From the list below, select the single most likely diagnosis for each of the five sets of results (reference range for TSH 0.3–5.5 IU/L and reference range for FT4 10–25 pmol/L):**

   1. A 43-year-old lady feeling generally unwell and tired:
       Serum FT4   8.0 pmol/L
       Serum TSH   1.3 mIU/L
   2. A 33-year-old female receiving 0.2 mg thyroxine/day:
       Serum FT4   31.0 pmol/L
       Serum TSH   0.1 mIU/L
   3. A 23-year-old female with heat intolerance, weight loss and oligomenorrhoea for 3 months
       Serum FT4   45.0 pmol/L
       Serum TSH   < 0.2 mIU/L
   4. A 34-year-old female with weight loss and sweating:
       Serum FT4   32.0 pmol/L
       Serum TSH   4.5 mIU/L
   5. A 55-year-old lady on anti-convulsant therapy and 0.2 mg thyroxine and complaining of tiredness and weight gain.
       Serum FT4   8.0 pmol/L
       Serum TSH   12.3 mIU/L

   A. Graves' Disease
   B. Hypopituitarism
   C. Over-replacement with thyroxine
   D. Primary hypothyroidism
   E. Under-replacement with thyroxine
   F. TSHoma

11. **Match the following nutrients with the conditions listed below:**
   1. Vitmain K.
   2. Folic acid.
   3. Iron.

    4.   Vitamin B_{12}.

    5.   Zinc.

A.   Hyperhomocysteinaemia.

B.   Peripheral neuropathy.

C.   Microcytic anaemia.

D.   Skin rash.

E.   Prolonged prothrombin time.

12. Match the following plasma enzymes to the conditions listed below:

    1.   Cholinesterase.

    2.   Alkaline phosphatase.

    3.   Amylase.

    4.   Gamma-glutamyl transferase(GGT).

    5.   Creatinine kinase.

A.   Hypothyroidism.

B.   Scoline apnoea.

C.   Osteomalacia.

D.   Acute pancreatitis.

E.   Treatment with anticonvulsants.

13. Match the following measurements to the with the corresponding conditions:

    1.   Serum 25-hydroxyvitamin D.

    2.   Activation of red cell transketolase.

    3.   Activation of red cell glutathione reductase.

    4.   Activation of red cell AST.

    5.   Serum methylmalonic acid.

A.   Vitamin B_{12} deficiency.

B.   Rickets.

C.   Vitamin B_6 deficiency.

D.   Vitamin B_2 (riboflavin) deficiency.

E.   Beriberi.

14. A patient is noted to have a plasma K of 2.7 mmol/L. Match each of the possible other laboratory findings with the likely diagnosis:

1.	Plasma bicarbonate	27 mmol/L	(23–32)
	Calcium	1.8 mmol/L	(2.20–2.55)
	Albumin	41 g/L	(37–47)
2.	Plasma bicarbonate	13 mmol/L	
	Urine potassium	52 mmol/L	

3. Plasma bicarbonate   14 mmol/L
    Urine potassium   10 mmol/L
4. Plasma bicarbonate   45 mmol/L
    Urine potassium   50 mmol/L
5. Plasma sodium   126 mmol/L   (135–145)
    Urine potassium   45 mmol/L

A. Primary hyperaldosteronism.
B. Hypomagnesaemia.
C. Renal tubular acidosis.
D. Diuretic therapy.
E. Laxative abuse.

Answers to Extended Matching Questions

1. 1. A–Primary hyperparathyroidism
    2. B–Osteomalacia
    3. C–Paget's disease of the bone
    4. E–Osteoporosis
    5. D–Pseudohypoparathyroidism

2. 1. C–Urine
    2. A–Gastric fluid
    3. B–Interstitial fluid
    4. E–CSF
    5. D–Pancreatic juice

3. 1. D–Diabetic ketoacidosis
    2. A–Diabetic hyperosmolar coma
    3. B–Chronic respiratory acidosis
    4. F–Water depletion
    5. E–Distal renal tubular acidosis

4. 1. D–Diabetic ketoacidosis
    2. B–Cardiac arrest
    3. C–Chronic obstructive airways disease
    4. F–Hypokalaemia
    5. G–Hysterical hyperventilation

5. 1. N–330
    2. G–40
    3. H–50

   4.   I–80
   5.   C–8
6.   1.   I–Glucose-6-phosphate dehydrogenase (G6PD)
     2.   K–Hypoxanthine–guanine phosphoribosyl transferase (HGPRT)
     3.   B–Alkaline phosphatase (ALP)
     4.   F–Creatine kinase (CK)
     5.   C–Amylase
7.   1.   O–Water intoxication
     2.   K–Syndrome of inappropriate ADH secretion (SIADH)
     3.   F–Diabetes mellitus
     4.   A–Addison's disease
     5.   B–Conn's syndrome
8.   1.   C–Dexamethasone suppression test
     2.   E–Glucose tolerance test
     3.   K–Water derivation test
     4.   F–HbA1c
     5.   I–Short Synacthen test
9.   1.   C–Osteomalacia
     2.   B–Hypoparathyroidism
     3.   D–Osteoporosis
     4.   F–Primary hyperparathyroidism
     5.   A–Carcinoma of breast
10. 1.   B–Hypopituitarism
     2.   C–Over-replacement with thyroxine
     3.   A–Graves' disease
     4.   F–TSHoma
     5.   E–Underreplacment with thyroxine
11. 1.   E–Prolonged prothrombin time
     2.   A–Hyperhomocysteinaemia
     3.   C–Microcytic anaemia
     4.   B–Peripheral neuropathy
     5.   D–Skin rash
12. 1.   B–Scoline apnoea
     2.   C–Osteomalacia
     3.   D–Acute pancreatitis
     4.   E–Treatment with anticonvulsants
     5.   A–Hypothyroidism

13. 1. B–Rickets
     2. E–Beriberi
     3. D–Vitamin B_2 (riboflavin) deficiency
     4. C–Vitamin B_6 deficiency
     5. A–Vitamin B_{12} deficiency

14. 1. B–Hypomagnesaemia
     2. C–Renal tubular acidosis
     3. E–Laxative abuse
     4. A–Primary hyperaldosteronism
     5. D–Diuretic therapy

Index

erythrocytes, *see* red blood cells

erythopoietic porphyria, 587

erythropoiesis, 385, 579, 589, 592, 690

esterase, pancreatic, 421, 423

ethanol, *see* alcohol

ethnicity, 10, 314

ethosuximide monitoring, 631

ethylene glycol poisoning, 95

exchange transfusion, neonatal jaundice, 692

exercise
amenorrhoea, 505
creatine kinase, 343
diabetes mellitus, 251

extracellular fluid
composition, 23
fluid loss effects, 37
hydrogen ion buffering, 92
osmolality, 25
volume regulation, 30

exudates, 40

factor VII, 39

faecal fat test, 428

faecal occult blood, 431

failure to thrive, 698

false negative, 12

false positive, 13

familial chylomicronaemia syndrome, 294–295

familial combined hyperlipidaemia, 292–293

familial defective apoB 100, 292

familial dysbetalipoproteinaemia, 295

familial hyperalphalipo-proteinaemia, 296

familial hypercholesterolaemia, 288–290

familial hypertriglyceridaemia, 293

familial hypocalciuric hypercalcaemia, 178

familial hypokalaemic periodic paralysis, 65

familial juvenile hyeruricaemia, 680

familial phytosterolaemia, 296

familial X-linked hypophosphataemia, 146

Fanconi syndrome, 147–148

fasting glucose levels, diabetes, 251

fasting hypoglycaemia, 266, 268, 269, 664

fat absorption, 428

fatty acid oxidation defects, 667–668

feminisation, chronic liver disease, 406

ferritin, 360, 591

fertility, *see* infertility

α-fetoprotein, *see* alphafetoprotein

fibrates, 293, 295, 296, 383

fibrinogen, 361, 382

fibronectin, 518

fibrosis
in cirrhosis, 400

fistula, 24, 66, 95, 427

fluid deprivation test, *see* water deprivation test

fluorescein dilaurate test, 421–422

fluoride
number, 341
sample tubes, 5